HYBRID ELECTRIC VEHICLE SYSTEM MODELING AND CONTROL

Automotive Series

Series Editor: Thomas Kurfess

HYBRID ELECTRIC VEHICLE SYSTEM MODELING AND CONTROL

Second Edition

Wei Liu
General Motors, USA

Registered Office(s)
John Wiley & Sons Ltd, The Atrium, Southern Gate, Chichester, West Sussex, PO19 8SQ, UK
John Wiley & Sons, Inc., 111 River Street, Hoboken, NJ 07030, USA

Editorial Office
The Atrium, Southern Gate, Chichester, West Sussex, PO19 8SQ, UK

For details of our global editorial offices, customer services, and more information about Wiley products visit us at www.wiley.com.

Wiley also publishes its books in a variety of electronic formats and by print-on-demand. Some content that appears in standard print versions of this book may not be available in other formats.

Library of Congress Cataloging-in-Publication Data

Names: Liu, Wei, 1960 August 30- author.
Title: Hybrid electric vehicle system modeling and control / Wei Liu.
Other titles: Introduction to hybrid vehicle system modeling and control
Description: 2nd edition. | Chichester, West Sussex, UK ; Hoboken, NJ, USA :
    John Wiley & Sons, Inc., 2017. | Series: Automotive series | Revised
    edition of: Introduction to hybrid vehicle system modeling and control. |
    Includes bibliographical references and index.
Identifiers: LCCN 2016045440 (print) | LCCN 2016048636 (ebook) | ISBN
    9781119279327 (cloth) | ISBN 9781119279334 (pdf) | ISBN 9781119278948 (epub)
Subjects: LCSH: Hybrid electric vehicles–Simulation methods. | Hybrid
    electric vehicles–Mathematical models.
Classification: LCC TL221.15 .L58 2017 (print) | LCC TL221.15 (ebook) | DDC 629.22/93–dc23
LC record available at https://lccn.loc.gov/2016045440

Cover design by Wiley
Cover image: Martin Pickard/ Henrik5000/ dreamnikon/ Gettyimages

Set in 10/12.5pt Times by SPi Global, Pondicherry, India
Printed and bound in Malaysia by Vivar Printing Sdn Bhd

10 9 8 7 6 5 4 3 2 1

To my wife Mei and son Oliver

Contents

Preface

With hybrid electric vehicle systems having undergone many great changes in recent years, hybrid electric vehicle modeling and control techniques have also advanced. Electrified powertrains are providing dramatic new opportunities in the automotive industry. Since hybrid vehicle systems naturally have nonlinear characteristics, exhibit fast parameter variation, and operate under uncertain and changing conditions, the associated modeling and control problems are extremely complex. Nowadays, hybrid vehicle system engineers must face head-on the challenge of mastering cutting-edge system modeling and control theories and methodologies in order to achieve unprecedented vehicle performance.

Hybrid electric vehicle systems, combining an internal combustion engine with one or more electric motors for propulsion, operate in changing environments involving different fuels, load levels, and weather conditions. They often have conflicting requirements and design objectives that are very difficult to formalize. Most hybrid controls are fundamentally multivariable problems with many actuators, performance variables, and sensors, but some key control variables are not directly measurable. To articulate these challenges, I published the first edition of this book in 2013 to meet the needs of those involved in hybrid vehicle system modeling and control development.

Continued advances in hybrid vehicle system technology make periodic revision of technical books in this area necessary in order to meet the ever-increasing demand for engineers to look for rigorous methods for hybrid vehicle system control design and analysis. The principal aims of this revision are to place added emphasis on advanced control techniques and to expand the various modeling and analysis topics to reflect recent advances in hybrid electric vehicle systems. Overall, many parts of the book have been revised. The most apparent change is that a chapter on noise and vibration has been added to present the unique control challenges arising in hybrid electric vehicle integration to meet driving comfort requirements.

The material assembled in this book is an outgrowth of my over fifteen years' work on hybrid vehicle research, development, and production at the National Research Council Canada, Azure Dynamics, and General Motors. The book is intended to contribute to a better understanding of hybrid electric vehicle systems, and to present all the major aspects of hybrid vehicle modeling, control, simulation, performance analysis, and preliminary design in the same book.

This revised edition retains the best of the first edition while rewriting some key sections. The basic structure of the book is unchanged. The book consists of ten main chapters and two appendices. Chapter 1 provides an introduction to hybrid vehicle system architecture, energy flow, and the controls of a hybrid vehicle system. Chapter 2 reviews the main components of a hybrid system and their characteristics, including the internal combustion engine, the electric motor/generator, the energy storage system, and hybrid electric transmission. This chapter also introduces the construction, basic materials, and requirements of Li-ion batteries for hybrid electric vehicle application.

Chapter 3 presents detailed mathematical models of hybrid system components for system design and simulation analysis, which include the internal combustion engine, the transmission system, the motor/generator, the battery system, and the vehicle body system, as well as the driver. One-mode and two-mode electrical continuously variable transmission system modeling and the lever analogy technique are introduced for hybrid transmission kinematic analysis in this chapter. The models presented in this chapter can be used either for individual component analysis or for building a whole vehicle simulation system.

Chapter 4 introduces the basics of power electronics and electric motor drives applied in hybrid electric vehicle systems. The characteristics of commonly used power electronic switches are presented first, followed by the introduction of the operational principles of the DC–DC converter and DC–AC inverter. Brushless DC motors and AC induction motors and their control principles are also introduced for hybrid vehicle applications. The techniques of plug-in charger design are presented in the last part of this chapter.

Chapter 5 addresses the modeling and controls of the energy storage system. Algorithms relating to the battery system play a very important role in hybrid electric vehicle systems because they directly affect the overall fuel economy and drivability and safety of a vehicle; however, due to the complexity of electrochemical reactions and dynamics as well as the availability of key variable measurements, hybrid vehicle system and algorithm engineers are facing head-on technical challenges in the development of the algorithms required for hybrid electric vehicles. In this chapter, the state of charge determination algorithms and technical challenges are first discussed. Then, the power capability algorithms and state of life algorithms with aging behavior and the aging mechanism are addressed, and the lithium metal plating issue and symptoms in Li-ion batteries are discussed as well. The cell-balancing algorithm necessary for hybrid vehicles, the battery cell core temperature estimation method, and the battery system efficiency calculation are also presented in this chapter.

Chapter 6 is concerned with the solution of energy management problems under different drive cycles. Both direct and indirect optimization methods are discussed. The methods presented in this chapter can be treated as the most general and practical techniques for the solution of hybrid vehicle energy management problems.

Chapter 7 elaborates on the other control problems in hybrid vehicle systems, including active engine fluctuation torque dumping control, voltage ripple control in the high-voltage bus, thermal control of the energy storage system, motor traction and anti-rollback control, and electric active suspension system control. For plug-in hybrid vehicles, the CS setpoint self-tuning control strategy and the CS lower bound real-time determination algorithm are presented to compensate for battery aging in this chapter.

Chapter 8 discusses the characteristics of AC-120, AC-240, and fast public plug-in charging for emerging plug-in hybrid and purely battery-powered vehicles. This chapter also presents plug-in charge control requirements and techniques for battery-powered electric vehicles. The impact of plug-in charging on battery life and safety as well as on the electric grid and power distribution system is presented in this chapter. In addition, the various plug-in charging strategies, including the optimal charging strategy, are introduced in this chapter.

Chapter 9 deals with noise and vibration issues. Noise and vibration have become an important aspect of hybrid powertrain development and the vehicle integration process, and there are stringent requirements to reduce HEV/PHEV/BEV vibration and noise levels. To articulate these challenges, this chapter first introduces the basics of vibration and noise, and then addresses the unique vibration and noise characteristics and issues associated with powertrain vibration, driveline vibration, gear rattle noise, and electrified-component-specific vibration and noise, such as accessory whine, motor/generator electromagnetic vibration and noise, and vibration and growl in the energy storage system, as well as vibration and noise pattern changes compared with traditional vehicles.

Chapter 10 presents typical cycles and procedures for fuel economy, emissions, and electric range tests, including FTP, US06, SC03, LA92, NEDC, and WLTP for hybrid electric vehicles, as well as single and multiple cycles for battery-powered electric vehicles. The necessary calculations and simulations for sizing/optimizing components and analyzing system performance at the concept/predesign stage of a hybrid vehicle system are addressed in this chapter.

Appendix A reviews the system identification, state and parameter estimation methods and techniques. Commonly used mathematical models are introduced for hybrid vehicle system control algorithm development. Recursive least squares and generalized least squares techniques are presented for parameter estimation. The Kalman filter and extended Kalman filter are introduced in this appendix to solve state and parameter estimation problems. In addition, the appendix also presents the necessary computational stability enhancement techniques of practical hybrid vehicle systems.

Appendix B briefly introduces some advanced control methods which are necessary to improve the performance of a hybrid electric vehicle system. These include system pole-placement control, objective-function-based optimal control, dynamic-programming-based optimal control, minimal variance, and adaptive control techniques for systems with stochastic behavior. To enhance the reliability and safety of a hybrid vehicle system, fault-tolerant control strategies are briefly introduced in this appendix.

In the hybrid electric vehicle system control field, there are many good practices that cannot be fully justified from basic principles. These practices are the 'art' of hybrid vehicle system control, and thus several questions arise for control engineers and researchers on the future control of hybrid vehicle systems: What form will scientific underpinnings take to allow control engineers to manage and control vehicle systems of unprecedented complexity? Is it time to design real-time control algorithms that address dynamic system performance in a substantial way? Is it feasible to develop a control methodology depending on ideas originating in other scientific traditions in addition to the dependence on mathematics and physics? Such questions provide strong evidence that control has a significant role to play in hybrid electric vehicle engineering.

This book has been written primarily as an engineering reference book to provide a text giving adequate coverage to meet the ever-increasing demand for engineers to look for rigorous methods for hybrid electric vehicle design and analysis. It should enable modeling, control, and system simulation engineers to understand the hybrid electric vehicle systems relevant to control algorithm design. It is hoped that the book's conciseness and the provision of selected examples illustrating the methods of modeling, control, and simulation will achieve this aim. The book is also suitable for a training course on hybrid electric vehicle system development with other supplemental materials. It can be used both on undergraduate and graduate-level hybrid vehicle modeling and control courses. I hope that my efforts here succeed in helping you to understand better this most interesting and encouraging technology.

I would like to express my gratitude to many present and former colleagues who have provided support and inspiration. Special thanks are due to Professor Yin Guodong who translated and introduced the first edition of this book to Chinese readers.

Wei Liu, PhD, PE, P Eng

List of Abbreviations

AC	Alternating current electricity
AER	All-electric range
AR	Autoregressive model
ARMAX	Autoregressive moving average exogenous model
ARX	Autoregressive exogenous model
ATDC	After top dead center
BDC	Bottom dead center
BDU	Battery disconnect unit
BEV	Battery-powered electric vehicle
BJT	Bipolar junction transistor
BLDC	Brushless DC motor
BTDC	Before top dead center
CAFE	Corporate average fuel economy
CC	Constant current plug-in charge
CD	Charge-depleting
CO	Carbon monoxide
CO_2	Carbon dioxide
CS	Charge-sustaining
CSC	Constant-speed driving cycle
CV	Constant-voltage plug-in charge
CVT	Continuously variable transmission
D	Sound energy density
DC	Direct current electricity
DCR	Driving cycle recognition
DoD	Depth of discharge
DOE	Department of Energy

DOE	Design of experiment
DOHC	Dual overhead cam engine
DP	Dynamic programming
DSR	Driving style recognition
ECE	Emission certificate Europe
ECU	Electronic/engine control unit
ECVT	Electronic continuously variable transmission
EDLC	Electrochemical double layer capacitor
EKF	Extended Kalman filter
EMF	Electromagnetic force
EMI	Electromagnetic interference
EOT	End-of-test criterion
EPA	Environmental Protection Agency of the USA
EREV	Electric range extended hybrid vehicle
ESS	Energy storage system
EUDC	Extra urban driving cycle
EV	Electric vehicle
EVT	Electrical variable transmission
FC	Fuel cell
FET	Field effect transistor
FTP	Federal test procedure
Genset	Engine-generator pair set
GM	General Motors
GVW	Gross vehicle weight
HC	Hydrocarbons
HD-UDDS	Urban dynamometer driving schedule for heavy-duty vehicles
HEV	Hybrid electric vehicle
HPPC	Hybrid pulse power characterization test
HWFET	Highway fuel economy test schedule
ICE	Internal combustion engine
IGBT	Insulated gate bipolar transistor
L_a	Acceleration level in decibels (dB)
L_I	Sound intensity level in decibels (dB)
L_P	Sound pressure level in decibels (dB)
L_V	Velocity level in decibels (dB)
L_W	Sound power in decibels (dB)
Li-ion	Lithium-ion battery
LQC	Linear quadratic control
MCT	Multi-cycle range and energy consumption test for battery-powered vehicles
MIMO	Multi-input, multi-output system
MOSFET	Metal oxide semiconductor field effect transistor

MPC	Model predictive control
MPG	Miles per gallon
MPGe	Miles per gallon equivalent
MRAC	Model reference adaptive control
NEC	Net energy change
NEDC	New European Driving Cycle
NHTSA	National highway traffic safety administration
NiCd	Nickel–cadmium battery
NiMH	Nickel–metal hydride battery
NiOx	Nickel oxyhydroxide
NOx	Nitrous oxides
NVH	Noise, vibration, and harshness
NYCC	New York City Driving Cycle
OCV	Open-circuit voltage
OP	Operational point
OVPU	Overvoltage protection unit
PEMFC	Proton exchange membrane fuel cell
PFC	Power factor correction
PG1	First planetary gear set
PG2	Second planetary gear set
PHEV	Plug-in hybrid electric vehicle
PID	Proportional integral derivative controller
PMR	Power to curb mass ratio
PNGV	Partnership for a New Generation of Vehicles
PST	Power split transmission
PVDF	Polyvinylidene fluoride
PWM	Pulse-width modulation
Rcda	Range of charge-depleting actual
Rcdc	Range of charge-depleting cycle
RESS	Rechargeable energy storage system
RMS	Root mean square
RPM	Revolutions per minute
RWR	Roadway recognition
SBR	Styrene butadiene copolymer
SCT	Single cycle range and energy consumption test cycle
SEI	Solid electrolyte interface
SFTP	Supplemental federal test procedure
SG	Specific gravity
SISO	Single-input, single-output system
SOC	State of charge
SOFC	Solid oxide fuel cell
SOH	State of health

SOHC	Single overhead cam engine
SOL	State of life
SP	Singular pencil model
SPL	Sound pressure level
TC	Torque converter of automatic transmission
TDC	Top dead center
TL	Sound transmission loss
TPIM	Traction power inverter module
TRC	Transient response characteristics
TWC	Test weight class
UDDS	Urban dynamometer driving schedule
USABC	US Advanced Battery Consortium
V_{oc}	Open-circuit voltage
VDV	Vibration dose value
VITM	Voltage, current, and temperature measurement unit
WLTP	Worldwide harmonized light vehicle test procedure
WOT	Wide open throttle

Nomenclature

A_d	Air mass density
$A_{H/C}$	Heating/cooling surface area between the battery pack and the heating/cooling channel
Ahr	Ampere hours
C_a	Air density correction coefficient for altitude
Cap_{BOL}	Battery Ahr capacity at the beginning of life
Cap_{EOL}	Designed battery Ahr capacity at the end of life
C_{bat_life}	Battery life cost weight factor
C_c	Specific heat of coolant
C_d	Vehicle aerodynamic drag coefficient
C_{diff}	Diffusion capacitance of the second-order electrical circuit battery model
C_{dl}	Double layer capacitance of the second-order electrical circuit battery model
C_{ds}	Source to drain capacitance of MOSFET
C_{dyn}	Dynamic capacitance of battery electrical circuit model
C_{ele}	Electrical power cost weight factor
$C_{energy\_balance}$	Imbalanced energy cost weight factor
C_{ESS}	Specific heat of the battery system
C_{fuel}	Fuel cost weight factor
C_{gd}	Parasitic capacitance from gate to drain of a MOSFET
C_{gs}	Parasitic capacitance from gate to source of a MOSFET
D	Duty cycle of PWM control method
D_{cf}	Distance between the center of gravity and the front wheel of a vehicle
D_{cr}	Distance between the center of gravity and the rear wheel of a vehicle
E_a	Back electromotive force
E_a	Activation energy of a battery

E_{ac}	AC recharge energy
E_{FD}	Full depletion DC discharge energy
F	Faraday constant, the number of coulombs per mole of electrons $(9.6485309 \times 10^4 \text{ C mol}^{-1})$
F_a	Frontal area of a vehicle
$F_{actuator\_max}$	Maximal output force of the actuator in an active suspension system
F_{wf}	Friction force acting on the front wheel of a vehicle
F_{wr}	Friction force acting on the rear wheel of a vehicle
$G_{actuator}(s)$	Transfer function of the actuator in an active suspension system
$G(Cap)$	Decline in the Ahr capacity of a battery system
$G(R)$	Increment of the internal resistance of a battery system
H_2	Hydrogen gas
H_{cg}	Height from the center of gravity of a vehicle to the road
$H_{gen'd}$	Battery heat generated
$I_{balancing\_max}$	Maximal balancing current of a battery system
I_{FAV}	Maximal average forward current
I_{FRMS}	Maximal RMS forward current
I_{FSM}	Maximal forward surge current
I_{GM}	Maximal peak positive gate current of a thyristor
I_H	Holding current of a thyristor
I_{max_chg}	Maximal allowable charge current of a battery system
I_{max_dischg}	Maximal allowable discharge current of a battery system
J	Motor rotor inertia
J_{axle}	Lumped inertia on the axle transferred from the powertrain
J_{eng}	Lumped engine inertia
J_{fd}	Final drive inertia
J_{gr}	Gearbox inertia
J_{mot}	Lumped motor inertia
J_{tc}	Lumped torque converter inertia
J_{wh}	Vehicle wheel inertia
K	Proportional gain of the PID controller
$K_{actuator}$	Gain of the actuator from input voltage to output force of an active suspension system
K_e	Voltage constant of a BLDC motor
K_m	Torque constant of a BLDC motor
L_m	Magnetizing inductance
L_r	Rotor phase inductance
L_r	Load rate of torque converter
L_s	Motor stator phase inductance
$LiCoO_2$	Lithium cobalt oxide
$LiFePO_4$	Lithium iron phosphate
LMO	Lithium manganese oxide

LTO	Lithium titanate
M_c	Total coolant mass of the energy storage system
M_{ESS}	Mass of the energy storage system
N_C	Number of teeth in the carrier of the planetary gear set
N_R	Number of teeth in the ring gear of the planetary gear set
N_S	Number of teeth in the sun gear of the planetary gear set
O_2	Oxygen gas
Pa	Pascal
P_{acc}	Lumped accessory power
P_{a_pct}	Acceleration pedal position as a percentage
P_{brake_pct}	Brake pedal position as a percentage
P_{bat}	Battery power
Pct	Percentage gradeability
P_{eng}	Engine power
$P_{max\_chg\_bat}$	Maximal allowable battery-charging power
$P_{max\_dischg\_bat}$	Maximal allowable battery-discharging power
$P_{max\_prop\_mot}$	Maximal allowable motor propulsive power
$P_{max\_regen\_mot}$	Maximal allowable motor regenerative power
P_{mot}	Motor power
P_{pump}	Operational power of heating/cooling system pump
P_s	Actual atmospheric pressure
P_{Veh}	Power demand of the vehicle
Q	Battery reaction quotient
Q_c	Surface-convection heat transfer
$\dot{Q}_{H/C}$	Energy transfer rate of heater/chiller
R	Electrical resistance
R_{AF}	Recharge allocation factor
R_{BOL}	Battery's internal resistance at the beginning of life
R_{EOL}	Battery's internal resistance at the end of life
R_{ct}	Charge transfer resistance of the second-order electrical circuit battery model
R_{diff}	Diffusion resistance of the second-order electrical circuit battery model
R_{dyn}	Dynamic resistance of the battery electrical circuit model
R_{ESS}	Internal resistance of the battery system
R_g	Universal gas constant: $R = 8.314472 \, \mathrm{J\,K^{-1}\,mol^{-1}}$
R_{int}	Battery cell internal resistance
R_{ohm}	Ohmic resistance of the battery electrical circuit model
R_r	Motor rotor phase resistance
R_s	Motor stator phase resistance
R_{wf}	Reaction force acting on the front wheel of a vehicle
R_{wr}	Reaction force acting on the rear wheel of a vehicle
SOC_{init}	Initial state of charge of the battery

SOC_{target}	Target state of charge of the battery
$T_{actuator}$	Time constant of the actuator from input voltage to output force
T_c	Coolant temperature
T_{c_init}	Initial coolant temperature
T_{c_sp}	Coolant temperature setpoint
T_d	Derivative time constant of the PID controller
T_{ESS}	Energy storage system temperature
T_{ESS_init}	Initial battery system temperature
T_{ESS_sp}	Temperature setpoint of the energy storage system
T_i	Integral time constant of the PID controller
T_J	Maximal junction temperature of the power electronics
V_{BE}	Base-emitter voltage of a bipolar transistor
V_{CE}	Collector-emitter voltage of a bipolar transistor
V_{DRM}	Peak repetitive forward blocking voltage of a thyristor
$V_{dynamic}$	Voltage on the dynamic component of the battery electrical circuit model
V_{GM}	Maximal peak positive gate voltage of a thyristor
V_{GS}	Gate voltage of a MOSFET
V_{max}	Maximal allowable battery system terminal voltage
V_{min}	Minimal allowable battery system terminal voltage
V^0	Standard cell potential
V_o	Potential of the battery electrical circuit model
V_R	Maximal reverse voltage of a power diode
V_{RRM}	Peak repetitive reverse blocking voltage of a thyristor
V_{RSM}	Non-repetitive peak reverse voltage of a thyristor
V_{RWM}	Maximal working peak reverse voltage of a power diode
$V_{terminal}$	Battery system terminal voltage
a	Acceleration
a_w	Frequency weighted acceleration in m/s^2
f_d	Final drive ratio
f_{emi_CO}	Carbon monoxide emissions
f_{emi_HC}	Hydrocarbon emissions
f_{emi_NOx}	Nitrogen oxide emissions
f_{emi_PM}	Particulate matter emissions
f_{fuel}	Fuel economy
g_{CO_hot}	Hot carbon monoxide emission rate
g_{fuel_hot}	Hot fuel economy rate
g_{HC_hot}	Hot hydrocarbon emission rate
g_m	MOSFET transconductance
g_{Nox_hot}	Hot nitrogen oxides emission rate
g_{PM_hot}	Hot particulate matter emission rate
g_r	Gear ratio
h	Heat transfer coefficient

h_{bat}	Battery heat transfer coefficient
i_{Dr}	Rotor direct-axis current of an AC induction motor
i_{Qr}	Rotor quadrature-axis current of an AC induction motor
i_{ds}	d-axis or air-gap flux current of an AC induction motor
i_{qs}	q-axis or torque current of an AC induction motor
i_{Ds}	Stator direct-axis current of an AC induction motor
i_{Qs}	Stator quadrature-axis current of an AC induction motor
k_{aero}	Aero drag factor
k_{chg}	Charging power margin factor
k_d	Distortion factor of plug-in charger
k_{rrc}	Rolling resistance coefficient
k_{sc}	Road surface coefficient
k_{split}	Split coefficient between the engine and the electric motor
$\dot{m}_c$	Mass flow rate of coolant
m_v	Manufacturer-rated gross vehicle mass
$m_v a$	Vehicle acceleration force
$m_v g$	Gross weight of vehicle
n_e	Number of electrons transferred in the cell reaction
r_C	Carrier radius of the planetary gear set
r_{ds}	Source to drain resistance of a MOSFET
r_o	MOSFET output resistance
r_R	Ring gear radius of the planetary gear set
r_S	Sun gear radius of the planetary gear set
r_{wh}	Effective wheel rolling radius
s_r	Speed ratio of the torque converter
v_{Dr}	Rotor direct-axis voltage of an AC induction motor
v_{Qr}	Rotor quadrature-axis voltage of an AC induction motor
v_{Ds}	Stator direct-axis voltage of an AC induction motor
v_{Qs}	Stator quadrature-axis voltage of an AC induction motor
ΔS	Delta entropy of reaction
ψ	Motor rotor magnetic flux
Ψ_{Dr}	Motor rotor direct-axis flux linkage
Ψ_{Qr}	Motor rotor quadrature-axis flux linkage
Ψ_{Ds}	Motor stator direct-axis flux linkage
Ψ_{Qs}	Motor stator quadrature-axis flux linkage
α	Road incline angle
λ	Forgetting factor of recursive least squares estimator
λ_{fuel}	Fuel economy temperature factor
λ_{CO}	Carbon monoxide emissions temperature factor
λ_{HC}	Hydrocarbon emissions temperature factor
λ_{NOx}	Nitrogen oxide emissions temperature factor
λ_{PM}	Particulate matter emissions temperature factor

$\delta(t)$	The Dirac delta function
ρ	Density of air
η_{bat}	Battery efficiency
η_{chg}	Battery charge efficiency
η_{fd}	Final drive efficiency
η_{gr}	Gearbox efficiency
$\eta_{H/C}$	Heater/chiller efficiency
η_{mot}	Motor efficiency
η_{pt_eng}	Engine power drivetrain efficiency
η_{pt_mot}	Electric motor power drivetrain efficiency
η_{tc}	Torque converter efficiency
μ_A	Membership function of fuzzy logic
τ_a	Acceleration torque
τ_{access}	Lumped torque of mechanical accessories
τ_C	Coulomb friction torque
τ_{cct}	The closed-throttle torque of the engine
τ_{com}	Compression torque
τ_{crank}	Cranking torque
τ_{demand}	Torque demand of the vehicle
τ_e	Electromagnetic torque
τ_{eng}	Engine torque
τ_{load}	Load torque
τ_{loss}	Lumped loss torque
τ_{mot}	Motor torque
τ_{regen}	Regenerative torque
τ_r	Torque ratio of the torque converter
τ_s	The static friction torque
τ_{trac}	Traction torque from the powertrain
τ_v	Viscous friction torque
ω	Vehicle wheel angular velocity
ω_c	Angular velocity of the carrier of the planetary gear set
ω_{eng}	Angular velocity of the engine
ω_{max_eng}	Maximal allowable angular velocity of the engine
ω_{mot}	Angular velocity of the motor
ω_{max_mot}	Maximal allowable angular velocity of the motor
ω_R	Angular velocity of the ring gear of the planetary gear set
ω_s	Synchronous speed of an AC induction motor
ω_s	Angular velocity of the sun gear of the planetary gear set

1

Introduction

In recent decades, hybrid electric technology has advanced significantly in the automotive industry. It has now been recognized that the hybrid is the ideal transitional phase between the traditional all-petroleum-fueled vehicle and the all-electric vehicles of the future. In popular concepts, a hybrid electric vehicle (HEV) has been thought of as a combination of an internal combustion engine (ICE) and an electric motor.

The most important feature of hybrid vehicle system technology is that fuel economy can be increased noticeably while meeting increasingly stringent emission standards and drivability requirements. Thus, hybrid vehicles could play a crucial role in resolving the world's environmental problems and the issue of growing energy insecurity. In addition, hybrid technology has been a catalyst in promoting the technology of electric motors, power electronics, and batteries to maturity (Powers and Nicastri, 1999; Chan, 2002).

An HEV is a complex system of electrical and mechanical components. Its powertrain control problems are complicated and often have conflicting requirements. Moreover, they are generally nonlinear, exhibit fast parameter variation, and operate under uncertain and changing conditions; for example, the vehicle has to run well on a cold January day in northern Ontario as well as on a sweltering day in Death Valley. Many control design objectives are very difficult to formalize, and many variables that are of the greatest concern are not measurable. The HEV system control is also fundamentally a multivariable problem with many actuators, performance variables, and sensors. It is often important to take advantage of these interactions with multivariable designs; however, multivariable designs may make control strategies less robust to parameter variation and uncertainties, and thus may be more difficult to calibrate. In this book, we will systematically introduce HEVs' control problems from powertrain architecture and modeling to design and performance analysis.

Hybrid Electric Vehicle System Modeling and Control, Second Edition. Wei Liu.
© 2017 John Wiley & Sons Ltd. Published 2017 by John Wiley & Sons Ltd.

1.1 Classification of Hybrid Electric Vehicles

In order to cover automotive needs, various hybrid electric vehicle concepts have been proposed and developed. According to the degree of hybridization, nowadays hybrid electric vehicles can be classified as micro hybrid, mild hybrid, full hybrid, or plug-in hybrid electric vehicles as well as fully electric vehicles. These hybrid electric vehicles are described briefly in the following sections and a classification summary is given in Table 1.1.

1.1.1 Micro Hybrid Electric Vehicles

Micro hybrid electric vehicles are normally operated at low voltages between 12 V and 48 V. Due to the low operational voltage, the electric power capability is often under 5 kW, and thus micro hybrid electric vehicles primarily have auto start–stop functionality. Under braking and idling circumstances, the internal combustion engine is automatically shut down, so fuel economy can be improved by 5–10% during city driving conditions. With the power capability increase of a 12 V battery, some micro hybrid vehicles even have a certain degree of regenerative braking capability and are able to store the recovered energy in the battery. Most micro hybrid electric systems are implemented through improving the alternator–starter system, where the conventional belt layout is modified and the alternator is enhanced to enable the engine to be started and the battery to be recharged. Valve-regulated lead–acid batteries (VRLAs) such as absorbent glass mat (AGM) batteries and gel batteries are widely used in micro hybrid electric vehicles. The biggest advantage of the micro hybrid vehicle is the lower cost, while the main drawback is the inability to recover all regenerative braking energy.

1.1.2 Mild Hybrid Electric Vehicles

Compared with micro hybrid electric vehicles, mild hybrid electric vehicles normally have an independent electric drivetrain providing 5–20 kW of electric propulsion power, and the electric drive system typically operates at voltages between 48 V and 200 V. Mild hybrid

Table 1.1 The main features and capabilities of various hybrid electric vehicles

Type of vehicle	Features and capabilities				
	Start–stop	Regenerative braking	Boost	Electric-only mode	Electric range (miles)
Micro hybrid	Yes	Possible	No	No	No
Mild hybrid	Yes	Yes	Yes	No	No
Full hybrid	Yes	Yes	Yes	Possible	Possible (<2)
Plug-in hybrid	Yes	Yes	Yes	Yes	Yes (20–60)
Pure electric	Yes	Yes	Yes	Yes	Yes (80–150)

electric vehicles can make use of an electric motor to assist the internal combustion engine during aggressive acceleration phases and enable the recovery of most regenerative energy during deceleration phases. Therefore, mild hybrid electric vehicles have great freedom to optimize vehicle fuel economy and vehicle performance, and improve driving comfort. Mild hybrid electric architecture is often implemented in several ways depending on the degree of hybridization. The belt starter–generator, mechanically coupled via the alternator belt in a similar manner to micro hybrids, and the starter–generator, mechanically coupled via the engine crankshaft, are typical implementations. Nickel–metal hydride and lithium-ion batteries are often employed in mild hybrid electric vehicles. One distinguishing characteristic of mild hybrid electric vehicles is that the vehicle does not have an exclusive electric-only propulsion mode. The fuel economy improvement is mainly achieved through shutting down the engine when the vehicle stops, using electrical power to initially start the vehicle, optimizing engine operational points, and minimizing engine transients. Typical fuel savings in vehicles using mild hybrid drive systems are in the range of 15 to 20%.

1.1.3 Full Hybrid Electric Vehicles

Full hybrid electric vehicles (HEVs) are also called strong hybrid electric vehicles. Here, the electric drive system normally has in excess of 40 kW of power and operates on a voltage level above 150 V for the sake of the operational efficiency of the electrical system and the component/wire size. The electric powertrain of a full hybrid electric vehicle is capable of powering the vehicle exclusively for short periods of time when the combustion engine runs with lower efficiency, and the energy storage system is designed to be able to store the free regenerative braking energy during various deceleration scenarios. These vehicles can also provide a purely electric driving range of up to two miles to meet some special requirements such as silent cruising in certain areas and zero emissions for driving in tunnels and indoors. The ideal application scenario for full hybrid electric vehicles is continuous stop-and-go operation; therefore, they are widely used as city buses and delivery trucks. Compared with traditional internal combustion engine vehicles, the overall fuel economy of a full hybrid electric vehicle in city driving could improve by up to 40%.

1.1.4 Electric Vehicles

Electric vehicles (EVs) are operated with electrical power only. Presently, most electric vehicles employ lithium-ion batteries as the energy storage system, with a plug to connect to the electric grid to charge the battery. The capacity of the energy storage system plays a crucial role in determining the electric driving range of the vehicle. However, enlarging the energy storage capacity would result in an increase in vehicle mass and volume, and would also require quite a long time to charge the battery without a fast-charging facility. Most electric vehicles on the market have an 80–150 mile electric range, while in the near future, 300–400 mile ranges could be achieved with a cutting-edge battery system with more than 80 kWh storage capability.

A major concern with such battery-powered electric vehicles (BEVs) is the range limitation, and technical challenges currently preventing progress with battery-powered electric vehicles include the need to reduce plug-in charging times significantly and to predict the energy remaining in the battery precisely. In the long term, a fuel-cell-powered electric vehicle could be a solution and could emerge on the automotive markets if the remaining technical and economic barriers are overcome and a hydrogen infrastructure established.

1.1.5 Plug-in Hybrid Electric Vehicles

Plug-in hybrid electric vehicles (PHEVs) share the characteristics of both full hybrid electric vehicles and all-electric vehicles with the capability of charging the battery through an AC outlet connected to the electric grid. The electric powertrain of PHEVs normally has an 80–150 kW electrical power capability that allows the vehicle to operate in exclusively electric mode with an electric range of 20–60 miles on most daily driving routes. Similar to BEVs, a PHEV also uses power from the grid to charge the battery. During a driving route, the vehicle normally first operates in electric mode using the energy stored in the battery; once the battery is depleted to a certain level, the internal combustion engine starts to propel the vehicle and the battery provides supplemental electric power and stores regenerative braking energy like a full HEV to improve fuel economy and dynamic performance and also to reduce emissions.

1.2 General Architectures of Hybrid Electric Vehicles

There are two fundamental architectures of hybrid electric vehicles:

1. The series hybrid vehicle, in which the engine, coupled with a generator, powers the generator for recharging the batteries and/or supplying electrical energy to the electric motor. The motor, in turn, provides all torque to the wheels.
2. The parallel hybrid vehicle is propelled by either an engine or an electric motor, or both. The electric motor works as a generator to recharge the batteries during regenerative braking or when the engine is producing more power than is needed to propel the vehicle.

Although possessing the advantageous features of both series and parallel HEVs, the series–parallel HEV is relatively more complicated and costly. Nevertheless, this system has been adopted by some modern HEVs, as advanced control and manufacturing technologies can be applied.

1.2.1 Series Hybrid

A series HEV, as shown in Fig. 1.1, has power sources in electromechanical series. The electric powertrain only provides propulsion power to the drive wheels, and an

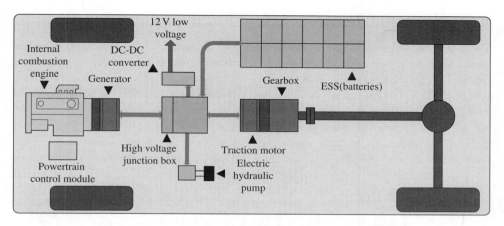

Figure 1.1 A rear-wheel-drive series hybrid electric vehicle layout

engine–generator pair unit (genset) provides electrical power and energy with a high-voltage bus. The energy storage system (ESS) is charged or discharged to achieve optimal fuel economy, while the electric motor propels the vehicle to realize vehicle performance requirements. Therefore, in simple terms, a series hybrid vehicle is an electric vehicle with a genset to supply electrical energy when the ESS lacks sufficient energy to power the vehicle.

Because of the simplicity of dynamic control, this type of hybrid vehicle has many practical uses, especially in the form of heavy/medium-duty delivery trucks and shuttle buses. In this type of system, the primary function of the genset is to extend the range of the electric vehicle beyond what is possible with the battery alone. The key technical challenge of this type of hybrid electric vehicle is to manage energy sources and power flow optimally.

1.2.2 Parallel Hybrid

In contrast to a series HEV, a parallel HEV essentially blends ICE power output with electric motor/generator power output. There are multiple potential points connecting these two power sources to the drivetrain depending on the availability of the components. In a parallel HEV configuration, as shown in Fig. 1.2, an electric powertrain system is added to the conventional powertrain system through a clutch that enables the vehicle to be driven by the electric motor or engine either separately or together. The maximal power rating of the electric powertrain is normally smaller than that of the engine powertrain in a parallel hybrid vehicle. The size of the electric powertrain is determined such that the electric motor and ESS can deliver the required power for a given drive cycle. In addition, the conventional powertrain must be able to provide sufficient flexible torque that can be smoothly and efficiently combined with the torque from the electric motor to meet the torque requirements to propel the vehicle. The engine may be turned on and off frequently in response to the system control strategy.

1.2.3 Series–Parallel Hybrid

The series–parallel architecture is a combination of the two described above. The electric motor, the electric generator, the internal combustion engine, and the wheels of the vehicle can be linked together through a device such as a planetary gear set. Figure 1.3 shows a conceptual layout of the series–parallel hybrid vehicle system, in which the power provided by the engine is split up and transmitted to the wheels through two paths: *series* and *parallel* paths. The *series path* leads through the electric generator jointed with the ESS to the electric motor to the wheels. In this path, the mechanical power of the engine is converted to electrical power through the generator, and the electrical power can partly flow to the ESS or

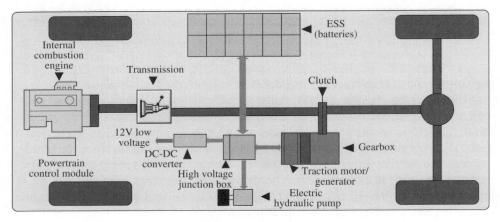

Figure 1.2   A rear-wheel-drive parallel hybrid electric vehicle layout

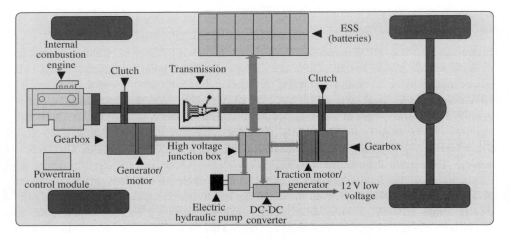

Figure 1.3   A rear-wheel-drive series–parallel hybrid electric vehicle layout

entirely to the wheels through the electric driveline. In the second path, the *parallel path*, the engine is connected through a gear set to the conventional drivetrain. In this path, the mechanical power of the engine is partly or entirely transmitted mechanically to the wheels, and the part not transmitted to the wheels is converted to electrical power through the electric motor to charge the battery. If the entire mechanical power of the engine cannot meet the vehicle's power demand, the electric motor drivetrain supplies supplemental power to the wheels. The series–parallel hybrid electric configuration acts at all times as a combination of the series and parallel configurations. It allows the electric motor drivetrain to adjust the engine load to achieve optimal fuel economy. The percentage of power flowing through the series and parallel paths is determined in real time to achieve the optimal vehicle performance. Although the power flow can be set by controlling the speeds of the planetary gear set, a sophisticated control system is needed to control the power flow to achieve the best fuel efficiency.

The comparison of series and parallel architectures described above leads to the conclusion that, in city driving conditions, series hybrid behavior is preferable, while in highway driving conditions, a parallel hybrid action is generally desired. Therefore, the series–parallel hybrid architecture combines the positive aspects of the series hybrid architecture – the independence of the engine operation from the driving conditions – with the advantage of the parallel hybrid architecture – efficient mechanical transmission. The complexity of the control task for the series–parallel configuration is the main distinct point compared to the individual series or parallel systems.

1.3 Typical Layouts of the Parallel Hybrid Electric Propulsion System

The basic parallel hybrid electric architecture shown in Fig. 1.2 is actually often implemented in several ways. There are several mechanical points in the drivetrain that can be used to implement hybridization. Figure 1.4 shows typical parallel hybrid arrangements. Based on the location of the hybrid joint points, these parallel hybrid architectures are called P0, P1 … P4, respectively from front to rear.

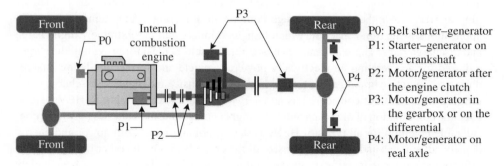

P0: Belt starter–generator
P1: Starter–generator on the crankshaft
P2: Motor/generator after the engine clutch
P3: Motor/generator in the gearbox or on the differential
P4: Motor/generator on real axle

Figure 1.4 Typical parallel hybrid electrified powertrain arrangements

The P0 hybrids are commonly implemented through an enhanced belt starter–generator system to handle frequent start–stop operation. The P1 hybrid joint point is on the crankshaft, where the starter in a traditional vehicle is replaced by a starter–generator which can support start–stop and recover some regenerative braking energy. P0 and P1 are typical architectures of micro hybrid electric vehicles and a 12 V battery system is normally employed as the energy storage device. The P2 hybrid joint is on the input gear shaft of the transmission; most mild hybrid electric vehicles use the P2 architecture, and a set of planetary gears is usually employed to distribute ICE power and electric motor power to the output shaft of the transmission. Compared with P2 architecture, the hybrid joint of P3 is on the transmission output shaft. Since the electric motor is mounted on the output shaft, the propulsion efficiency is improved. P2 and P3 are often combined in strong hybrid electric vehicles, and multiple operational modes can be implemented to achieve superior fuel economy and performance in either city driving or highway driving. Since the motor/generators are mounted on the rear axle, P4 architecture is the most efficient implementation to recover regenerative braking energy and boost the vehicle. This type of parallel hybrid is also called parallel-through-road, since the ICE power and electric motor power are coupled through the road. Therefore, the battery can only be charged when the vehicle is running; that is, the ability to recharge the battery is limited.

1.4   Hybrid Electric Vehicle System Components

Compared with traditional vehicles, the ESS, electric motors, transmission systems, and power-electronics-related components such as converters and inverters are key components in hybrid vehicle systems. In order to size these components and analyze hybrid system performance, it is necessary to establish their models based either on physical principles or test data.

- **The ESS:** This is one of the most important subsystems in hybrid vehicles, which directly affects the efficiency and other performance factors of the vehicle. In hybrid vehicle applications, the batteries need to have high energy density, low internal resistance, and long cycle and calendar life. Depending on the design objective, higher power density batteries are generally used for traditional HEVs and higher energy density batteries are needed for plug-in HEVs. Another energy storage component attracting R&D attention for HEV applications is the ultracapacitor, which lasts indefinitely and has extremely high charge and discharge rates. These advantages make ultracapacitors ideal for providing the surges required for accelerating an electrically powered vehicle and for accumulating charges during regenerative braking. Due to their low energy density and high self-discharge rates, ultracapacitors are not considered as an energy storage device for plug-in HEVs. However, the combination of ultracapacitors and higher energy density batteries may have considerable potential for all types of HEV, as this combination has both power and energy density advantages and decreases the size of the entire ESS. On the other hand, with significant reductions in manufacturing cost, lithium-ion (Li-ion) batteries have been widely regarded as the best choice for hybrid and purely electric vehicles.

- **Transmission:** Hybrid vehicle systems carry some specific demands for transmission design. Generally speaking, the hybrid vehicle transmission must be able to manage ICE driving, electric-only driving, and combinations of the two. Functionally, it has to support functions of stop–start, regenerative braking, and shifting the operational zone of the ICE; furthermore, the transmission must also be able to adjust its parameters to match the actual drive scenarios. That is, a hybrid vehicle system mainly relies on the transmission to implement an optimal performance for multiple types of drive cycle rather than a particular cycle. Other challenges for hybrid transmission design include minimizing additional weight, cost, and packaging.
- **Electric motors:** Efficient, light, powerful electric motors also play a key role in hybrid technology. Depending on the architecture of an HEV, the electric motor can be used as a peak power regulation device, a load-sharing device, or a small transient source of torque. Electric motors also operate well in two modes – normal mode and extended mode. In the 'normal' mode, the motor exerts constant torque throughout the rated speed range. Once past the rated speed, the motor enters its 'extended' mode, in which torque decreases with speed. In HEVs, the electric motor is primarily designed to deliver the necessary torque for adequate acceleration during its normal mode before it changes to its extended mode for steady speeds. Depending on the design objectives, direct current (DC), brushless DC, and alternating current (AC) induction motors can be selected for HEVs.

  The second function of electric motors is to capture the energy from regenerative braking. The electric motors for HEV applications need to have the capacity to operate equally well as a generator when driven by some external rotational force. Applying the brake pedal in an HEV normally signals to the control system for the motor to generate negative torque, switch off the ICE or let the vehicle's momentum drive the electric motor via the drivetrain. In the case where the electric motor generates negative torque, the mechanical energy of the vehicle will be converted to AC electrical energy by the motor, and then the inverter system on the motor assembly will invert the AC to DC to recharge the battery system. The control system tasks include optimizing the regenerative braking strength in combination with activating the conventional hydraulic braking system in accordance with the pressure applied to the brake pedal. Gentle deceleration generally maximizes the use of the regenerative system, but emergency braking sometimes needs to utilize the conventional braking system. As stop–start urban driving involves frequent acceleration and deceleration, the regenerative braking system and control strategy are crucial technologies for the improvement of the fuel efficiency of a hybrid vehicle.
- **Power electronic components:** In addition to batteries, electric motors, and the transmission, DC–DC converters and DC–AC inverters are key components in hybrids. The function of a DC–DC converter in HEVs/EVs is to convert the high voltage supplied by the ESS to a lower voltage, which normally supplies 12 V electrical power to various accessories such as headlamps and wipers. The function of the inverter in HEVs/EVs is to convert the DC voltage of the ESS to a high AC voltage to power the AC electric propulsion motor. Under regenerative braking, this process is reversed; the output AC power of the motor, operating as a generator, is converted to DC power to charge the battery. The efficiencies of these power electronic components have significant impact on the overall efficiency of the vehicle.

1.5 Hybrid Electric Vehicle System Analysis

1.5.1 Power Flow of Hybrid Electric Vehicles

Different types of HEV configuration have different power flow paths. In series hybrid power flow, as shown in Fig. 1.5, the propulsion power comes from an electric motor which converts electrical energy into the mechanical energy required by the vehicle, while the motor can be powered by either a generator or the ESS. The engine and generator pair can either power the electric motor or charge the ESS. During regenerative braking, the motor works as a generator which converts braking mechanical energy into electrical energy to charge the ESS. When cranking the engine, the battery will provide electrical energy to the generator. In parallel hybrid power flow, the vehicle can be powered by either the engine or the electric motor or both, depending on the system state and the control objectives. During regenerative braking, the captured braking energy will be converted into electrical energy by the electric motor and stored in the ESS. The ESS will power the motor/generator to crank the engine when the key starts. The power flow path of a parallel hybrid system is shown in Fig. 1.6.

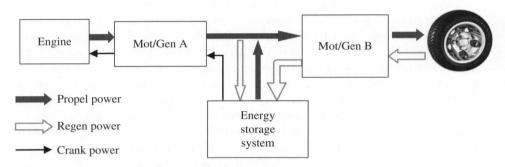

Figure 1.5 Power flow of a series hybrid electric vehicle

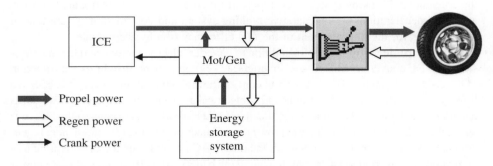

Figure 1.6 Power flow of a parallel hybrid electric vehicle

1.5.2   Fuel Economy Benefits of Hybrid Electric Vehicles

Figure 1.7 illustrates the fuel economy improvement of a P2-type hybrid electric vehicle from an ICE-only conventional vehicle in typical urban cycles. For one unit of energy to be delivered at the wheels, if we assume that the final drive efficiency is 92%, the transmission efficiency is 85%, and the conversion efficiency of fossil fuel energy to mechanical energy is about 28% for an ICE in the conventional powertrain, 4.6 units of fossil energy are required to propel the vehicle in a typical urban cycle, shown by the dotted lines in Fig. 1.7. Typically, 60% of the energy at the wheels is consumed by aero drag and rolling resistance and the other 40% is converted into kinetic energy and eventually consumed by braking in urban cycles. Since the overall vehicle mass increases with hybridization, to achieve the same performance as a conventional vehicle, the hybrid vehicle requests 1.05 units of energy at the wheels. However, most kinetic energy consumed by braking in a conventional vehicle can be recovered by regenerative braking in a hybrid electric vehicle.

As shown by the dashed line in Fig. 1.7, there is 0.38 of a unit of energy available at the wheels for regenerative braking; this amount of energy is reduced to 0.35 of a unit at the output shaft of the transmission, and further reduced to 0.3 of a unit when it reaches the motor/generator shaft due to efficiency loss. If we assume that the one-way efficiencies of the motor/invertor and the battery are 92% and 96% respectively, the 0.38 of a unit of regenerative energy at the wheels will be converted into 0.26 of a unit of chemical energy stored in the battery, which can contribute 0.23 of a unit of mechanical energy at the input shaft of transmission to propel the vehicle afterwards.

On the other hand, the second electrical power source provides the opportunity to operate the ICE in optimal states and shut down the ICE when the vehicle is still, which saves about 0.68 of a unit of fossil fuel energy. Overall, the hybridization shown can improve fuel economy by about 24% in typical urban cycles, shown by the pale grey line in Fig. 1.7.

1.5.3   Typical Drive Cycles

Since a vehicle's fuel economy and emissions are strongly affected by environmental factors such as road condition, traffic, driving style, weather, etc., it is not a good idea to try and judge whether a vehicle really improves fuel economy or emissions based on actual fuel consumption and emissions measured on the road. To get around this problem, the automobile industry and governments have developed a series of standard tests whereby the fuel consumption and emissions of a vehicle can be measured under completely repeatable conditions, and different vehicles can be compared fairly to each other. These tests are described as drive cycle tests and are conducted as a matter of routine on all new car designs. Most drive cycle tests will be described in Chapter 10.

1.5.4   Vehicle Drivability

Drivability can be understood as the capacity of a vehicle to deliver the torque requested by the driver at the time expected. It is often evaluated subjectively but can also be quantified

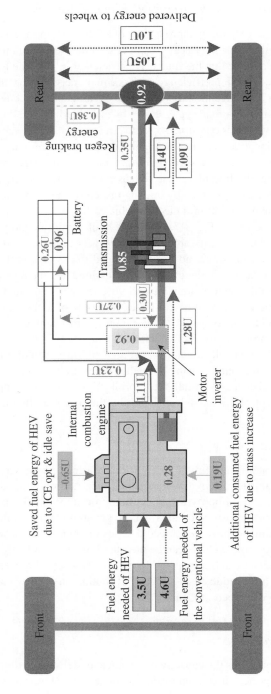

Figure 1.7 Typical urban cycle energy flows of a conventional powertrain and a hybrid electrified powertrain

objectively through accelerometers. Problems such as hesitation, powertrain excitation during acceleration (acceleration pedal tip-in) and deceleration (acceleration pedal back-out) maneuvers are identified in this attribute. Compared with conventional vehicles, hybrid vehicles have more operational modes. The delivered torque is associated not only with the states of the internal combustion engine (ICE), the electric motor/converter and the ESS, but also with the energy management strategy determining how to split the vehicle's required power between the ICE and the electric motor. In order to achieve the maximal fuel economy and meet emission standards under different driving situations, an HEV has to employ more complex control strategies to meet the drivability requirements. The complexity of the control and powertrain systems makes it a challenge to analyze an HEV's drivability.

1.5.5   Hybrid Electric Vehicle Fuel Economy and Emissions

The actual fuel consumption and emissions of ICE-driven vehicles can be measured directly. Since HEVs, especially plug-in HEVs, can make use of an external electrical source (such as the public grid), the electrical energy withdrawn from that source must be separately accounted for when performing fuel consumption and emissions calculations.

1.6   Controls of Hybrid Electric Vehicles

Since a hybrid vehicle is a complex system of electrical and mechanical components which contains multidisciplinary technologies, modern control system techniques and methodologies are playing important roles in hybrid technology (Powers and Nicastri, 1999). An HEV's performance is affected by many interrelated multidisciplinary factors; therefore, advanced control strategies could significantly improve its performance and lower its costs. The overall control objective of a hybrid vehicle is to maximize fuel economy and minimize emissions. In order to achieve the objectives, some key system variables must be optimally governed; these primarily include the energy flow of the system, the availability of energy and power, the temperatures of subsystems, and the dynamics of the engine and the electric motor. Some typical HEV control issues are as follows:

- **Make sure the ICE works at the optimal operating points:** Each ICE has optimal operating points on its torque–speed plane in terms of fuel economy and emissions. If the ICE operates at these points, maximal fuel economy, minimal emissions, or a compromise between fuel economy and emissions can be achieved. Ensuring that the HEV's ICE operates at these points under various operating conditions is a challenging control objective.
- **Minimize ICE dynamics:** As an ICE has inertia, additional energy is consumed to generate the related kinetics whenever the operating speed changes. Therefore, the operating speed of the ICE should be kept constant as much as possible and any fast fluctuations should be avoided. HEVs make it possible to minimize the dynamics under changing load, road, and weather conditions.

- **Optimize ICE operational speed:** According to the working principle of an ICE, its fuel efficiency is low if the ICE operates at low speed. The ICE speed can be independently controlled with the vehicle speed and can even be shut down when its speed is below a certain value, in order to achieve maximal benefits.
- **Minimize ICE turn on/off times:** The ICE in an HEV can be turned on and off frequently as it has a secondary power source; furthermore, the times at which the ICE is turned on/off can be determined based on an optimal control method to minimize fuel consumption and emissions.
- **Optimally manage the battery's state of charge (SOC):** The battery's SOC needs to be controlled optimally so that it is able to provide sufficient energy to power the vehicle and accept regenerative energy during braking or while traveling downhill as well as maximizing its service life. The simplest control strategy is to turn the ICE off if the battery's SOC is high and turn the ICE on if the SOC is too low. A more advanced control strategy will be able to regulate the output power of the ICE based on the actual SOC level of the ESS.
- **Optimally control the voltage of the high-voltage bus:** The actual voltage of the high-voltage bus of an HEV has to be controlled during discharging and charging to avoid being over or under limits; otherwise, the ESS or other components may be permanently damaged.
- **Optimize power distribution:** Since there are two power sources in an HEV, the most challenging and important control task is to split the vehicle's power demand between the ICE and the electric motor based on the driving scenario, road and weather conditions, as well as the state of the ESS, to achieve the best fuel economy, minimal emissions, and maximal service life of the ESS.
- **Follow zero emissions policy:** In certain areas such as tunnels or workshops, some HEVs may need to be operated in the purely electric mode.
- **Optimally control the HEV transmission system:** The most recent HEV systems not only possess the features of the parallel hybrid but also incorporate unique advantages of the series hybrid. The key for this implementation is to employ an advanced transmission system that provides at least two mechanical transmission channels through the clutch control. In city driving, the HEV system maximally uses the advantage of a series hybrid. If full-throttle acceleration is needed, the required power is simultaneously delivered by the ICE and the electric motor, but the ICE is operated at steady speed as much as possible. While the vehicle is driving normally, the power is collaboratively fed by the ICE and the electric motor to achieve the maximal fuel economy.

References

Chan, C. C. 'The State of the Art of Electric and Hybrid Vehicles,' *Proceedings of the IEEE*, **90**(2), 248–275, February 2002.

Powers, W. F. and Nicastri, P. P. 'Automotive Vehicle Control Challenges in the 21st Century,' *Control Engineering Practice*, **8**(2000), 605–618, 1999.

2

Basic Components of Hybrid Electric Vehicles

Regardless of the type of hybrid electric vehicle, the propulsion system is composed of at least the following components:

- A prime mover;
- An electric motor with a DC–DC converter, a DC–AC inverter, and the controller;
- An energy storage system;
- A transmission system.

2.1 The Prime Mover

The prime mover of a hybrid vehicle is its main energy source, which generally is one of the following: a gasoline engine, a diesel engine, or fuel cells. The selection of the prime mover is based mainly on the requirements of drivability, fuel economy, and emissions.

2.1.1 Gasoline Engines

A gasoline engine is a highly developed machine that converts natural fossil energy into mechanical work to propel a vehicle. The main advantages of this type of engine are its high specific power (power/weight ratio), the wide range of rotational speeds, and higher mechanical efficiency. For hybrid system design and performance analysis, it is necessary to understand the curves of an ICE's torque/power versus speed and graphs of fuel consumption and emissions. An example curve of torque/power versus speed for a gasoline engine is shown in Fig. 2.1, and the fuel consumption contour is shown in Fig. 2.2.

Hybrid Electric Vehicle System Modeling and Control, Second Edition. Wei Liu.
© 2017 John Wiley & Sons Ltd. Published 2017 by John Wiley & Sons Ltd.

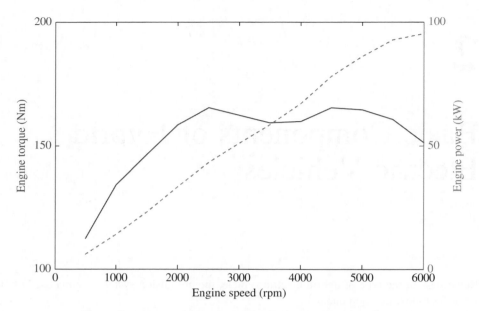

Figure 2.1   Torque/power vs. speed curve of a gasoline engine

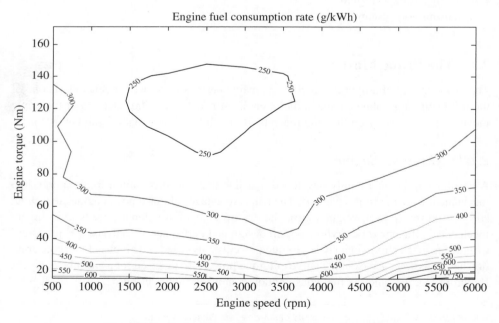

Figure 2.2   Fuel consumption contour of a gasoline engine

2.1.2　Diesel Engines

A diesel engine is a reciprocating piston engine that automatically ignites the injected fuel through heat during the final stage of compression. This type of ignition makes diesel engines burn fuel at much higher temperatures than gasoline engines, and the efficiency is, therefore, inherently higher. A diesel engine also has longer maintenance intervals and lifetime than a gasoline engine. A typical curve of torque/power versus speed for a diesel engine is shown in Fig. 2.3, and the corresponding fuel efficiency contour is shown in Fig. 2.4. From these curves it can be seen that the operational speed of a diesel engine is lower than that of a gasoline engine and the torque curve is flatter. These features enable a diesel engine to be a better prime mover for medium-duty hybrid vehicles. The disadvantages of a diesel engine include noise, vibration, harshness, smell, weight, higher maintenance costs, and an inability to warm up quickly in cold weather, especially if used for short hauls and stop-and-go work.

2.1.3　Fuel Cells

Fuel cells (FCs) are promising candidates to reduce emissions and energy consumption to improve the global climate situation. Although many types of fuel cell have been successfully developed, proton exchange membrane fuel cells (PEMFCs) and solid oxide fuel cells (SOFCs) have received considerable attention as prime movers for hybrid vehicles, with

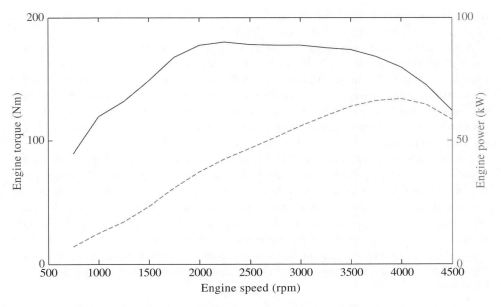

Figure 2.3　Typical torque/power vs. speed curve of a diesel engine

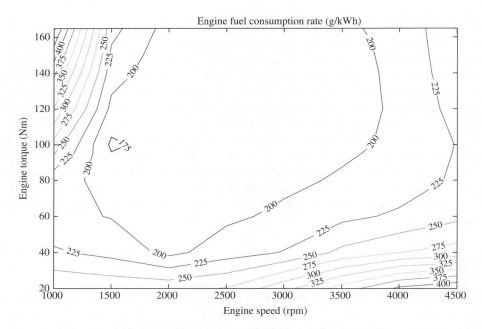

Figure 2.4   Typical fuel efficiency contour of a diesel engine

improvements in cost, transient response, and cold performance of FC systems (Markel *et al.*, 2005; Rechberger and Prenninger, 2007).

A fuel cell is an electrochemical component that can continuously convert fuel's chemical energy into electrical energy. The working principles of a fuel cell are similar to electrochemical batteries. The main difference is that the chemical energy converted by a battery is stored inside the battery. Thus, once the chemical energy has been converted into electrical energy, the battery must be recharged or its life ends. But for fuel cells, the converted chemical energy is stored externally, so the chemical reaction to supply electrical energy can take place continuously as long as fuel and oxygen are supplied.

The fundamental construction of a PEM fuel cell can be illustrated by Fig. 2.5; it consists mainly of an anode, a cathode, and an electrolyte between the electrodes. Hydrogen gas is fed into the anode, and oxygen (usually from air) enters at the cathode. The catalyst on the anode breaks down the hydrogen into electrons and protons. The electrons travel through the external wire/load, producing current to the cathode, while the protons pass through the electrolyte to the cathode. Once the electrons and protons reach the cathode, they are reunited and react with oxygen on the cathode catalyst to form a waste product – water. Figure 2.6 shows the terminal voltage characteristics versus current density for typical PEM fuel cells at 25 °C.

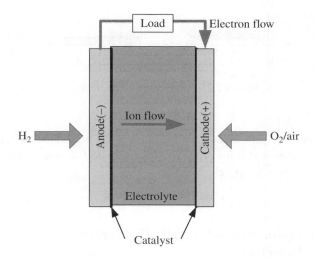

Figure 2.5 Working principle of PEM fuel cells

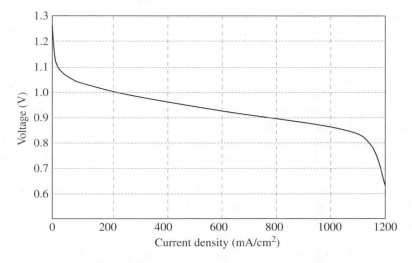

Figure 2.6 Fuel cell voltage vs. current density curve

In addition, to be a fuel-cell-based mover, the fuel cell subsystem needs to be integrated with a DC–DC converter or a DC–AC inverter and an electric motor, shown in Fig. 2.7. Since a fuel cell mover consists of many subsystems, its detailed characteristics are very complicated. From a control point of view, the operating temperature, internal pressure, and the humidification of the fuel cell stack are the most important and challenging control

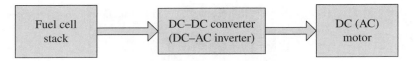

Figure 2.7   Fuel-cell-based prime mover

variables affecting the performance and efficiency of the whole system. The biggest advantage of using fuel cells as the prime mover for a hybrid vehicle is that the implemented vehicle is a true zero-emission vehicle.

2.2   Electric Motor with a DC–DC Converter and a DC–AC Inverter

An electric motor is one of the most important components in a hybrid electric vehicle. The brushless DC motor and the AC induction motor are widely used for this application due to their higher efficiency, lower costs, less maintenance, and longer lifetimes. Induction motors might seem unsuitable for a DC source of a hybrid electric vehicle as they require an AC supply, but AC can easily be inverted from a DC source with the advances in modern power electronics.

The rotor current of an induction motor is induced through the rotating magnetic field created by the stator winding, and, in turn, this rotor current will generate the rotor magnetic field. If there is a difference between the spinning speed of the rotor and the rotating speed of the magnetic field in the stator, the rotor magnetic field and rotating magnetic field in the stator will interact with each other, resulting in spin of the rotor. Since it is necessary to have a speed difference between the two fields in order for the rotor to turn, the induction motor is also called an asynchronous machine. On the other hand, the brushless direct current (BLDC) motor is a special DC motor in which the mechanical brush/commutator system is replaced by a motionless electronic controller. In a BLDC motor, current and torque, voltage and turning speed are linearly related.

Since DC voltage varies with battery SOC and operational conditions in hybrid electric vehicles, either a BLDC motor or an AC induction motor has to work with a DC–DC converter or a DC–AC inverter, which converts or inverts the DC voltage to the required operational voltage level of the electric motor. The required efficiency of the DC–DC converter or the DC–AC inverter is normally above 95% in hybrid vehicle applications.

In hybrid electric vehicle system design, a system design engineer needs to know the curves of torque/power versus speed and the efficiency contour of the electric motor to size the components and select the optimal operational points. The typical curves of an AC induction motor and a BLDC motor are illustrated in Fig. 2.8, Fig. 2.9, Fig. 2.10, and Fig. 2.11.

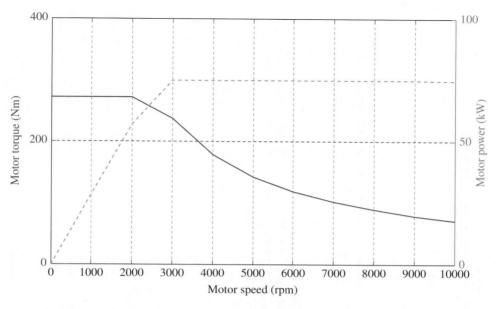

Figure 2.8 Typical curve of torque/power vs. speed of an induction motor system

2.3 Energy Storage System

2.3.1 *Energy Storage System Requirements for Hybrid Electric Vehicles*

The energy storage system, one of the most important subsystems in hybrid electric vehicles and all-electric vehicles, consists of an energy storage pack, a voltage, current, and temperature measurement module (VITM), a cell-balancing circuit, and a cooling/heating system; in addition, battery-related estimation algorithms also play a key role. The function of the ESS is to convert chemical energy into electrical energy and vice versa by electrochemical oxidation or reduction reactions to provide/capture electrical energy for/from the vehicle. The stored energy may be captured from regenerative braking or charged from the electric grid through a plug-in charger. The basic requirements of the energy storage system include safety, reliability, high efficiency, and low cost, but the actual requirements for a hybrid electric vehicle application vary with the architecture of the hybrid vehicle system, maximal speed, acceleration time, electric mile range requirements, and designed operational modes. For a conventional hybrid propulsion system, the requirements for the ESS substantially depend on the design of the driving power distribution between the engine and the traction motor, and the power capability of the ESS is the main consideration factor; however, both power capability and energy capacity of the ESS should be taken into account for a plug-in hybrid electric

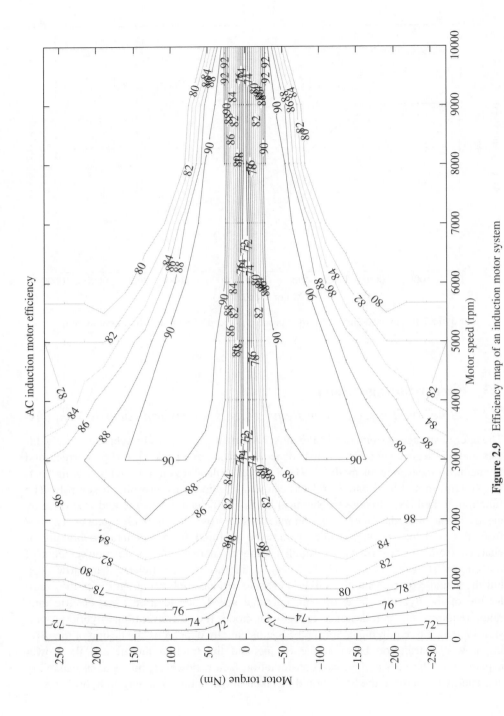

AC induction motor efficiency

Figure 2.9 Efficiency map of an induction motor system

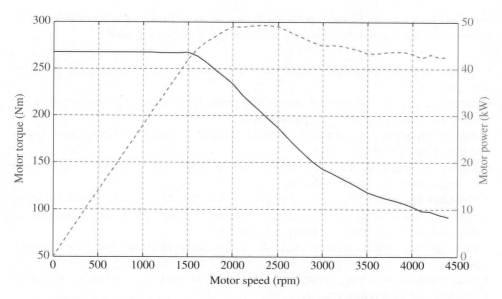

Figure 2.10   Typical curve of torque/power vs. speed of a BLDC motor system

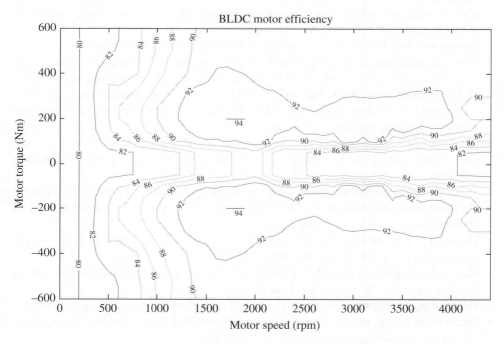

Figure 2.11   Efficiency map of a BLDC motor system

vehicle (PHEV) or a full battery-powered electric vehicle (BEV). In hybrid electric vehicle applications, the following battery terminology is commonly used:

- **Anode:** Also called a negative electrode, which donates electrons during the discharge process.
- **Battery:** A device converting chemical energy into electrical energy, consisting of one or many electrochemical cells connected together.
- **Battery capacity:** The amount of electrical charge that a battery contains, expressed in ampere-hours (Ahr).
- **Binder:** A component of a Li-ion battery to hold the active material particles together and in contact with the current collectors.
- **Calendar life (year):** The time that a battery can be stored in an inactive state before its capacity fades to the designed percentage of the initial capacity; temperature and time are the two key factors affecting battery calendar life.
- **Cathode:** Also called a positive electrode, which accepts electrons during the discharge process.
- **Cold-cranking power capability (kW):** The maximal short-term discharge power at the minimal operational temperature.
- **Cycle life (#):** The number of complete charge–discharge cycles before the battery's capacity falls to the designed percentage of the initial capacity.
- **Electrolyte:** A chemical medium in a Li-ion battery cell that allows the lithium ions to travel between the cathode and the anode; high-purity electrolytes are a core component of Li-ion batteries.
- **Energy density:** Energy per unit weight of a device, expressed in Wh/kg.
- **Maximal continuous charge power (kW):** The maximal allowable long-term charge power at a given temperature, which is usually set as the plug-in charging power limit.
- **Maximal continuous discharge power (kW):** The maximal allowable long-term discharge power at a given temperature, which is usually selected based on the power necessary for highway cruising performance.
- **Maximal operating current (A):** The maximal allowable regenerative or plug-in charging current, which is the charge current with a positive sign from the battery's perspective.
- **Maximal operating temperature (°C):** The highest allowable operating temperature.
- **Maximal operating voltage (V):** The maximal allowable battery terminal voltage during plug-in charging and regenerative events.
- **Maximal self-discharge rate:** The maximal allowable self-discharge rate in the storage condition, expressed as energy loss per day (Wh/day).
- **Minimal operating current (A):** The minimal allowable discharging current, which has a negative sign from the battery's perspective.
- **Minimal operating temperature (°C):** The lowest allowable operating temperature.
- **Minimal operating voltage (V):** The minimal allowable battery terminal voltage during motoring or cranking engine events.

- **Peak charge power (kW):** The maximal allowable short-term charge power at a given temperature, which is usually set as the regenerative limit.
- **Peak discharge power (kW):** The maximal allowable short-term discharge power at a given temperature, which is usually set as the motoring power limit for the wide-open-throttle (WOT) event.
- **Power density:** The output power capability per unit weight of a device, expressed in W/kg.
- **Power energy ratio:** The ratio of output power capability over the energy capacity of a battery.
- **Round-trip energy efficiency (%):** The ratio of the energy put in and the energy pulled out at a specified C-rate.
- **Separator:** A component of a Li-ion battery which separates the cathode from the anode.
- **Specific energy:** Energy per unit volume of a device, expressed in Wh/L.
- **Specific power:** Output power per unit volume of a device, expressed in W/L.

2.3.2   Basic Types of Battery for Hybrid Electric Vehicle System Applications

Over the past decade, battery technology has advanced dramatically and many high-performance batteries have been developed. As a result, a variety of batteries can be used for hybrid vehicle applications. This section briefly introduces some typical batteries that are used in hybrid electric vehicles, and the interested reader should refer to the *Handbook of Batteries* for more detail (Linden and Reddy, 2002).

2.3.2.1   Lead–Acid Batteries

The lead–acid battery is one of the oldest rechargeable batteries. Two types of lead–acid battery, starting and deep-cycle, are used for vehicle applications. The starting type of lead–acid battery consists of many thin plates to achieve maximal surface area to obtain maximal current output, while deep-cycle lead–acid batteries have much thicker plates to achieve a longer lifetime. The starting type of lead–acid battery is used to start the engine in a conventional vehicle, and for most of the rest of the time it is kept in a float charging state. If this type of battery is repeatedly charged/discharged, as is the case for a battery used in a hybrid electric vehicle, it will ultimately result in damage. For hybrid vehicle applications, specially designed deep-cycle lead–acid cells are needed to withstand frequent charging/discharging, and the basic technical requirements for this type of battery are listed in Table 2.1. The major advantages of the lead–acid battery include:

- Lower cost compared with other batteries;
- High cell open-circuit voltage;

Table 2.1 Basic technical requirements for a lead–acid cell in HEV/EV applications

Energy density	35–40 Wh/kg
Specific energy	60–75 Wh/L
Power density	150–200 W/kg
Round-trip energy efficiency	70–90%
Self-discharge rate	5–15%/month
Cycle durability	500–800 cycles
Nominal cell voltage	~2.1 V

Table 2.2 Basic technical requirements for an NiMH cell in HEV/EV applications

Energy density	60–80 Wh/kg
Specific energy	160–200 Wh/L
Power density	240–300 W/kg
Round-trip energy efficiency	75–95%
Self-discharge rate	10–20%/month
Cycle durability	800–1000 cycles
Nominal cell voltage	~1.2 V

- Easy recycling of the cell components;
- Accurate SOC indication due to the electrolyte taking part in the reaction, as a result of which, the SOC can be determined by measuring the specific gravity of the electrolyte.

The major disadvantages include:

- Relatively low cycle life (this normally amounts to only 500–800 full cycles);
- Low energy density, typically 30–40 Wh/kg;
- High self-discharge rates;
- Low charge/discharge efficiency.

2.3.2.2 Nickel–Metal Hydride (NiMH) Batteries

A nickel–metal hydride (NiMH) battery uses hydrogen absorbed in a metal alloy for the active negative material. Since a metal hydride electrode has a higher energy density than a cadmium electrode, NiMH batteries have a higher capacity and a longer life than NiCd batteries. Also, an NiMH battery is free from cadmium, thus it is considered an environmentally friendly battery. The higher specific energy and good cycle life make an NiMH battery very suitable for HEV/EV applications. The basic technical requirements for an NiMH cell are listed in Table 2.2.

2.3.2.3 Lithium-ion (Li-ion) Batteries

Due to the outstanding features of the metal lithium, lithium-ion batteries have rapidly penetrated the field of hybrid electric and all battery-powered electric vehicles. Li-ion batteries have high efficiency, specific power, and energy densities, low self-discharge rates, and long lifetimes. They are also environmentally friendly, as their components can be recycled. A lithium-ion battery cell consists mainly of a positive electrode (cathode), a negative electrode (anode), electrolytes, separators, binders and two current collectors.

Positive Electrode Materials and Characteristics
The cathode is the key element that limits the performance of present-day Li-ion batteries. The cathode is also the most expensive component of Li-ion batteries, as it comprises mainly crystalline cobalt, nickel, and manganese, which form a multi-metal oxide material to which lithium is added. Since the 1990s, continuous efforts have been devoted to improving performance, enhancing safety, and lowering costs. In order to maximize energy density, the cathode materials need to have higher average potentials. Furthermore, to meet the high specific energy requirements for hybrid electric and all-electric vehicle applications, researchers and scientists have been looking for transition-metal element-based compounds with the crystalline structures that favor high mobility of lithium ions.

The Li-ion battery performance requirements for hybrid electric vehicle applications are for the cathode materials to be good conductors of both lithium ions and electrons. In order to meet such requirements, today's Li-ion battery cathodes have been developed using various approaches, and can essentially be classified into three different families from a crystalline structure perspective (Doeff, 2012; Julien *et al.*, 2014). The three families are: a layered structure typified by $LiCoO_2$; a spinel structure such as $LiMn_2O_4$ (LMO); and an olivine structure such as $LiFePO_4$ (LFP). The crystalline structure characteristics of these cathode materials are shown in Fig. 2.12 and their general properties and characteristics are summarized in Table 2.3. In general, the layered structure typified by $LiCoO_2$, has a cubic-close-packed oxygen array providing a two-dimensional network of edge-shared CoO_6

(a) The layered structure   (b) The cubic structure   (c) The olivine structure

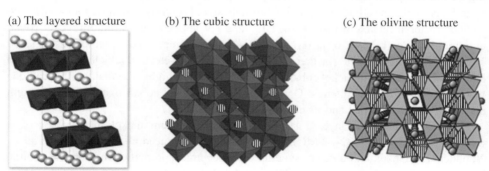

Figure 2.12 Common crystalline structures of cathode materials for Li-ion batteries. Reproduced from *Encyclopedia of Sustainability Science and Technology*, edited by Robert A. Meyers, p. 711, with permission from Springer

Table 2.3 Characteristics of common Li-ion battery cathode materials

Structure	Compound	Average potential (V vs. Li^0/Li^+)	Specific capacity (mAh/g)	Specific energy Wh/kg
Layered	$LiCoO_2$ (LCO)	4.2	120–140	520–570
Layered	$LiNi_{1/3}Mn_{1/3}Co_{1/3}O_2$ (NMC)	4.0	160–180	610–660
Layered	$LiNi_{0.8}Co_{0.15}Al_{0.05}O_2$ (NCA)	4.0	180–200	700–760
Spinel	$LiMn_2O_4$ (LMO)	4.1	100–120	420–500
Spinel	$LiMn_{1.5}Ni_{0.5}O_4$ (LMN)	4.7	100–120	460–520
Olivine	$LiFe_{0.5}Mn_{0.5}PO_4$ (LFP)	3.4	150–170	520–580

octahedra (dark gray) for the lithium ions; the spinel structure such as $LiMn_2O_4$ also has a cubic-close-packed oxygen array but provides a three-dimensional array of edge-shared MnO_6 octahedra (mid-gray) for the lithium ions; the olivine structure such as $LiFePO_4$ has a hexagonally-close-packed oxygen array in which there are corner-shared FeO_6 octahedra (light gray) and PO_4 tetrahedra (striped).

Since lithium cobalt oxide ($LiCoO_2$) is able to offer the highest energy density, the first commercialized Li-ion battery used it as the cathode, and it is still used today for consumer devices. However, the $LiCoO_2$ battery is not acceptable for applications demanding high power and energy such as hybrid electric and all-electric vehicles due to the high material cost of using cobalt and the safety risks resulting from poor thermal stability in higher operating temperatures or overcharge situations.

Since nickel is much cheaper than cobalt and has the potential for higher energy density, many scientists and researchers have extensively explored $LiNiO_2$-based layered compounds as the cathode in battery applications, while the head-on challenge now is how to improve the thermal stability at high SOCs. Although the developed Mg coating and doping techniques have improved the characteristics of $LiNiO_2$, concerns over safety when this material is used as a cathode have still not disappeared. Presently, the most commonly used cathode material related to $LiNiO_2$ for HEV and EV applications is lithium nickel cobalt aluminum oxide ($LiNi_{0.8}Co_{0.15}Al_{0.05}O_2$, NCA). For example, Tesla Motors use 18,650 NCA cells from Panasonic for their 200-mile model S all-electric vehicle. In the layered structure family, lithium nickel cobalt aluminum oxide is a highly thermally stable cathode material used in Li-ion batteries. Doping the lithium nickel cobalt oxide with aluminum stabilizes both its thermal and its charge transfer resistance.

The archetype of the layered structure family is lithium nickel manganese cobalt oxide ($LiNi_{1/3}Mn_{1/3}Co_{1/3}O_2$, NMC), which was first introduced as a cathode material by Yabuuchi and Ohzuku (2003). The superior electrochemical properties and better structural, chemical, and thermal stability of the NMC material compared to $LiCoO_2$ and NCA make it an attractive replacement cathode material for consumer batteries and it has gradually become the material of choice in batteries for HEV and EV applications due to its longer lifespan and inherent safety.

The archetype of the spinel structure family is lithium manganese oxide ($LiMn_2O_4$, LMO). Since manganese is five times cheaper than cobalt and found in abundance in nature, the LMO material has been extensively studied as a cathode element for HEV and EV applications, but it also suffers from the dissolution of manganese in the electrolyte, especially upon use above room temperature, which reduces the calendar life. However, overall, although the cycling behavior of Li-ion cells with spinel electrodes is inferior to cathodes made from other commercial materials, the good rate capability and safety, the wide availability of manganese precursors, and the potential for low cost still make LMO attractive for vehicular applications.

With the advances in hybrid electric vehicle technology, the requirements for battery performance, life, and cost are becoming more and more rigorous. Over the past decade, the performance of lithium iron phosphate chemistry ($LiFePO_4$, LFP) has been significantly improved due to the better control of synthesis parameters and the use of conductive coatings and nano-structuring, so that LFP has become one of the most attractive cathodes for Li-ion batteries (Padhi, Nanjundaswamy, and Goodenough, 1997). From a crystalline structure point of view, LFP belongs to the olivine family. The advantages of LFP are not only that it contains no noble elements such as cobalt, making the price of the raw material lower and the availability of the raw material higher, as both phosphorus and iron are abundant on Earth, but also that LPF chemistry is environmentally friendly, has better thermal stability, and is safer. Batteries using this cathode material have a moderate operating voltage (3.3 V), high energy storage capacity (170 mAh/g), high discharge power, fast charging, a long cycle life, and high stability when placed under high temperatures. All these superior features make LFP particularly well suited to hybrid vehicle applications.

The advantages and disadvantages of Li-ion batteries made from different cathode materials are listed in Table 2.4, and the characteristics of open-circuit-voltage (OCV) vs. SOC and 1-C discharge profiles are shown in Fig. 2.13 and Fig. 2.14, respectively. In general, NMC materials have gradually become mainstream in terms of active cathode materials for hybrid electric propulsion applications, since they have higher performance, higher energy, higher density, and higher capacity than existing cathode materials.

Negative Electrode Materials and Characteristics

The negative electrode, or anode, in a Li-ion battery is typically made by mixing an active material, binder powder, a solvent, and additives into a slurry. Graphite is commonly used as the active material in anodes because it has high conductivity, low cost, and can reversibly place lithium ions between its many layers due to its unique electronic structure. Graphite is a crystalline solid with a black/gray color and a metallic sheen. Furthermore, the graphite materials used in practice can be classified as natural graphite, artificial graphite, or modified natural graphite. Natural graphite materials generally have excellent first charge efficiency, and artificial graphite materials are normally designed to improve rate capabilities, while modified graphite materials may have both advantages and may also provide superior cell capacity. The reversible electrochemical capability of a graphite anode needs to be over several thousands of cycles.

Table 2.4 Advantages and disadvantages of common Li-ion batteries with different cathode materials

Compound	Advantages	Disadvantages
$LiNi_{1/3}Mn_{1/3}Co_{1/3}O_2$ (NMC)	High energy capacity Slow reaction with electrolytes Ready SOC estimation Moderate safety (oxygen release)	High cost of Ni and Co Potential resource limitations Poor high rate performance
$LiNi_{0.8}Co_{0.15}Al_{0.05}O_2$ (NCA)	Slow reaction with electrolytes High energy capacity High operating voltage Excellent high rate performance	High cost of Ni and Co Potential resource limitations Poor safety
$LiMn_2O_4$ (LMO)	Moderately low cost Excellent high rate performance High operating voltage No resource limitations Moderate safety (oxygen release)	Low cycle life due to Mn solubility issue Low energy capacity
$LiFe_{0.5}Mn_{0.5}PO_4$ (LFP)	Moderately low cost Excellent high rate performance No resource limitations Very slow reaction with electrolyte Excellent safety (no oxygen release)	Low operating voltage Low energy density Low power density Very challenged for SOC estimation

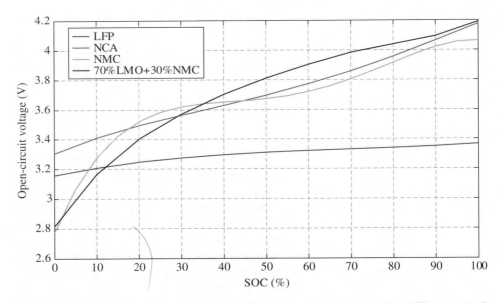

Figure 2.13 The characteristics of OCV vs SOC of Li-ion batteries made from different cathode materials

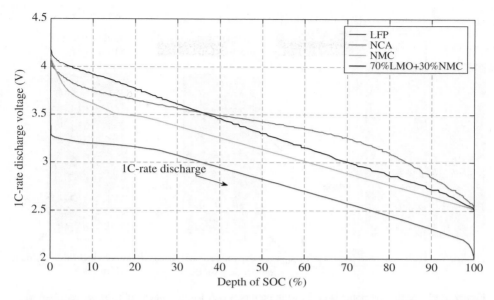

Figure 2.14   1C discharge profiles of Li-ion batteries made from different electrode materials

Other negative electrode materials used in practice or evaluated for future use include lithium titanate ($Li_4Ti_5O_{12}$, LTO), hard carbon, a tin/cobalt alloy, and silicon. The LTO anode has better durability and safety, and the operating temperature is between −50 °C and 70 °C; hard carbon and tin/cobalt alloy anodes have larger storage capacity, while silicon-based anodes are capable of improving energy density significantly.

Other Main Component Materials and Characteristics

Other components of a Li-ion battery include a binder, an electrolyte, a separator, and foils; the structure of a Li-ion battery cell (pouch cell) is shown in Fig. 2.15. The binder in Li-ion batteries holds the active material particles together and in contact with the current collectors. At present, most binder materials are styrene butadiene copolymer (SBR) and polyvinylidene fluoride (PVDF). The characteristics of the binder material used are critical for hybrid electric vehicle applications, and the material must have the following properties:

- Ionic conductivity;
- Tensile strength;
- Water absorption;
- Adhesion property;
- Swelling in electrolyte;
- Melting point and crystallinity;
- Dissolution properties;
- Purity.

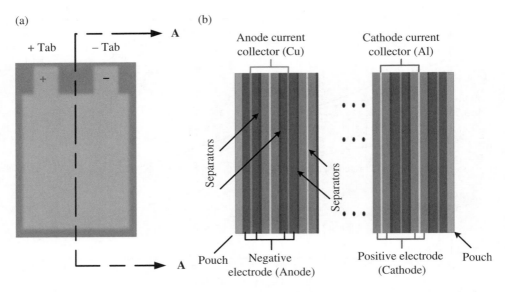

Figure 2.15   Main components of a Li-ion battery. (a) A Li-ion pouch cell; (b) section A–A view: structure and main components of a Li-ion cell

The battery's electrolyte is a core component of Li-ion batteries and plays a key role in transporting lithium ion particles between the cathode and the anode. The most commonly used electrolyte is composed of lithium salt, such as LiPF6, in an organic solution. In addition to lithium salt, several additives are also included to give the required properties for the electrolyte solution. These additives are also used to improve the stability, preventing dendritic formation and degradation of the solution, and the specific additives have a great impact on the overall performance of the battery, especially in high-energy EV applications. The specific electrolyte formulation varies depending on the specific anode and cathode materials being used. The principal criteria for selecting an electrolyte product in hybrid electric propulsion applications are as follows:

- Good cycle life;
- Excellent thermal and hydrolytic stability;
- High conductivity;
- High discharge rate;
- Notable improvement in large current discharge performance;
- Good performance at high temperatures;
- Good performance at low temperatures;
- Excellent anti-overcharging performance.

Battery separators play a critical role in separating the cathode from the anode so that electricity can be generated. Battery separators are power-driven spacers and can be produced with fiberglass cloth or flexible plastic films made from nylon, polyethylene, or polypropylene. The separator must be permeable and thin to allow the charged lithium ions to pass through without obstruction, and they must occupy minimal space, leaving space for the cathode/anode active elements. During the charging cycle, the charged lithium ions move from the cathode, through the separator to the anode. By contrast, during discharge, the charged lithium ions move from the anode, through the battery separators to the cathode. During either charge or discharge cycle, electrons move through the external load from the cathode/anode to the anode/cathode, resulting in a current through the load. The lithium ions move through the separator via the electrolyte solution.

Requirements for battery separators depend on the required performance of the battery in terms of the operational temperature, charge rate, safety, and cost. The main considerations when selecting a separator include:

- Electronic insulation;
- Electrolyte resistance;
- Mechanical and dimensional stability;
- Chemical resistance to degradation by electrolyte, impurities, and electrode reactants;
- Effectiveness in preventing migration of particles or colloidal or soluble species between the two electrodes;
- Readily wetted by the electrolyte.

The most commonly used foil materials in Li-ion batteries are aluminum, copper, and nickel. Typically, copper foil is used for the anode current collector connected to the negative electrode and aluminum foil is used for the cathode current collector connected to the positive electrode. Nickel foil is electroconductive and has a higher cost; therefore, it is generally used when there are requirements for its high anti-corrosive and ultra-thin properties. Compared with copper foil, nickel foil is produced wider and thinner. The key features of foil are:

- High conductivity to minimize cell impedance;
- Excellent for high-power applications – electric propulsion and power tools;
- Heat resistance – offering high heat tensile and ductile strength;
- Consistently smooth surface – no large protrusions;
- Clean – free of oxides and oils;
- Surface treatment for adhesion enhancements.

Basic Technical Requirements for a Li-ion Cell in HEV/EV Applications
Due to the outstanding features of lithium metal, Li-ion batteries have rapidly penetrated the hybrid electric propulsion systems market recently. Li-ion batteries have high efficiency, specific power and energy densities, low self-discharge rates, and long lifetimes. They are also environmentally friendly, as their components can be recycled. For HEV/EV applications, the basic technical requirements of Li-ion cells are listed in Table 2.5.

Table 2.5   Basic technical requirements for a Li-ion cell in HEV/EV applications

Specifications	Unit	HEV application	EV application
Energy density (1C rate at 25 °C)	Wh/L	70–200	250–550
Specific energy (1C rate at 25 °C)	Wh/kg	50–100	150–230
Power density (10 second, at 50% SOC and 25 °C)	W/L	2000–9000	2000–4000
Specific power (10 second, at 50% SOC and 25 °C)	W/kg	1000–4500	1000–3000
Round-trip energy efficiency (1C rate at 25 °C)		92–97%	94–98%
Self-discharge rate (at 50% SOC and 25 °C)		<5%/month	<3%/month
Power/Energy (P/E) ratio (at 50% SOC and 25 °C)	hr^{-1}	25–50	10–25
Cycle durability		2000–4000 cycles	1500–3000 cycles
Nominal cell voltage	V	~3.75	~3.75

2.3.3   Ultracapacitors for Hybrid Electric Vehicle System Applications

Ultracapacitors, also called supercapacitors, are electrochemical double-layer capacitors. Unlike batteries, ultracapacitors directly store electrical energy by physically storing separated positive and negative charges. Conventional capacitors stored electrical energy by storing a number of charges on the two metal plates with a certain potential. In order to increase the capacity, various materials, called dielectric materials, are generally inserted between the plates so that higher voltages can be stored. Unlike conventional capacitors, ultracapacitors structurally use an electrical double layer to form a very large surface area to allow the storage of a greater number of charges. For HEV applications, ultracapacitors have the following advantages over batteries:

- **Significantly higher power density:** Unlike a battery, an ultracapacitor can be charged or discharged at a very high C-rate, and the temperature of electrodes heated by the current is the only limiting factor. It is not difficult for an ultracapacitor to have a 5000 W/kg power density.
- **Excellent cycle life:** Compared to batteries, ultracapacitors can endure millions of charge/discharge cycles, which makes it possible to capture all available regenerative energy in an HEV.
- **More environmentally friendly:** due to the lack of unrecyclable parts during the entire service life.
- **Higher efficiency:** Ultracapacitors have very low internal resistance, resulting in little heat being generated during operation. The efficiency of an ultracapacitor can be over 97%, which is higher than a battery.

There are two main disadvantages of ultracapacitors for HEV applications:

- The energy density is substantially lower than for batteries – generally only 1/10th that of a same-size battery.
- There is low resistance to leaks, resulting in a higher self-discharge rate.

2.4 Transmission System in Hybrid Electric Vehicles

Transmission is another important subsystem of a hybrid electric vehicle and performs the following functions:

- Achieves transition from a stationary to a mobile state;
- Converts torque and rotational speed from the mover to meet the vehicle's momentary traction requirements;
- Provides for forward and reverse motion;
- Gives the vehicle maximal fuel economy and minimal emissions while meeting drivability requirements.

Since a hybrid vehicle has two or more prime movers with different characteristics, the transmission plays a more important role than in a conventional vehicle. In order to obtain the maximal efficiency and optimal performance, a specially designed transmission is necessary for a given hybrid vehicle system configuration. The electrical continuously variable transmission (ECVT) and power split transmission (PST) are two commonly used transmissions in HEVs. These transmissions generally have more than two planetary gear sets, which provide an additional mechanical power path so that the electric motor/generator can directly power or regeneratively brake the vehicle.

A typical hybrid electric powertrain system includes an ICE and two electric motors (MG_1 and MG_2) which are interconnected by a planetary gear set that provides various power flow configurations for different modes of operation. MG_2 is the primary electric motor assisting the ICE in providing mechanical drive power for the vehicle, and it also acts as a generator to recharge the battery during regenerative braking. MG_1 is the secondary electric motor, cranking the ICE during the key start period; it also functions as a generator to transfer power from the ICE to recharge the battery and it provides supplemental power with the powertrain so that the global optimal power flow is achieved while meeting the vehicle's power demands. A simple schematic diagram of ECVT is given in Fig. 2.16.

In order for the vehicle to operate in the most efficient state at all times, a practical hybrid electric vehicle normally has an additional mechanical power path so that the vehicle can operate in purely electric mode, purely mechanical mode, and hybrid mode. For example, the Allison two-mode hybrid transmission consists of three planetary gear sets and four controllable engaging friction clutches. This transmission further incorporates two BLDC

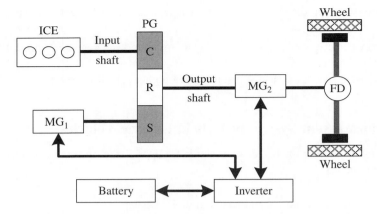

Figure 2.16   A simple ECVT schematic diagram

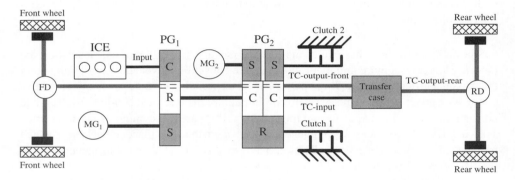

Figure 2.17   Mechanical diagram of an ECVT gear system

motors so that the vehicle controller is able to select, amplify, and transfer output torques of the ICE and electric motors to the wheels based on the real driving demand to achieve optimal performance and efficiency. Another example of ECVT is composed of motor, generator, and gearbox subassemblies; its mechanical diagram is shown in Fig. 2.17. This set-up utilizes a simple planetary gear set to coordinate ICE output power and MG_1 output power, which can operate as a motor (positive) or generator (negative); that is, the sun gear of the planetary set (PG1) receives input from MG_1, the ICE is connected to the carrier of PG1, and the ring of PG1 connects to the carrier of a Ravigneaux planetary gear set (PG2). Compared with a simple planetary gear set, a Ravigneaux gear set has two sun gears, two sets of planets on a single carrier. One advantage of using a Ravigneaux gear set is that the gear ratio can be selected through clutch-2, depending on the demand power or available regenerative power of the vehicle to achieve optimal performance. The carrier of the Ravigneaux gear set mates

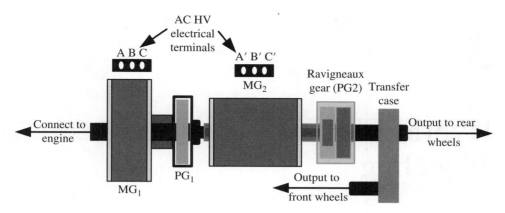

Figure 2.18 Section diagram of an ECVT housing

with the input of the transfer case, which distributes total power to the front and rear wheels appropriately through a small planetary gear set.

The section diagram of an ECVT housing is shown in Fig. 2.18. As shown in the figure, the primary sections of an ECVT include the generator (MG_1), a power split planetary gear set (PG1), a motor (MG_2), a Ravigneaux gear set (PG2), and a transfer case. Mechanically, the engine mounts directly on the ECVT and supplies power through a splined shaft. The MG_1 rotor shaft connects to the sun of the planetary gear, while the engine shaft passes through the hollow shaft of MG_1 and connects to the planetary carrier. The ring gear and ring gear shaft of PG1 passing through the MG_2 rotor connect to the carrier of PG2, while the MG_2 rotor shaft connects to the ring gear shaft of PG2. The carrier shaft outputs power to the transfer case, which distributes the power to the front and rear wheels.

References

Doeff, M. M. 'Battery Cathode's. In *Encyclopedia of Sustainability Science and Technology*, edited by Robert A. Meyers, pp. 708–739, Springer, 2012.

Julien, M. C., Mauger, A., Zaghib, K., and Groult, H. 'Comparative Issues of Cathode Materials for Li-Ion Batterie's, *Inorganics*, **2**, 132–154, 2014.

Linden, D. and Reddy, B. T. *Handbook of Batteries*, 3rd edition, McGraw-Hill, 2002.

Markel, T., Pesaran, A. A., Zolot, M. *et al*. 'Energy Storage Fuel Cell Vehicle Analysis,' Conference paper, NREL/CP-540-37567, April 2005.

Padhi, A. K., Nanjundaswamy, K. S., and Goodenough, J. B. 'Phospho-olivines as positive-electrode materials for rechargeable lithium batteries,' *Journal of the Electrochemical Society*, **144**, 1188–1194, 1997.

Rechberger, J. and Prenninger, P. 'The Role of Fuel Cells in Commercial Vehicles,' SAE technical paper, 2007-01-4273.

Yabuuchi, N. and Ohzuku, T. 'Novel lithium insertion material of $LiCo_{1/3}Ni_{1/3}Mn_{1/3}O_2$ for advanced lithium-ion batteries,' *Journal of Power Sources*, **119–121**, 171–174, 2003.

3

Hybrid Electric Vehicle System Modeling

The modeling objectives generally determine the accuracy and architecture requirements of a mathematical model, the employed methodology, and the time required to build the model. For instance, the mathematical model for analysis, design, and diagnosis should be more accurate than the model for prediction. Since an HEV has at least two power sources, its analysis, design, and calibration tasks are much more complex and challenging than those for a conventional vehicle; therefore, it is important for a hybrid vehicle system engineer to be familiar with mathematical models of the major components or subsystems. The objective of this chapter is to provide engineers with more practical, up-to-date, and comprehensive models of a hybrid electric vehicle system. The mathematical models given in this chapter include those for the engine, motor, energy storage subsystem, transmission, driveline, vehicle body, and driver.

3.1 Modeling of an Internal Combustion Engine

An ICE converts fuel's chemical energy into mechanical energy through the internal combustion process, which creates high temperature and pressure gases that act directly to cause movement of the solid parts such as pistons. The working process of an ICE is very complex, and various models have been developed for different purposes. The model presented in this section exclusively describes the input–output static mechanical characteristics of an ICE and is limited to hybrid vehicle system performance analysis. The internal combustion process and thermal dynamics are not covered. Since an ICE usually has four operational states: cranking (key start), idle, engine on, and engine off, each individual model corresponding to these states is presented below.

Hybrid Electric Vehicle System Modeling and Control, Second Edition. Wei Liu.
© 2017 John Wiley & Sons Ltd. Published 2017 by John Wiley & Sons Ltd.

3.1.1   Cranking (Key Start)

During this state, the engine provides negative torque, and a starter has to overcome this torque to start the engine, with the driveshaft clutch in the disengaged state. The engine model in this state can be developed based on Newton's laws of motion, as follows:

$$\tau_{crank} = J_{eng}\frac{d\omega}{dt} + \tau_{access} + \tau_{cct}$$

$$\omega_{eng} = \frac{1}{J_{eng}}\int_0^t (\tau_{starter} - \tau_{access} - \tau_{cct})dt \tag{3.1}$$

where τ_{crank} is the required torque to crank the engine, J_{eng} is the engine's inertia (kgm^2), ω_{eng} is the shaft angular velocity of the engine (rad/s), τ_{access} is the lumped torque from the mechanical accessories requiring constant torque (Nm), and τ_{cct} is the closed throttle torque of the engine (Nm).

Since the closed throttle torque is generated by static friction, viscous friction, Coulomb friction, and retarding compression torques, the torque τ_{cct} can be described by the following equation:

$$\tau_{cct} = \alpha_1(T)d \cdot \delta(t) + \alpha_2(T)\mathrm{sgn}(\omega) + \alpha_3(T)\left(\frac{\omega}{\omega_{max\_eng}}\right) + \alpha_4(T)\left(\frac{\omega}{\omega_{max\_eng}}\right)^2 \tag{3.2}$$

where d is the engine's displacement (L), $\delta(t)$ is the Dirac delta function, ω is the angular velocity (rad/s), ω_{max_eng} is the maximal allowable angular velocity (rad/s) of the engine, T is temperature (°C), and $\alpha_1(T)$, $\alpha_2(T)$, $\alpha_3(T)$, $\alpha_4(T)$ are the static friction coefficient, Coulomb friction coefficient, viscous friction coefficient, and air compression torque coefficient, respectively. These coefficients can be estimated based on experimental data using the methods introduced in Appendix A.

Static friction torque represents a retarding torque that tends to prevent rotation in a stationary object. This static friction torque only exists when the body is stationary but tends to rotate. It is expressed by the following equation and is illustrated in Fig. 3.1(a). It should be noted that the static friction torque vanishes once rotation begins, and other types of friction take over.

$$\tau_s(t) = \pm(T_s)_{\varpi=0} = \alpha_1(T)d \tag{3.3}$$

Coulomb friction represents a retarding torque that has constant amplitude with respect to the change in angular velocity, but the sign of the frictional torque changes with a reversal of the rotation direction. The functional description of Coulomb friction torque is shown in Fig. 3.1(b), and the mathematical expression is given by the following equation:

$$\tau_C(t) = \alpha_2(T)\left(\frac{d\theta}{dt} \Big/ \left|\frac{d\theta}{dt}\right|\right) = \alpha_2(T)\left(\omega/|\omega|\right) = \alpha_2(T)\ \mathrm{sgn}(\omega) \tag{3.4}$$

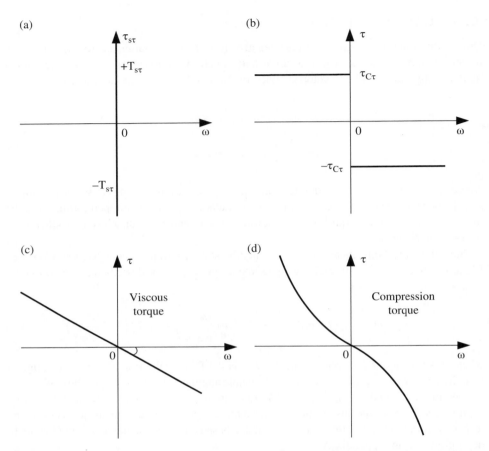

Figure 3.1 Frictional and compression torque of an internal combustion engine. (a) Static friction torque; (b) Coulomb friction torque; (c) the functional relationship between viscous friction torque and angular velocity; (d) the functional relationship between the compression torque and angular velocity

Viscous friction torque represents a retarding torque which is a linear relationship between applied torque and angular velocity. Figure 3.1(c) shows the functional relationship between viscous friction torque and angular velocity. Viscous friction torque is described by the following equation:

$$\tau_V(t) = k(T)\frac{d\theta}{dt} = \alpha_3(T)\left(\frac{\omega}{\omega_{max\_eng}}\right) \tag{3.5}$$

Air compression torque represents a retarding torque generated inside the cylinder by compressed air and gas before combustion. Figure 3.1(d) shows the functional relationship

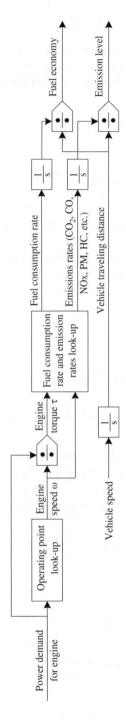

Figure 3.4 Diagram of the fuel consumption and emissions calculation

3.1.5   Engine Fuel Economy and Emissions

There are various methods and approaches to predict and calculate fuel economy and emissions in real time. For hybrid electric vehicle design and analysis, fuel economy and emissions are normally calculated based on numerical models or look-up tables extracted from engine-mapping data, as follows:

$$\text{fuel consumption}: \ f_{\text{fuel}}(\tau, \omega, T) = \lambda_{\text{fuel}}(T) g_{\text{fuel\_hot}}(\tau, \omega) \tag{3.15}$$

$$\text{emissions} - \text{CO}: \ f_{\text{emi\_CO}}(\tau, \omega, T) = \lambda_{\text{CO}}(T) g_{\text{CO\_hot}}(\tau, \omega) \tag{3.16}$$

$$\text{emissions} - \text{HC}: \ f_{\text{emi\_HC}}(\tau, \omega, T) = \lambda_{\text{HC}}(T) g_{\text{HC\_hot}}(\tau, \omega) \tag{3.17}$$

$$\text{emissions} - \text{NOx}: \ f_{\text{emi\_NOx}}(\tau, \omega, T) = \lambda_{\text{NOx}}(T) g_{\text{Nox\_hot}}(\tau, \omega) \tag{3.18}$$

$$\text{emissions} - \text{PM}: \ f_{\text{emi\_PM}}(\tau, \omega, T) = \lambda_{\text{PM}}(T) g_{\text{PM\_hot}}(\tau, \omega) \tag{3.19}$$

where $g_{\text{x\_hot}}(\tau, \omega)$ are fuel consumption (g/s) and emissions (g/s) at the engine coolant thermostat set temperature, which are functions of the torque (Nm) and the speed (rpm) of the engine, $\lambda_x(T)$ is a temperature factor, which is used to modulate hot temperature fuel consumption and emissions to the value at a given temperature. A diagram showing the vehicle fuel economy and emissions calculations is given in Fig. 3.4.

Example 3.1
Consider a 1.8 L, 85 kW diesel engine attached to a 500 W mechanical accessory. The closed throttle torque is described by the following empirical equation at room temperature:

$$\tau_{\text{cct}} = -30\delta(t) - 17.27 - 28.99 \left(\frac{\omega}{\omega_{\text{max\_eng}}} \right) - 98.63 \left(\frac{\omega}{\omega_{\text{max\_eng}}} \right)^2 \tag{3.20}$$

The engine inertia, $J_{\text{eng}} = 0.2$ (kgm^2), and the model of the engine cranking process is shown in Fig. 3.5. If the cranking torque from the motor/generator set is 50 Nm, the engine is fired at 350 rpm, and the characteristics of the mechanical accessory are as shown in Fig. 3.6(a), the dynamic response of the cranking is as shown in Fig. 3.6(b).

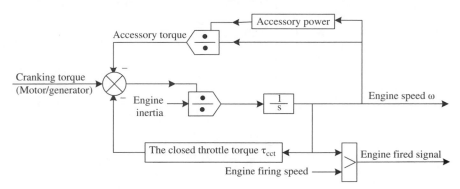

Figure 3.5   Engine cranking model

(a)

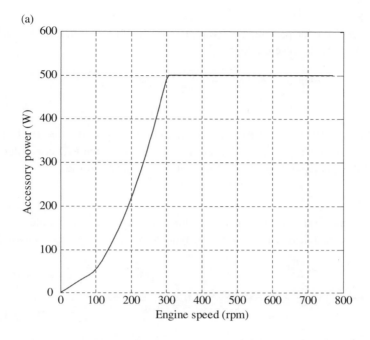

(b)

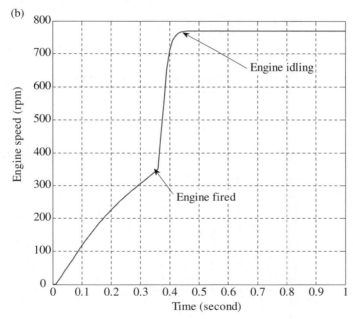

Figure 3.6 Cranking speed response of the engine. (a) The power characteristics of the engine accessory; (b) the cranking speed response of the engine

Example 3.2

Consider the same engine presented in Example 3.1. The PID control, Eq. 3.9, can be implemented in the following discrete form:

$$\tau_{\text{idle}}(k) = k_{\text{p}}e(k) + k_{\text{i}}T_{\text{s}}\sum_{i=0}^{k}e(i) + k_{\text{d}}\frac{e(k)-e(k-1)}{T_{\text{s}}} \tag{3.21}$$

$$e(k) = \omega_{\text{desired\_idle}} - \omega_{\text{actual}}(k)$$

If the idling speed is set at 770 rpm and the sampling time period is $T_{\text{s}} = 0.01\text{s}$, the engine idling speed is as shown in Fig. 3.6(b) and the parameters of the PID controller are set as $k_{\text{p}} = 4$, $k_{\text{i}} = 0.01$, and $k_{\text{d}} = 0.001$. The control diagram is given in Fig. 3.7.

Example 3.3

Consider a series hybrid electric vehicle using the same engine as in Example 3.1 as the main mover. When the vehicle runs the EPA urban dynamometer driving cycle, shown in Fig. 3.8,

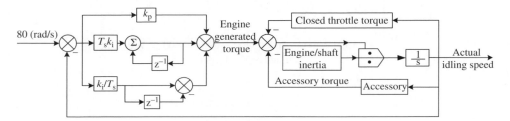

Figure 3.7   Diagram of engine idle control

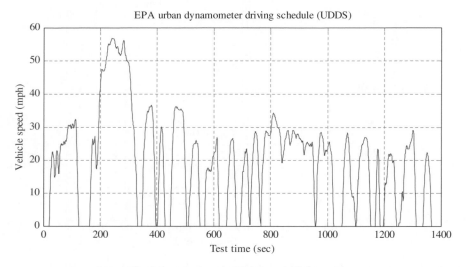

Figure 3.8   EPA urban dynamometer driving schedule

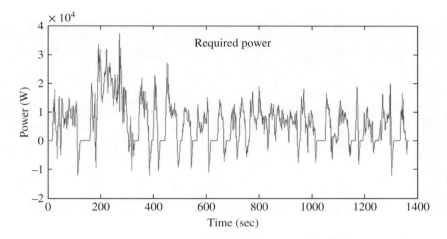

Figure 3.9 The power demand on the engine shaft over the EPA urban cycle

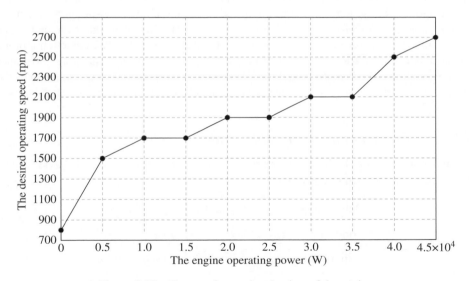

Figure 3.10 The set of operational points of the engine

the power demand on the engine driveshaft is as shown in Fig. 3.9. If the engine's operational points are set as in Fig. 3.10 and the fuel consumption data of the engine are as listed in Table 3.1, the average fuel consumption over the driving cycle is 1.06 (L/12 km) = 8.85 (L/100 km) based on the calculation principles shown in Fig. 3.4.

Table 3.1   The fuel consumption mapping data of a 1.8 L diesel engine (mL/s)

	1600 rpm	1800 rpm	2000 rpm	2200 rpm	2400 rpm	2600 rpm	2800 rpm	3000 rpm	3200 rpm	3400 rpm
0 kW	0.092	0.183	0.183							
5 kW	0.275	0.275	0.366	0.366	0.366					
10 kW	0.527	0.527	0.618	0.710	0.710	0.801	0.801	0.985	1.076	1.168
15 kW	0.710	0.710	0.801	0.893	0.893	0.985	1.076	1.168	1.259	1.420
20 kW		1.076	1.076	1.168	1.168	1.214	1.351	1.420	1.511	1.649
25 kW		1.351	1.420	1.466	1.420	1.466	1.603	1.695	1.786	1.878
30 kW				1.695	1.786	1.786	1.878	1.992	2.084	2.153
35 kW						2.153	2.198	2.290	2.359	2.404

3.2   Modeling of an Electric Motor

As mentioned in Chapter 2, BLDC and AC induction motors are extensively used in hybrid vehicle applications. BLDC motors have several apparent advantages including higher efficiency and reliability, longer lifetimes, reduced noise, the elimination of sparks from the commutator, and less electromagnetic interference (EMI). However, due to higher costs, BLDC motors are generally used in higher-end HEVs/PHEVs/EVs. AC induction motors, on the other hand, have lower cost and are suitable for cost-sensitive applications. No matter what type of motor is used, the model for system performance analysis and simulation is the same and is used primarily to describe the external electrical and mechanical behaviors.

3.2.1   Operation in the Propulsion Mode

When a motor operates in this mode, it provides propulsion torque, and its behavior can be described by the following equations:

$$\text{Output torque}: \ \tau_{\text{mot}} = \tau_{\text{demand}} + \tau_{\text{spin\_loss}} + J_{\text{mot}}\frac{d\omega}{dt} \tag{3.22}$$

$$\text{Constraint}: \quad \tau_{\text{mot}} \le \text{Max}(\tau_{\text{mot}}) = f(\omega)$$

$$\text{Lost spin torque}: \ \tau_{\text{spin\_loss}} = \alpha_1 \delta(t) + \alpha_2 \omega + \alpha_3 \ \text{sgn}(\omega) \tag{3.23}$$

$$\text{Req'd electrical power}: \ P_{\text{elec}} = \frac{P_{\text{mech}}}{\eta_{\text{mot}}} = \frac{\tau_{\text{mot}} \cdot \omega}{\eta(\tau_{\text{mot}}, \omega)} \tag{3.24}$$

$$\text{Motor efficiency}: \ \eta_{\text{mot}} = \eta(\tau_{\text{mot}}, \omega) \tag{3.25}$$

$$\text{Motor voltage}: \ V_{\text{mot}} = V_{\text{bus}} \tag{3.26}$$

$$\text{Motor req'd current}: \ I_{\text{mot}} = \frac{P_{\text{elec}}}{V_{\text{bus}}} \tag{3.27}$$

where J_{mot} is the motor's inertia, τ_{mot} is the propulsion torque provided by the motor (Nm), τ_{spin_loss} is the lost torque due to friction (Nm), τ_{demand} is the torque demanded by the vehicle (Nm), ω is the motor's angular velocity (rad/s), $\max(\tau_{mot}) = f(\omega)$ is the motor's maximal physical torque, α_1, α_2, and α_3 are the static friction coefficient, the viscous friction coefficient, and the Coulomb friction coefficient, which can be estimated based on test data, η_{mot} is the lumped efficiency of the motor, the inverter, and the controller, which may be found from a look-up table, and V_{bus} is the high-voltage bus voltage. The diagram of the electric motor model in propulsion mode is given in Fig. 3.11.

3.2.2   Operation in the Regenerative Mode

When a motor operates in the regenerative mode, it works as a generator and provides negative (brake) torque for the vehicle. In this mode, the motor's behavior can be described by the following equations:

$$\text{Brake torque}: \quad |\tau_{regen}| = |\tau_{demand}| - \tau_{spin\_loss} + J_{mot}\frac{d\omega}{dt} \tag{3.28}$$

$$\text{Constraint}: \quad |\tau_{regen}| \leq \text{Max}(\tau_{regen}) = g(\omega)$$

$$\text{Lost spin torque}: \quad \tau_{spin\_loss} = \alpha_1 + \alpha_2\omega + \alpha_3 \ \text{sgn}(\omega) \tag{3.29}$$

$$\text{Generated electrical power}: \quad P_{elec} = \eta_{regen} \cdot P_{mech} = \eta(\tau_{regen}, \omega) \cdot \tau_{regen} \cdot \omega \tag{3.30}$$

$$\text{Motor efficiency}: \quad \eta_{mot} = \eta(\tau_{regen}, \omega) \tag{3.31}$$

$$\text{Motor voltage}: \quad V_{mot} = V_{bus} \tag{3.32}$$

$$\text{Motor generated current}: \quad I_{regen} = \frac{P_{elec}}{V_{bus}} \tag{3.33}$$

where J_{mot} is the motor's inertia, τ_{regen} is the motor's negative brake torque (Nm), τ_{demand} is the torque demanded by the vehicle (Nm), ω is the motor's angular velocity (rad/s), max (τ_{regen}) is the motor's maximal regenerative torque, and η_{mot} is the lumped efficiency of the motor, the inverter, and the controller in regenerative mode. The diagram of the motor model in regenerative mode is given in Fig. 3.12.

3.2.3   Operation in Spinning Mode

When a motor operates in this mode, it is passively spun and provides a small negative frictional torque for the vehicle, as follows:

$$\tau_{mot} = \tau_{spin\_loss} = \alpha_1 \cdot \delta(t) + \alpha_2\omega + \alpha_3 \ \text{sgn}(\omega) \tag{3.34}$$

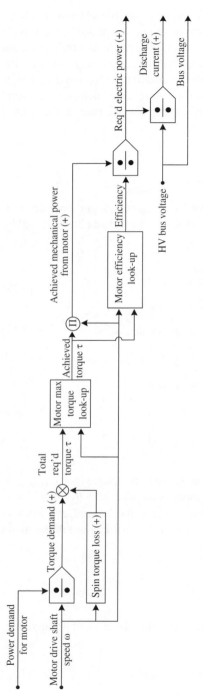

Figure 3.11 Diagram of the electric motor model in propulsion mode

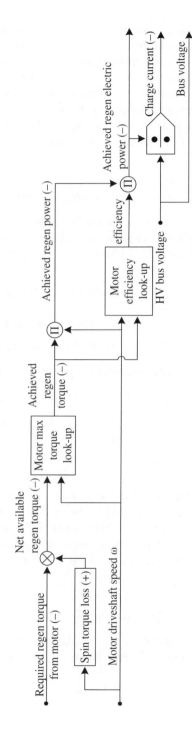

Figure 3.12 Diagram of the electric motor model in regenerative mode

Example 3.4

Consider a series hybrid electric vehicle running on the US06 driving cycle shown in Fig. 3.13. The motor shaft speed is as shown in Fig. 3.14, and the demand torque on the driveshaft is as shown in Fig. 3.15. The maximal torque of the motor/generator set is shown in Fig. 3.16, while the efficiency is listed in Table 3.2. If we neglect the spin loss and assume a nominal high-voltage bus voltage of 375 V, we can determine the current on the high-voltage bus as follows.

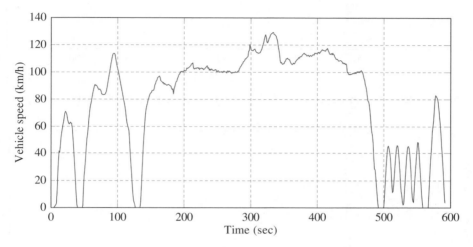

Figure 3.13 US06 drive cycle

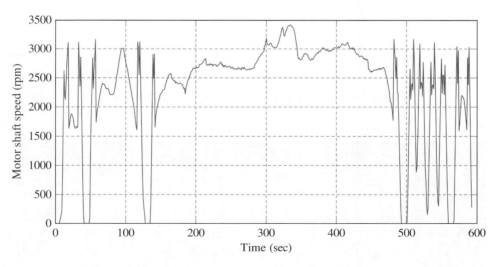

Figure 3.14 The electric motor speed over the US06 drive cycle

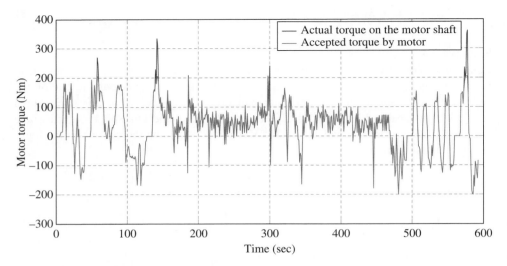

Figure 3.15 The demand torque on the electric motor driveshaft over the US06 drive cycle

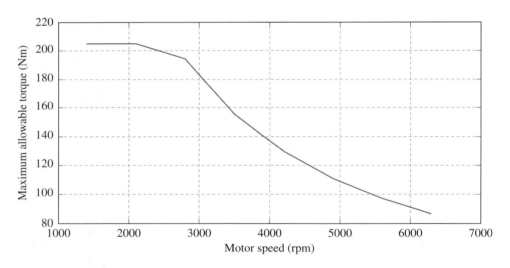

Figure 3.16 The maximal allowable torque of an electric motor

Table 3.2 The efficiency (%) of the motor/generator set

	34 Nm	68 Nm	103 Nm	137 Nm	162 Nm	170 Nm	205 Nm
1400 rpm	94	94	92	91	89	89	89
2100 rpm	95	95	94	93	92	92	91
2800 rpm	94	96	95	94	94	93	93
3500 rpm	94	96	96	95	94	94	94
4200 rpm	94	96	96	96	95	95	94
4900 rpm	93	96	97	96	96	96	95
5600 rpm	93	96	96	96	96	96	96
6300 rpm	93	96	96	96	96	96	96

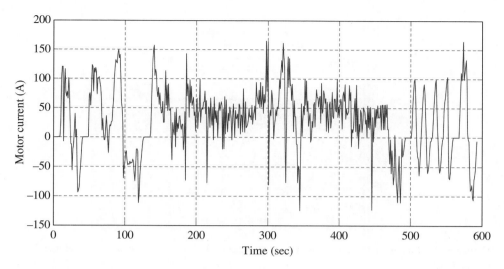

Figure 3.17   The current demand on the high-voltage bus over the US06 drive cycle

Based on Figs 3.11 and 3.12, the electric power demand on the high-voltage bus is:

$$P_e(t) = \begin{cases} \dfrac{\min\{\tau_{\text{demand}} \quad \tau_{\max}^{\text{mot}}\}}{\eta_{\text{mot}}}\omega_{\text{mot\_shaft}} & \tau_{\text{demand}} > = 0 \\ \max\{\tau_{\text{demand}} \quad -\tau_{\max}^{\text{mot}}\}\omega_{\text{mot\_shaft}}\eta_{\text{mot}} & \tau_{\text{demand}} < 0 \end{cases} \qquad (3.35)$$

The current demand on the high-voltage bus is:

$$I(t) = \frac{P_e(t)}{V_{\text{bus}}} \qquad (3.36)$$

In this example, the corresponding current profile is shown in Fig. 3.17.

3.3   Modeling of the Battery System

A battery system consists of battery cells, the working principle of which is shown in Fig. 3.18. Depending on the requirements for output voltage, power, and energy capacity for the hybrid electric vehicle, a battery pack is configured by many cells connected in series or parallel, or both.

An electrical circuit, illustrated in Fig. 3.19, is normally used to model the relationship between the current and voltage observed at the battery terminals. The state of charge of the battery system can be calculated based on the battery's Ahr capacity, the current history, and the self-discharge and charge/discharge efficiency. The parameters of the electrical

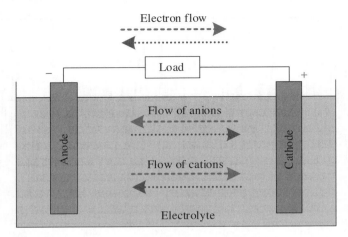

Figure 3.18 Electrochemical operation of a battery cell (dashed lines: discharge flow; dotted lines: charge flow)

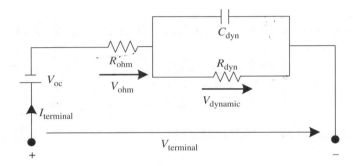

Figure 3.19 The electrical circuit equivalent model of a battery

circuit model are generally identified offline based on test data from a test such as the hybrid pulse power characterization test (HPPC). The HPPC test is designed to demonstrate the charge and discharge power capabilities of an HEV battery over a short period of time such as two seconds or ten seconds at various SOC levels under different operational temperatures. An HPPC test profile is illustrated in Fig. 3.20, where the voltages at the end point of each rest period establish the open-circuit voltage (V_{oc}).

3.3.1   Modeling Electrical Behavior

According to the equivalent electrical circuit, the relationship between terminal voltage and current can be expressed by the following equations:

$$\text{Terminal voltage}: V_{\text{terminal}} = V_{oc} + V_{\text{ohm}} + V_{\text{dynamic}} \tag{3.37}$$

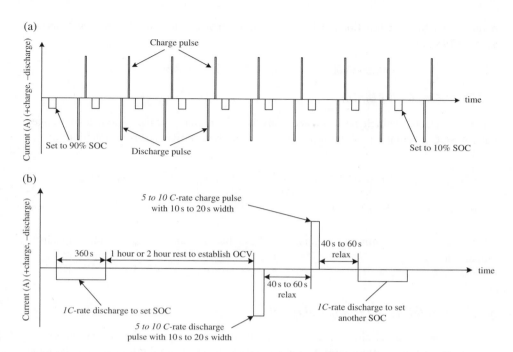

Figure 3.20   (a) Entire HPPC test sequence; (b) the detailed test pulse at each SOC level

$$\text{Open-circuit voltage}: \ V_{oc} = f(SOC, \ T) \tag{3.38}$$

$$\text{Voltage on the ohmic resistor}: \ V_{ohm} = I \cdot R_{ohm} = I \cdot R_{ohm}(SOC, T) \tag{3.39}$$

$$\text{Dynamic voltage}: \ I_{R_{dyn}} = \frac{V_{dynamic}}{R_{dyn}(SOC, T)};$$

$$I_{C_{dyn}} = C_{dyn}(SOC, T)\frac{dV_{dynamic}}{dt} \tag{3.40}$$

$$I = I_{R_{dyn}} + I_{C_{dyn}}$$

That is, the dynamic voltage can be described by the following differential equation:

$$\frac{dV_{dynamic}}{dt} + \frac{V_{dynamic}}{R_{dyn}(SOC, \ T)C_{dyn}(SOC, \ T)} = \frac{I}{C_{dyn}(SOC, \ T)} \tag{3.41}$$

The overall differential equation of the electrical circuit is:

$$\frac{dV_{terminal}}{dt} + \frac{V_{terminal}}{R_{dyn}C_{dyn}} = R_{ohm}\frac{dI}{dt} + \frac{R_{dyn} + R_{ohm}}{R_{dyn}C_{dyn}}I + \frac{V_{oc}}{R_{dyn}C_{dyn}} \tag{3.42}$$

where I is the battery system terminal current (A), $V_{terminal}$ is the battery system terminal voltage (V), T is the battery's temperature, V_{oc}, R_{ohm}, R_{dyn}, and C_{dyn} are circuit model parameters which are functions of T and SOC and can be elicited based on special tests such as the HPPC test.

3.3.2 SOC Calculation

For hybrid electric vehicle performance analysis and simulation at the system design stage, a battery SOC calculation can be implemented from the following current integration:

$$\text{SOC}: \quad SOC(t) = SOC(t_i) + \frac{1}{Cap_{Ahr} \cdot 3600} \int_{t_i}^{t} I(t)\eta_{bat}(SOC, T, sign(I(t)))dt \tag{3.43}$$

$$\text{Initial SOC}: \quad SOC(t_i) = SOC_i$$

where η_{bat} is the coulombic efficiency of the battery and Cap_{Ahr} is the capacity in ampere-hours.

3.3.3 Modeling Thermal Behavior

Based on the electrical circuit model and the thermal dynamics of the battery, the power-generating heat and battery temperature can be calculated from the following equations:

$$\text{Heat generation [w]}: \quad H_{gen'd} = H_{Rohm} + H_{dynamic} + H_{react} \tag{3.44}$$

$$\text{Ohmic resistance heat [w]}: \quad H_{Rohm} = |I| \cdot V_{ohm} = I^2 \cdot R_{ohm}(SOC, T) \tag{3.45}$$

$$\text{Dynamic heat [w]}: \quad H_{dynamic} = |I| \cdot |V_{dynamic}|$$

$$= |I| \cdot \left| \int_{0}^{t} \left[\frac{I(t)}{C_{dyn}(SOC, T)} - \frac{V_{dynamic}(t)}{R_{dyn}(SOC, T)C_{dyn}(SOC, T)} \right] dt \right| \tag{3.46}$$

$$\text{Initial condition}: \quad V_{dynamic}|_{t=0} = 0$$

$$\text{Reaction heat [w]}: \quad H_{react} = \frac{\Delta S \cdot I \cdot \eta_{bat} \cdot T}{F} \tag{3.47}$$

$$\text{Dissipated heat [w]}: \quad H_{dissip'd} = (T_{coolant} - T)h_{bat} \tag{3.48}$$

$$\text{Battery temperature}: \quad T(t) = T(t_i) + \int_{t_i}^{t} \frac{H_{gen'd} + H_{dissip'd}}{C_{ESS}M_{ESS}}dt \tag{3.49}$$

where ΔS is the delta entropy of reaction [J/mol-K], η_{bat} is the battery efficiency, and F is Faraday's constant $= 96487$ [C/mol], C_{ESS} is the specific heat capacity (J/K-kg) of the battery system, M_{ESS} is the mass (kg) of the battery system, $T_{coolant}$ is the temperature of the coolant in kelvin, and h_{bat} is the heat transfer coefficient of the battery (W/K).

Based on the above equations, an electrical model of a hybrid electric vehicle battery system can be implemented by the diagram shown in Fig. 3.21.

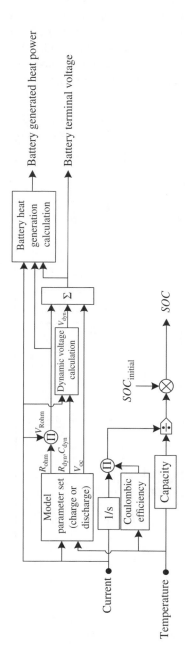

Figure 3.21 Hybrid pulse power characterization test

Table 3.3 Battery model parameters of a 4.4 Ahr battery system at 25 °C at the beginning of the system's life

	SOC 0%	SOC 10%	SOC 20%	SOC 30%	SOC 40%	SOC 50%	SOC 60%	SOC 70%	SOC 80%	SOC 90%	SOC 100%
V_{oc} (V)	106.24	116.36	125.04	132.44	138.72	144.04	148.6	152.52	155.96	159.08	162.08
R_{ohm_chg} (Ω)	0.0988	0.0944	0.0904	0.0876	0.0852	0.0836	0.0820	0.0812	0.0804	0.0800	0.0792
R_{ohm_dischg} (Ω)	0.1128	0.1024	0.0956	0.0916	0.0888	0.0868	0.0852	0.0836	0.0824	0.0816	0.082
R_{dyn_chg} (Ω)	0.8488	0.7076	0.5852	0.4816	0.3956	0.3276	0.2764	0.2420	0.2240	0.2220	0.2356
R_{dyn_dischg} (Ω)	0.4396	0.4280	0.3904	0.3408	0.2892	0.2436	0.2088	0.1856	0.1724	0.1652	0.156
C_{dyn_chg} (F)	73.6	86.1	97.6	108.0	117.5	125.7	132.9	139.0	143.9	147.6	150.1
C_{dyn_dischg} (F)	43.8	63.8	80.5	94.3	105.5	114.7	122.0	128.0	132.9	137.2	141.2

Example 3.5

Consider an HEV battery system consisting of 40 4.4 Ahr Li-ion cells connected in series. At 25 °C at the beginning of the system's life, the model parameters are as listed in Table 3.3. If a battery usage profile is given (Fig. 3.22), the battery terminal voltage can be calculated by the above model. Figures 3.23 and 3.24 show the model-predicted terminal voltage, the actual, the measured terminal voltage, and the corresponding SOC varying range.

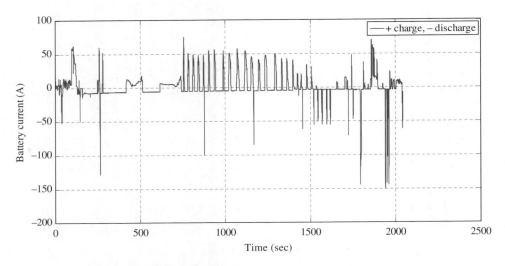

Figure 3.22   The battery test current profile

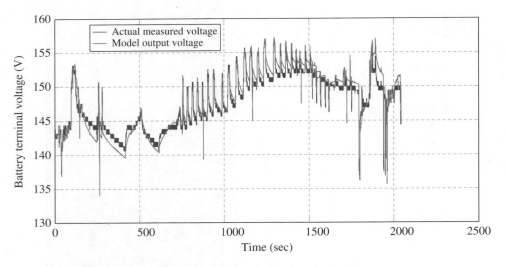

Figure 3.23   Actual and model-predicted battery terminal voltages

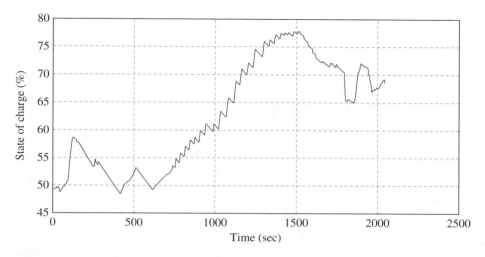

Figure 3.24 Battery SOC varying range over the battery usage test profile

3.4 Modeling of the Transmission System

Unlike a conventional vehicle, in which the transmission just matches the ICE characteristics with that particular vehicle's required characteristics, an HEV transmission also needs to perform the task that controls the direction and amount of power flow on the hybrid power-train to achieve the optimal overall efficiency of the hybrid electric vehicle. Therefore, an HEV transmission must be able to handle engine-only driving, electric-only driving, and a combination of the two. In addition, it also has to support stop–start functionality, the use of regenerative braking, and shifts in the ICE's operational zone. Since the transmission plays a very important role in a hybrid vehicle system, various advanced transmissions have been developed to make use of two movers efficiently. The electronic continuously variable transmission (ECVT) has been globally accepted as the standard transmission for hybrid vehicle systems.

The transmission system model presented in this section is based on the assumption that the vehicle's power demand is split between the ICE and the electric motor at the input shaft of the transmission through clutches. A conceptual architecture of such a hybrid vehicle system is illustrated by Fig. 3.25(a), where the power split device is the key component, and this generally consists of one or two planetary gear sets, as shown in Fig. 3.25(b). For a detailed analysis of such a power split system, the interested reader may refer to Miller's paper (Miller, 2006).

Since most hybrid electric transmissions are designed based on an existing automatic transmission, this section introduces a generic model of automatic transmission which includes a torque convertor, clutches, a gearbox, and a controller with a complex control strategy. The diagram of this transmission model is shown in Fig. 3.26.

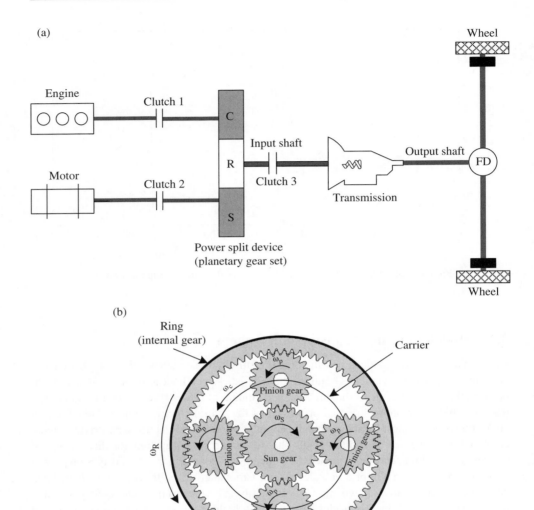

Figure 3.25 The conceptual architecture of an input split hybrid electric vehicle. (a) The conceptual architecture; (b) a schematic of the planetary gear set

3.4.1 *Modeling of the Clutch and Power Split Device*

In order to optimize two power sources, several ON–OFF clutches are needed in a hybrid electric vehicle transmission. A vehicle energy management algorithm will determine whether these clutches are ON or OFF. For the conceptual hybrid electric vehicle shown

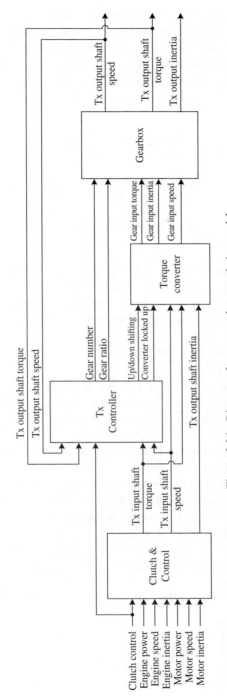

Figure 3.26 Diagram of an automatic transmission model

in Fig. 3.25(a), the splitting or combination of power is implemented through three clutches and one planetary gear set. The basic planetary gear set consists of a sun gear, a carrier with several pinion gears, and a ring gear. Each gear can act as the input or output, or be held stationary. If we assume that the radius of the sun gear is r_S with N_S teeth, the radius of the ring gear is r_R with N_R teeth, and the radius of the carrier is r_C, the basic equation of the planetary gear set is:

$$\omega_S + \frac{N_R}{N_S}\omega_R - \left(1 + \frac{N_R}{N_S}\right)\omega_C = 0 \tag{3.50}$$

or

$$\omega_S + \frac{r_R}{r_S}\omega_R - \left(1 + \frac{r_R}{r_S}\right)\omega_C = 0 \tag{3.51}$$

where ω_S is the angular velocity (rad/s) of the sun gear, ω_C is the angular velocity (rad/s) of the carrier, and ω_R is the angular velocity (rad/s) of the ring gear.

Corresponding with the states of clutches in the system shown in Fig. 3.25, the operational mode, transmission ratios, and operational states of the engine and motor are listed in Table 3.4.

The detailed clutch control logic and signals are:

Clutch Ctrl – Clutch control signals come from energy management.

Clutch Ctrl = 011 – This combination is the control command for the vehicle to operate in the regenerative braking state or the electric-only drive mode; the corresponding power flows are $P_{veh} \rightarrow P_{mot} \rightarrow P_{bat}$ or $P_{bat} \rightarrow P_{mot} \rightarrow P_{veh}$.

Clutch Ctrl = 101 – This combination is the control command for the vehicle to operate in ICE-only drive mode; the power flow is $P_{eng} \rightarrow P_{veh}$.

Clutch Ctrl = 110 – This combination means that clutch #1 and clutch #2 are engaged but clutch #3 is disengaged. In this combination, either the motor is cranking the ICE or the vehicle is stationary but the ICE is powering the motor (generating) to charge the battery system. The corresponding power flows are $P_{bat} \rightarrow P_{mot} \rightarrow P_{eng}$ or $P_{eng} \rightarrow P_{mot} \rightarrow P_{bat}$.

Clutch Ctrl = 111 – This combination is the control command for the vehicle to operate in hybrid drive mode, or for the ICE to propel the vehicle and power the motor (generating) to charge the battery; the power flows are $\begin{bmatrix} P_{eng} \\ P_{bat} \rightarrow P_{mot} \end{bmatrix} \rightarrow P_{veh}$ or $P_{eng} \rightarrow \begin{bmatrix} P_{veh} \\ P_{mot} \rightarrow P_{bat} \end{bmatrix}$.

Table 3.4 The states of the clutches and the power split device in the hybrid vehicle system illustrated in Fig. 3.25

| The state of clutches | | | Sun gear (S) | Planet & Carrier (C) | Ring gear (R) | Input–output relationship | Operational mode | Sign of motor torque | Sign of engine torque | Sign of Tx input shaft torque |
1	2	3								
OFF	ON	ON	Output	Stationary	Input	$\omega_S = -\dfrac{N_R}{N_S}\omega_R$	Regenerative braking	(−)	0	(+)
OFF	ON	ON	Input	Stationary	Output	$\omega_R = -\dfrac{N_S}{N_R}\omega_S$	Electric-only driving mode	(+)	0	(−)
ON	OFF	ON	Stationary	Input	Output	$\omega_R = \dfrac{N_S+N_R}{N_R}\omega_C$	Engine-only driving mode	0	(+)	(−)
ON	ON	OFF	Input	Output	Stationary	$\omega_C = \dfrac{N_S}{N_R+N_S}\omega_S$	Cranking	(+)	(−)	0
ON	ON	OFF	Output	Input	Stationary	$\omega_S = \dfrac{N_R+N_S}{N_S}\omega_C$	Stationary charging ESS	(−)	(+)	0
ON	ON	ON	Input	Input	Output	$\omega_R = \dfrac{N_S}{N_R}(\omega_C-\omega_S)+\omega_C$	Hybrid drive mode	(+)	(+)	(−)
ON	ON	ON	Output	Input	Output	$\omega_R + \dfrac{N_S}{N_R}\omega_S = \dfrac{N_S+N_R}{N_R}\omega_C$	Engine driving and motor charging	(−)	(+)	(−)

If we neglect the efficiency loss, the power split device can be modeled based on the clutch control signals as follows.

3.4.1.1 Clutch Control Signal = 011

In this case, the vehicle is exclusively driven by the motor, or is applying regenerative braking. The model inputs are:

ω_{mot} = angular velocity on the motor shaft (rad/s)
τ_{mot} = motor shaft torque (Nm)
J_{mot} = lumped inertia on the motor shaft (kg.m^2)

and the model outputs are:

ω_{tx_input} = angular velocity on the input shaft of the torque converter (rad/s)
τ_{tx_input} = the torque on the transmission input shaft (Nm)
J_{tx_input} = lumped inertia on the transmission input shaft (kg.m^2).

The relationships between input and output are:

$$\omega_{tx\_input} = -\frac{r_S}{r_R}\omega_{mot} \tag{3.52}$$

$$\tau_{tx\_input} = -\frac{r_R}{r_S}\tau_{mot} \tag{3.53}$$

$$J_{tx\_input} = \left(\frac{r_R}{r_S}\right)^2 J_{mot} + J_{tx\_shaft} \tag{3.54}$$

The corresponding model diagram is shown in Fig. 3.27(a).

3.4.1.2 Clutch Control Signal = 101

In this case, the vehicle is exclusively driven by the ICE. The model inputs are:

ω_{eng} = angular velocity on the engine shaft (rad/s)
τ_{eng} = engine shaft torque (Nm)
J_{eng} = lumped inertia on the engine shaft (kg.m^2)

The model outputs are:

ω_{tx_input} = angular velocity on the input shaft of the torque converter (rad/s)
τ_{tx_input} = torque on the transmission input shaft (Nm)
J_{tx_input} = lumped inertia on the transmission input shaft (kg.m^2)

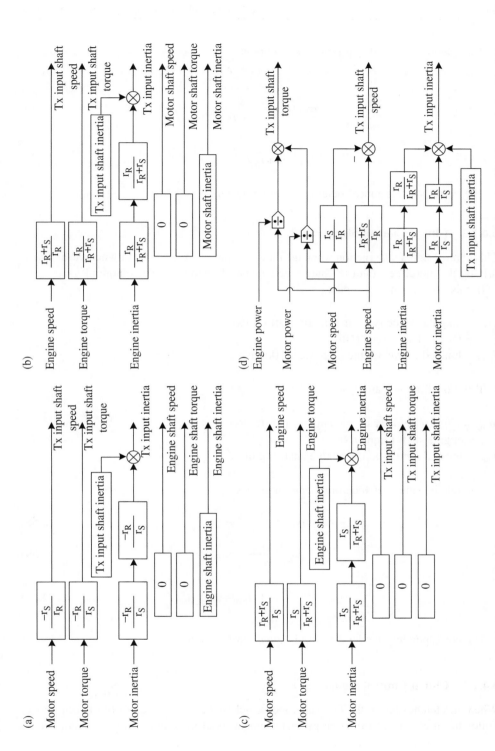

Figure 3.27 Model diagrams of the power split device with clutches in a hybrid electric vehicle transmission. (a) Clutch control signal is 101; (b) clutch control signal is 110; (c) clutch control signal is 110; (d) clutch control signal is 111

The relationships between input and output are:

$$\omega_{\text{tx\_input}} = \frac{r_S + r_R}{r_R}\omega_{\text{eng}} \qquad (3.55)$$

$$\tau_{\text{tx\_input}} = \frac{r_S}{r_R + r_S}\tau_{\text{eng}} \qquad (3.56)$$

$$J_{\text{tx\_input}} = \left(\frac{r_S}{r_R + r_S}\right)^2 J_{\text{eng}} + J_{\text{tx\_shaft}} \qquad (3.57)$$

The corresponding model diagram is shown in Fig. 3.27(b).

3.4.1.3   Clutch Control Signal = 110

This control signal takes action when the motor is cranking the engine or when the engine is driving the motor/generator to charge the battery while the vehicle is motionless.

In this case, the model inputs are:

ω_{mot} = angular velocity on the motor shaft (rad/s)
τ_{mot} = motor shaft torque (Nm)
J_{mot} = lumped inertia on the motor shaft (kg.m^2)

The model outputs are:

ω_{eng} = angular velocity on the engine shaft (rad/s)
τ_{eng} = engine shaft torque (Nm)
J_{eng} = lumped inertia on the engine shaft (kg.m^2)

The relationships between input and output are:

$$\omega_{\text{eng}} = \frac{r_R + r_S}{r_S}\omega_{\text{mot}} \qquad (3.58)$$

$$\tau_{\text{eng}} = \frac{r_S}{r_R + r_S}\tau_{\text{mot}} \qquad (3.59)$$

$$J_{\text{eng}} = \left(\frac{r_S}{r_R + r_S}\right)^2 J_{\text{mot}} + J_{\text{eng\_shaft}} \qquad (3.60)$$

The corresponding model diagram is shown in Fig. 3.27(c).

3.4.1.4   Clutch Control Signal = 111

When all clutches are in the ON state, the vehicle is being operated in full hybrid mode. Following a command from energy management, both the engine and the motor propel the vehicle, or the engine drives the vehicle and powers the motor to charge the battery. In this mode, the model inputs are:

ω_{eng} = angular velocity on the engine shaft (rad/s)
ω_{mot} = angular velocity on the motor shaft (rad/s)
P_{eng} = engine shaft power (W)
P_{mot} = motor shaft power (W)
J_{mot} = lumped inertia on the motor shaft (kg.m^2)
J_{eng} = lumped inertia on the engine shaft (kg.m^2)

The model outputs are:

ω_{tx_input} = angular velocity on the input shaft of the torque converter (rad/s)
τ_{tx_input} = torque on the transmission input shaft (Nm)
J_{tx_input} = lumped inertia on the transmission input shaft (kg.m^2)

The relationships between input and output are:

$$P_{tx\_input} = P_{mot} + P_{eng} \tag{3.61}$$

$$\omega_{tx\_input} = \left(\frac{r_R + r_S}{r_R} \right) \omega_{eng} - \frac{r_S}{r_R} \omega_{mot} \tag{3.62}$$

$$\tau_{tx\_input} = \frac{P_{mot}}{\omega_{mot}} + \frac{P_{eng}}{\omega_{eng}} \tag{3.63}$$

$$J_{tx\_input} = \left(\frac{r_R}{r_R + r_S} \right)^2 J_{eng} + \left(\frac{r_R}{r_S} \right)^2 J_{mot} + J_{tx\_shaft} \tag{3.64}$$

The corresponding model diagram is shown in Fig. 3.27(d).

3.4.2   Modeling of the Torque Converter

The function of the torque converter is to transfer torque smoothly from the engine to the gearbox through the fluid path and the mechanical clutch path. These two paths are in parallel. In the fluid path, the engine-generated torque is hydraulically transmitted to the gearbox, and the torque converter works as a fluid coupler which allows the engine to rotate fairly independently of the speed of the transmission output shaft. Physically, the torque converter is filled with oil and transmits the input shaft torque to the output shaft by means of the viscosity of the oil.

The parameters of a torque converter include the speed ratio, the torque ratio, the K-factor, and the C-factor, which are defined as:

$$\text{Speed ratio} = \frac{\text{Turbine speed}}{\text{Pump speed}} \tag{3.65}$$

$$\text{Torque ratio} = \frac{\text{Turbine torque}}{\text{Pump torque}} \tag{3.66}$$

$$\text{K-factor} = \frac{\text{Pump speed (rpm)}}{\sqrt{\text{Pump torque (Nm)}}} \tag{3.67}$$

$$\text{C-factor} = \frac{\text{Pump speed (rad/s)}}{\sqrt{\text{Pump torque (Nm)}}} \tag{3.68}$$

The torque converter is an important part of automatic transmission, and generally consists of a converting pump, a turbine, a stator, a lock-up clutch, and a torsional damper assembly. Unlike a traditional powertrain, the input torque of the torque converter in a hybrid electric vehicle is engine torque, motor torque, or the combined torque from these two sources. A diagram of the torque converter model is shown in Fig. 3.28.

The model inputs include:

lockedup = the state in which the torque converter is locked up
shifting = the state in which the transmission gears are shifting
coasting = the state in which the vehicle is coasting
$\tau_{\text{tx\_in}}$ = the torque on the input shaft of the transmission (Nm)
$\omega_{\text{tx\_in}}$ = the angular velocity of the input shaft of the transmission (rad/s)
$J_{\text{tx\_in}}$ = the inertia on the input shaft of the transmission (kg.m^2)

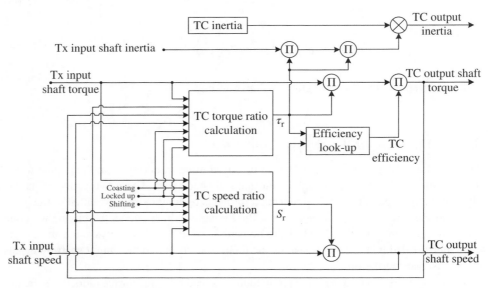

Figure 3.28   A diagram of the torque converter model

The model outputs are:

$\tau_{\text{tc\_out}}$ = the torque on the output shaft of the torque converter (Nm)
$\omega_{\text{tc\_out}}$ = the angular velocity of the output shaft of the torque converter (rad/s)
$J_{\text{tc\_out}}$ = the inertia on the output shaft of the torque converter (kg.m^2)

The relationships between inputs and outputs are:

$$\tau_{\text{tc\_out}} = \tau_{\text{tx\_in}} \cdot \tau_r \cdot \eta_{\text{tc}} \tag{3.69}$$

$$\omega_{\text{tc\_out}} = \omega_{\text{tx\_in}} \cdot s_r \tag{3.70}$$

$$J_{\text{tc\_out}} = J_{\text{tx\_in}} \cdot \tau_r^2 + J_{\text{tc}} \tag{3.71}$$

where s_r and τ_r are the speed ratio and the torque ratio of the torque converter, which are calculated from the equations below, and η_{tc} is the efficiency of the torque converter, which is a calibratable look-up table indexed by s_r.

$$s_r = \begin{cases} \min\left\{ \dfrac{\omega_{\text{tx\_in}}}{\sqrt{\tau_{\text{tx\_in}}}}, \dfrac{\omega_{\text{tc\_out}}}{(\omega_{\text{tc\_out}} + k_\omega \cdot \Delta t)} \right\} & \text{if } \omega_{\text{tx\_in}} \geq \omega_{\text{coasting}} \text{ and TC is unlocked} \\ s_{r|\text{coasting}} = lookup(\tau_{\text{tx\_in}}, \omega_{\text{tx\_in}}) & \text{if } \omega_{\text{tx\_in}} < \omega_{\text{coasting}} \text{ and TC is unlocked} \\ 1 & \text{if TC is locked} \end{cases}$$

$$\tag{3.72}$$

$$\tau_r = \begin{cases} f(s_r) = lookup(s_r) & \text{if TC is unlocked} \\ \tau_{r|\text{coasting}} = f\left(s_{r|\text{coasting}}\right) = lookup\left(s_{r|\text{coasting}}\right) & \text{if TC is coasting} \\ 1 & \text{if TC is locked} \end{cases} \tag{3.73}$$

where $s_{r|\text{coasting}}$ is the speed ratio when coasting, $\tau_{r|\text{coasting}}$ is the torque ratio when coasting, and k_ω is the calibratable speed change rate.

3.4.3 Modeling of the Gearbox

The function of the gearbox is to make the prime mover output match the vehicle's momentary traction requirements with maximal efficiency and optimal performance. A model diagram of a gearbox is shown in Fig. 3.29, whereby the inputs are the outputs of the torque converter model, as follows:

$\tau_{\text{gr\_in}} = \tau_{\text{tc\_out}}$ = the torque on the output shaft of the torque converter (Nm)
$\omega_{\text{gr\_in}} = \omega_{\text{tc\_out}}$ = angular velocity of the output shaft of the torque converter (rad/s)
$J_{\text{gr\_in}} = J_{\text{tc\_out}}$ = the inertia on the output shaft of the torque converter (kg.m^2)

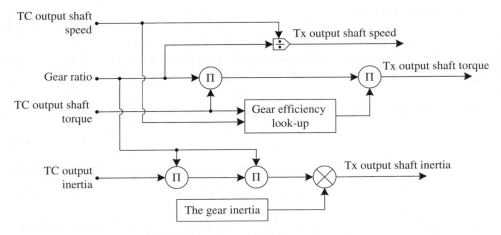

Figure 3.29 The model diagram of a gearbox

The model outputs are also the outputs of the entire transmission model:

$\tau_{tx\_out} = \tau_{gr\_out}$ = the torque on the output shaft of the transmission (Nm)
$\omega_{tx\_out} = \omega_{gr\_out}$ = angular velocity of the output shaft of the transmission (rad/s)
$J_{tx\_out} = J_{gr\_out}$ = the inertia on the output shaft of the transmission (kg.m^2)

The relationships between the inputs and outputs are expressed as follows:

$$\omega_{gr\_out} = \frac{\omega_{gr\_in}}{g_r} \tag{3.74}$$

$$\tau_{gr\_out} = g_r \cdot \left(\tau_{gr\_in} - \tau_{gr\_loss} \right) = g_r \cdot \eta_{gr} \cdot \tau_{gr\_in} \tag{3.75}$$

$$J_{gr\_out} = g_r^2 \cdot J_{gr\_in} + J_{gr} \tag{3.76}$$

where g_r is the gear ratio and η_{gr} is the gear efficiency.

3.4.4 Modeling of the Transmission Controller

The transmission controller is an electronic device with algorithms to control the action of the transmission. Its major function is to determine how and when to lock up the torque converter and change transmission gears to achieve the desired drivability, optimal fuel economy, and shifting quality. The decision to change gears is based on sensed data regarding the state of the transmission and data provided from the powertrain control unit regarding the state of the vehicle; therefore, the task of modeling the controller of an automatic transmission is actually to develop a simple transmission control algorithm for which the inputs and outputs are listed as follows.

The model inputs are:

Clutch Ctrl = control signal from the vehicle controller to engage or disengage the clutches (ON/OFF)

Input_shaft_spd = angular velocity on the input shaft of the torque converter (rad/s)

Input_shaft_trq = total torque on the input shaft of the torque converter (Nm)

Ouput_shaft_spd = angular velocity on the output shaft of the torque converter (rad/s)

Output_shaft_trq = total torque on the output shaft of the torque converter (Nm)

The model outputs are:

Gear # = control signal to determine which gear to engage

Gear ratio = actual gear ratio of the currently engaged gear

Shifting = gear shifting state; the gear is shifting when the shifting signal = 1

Lockedup = control signal to lock the torque converter; when *Lockedup* = 1, the TC is locked up.

The controls of an automatic transmission need to perform the following subtasks.

1. Estimate the input shaft speed of the torque converter

Based on the actual speed and torque on the output shaft of the transmission, the input shaft speed of the torque converter has to be estimated before sending a gear change command. The following equation can be used to estimate the speed:

$$\omega_{tc\_out} = \omega_{gr\_in} = g_r \cdot \omega_{gr\_out} \tag{3.77}$$

$$\hat{\omega}_{tc\_in} = \begin{cases} \omega_{tc\_out} & \text{if TC is locked} \\ \omega_{tc\_out}/s_r & \text{if TC is unlocked} \end{cases} \tag{3.78}$$

$$s_r = \frac{\omega_{tc\_out}}{\omega_{tc\_in}} = f(C\text{-}factor) = lookup(C\text{-}factor) \tag{3.79}$$

where s_r is the speed ratio of the torque converter; it is a look-up vector indexed by the C-factor defined in Eq. 3.68.

2. Calculate the load rate

Based on the power/torque capacities of movers, the controller needs to know the current load level relative to the capacity; then, a gear change decision can be made in order to best match the movers' output with the vehicle's demands. The load rate is calculated from the following equations:

$$L_r = \begin{cases} \dfrac{\tau_{tc\_in}}{\tau_{mot\_max}(\omega_{tc\_in})} & \text{if the vehicle is operating in EV mode} \\[3mm] \dfrac{\tau_{tc\_in}}{\tau_{eng\_max}(\omega_{tc\_in})} & \text{if the vehicle is operating in ICE only mode} \\[3mm] \dfrac{\tau_{tc\_in}}{\tau_{eng\_max}(\omega_{tc\_in}) + \tau_{mot\_max}(\omega_{tc\_in})} & \text{if the vehicle is operating in hybrid mode} \end{cases}$$

$$\tag{3.80}$$

where τ_{tc_in} is the actual torque on the input shaft of the transmission, τ_{x_max} is the maximal deliverable torque of movers on the input shaft of the transmission.

3. Shift up command

The command to shift the gear up is made based on a comparison between the current load rate and the predicted load rate once the gear is shifted up, and the actual decision can be made based on the following equation:

$$Ctrl_{up\_shift} = \begin{cases} 1 & \text{if } \dfrac{L_r(\omega_{tc\_in})}{L_{up\_f}(gr)} \geq 1 \\[4mm] 0 & \text{if } \dfrac{L_r(\omega_{tc\_in})}{L_{up\_f}(gr)} < 1 \end{cases} \tag{3.81}$$

where L_r is the actual load rate calculated by Eq. 3.81, L_{up_f} is the predetermined up-shifting torque fraction, $Ctrl_{up\_shift} = 1$ means that the shift up command is valid, $Ctrl_{up\_shift} = 0$ means that the shift up command is invalid.

4. Shift down command

As for the shift up determination, the command to shift down a gear is also determined based on a comparison between the current load rate and the predicted load rate if the gear is shifted down, and the following equation can be used to determine the command:

$$Ctrl_{dn\_shift} = \begin{cases} 1 & \text{if } \dfrac{L_r(\omega_{tc\_in})}{L_{dn\_f}(gr)} \geq 1 \text{ and } Ctrl_{up\_shift} = 0 \\[4mm] 0 & \text{if } \dfrac{L_r(\omega_{tc\_in})}{L_{dn\_f}(gr)} < 1 \text{ or } Ctrl_{up\_shift} = 1 \end{cases} \tag{3.82}$$

where L_r is the actual load rate calculated by Eq. 3.80, L_{dn_f} is the predetermined down-shifting torque fraction, $Ctrl_{dn\_shift} = 1$ means that the shift down command is valid, $Ctrl_{dn\_shift} = 0$ means that the shift down command is invalid.

5. Shifting duration

When gears are being changed in a transmission, it is necessary to take a short period of time to allow the engaged gears to synchronize with the driveshaft. The complete gear-shifting signal can be modeled based on the accumulated time as:

$$G_{shifting} = \begin{cases} 1 & \text{if } t_{shifting} < t_{gsh} \\[2mm] 0 & \text{if } t_{shifting} \geq t_{gsh} \end{cases} \tag{3.83}$$

where $G_{shifting} = 1$ means that the gear is shifting, $G_{shifting} = 0$ means that the gear-shifting process has been completed, $t_{shifting}$ is the accumulated shifting time since the gear started to shift, and t_{gsh} is the time needed to complete the gear shift.

6. Lock torque converter command

In order to improve fuel economy and drivability, the torque converter needs to be locked up when the vehicle is cruising or when gears are shifting. The TC locked-up command can be determined simply from the following logic:

$$TC_{\text{locked\_up}} = \begin{cases} 1 & \text{if} \left(G_{\text{shifting}} = 0 \right) \text{ or } (\omega_{\text{tx\_out}} = \text{constant}) \\ 0 & \text{if} \ \ G_{\text{shifting}} = 1 \end{cases} \tag{3.84}$$

where $TC_{\text{locked\_up}} = 1$ means that the torque converter is locked up.

3.5 Modeling of a Multi-mode Electrically Variable Transmission

Modern hybrid electric vehicles commonly exploit continuously variable transmission (CVT) and electronic continuously variable transmission (ECVT) to combine/split two power sources and propel the vehicle. The ratio between the input shaft and the output shaft of a CVT can continuously vary within a given range, so that the vehicle controller is able to select the optimal relationship between the speed of the engine and the speed of the wheels within a continuous range; therefore, the CVT can provide even better fuel economy if the engine runs constantly at a single speed. On the other hand, an ECVT normally consists of at least one set of planetary gears and several clutches to combine or split the power from the internal combustion engine (ICE) and the electric motor; one of the distinct characteristics is the multiple operational modes, which help the vehicle to operate most efficiently in most driving scenarios. ECVTs are capable of continuously adjusting effective output/input speed ratios like mechanical CVTs, but they offer the distinct benefit of also being able to apply power from two different sources to one output, as well as potentially reducing overall complexity. In the simple ECVT implementation shown in Fig. 3.25, an ICE is connected to the planetary carrier through a clutch, and an electric motor/generator is connected to the central sun gear through another clutch, while the output shaft is directly connected to the ring gear.

3.5.1 Basics of One-mode ECVT

A simple ECVT operates in a single mechanical configuration, as shown in Fig. 3.30. The kinematics of this ECVT are quite simple. As described by Eq. 3.50 in the previous section, for the planetary gear set, the speed of the carrier is weighted by the average of the ring gear speed and the sun gear speed. Thus, for this particular ECVT arrangement, which maximizes output torque, the speed of the output is the weighted average of the speed of the input and the speed of motor/generator 1 (MG_1). In Fig. 3.30, motor/generator 2 (MG_2) has the same speed as the output.

Since MG_2 is coupled to the output shaft of the transmission in this example arrangement, MG_1 is typically used to balance the torques of the planetary gear set, and the design objective is to have the ICE power transmit directly to the output without using the battery in most normal operating times. MG_2 is used mainly to capture regenerative energy if the vehicle needs to be slowed down by electric braking. MG_2 also provides additional power to boost the vehicle in the event of a wide open throttle to give the vehicle superior performance with a small ICE. Through this ECVT, the vehicle can operate in the following three modes:

1. **Load-sharing mode:** The vehicle is powered by the ICE and MG_2; MG_1 serves as a starter to start the ICE.
2. **Series mode:** The ICE powers MG_1 to generate electricity for MG_2 to propel the vehicle and/or to charge the battery.
3. **Power split mode:** A fraction of the ICE power can be transmitted to the sun gear to charge the battery through MG_1 and to the ring gear to propel the vehicle mechanically, depending on the detailed speeds of the gear shafts. In this operational mode, by varying the speed of MG_1, the ICE can hold its optimal fuel economy speed constantly over a long period of time, independent of the vehicle speed.

Because the ICE shaft is connected to a shaft of the planetary gear set, the ECVT arrangement in Fig. 3.30 is sometimes called an *input-split ECVT*, where the power flow through the transmission is effectively split by the gearing at the input. Typically, some of the ICE power flows to MG_1, which acts as a generator and turns the power into electricity, and the rest of the ICE power flows along to the output shaft. The output shaft power is simultaneously added from MG_2, which consumes the electrical power generated from MG_1. Therefore, this type of ECVT has two power paths through the transmission from input to output: a completely mechanical path from input to output through gears, and an electrical–mechanical path from MG_1 to MG_2 by means of a mechanically balanced gearing.

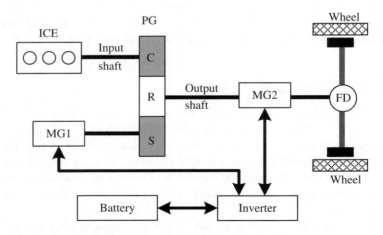

Figure 3.30   A simple ECVT schematic diagram

It should be noted that although the electrical–mechanical path enables the ICE to operate in the optimal efficiency zone and minimize the mechanical transient energy lost, the mechanical–electrical and electrical–mechanical conversion losses need to be taken into account in this ECVT arrangement. For the example transmission shown in Fig. 3.30, the direction of MG_1 power flow can be indicated by the speed ratio. With a change in the rotational direction of the sun gear, MG_1's functionality is altered from motoring to generating, and vice versa. The speed ratio of the transmission is defined as:

$$S_r = \frac{\omega_{\text{input}}}{\omega_{\text{output}}} = \frac{1}{1+R} + \frac{R}{1+R}\frac{\omega_{MG1}}{\omega_{\text{output}}} \tag{3.85}$$

where R is the planetary gear set ratio, which equals $\frac{N_{\text{sun}}}{N_{\text{ring}}}$.

From Eq. 3.85, a unique speed ratio $S_{rs} = \frac{1}{1+R}$ can be found, in which MG_1 is stationary. Since the power flow from input to output all stays in mechanical form in this operational condition, this particular speed ratio is often called the *mechanical point* of the ECVT transmission. As there is no mechanical power converted into electricity and back again, the mechanical point tends to be the most efficient ratio for mechanical power flow through the transmission. Because the ECVT shown in Fig. 3.30 has only one mechanical point, it is often called *one-mode ECVT*.

Example 3.6
A one-mode hybrid vehicle utilizes the ECVT shown in Fig. 3.30. The ECVT planetary gear set has a 1:2.6 ratio, the lumped ratio between the vehicle speed (km/h) and the ECVT output shaft rotation (rpm) is 1:35.6, and the ICE operational speed is within [700, 5500] rpm. Determine the mechanical point and the ICE operational speed vs. vehicle speed, and analyze the power flow during highway cruising and high power demand scenarios.

Solution:
With the given $R = 0.385$, if ω_{MG1} is respectively replaced by the maximal, zero, and minimal operational speed of MG_1, the operational speed of the ICE will be limited by:

$$\omega_{\text{ICE}} = \frac{1}{1+R}\omega_{\text{output}} + \frac{R}{1+R}\omega_{MG1} \tag{3.86}$$

Furthermore, given that the ratio between the speed (rpm) of the transmission output shaft and the vehicle speed (km/h) is 35.6:1 and the ICE operational speed is within [700, 5500] rpm, the ICE will operate within the zone shown in Fig. 3.31.

Some operational speed setpoints of this ECVT are determined in Table 3.5; for example, during light acceleration, when the output shaft speed (also MG_2) reaches 1500 rpm, the ICE operational speed is set at 2500 rpm and MG_1 operates at 5100 rpm. This example demonstrates that the speed ratio of this ECVT is variable. Table 3.5 also shows that the mechanical

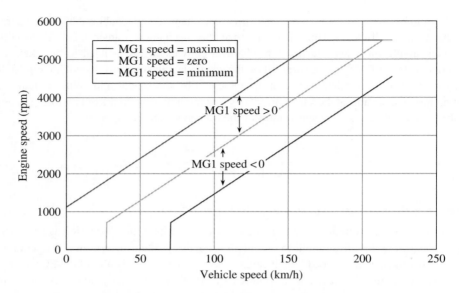

Figure 3.31   Curve of ICE operational speed vs. vehicle speed

Table 3.5   Some operational speeds of the given ECVT

Operational mode	Vehicle speed (km/h)	Input (ICE)	MG_1	Output (MG_2)
		Carrier	Sun	Ring
Engine warm-up (idle vehicle stopped, stationary charging)	Vspd = 0	1000 rpm	3600 rpm	0 rpm
Acceleration (hybrid)	Vspd > 0	2000 rpm	3300 rpm	1500 rpm
Cruising (MG_1 off, mechanical point)	Vspd = 105	2700 rpm	0 rpm	3738 rpm
Electric driving (engine off)	Vspd < = 55	0 rpm	−5200 rpm	2000 rpm
Series hybrid	55 < Vspd < = 82	1800 rpm	−1320 rpm	3000 rpm

point of this system occurs when the vehicle runs at 105 km/h corresponding with the ICE turning at 2700 rpm, MG_1 being stationary, and the output shaft turning at 1000 rpm.

Figures 3.32 and 3.33 demonstrate the power distribution among the ICE (input of transmission), MG_1, MG_2, and the output of transmission in two particular operational scenarios: highway cruising and maximal vehicle power demand. Under low power demand operational conditions such as cruising, the hybrid vehicle system is designed such that the ICE operates with approximately constant speed, the output shaft varies with the vehicle speed, and no battery energy is used. Figure 3.32 shows the power flowing through this input-split ECVT. Figure 3.33 demonstrates the mechanical power flow through this input-split ECVT without battery assistance under the maximal vehicle power demand

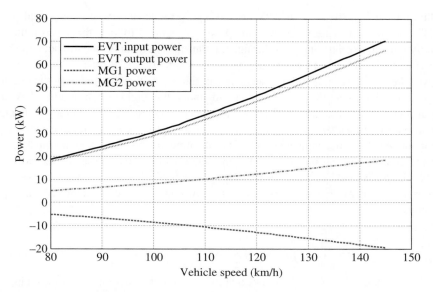

Figure 3.32 One-mode ECVT power flow chart under a low power demand operational condition (highway cruising)

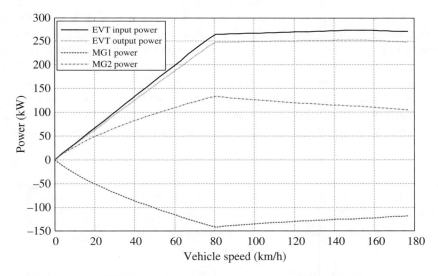

Figure 3.33 One-mode ECVT power flow chart under a high power demand operational condition (WOT acceleration)

condition. It can be seen that both MG_1 and MG_2 power reach a very large magnitude for this ECVT, even without battery power. This power is required to vary the speed ratio of the transmission and to transmit power through the transmission, not for the fundamental hybrid requirement of delivering power to the battery.

If we take Figs 3.32 and 3.33 together, we can find the critical limitation of the one-mode ECVT. In order to meet the vehicle's power demand, the engine often operates beyond its most efficient range, and the increased engine speed may push the speed ratio of the transmission out of its mechanical point. Since a hybrid electric vehicle with one-mode ECVT cannot operate at the mechanical point most of the time, the mechanical point must be determined to compromise propulsion system efficiency with the capacities of electric motors. For the given system, Fig. 3.32 shows that the mechanical point will restrain continuous motor power during cruising if it is chosen for low engine speeds to achieve highway fuel economy.

3.5.2   Basics of Two-mode ECVT

Since the mechanical point of the one-mode ECVT cannot be determined for both highway fuel economy and high power demand scenarios, it must compromise between fuel economy and power. Therefore, this type of ECVT is normally used in small hybrid and electric vehicles such as Toyota's Prius and GM's Chevy Volt. In order to meet the increased stringent emission standards and drivability requirements and achieve superior performance, many vehicle manufacturers have created advanced hybrid ECVT architectures. A characteristic of present-day power-split architectures is that the ECVT consists of multiple planetary gear sets and clutches and supports multiple operational modes. A well-known example of such an ECVT is the GM Allison Hybrid System II (GM-AHSII), commonly called a two-mode hybrid ECVT. A two-mode ECVT is capable of operating in input-split and compound-split modes, the system diagram for which is shown in Fig. 3.34 and the schematic cross-section is shown in Fig. 3.35. In principle, the two-mode ECVT can be implemented through two planetary gear sets. The first power-split device is a compound

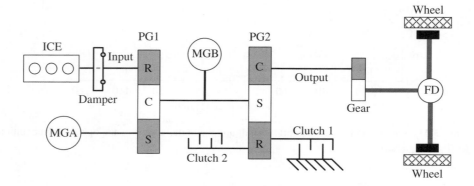

Figure 3.34   A two-mode EVT schematic diagram

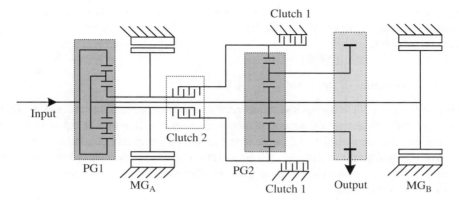

Figure 3.35 Schematic cross-section of the two-mode ECVT

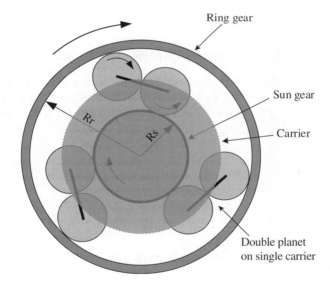

Figure 3.36 Compound planetary gear set with double planetary gears

planetary gear set with three sets of double planetary gears mounted on a single carrier, as shown in Fig. 3.36. The ICE directly connects to the ring gear of the first planetary gear set through a damper. Motor/generator A (MG_A) and the sun gear of the first planetary gear set are conjoined and selectively connected to the ring gear of the second planetary gear set through a rotating clutch. The ring gear of the second planetary gear set is also selectively connected to the ground through a stationary clutch. The carrier of the first planetary gear set, motor/generator B (MG_B), and the sun gear of the second planetary gear set are conjoined. The output shaft of the ECVT is connected to the carrier of the second planetary gear set.

3.5.2.1 Kinematics of the Two-mode ECVT

This two-mode ECVT achieves its first operational mode (input-split operational mode) by grounding the ring gear of the second planetary gear set (PG2), which is equivalent to the one-mode operation described in the previous section. During this operational mode, the ICE power is split by the first planetary gear set (PG1) while PG2 provides torque multiplied by MG_B as a conventional two-terminal gear set. The constraining speed and torque equations in input-split mode are:

$$\omega_{c_1} = \frac{1}{1-R_1}\omega_{r_1} - \frac{R_1}{1-R_1}\omega_{s_1} \tag{3.87}$$

$$-\tau_{c_1} = (1-R_1)\tau_{r_1} = \frac{R_1-1}{R_1}\tau_{s_1} \tag{3.88}$$

where $R_1 = \dfrac{N_{sun_1}}{N_{ring_1}}$ (the ratio of the first planetary gear set), $\omega_{r_1} = \omega_{ICE}$, $\omega_{s_1} = \omega_{MG_A}$.

The second operational mode, the compound-split mode, is achieved by locking the ring gear of PG2 to the sun gear of PG1 by the rotating clutch 2. During this operational mode, the split power by PG1 at the input is combined with that of PG2 at the output. The constraining speed and torque equations in compound-split mode are:

$$\omega_{c_2} = \frac{1}{1+R_2}\omega_{r_2} + \frac{R_2}{1+R_2}\omega_{s_2} \tag{3.89}$$

$$\tau_{c_2} = -(1+R_2)\tau_{r_2} = -\frac{1+R_2}{R_2}\tau_{s_2} \tag{3.90}$$

where $R_2 = \dfrac{N_{sun_2}}{N_{ring_2}}$ (the ratio of the second planetary gear set), $\omega_{r_2} = \omega_{MG_A}$, $\omega_{s_2} = \omega_{MG_B}$.

3.5.2.2 Operational Modes

As described in the previous section, the two-mode ECVT achieves its first operational mode (input-split operational mode) by grounding the ring gear of the second planetary gear set. During this operational mode, the rotational speed of MG_B is always proportional to the output shaft speed, while the rotational speed of MG_A is not proportional to the input shaft (ICE shaft) speed. As MG_B is directly connected to the output shaft, it is possible to propel the vehicle without the combustion engine. At low speeds, the vehicle can move using MG_A and MG_B, the ICE, or both, making it a so-called full hybrid vehicle. All accessories will still function under electrical power, and the ICE can restart instantly if necessary. This operational mode is often called the *input-split mode* because the power flow is split only at the input. In this mode, MG_A can act as a generator, while MG_B operates as a motor. If the power generated by MG_A equals the propelling power of MG_B, the battery operates in charging neutral and the vehicle is fully powered by the ICE.

Using Eqs 3.87 and 3.89 with $\omega_{input} = \omega_{ICE} = \omega_{r_1}$, $\omega_{output} = \omega_{c_2}$ and $\omega_{r_2} = 0$, we can obtain the speeds of MG_A and MG_B as follows:

$$\omega_{c_1} = \frac{1}{1-R_1}\omega_{input} - \frac{R_1}{1-R_1}\omega_{MG_A} \tag{3.91}$$

$$\omega_{output} = \frac{R_2}{1+R_2}\omega_{MG_B} \tag{3.92}$$

where $R_1 = \frac{N_{sun_1}}{N_{ring_1}}$, $R_2 = \frac{N_{sun_2}}{N_{ring_2}}$, $\omega_{r_1} = \omega_{ICE}$, $\omega_{s_1} = \omega_{MG_A}$, $\omega_{c_1} = \omega_{MG_B}$.

Thus, MG_A and MG_B speeds can be obtained based on the input shaft speed and the output shaft speed of the ECVT as follows:

$$\omega_{MG_A} = \frac{1}{R_1}\omega_{input} - \frac{(1-R_1)(1+R_2)}{R_1 R_2}\omega_{output} \tag{3.93}$$

$$\omega_{MG_B} = \frac{1+R_2}{R_2}\omega_{output} \tag{3.94}$$

From Eq. 3.93, there is a special input–output speed ratio of the ECVT in which MG_A does not rotate, and this is determined by the ratios of both planetary gear sets as follows:

$$\frac{\omega_{output}}{\omega_{input}} = \frac{R_2}{(1-R_1)(1+R_2)} \tag{3.95}$$

As described in the previous section, this speed ratio is called the mechanical point, and it corresponds to the point at which the power required by the vehicle is provided entirely by the ICE through the mechanical path without any mechanical–electrical–mechanical conversion.

The other operational mode of this ECVT is called *compound-split mode*; here, MG_B is not proportional to the output shaft speed, similarly, MG_A is not proportional to the input shaft speed. In this operational mode, the power flow is split at both the input and output shafts. The speed relationships between MG_A, MG_B, the input shaft, and the output shaft can be obtained by applying $R_1 = \frac{N_{sun_1}}{N_{ring_1}}$, $R_2 = \frac{N_{sun_2}}{N_{ring_2}}$, $\omega_{r_1} = \omega_{ICE} = \omega_{input}$, $\omega_{c_2} = \omega_{output}$, $\omega_{s_1} = \omega_{r_2} = \omega_{MG_A}$, $\omega_{c_1} = \omega_{s_2} = \omega_{MG_B}$ to Eqs 3.87 and 3.89:

$$\omega_{MG_B} = \frac{1}{1-R_1}\omega_{input} - \frac{R_1}{1-R_1}\omega_{MG_A} \tag{3.96}$$

$$\omega_{\text{output}} = \frac{1}{1+R_2}\omega_{MG_A} + \frac{R_2}{1+R_2}\omega_{MG_B} \tag{3.97}$$

Thus, we have the following relationships:

$$\omega_{MG_A} = -\frac{R_2}{1-R_1-R_1R_2}\omega_{\text{input}} + \frac{(1+R_2)(1-R_1)}{1-R_1-R_1R_2}\omega_{output} \tag{3.98}$$

$$\omega_{MG_B} = \frac{1}{1-R_1-R_1R_2}\omega_{\text{input}} - \frac{R_1(1+R_2)}{1-R_1-R_1R_2}\omega_{output} \tag{3.99}$$

From the above equations, there are the following two mechanical points at which neither MG_A nor MG_B rotate:

$$\frac{\omega_{\text{output}}}{\omega_{\text{input}}} = \begin{cases} \dfrac{R_2}{(1-R_1)(1+R_2)} & \omega_{MG_A} = 0 \\[3mm] \dfrac{1}{R_1(1+R_2)} & \omega_{MG_B} = 0 \end{cases} \tag{3.100}$$

Similarly, the torque relationships between MG_A, MG_B, the input shaft, and the output shaft are as follows:

$$
\begin{aligned}
\tau_{\text{input}} &= \tau_{\text{ICE}} = \frac{1}{R_1-1}\tau_{c_1} = -\frac{1}{R_1}\tau_{s_1} \\[2mm]
\tau_{\text{output}} &= -(1+R_2)\tau_{r_2} = -\frac{1+R_2}{R_2}\tau_{s_2} \\[2mm]
\tau_{MG_A} &= \tau_{s_1} - \tau_{r_2} = -R_1\cdot\tau_{\text{input}} + \frac{1}{1+R_2}\tau_{\text{output}} \\[2mm]
\tau_{MG_B} &= \tau_{c_1} - \tau_{s_2} = -(1-R_1)\cdot\tau_{\text{input}} + \frac{R_2}{1+R_2}\tau_{\text{output}}
\end{aligned} \tag{3.101}
$$

Example 3.7

A two-mode hybrid ECVT is shown in Fig. 3.34. The first planetary gear set is a compound planetary gear set with 44:60 sun to planetary and 60:104 planetary to ring ratios. The second planetary gear set is a simple planetary gear set with a 37:83 sun to ring ratio and a lumped ratio between the vehicle speed (km/h) and the ECVT output shaft rotation (rpm) of 1:23.2. Both MG_A and MG_B have a 10,000 rpm maximal speed limit and an 85 kW power capability. If the low and high power profiles are given by Fig. 3.37, determine the power flows of the ICE, MG_A and MG_B without battery power assistance.

Solution:

Given that $R_1 = \dfrac{44}{104} = 0.423$ and $R_2 = \dfrac{37}{83} = 0.446$, the speeds of MG_A and MG_B relative to the speeds of the transmission input shaft and the output shaft, and the corresponding torque relationships are as follows:

$$
\begin{aligned}
\omega_{MG_A} &= -1.148\cdot\omega_{\text{input}} + 2.148\cdot\omega_{output} \\[2mm]
\omega_{MG_B} &= 2.575\cdot\omega_{\text{input}} - 1.575\cdot\omega_{output}
\end{aligned} \tag{3.102}
$$

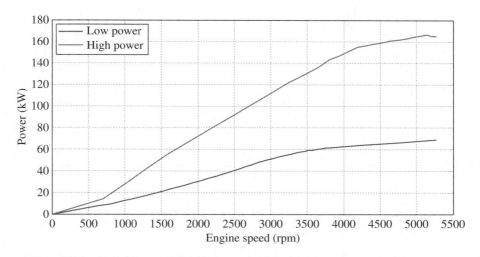

Figure 3.37   Optimal operation curves of the ICE under low and high power conditions

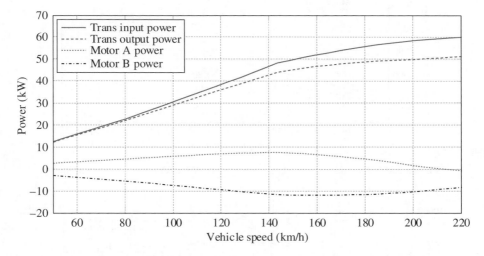

Figure 3.38   Operational power of two-mode ECVT under low power conditions

$$\tau_{MG_A} = -0.423 \cdot \tau_{\text{input}} + 0.692 \cdot \tau_{\text{output}}$$
$$\tau_{MG_B} = -0.577 \cdot \tau_{\text{input}} + 0.384 \cdot \tau_{\text{output}}$$
(3.103)

If the optimal operation curves of the ICE under low and high power operational conditions are as shown in Fig. 3.37, the operational power of MG_A, MG_B, the input shaft, and the output shaft are as shown in Fig. 3.38 and Fig. 3.39, respectively. The corresponding operational speeds are as shown in Fig. 3.40 and Fig. 3.41.

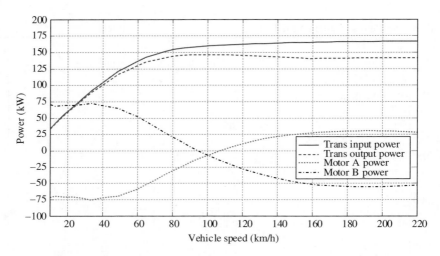

Figure 3.39 Operational power of two-mode ECVT under high power conditions

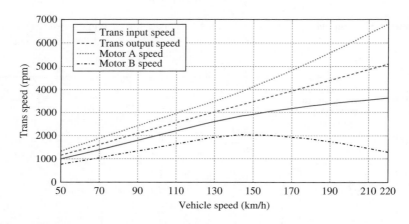

Figure 3.40 Operational speed of two-mode ECVT under low power conditions

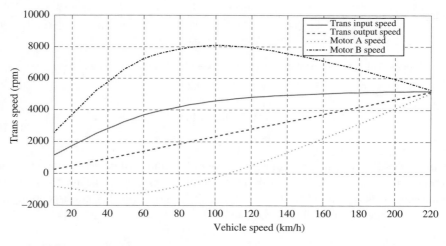

Figure 3.41 Operational speed of two-mode ECVT under high power conditions

3.6 Lever Analogy as a Tool for ECVT Kinematic Analysis

As described in the previous section, planetary gear sets are fundamental components of an ECVT, and it is quite cumbersome to analyze their kinematics by means of the traditional torque and speed calculation method. The lever analogy diagram introduced in this section is a very useful tool when analyzing a gear train that has more than two connected planetary gear sets. The lever analogy diagram is a translational system representing the rotating parts of a planetary gear. Hybrid vehicle engineering practices show that this tool can easily visualize the essential functions of ECVTs without addressing the complexities of planetary gear kinematics (Benford and Leising, 1981; Conlon, 2005; Hall and Dourra, 2009).

3.6.1 Lever System Diagram Set-up

The lever diagram is a basic building block of the analogy which replaces the planetary gear set, and its proportions are determined by the numbers of teeth on the sun and ring gears. In the lever analogy diagram, all planetary gear sets connected in an ECVT are denoted by a single vertical lever, in which the ECVT input, output, and reaction torques are represented by horizontal forces on the lever, and the rotational velocities relative to the reaction points are represented by the lever motion.

The lever analogy diagram of a simple planetary gear set can be set up as shown in Fig. 3.42. For case A, shown in Fig. 3.42(b), it is easy to obtain the following relationships from the lever analogy diagram:

$$\tau_C = -(\tau_R + \tau_S), \quad \tau_R = -\tau_C \cdot \frac{N_R}{N_R + N_S}, \quad \tau_S = -\tau_C \cdot \frac{N_S}{N_R + N_S}$$

$$\frac{\omega_S - \omega_R}{N_R + N_S} = \frac{\omega_C - \omega_R}{N_S} \quad \Rightarrow \quad \omega_C = \frac{N_S}{N_R + N_S} \cdot \omega_S + \frac{N_R}{N_R + N_S} \cdot \omega_R \tag{3.104}$$

Grounding the carrier can result in a special case B, shown in Fig. 3.42(c), and from the diagram, the following speed and torque relationships between the sun gear and the ring gear can be obtained:

$$\frac{\omega_S}{N_R} = -\frac{\omega_R}{N_S} \quad \Rightarrow \quad \omega_S = -\frac{N_R}{N_S} \cdot \omega_R$$

$$\tau_S \cdot N_R = \tau_R \cdot N_S \quad \Rightarrow \quad \tau_S = \frac{N_S}{N_R} \cdot \tau_R \tag{3.105}$$

For a compound gear set, the lever analogy diagram can be set up as shown in Fig. 3.43. The speed and torque relationships among ring, sun, and carrier shafts of the case A, shown in Fig. 3.43(b), are as follows:

$$\tau_C = -(\tau_R + \tau_S), \quad \tau_R = -\frac{N_R}{N_R - N_S} \cdot \tau_C, \quad \tau_S = \frac{N_S}{N_R - N_S} \cdot \tau_C, \quad \tau_S = -\frac{N_S}{N_R} \cdot \tau_R$$

$$\frac{\omega_S - \omega_C}{N_R} = \frac{\omega_R - \omega_C}{N_S} \quad \Rightarrow \quad \omega_C = \frac{N_R}{N_R - N_S} \cdot \omega_R - \frac{N_S}{N_R - N_S} \cdot \omega_S \tag{3.106}$$

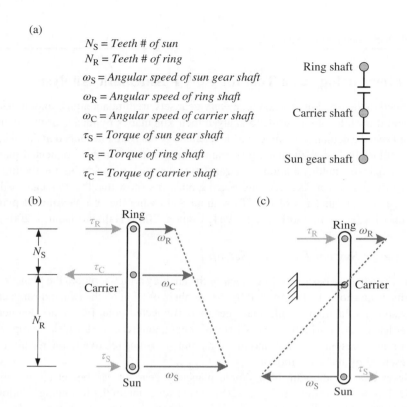

Figure 3.42 Simple planetary gear set lever analogy diagram. (a) Stick diagram, (b) Lever diagram case A, (c) Lever diagram case B

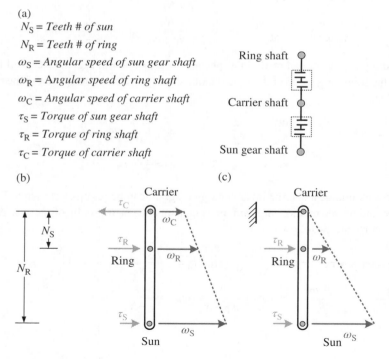

Figure 3.43 Compound planetary gear set lever analogy diagram. (a) Stick diagram, (b) Lever diagram case A, (c) Lever diagram case B

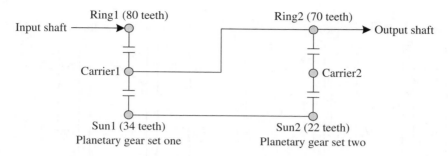

Figure 3.44 Diagram of two interconnected planetary gear sets

Grounding the carrier can result in a special case *B*, shown in Fig. 3.43(c), and from the diagram, there are the following speed and torque relationships between the sun gear and the ring gear:

$$\frac{\omega_S}{N_R} = \frac{\omega_R}{N_S} \quad \Rightarrow \quad \omega_S = \frac{N_R}{N_S} \cdot \omega_R$$

$$\tau_S \cdot N_R = \tau_R \cdot N_S \quad \Rightarrow \quad \tau_S = \frac{N_S}{N_R} \cdot \tau_R \tag{3.107}$$

For a planetary system interconnected by two planetary gear sets, as shown in Fig. 3.44, the lever analogy diagram can be set up by the following two procedures (shown in Fig. 3.45).

1. Set up the lever analogy diagram for each planetary gear set and insert horizontal links between the places associated with the interconnections between planetary gear sets, as shown in Fig. 3.45(a).
2. Keep the levers in parallel and rescale the constants of the diagram to make the links horizontal. Since levers connected by a pair of horizontal links remain parallel, they can be replaced functionally by a single lever with the same vertical dimension between points, as shown in Fig. 3.45(b).

3.6.2 Lever Analogy Diagram for ECVT Kinematic Analysis

Since a planetary gear system plays an important role in distributing the power between the ICE and electric motors in the ECVT, a lever analogy diagram provides an effective method for hybrid electric vehicle engineers to analyze various planetary arrangements. This section takes the two-mode ECVT shown in Fig. 3.34 as an example to illustrate the procedures when performing ECVT kinematic analysis.

The chosen ECVT runs in the first mode by grounding the ring gear of PG2 and disengaging the rotating clutch 2. In this mode, the ICE power is split by PG1; PG2 becomes a fixed gear outputting torque by multiplying the sum of the PG1 output shaft torque and the MG_B torque; and the corresponding lever analogy diagram is shown in Fig. 3.46.

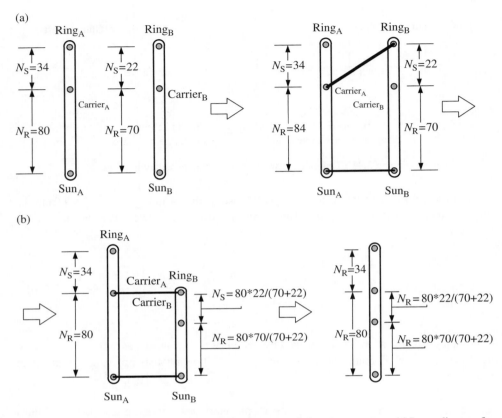

Figure 3.45 Lever analogy diagram for two interconnected planetary gear sets. (a) Lever diagram for procedure 1; (b) lever diagram for procedure 2

Based on the speed diagram in Fig. 3.46(d), the MG_B and output shaft speeds in the first mode are derived as:

$$\omega_{MG_B} = \frac{N_{sun_2} + N_{ring_2}}{N_{sun_2}} \cdot \omega_{output} = \left(1 + \frac{1}{R_2}\right) \cdot \omega_{output} \cong 3.24 \cdot \omega_{output}$$

$$\omega_{output} = \frac{N_{ring_1} + N_{ring_2}}{N_{ring_1} - N_{sun_1}} \cdot \omega_{input} - \frac{N_{sun_1} + N_{ring_2}}{N_{ring_1} - N_{sun_1}} \cdot \omega_{MG_A} \qquad (3.108)$$

$$= \frac{104 + 83}{104 - 44} \cdot \omega_{input} - \frac{44 + 83}{104 - 44} \cdot \omega_{MG_A} \cong 3.12 \cdot \omega_{input} - 2.12 \cdot \omega_{MG_A}$$

The chosen ECVT runs in the second mode, called compound-split mode, by disengaging clutch 1 and engaging clutch 2 to place the sun gear of the first planetary gear set and the ring

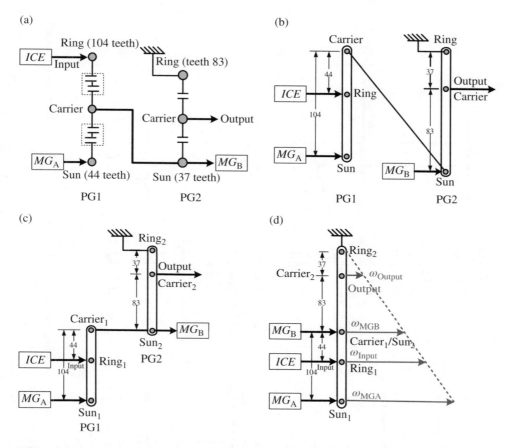

Figure 3.46 Lever analogy diagram of the first-mode ECVT. (a) Stick diagram; (b) lever analogy diagram; (c) rescale the lever analogy diagram; (d) combined lever analogy diagram with gear-shafts speed

gear of the second planetary gear set together. In this mode, the power flow is split at the input and at the output. The corresponding lever analogy diagram with compound split mode is shown in Fig. 3.47.

Based on the speed diagram in Fig. 3.47(d), the speed relationships between the input shaft, MG_A, MG_B, and the output shaft in the compound split mode are derived as:

$$\frac{\omega_{MG_A} - \omega_{\text{output}}}{N_{\text{ring}_1} \cdot N_{\text{sun}_2} / \left(N_{\text{sun}_2} + N_{\text{ring}_2}\right)} = \frac{\omega_{\text{output}} - \omega_{MG_B}}{N_{\text{ring}_1} \cdot N_{\text{ring}_2} / \left(N_{\text{sun}_2} + N_{\text{ring}_2}\right)}$$

$$\Rightarrow \quad \omega_{\text{output}} \cdot \left(N_{\text{sun}_2} + N_{\text{ring}_2}\right) = \omega_{MG_A} \cdot N_{\text{ring}_2} + \omega_{MG_B} \cdot N_{\text{sun}_2} \tag{3.109}$$

$$\Rightarrow \quad \omega_{\text{output}} = \frac{1}{1+R_2} \omega_{MG_A} + \frac{R_2}{1+R_2} \omega_{MG_B} \cong 0.69 \omega_{MG_A} + 0.31 \omega_{MG_B}$$

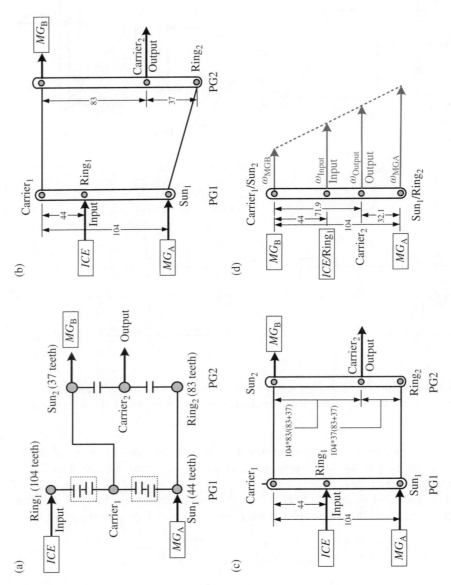

Figure 3.47 Lever analogy diagram of the second-mode ECVT. (a) Stick diagram; (b) lever analogy diagram; (c) rescale the lever analogy diagram; (d) combined lever analogy diagram with gear-shaft speed

$$\frac{\omega_{\text{output}} - \omega_{\text{input}}}{N_{\text{ring}_1} \cdot N_{\text{ring}_2} / (N_{\text{sun}_2} + N_{\text{ring}_2}) - N_{\text{sun}_1}} = \frac{\omega_{\text{input}} - \omega_{MG_B}}{N_{\text{sun}_1}}$$

$$\Rightarrow \quad \omega_{\text{output}} \cdot (N_{\text{sun}_1} \cdot N_{\text{ring}_2} + N_{\text{sun}_1} \cdot N_{\text{sun}_2}) = N_{\text{ring}_1} \cdot N_{\text{ring}_2} \cdot \omega_{\text{input}}$$

$$+ (N_{\text{sun}_1} \cdot N_{\text{sun}_2} + N_{\text{sun}_1} \cdot N_{\text{ring}_2} - N_{\text{ring}_1} \cdot N_{\text{ring}_2}) \cdot \omega_{MG_B}$$

(3.110)

$$\Rightarrow \quad \omega_{\text{output}} = \frac{1}{R_1(1 + R_2)} \cdot \omega_{\text{input}} - \frac{1 - R_1 - R_1 \cdot R_2}{R_1(1 + R_2)} \cdot \omega_{MG_B} \cong 1.63 \cdot \omega_{\text{input}} - 0.63 \omega_{MG_B}$$

3.7 Modeling of the Vehicle Body

The vehicle body, shown in Fig. 3.48, provides rolling resistance and aero-drag resistance forces. The retarding force of the vehicle when motionless is equal to $F_{\text{fric}} + F_{\text{grade}}$. The required propulsion force at a given speed is described by the equation:

$$F_{\text{wh}} = F_{\text{rolling}} + F_{\text{aero}} + F_{\text{grade}} + m_{\text{v}} a$$

(3.111)

where F_{wh} is the traction force on the wheel from the powertrain, m_{v} is the vehicle's mass (kg), a is the acceleration of the vehicle (m/s^2), F_{rolling}, F_{aero}, F_{grade} are the rolling resistance, the aero-drag resistance, and the grade weight forces, respectively, and are expressed as:

$$F_{\text{rolling}} \approx k_{\text{rrc}} k_{\text{sc}} m_{\text{v}} g \cos(\alpha) = k_{\text{r}} m_{\text{v}} g \cos(\alpha)$$

$$k_{\text{r}} = k_{\text{rrc}} k_{\text{sc}}$$

(3.112)

$$F_{\text{aero}} = k_{\text{aero}} v^2$$

$$k_{\text{aero}} = \frac{1}{2} C_{\text{a}} \cdot A_{\text{d}} \cdot C_{\text{d}} \cdot F_{\text{a}}$$

(3.113)

$$F_{\text{grade}} = m_{\text{v}} g \sin(\alpha)$$

(3.114)

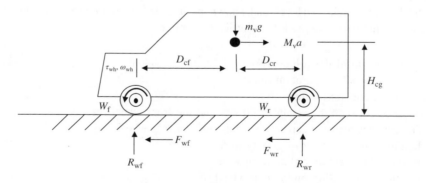

Figure 3.48 Free body diagram of a vehicle

where k_{rrc} is the rolling resistance coefficient, k_{sc} is the road surface coefficient, k_r is the effective rolling resistance coefficient, k_{aero} is the aero drag factor, C_a is the air density correction coefficient at a given altitude, A_d is the air mass density (kg/m^3), C_d is the aerodynamic drag coefficient of the vehicle (Ns^2/kgm), F_a is the frontal area of the vehicle (m^2), and α is the road incline angle (grade angle).

3.8   Modeling of the Final Drive and Wheel

3.8.1   Final Drive Model

The final drive is a set of rigid mechanical devices which transfers power from a transmission to the wheels. The following simple models can be used to simulate and analyze system performance:

$$\text{Output speed}: \ \omega_{fd\_out} = \frac{1}{f_d}\omega_{fd\_in} \qquad (3.115)$$

$$\text{Output torque}: \ \tau_{fd\_out} = f_d \cdot \eta_{fd}(\omega_{fd\_in}, \tau_{fd\_in})\tau_{fd\_in} \qquad (3.116)$$

$$\text{Output inertia}: \ J_{fd\_out} = f_d^2 \cdot J_{fd\_in} + J_{fd} \qquad (3.117)$$

where f_d is the final drive ratio, $\eta_{fd}(\omega_{fd\_in}, \tau_{fd\_in})$ is the final drive efficiency, which is a function of input speed and torque and can be obtained through actual tests, J_{fd_out} is the output inertia of the final drive, J_{fd_in} is the input inertia of the transmission, and J_{fd} is the inertia of the final drive itself.

3.8.2   Wheel Model

A vehicle wheel is a deformable part in contact with the road. When mover-generated torque is applied to the wheel axle, the tire deforms and pushes on the ground against friction, resulting in the vehicle moving forwards or backwards. From a free body diagram of a rear-wheel-drive vehicle on a level road, shown in Fig. 3.48, the friction forces on the wheels can be calculated based on the following list of parameters:

$m_v g$ = gross weight of the vehicle (N)
$g = 9.81$ gravitational acceleration (m/s^2)
$m_v a$ = vehicle acceleration force (N)
D_{cf} = distance between the center of gravity and the front wheel (m)
D_{cr} = distance between the center of gravity and the rear wheel (m)
H_{cg} = height from the center of gravity to the road (m)
R_{wf} = reaction force acting on the front wheel (N)
R_{wr} = reaction force acting on the rear wheel (N)
F_{wf} = friction force acting on the front wheel (N)

F_{wr} = friction force acting on the rear wheel (N)
τ_{trac} = traction torque from the powertrain (Nm)
ω_{wh} = wheel angular velocity (rad/s)
V_{veh} = vehicle velocity (m/s)
r_{wh} = effective wheel rolling radius (m)
J_{axle} = lumped inertia on the axle transferred from the powertrain (kg.m^2)
J_{wh} = wheel inertia (kg.m^2)
R_{L} = road load (N)

On a level road, taking the moment about the front wheel and the road contact point, we get:

$$R_{wr} = \frac{m_v g \cdot D_{cf} + m_v a \cdot H_{cg}}{D_{cf} + D_{cr}} \tag{3.118}$$

$$R_{wf} = m_v g - R_{wr} \tag{3.119}$$

Equations (3.118) and (3.119) give the reaction forces on the rear wheel and the front wheel, respectively. The frictional forces on the front and rear wheels are obtained by the following equations:

$$F_{wf} = \mu \cdot R_{wf} \tag{3.120}$$

$$F_{wr} = \mu \cdot R_{wr} \tag{3.121}$$

The total frictional force is:

$$F_{fric} = F_{wf} + F_{wr} \tag{3.122}$$

where μ is the frictional coefficient between the wheel tire and the road, and it is normally given as a function of tire slip in automobile handbooks.

The traction force from the powertrain is:

$$F_{wh} = \frac{\tau}{r_{wh}} = \frac{\tau_{trac} - \tau_{brk}}{r_{wh}} \tag{3.123}$$

The vehicle velocity is:

$$V_{veh} = r_{wh} \omega_{wh} \tag{3.124}$$

The transferred equivalent mass from the powertrain inertia is:

$$m_{equ} = \frac{J_{axle} + J_{wh} \cdot 4}{r_{wh}^2} \tag{3.125}$$

The road load is an important characteristic or property of any vehicle, and it refers to the minimal propulsive force requirement on the tires to move the vehicle in the absence of any

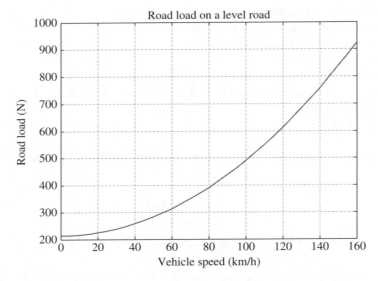

Figure 3.49   The road load of the vehicle given in Example 3.8

ambient wind. The road load is normally determined by performing coastdown tests and is described by the following equation on a level road:

$$R_L = F_{\text{rolling}} + F_{\text{aero}} = k_r m_v g + \frac{1}{2} C_a \cdot A_d \cdot C_d \cdot F_a \tag{3.126}$$

Furthermore, the following equation can be used to describe the road load for a vehicle on a graded road:

$$R_L = F_{\text{rolling}} + F_{\text{aero}} + F_{\text{grade}} = k_r m_v g \cos(\alpha) + \frac{1}{2} C_a \cdot A_d \cdot C_d \cdot F_a + m_v g \sin(\alpha) \tag{3.127}$$

Example 3.8
If we assume that a vehicle has a mass of 1450 kg, an effective wheel rolling radius of 0.33 m, a frontal area of 2.0 m^2, an aero drag coefficient of 0.3 N $\cdot$ s^2/kg $\cdot$ m, and an effective coefficient of rolling resistance of 0.015, then the road load of the given vehicle on a straight, level road under windless conditions is shown in Fig. 3.49.

3.9   PID-based Driver Model

When simulating vehicle performance, it is necessary for a driver to operate the vehicle according to a given cycle. The driver can be modeled by a PID controller which outputs the positions of the acceleration pedal and the brake pedal.

3.9.1 Principle of PID Control

The principle of the PID control algorithm can be described as follows:

$$u(t) = K\left(e(t) + \frac{1}{T_i}\int_0^t e(\tau)d\tau + T_d\frac{de(t)}{dt}\right) \tag{3.128}$$

where $e(t) = V_{sch}(t) - V_{veh}(t)$ is the control error, $u(t)$ is the control variable which is a sum of three terms: the *P term: $Ke(t)$*, which is proportional to the error; the *I term: $K\frac{1}{T_i}\int_0^t e(\tau)d\tau$*, which is proportional to the integral of the error; and the *D term: $K\cdot T_d\frac{de(t)}{dt}$*, which is proportional to the derivative of the error. The PID control algorithm given by Eq. 3.128 can be represented by the following transfer function form:

$$G(s) = \frac{U(s)}{E(s)} = K\left(1 + \frac{1}{T_i s} + T_d s\right) \tag{3.129}$$

The PID control algorithm diagram is shown in Fig. 3.50. The parameters of the controller are the proportional gain, K, the integral time, T_i, and the derivative time, T_d. The action of the proportional *P* term is simply proportional to the control error. The integral *I* term gives a control action proportional to the time integral of the error, which ensures that the steady-state error becomes zero.

In practice, some nonlinear effects must be accounted for, such as the traction speed of the motor has a limit. From a control point of view, when a control variable reaches the max/min limit, the feedback loop is broken, and the system actually runs as an open loop. In this case, if the controller has an integrating action and the control variable reaches the actuator limits, the control error will be integrated continuously so that the control action will be accumulated. Therefore, it will take a long period of time to obtain an opposite sign for the error in order to return the control action to normal. In order to cope with this problem, it is necessary to provide the PID algorithm with an anti-wind-up function. For the detailed implementation and analysis of PID control strategies with an anti-wind-up function, the interested reader should refer to Åstrom and Hägglund (1995).

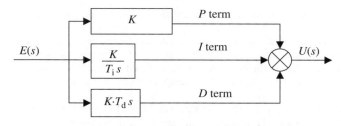

Figure 3.50 Block diagram of a PID control algorithm

3.9.2   Driver Model

The inputs of the driver model are the vehicle speed and the scheduled vehicle speed (the driver's desired speed), and the outputs are the percentage usages of the acceleration and brake pedals. The driver model here is implemented by a PID controller and takes the vehicle's physical limitations into account, as follows:

$$\tau_{\text{demand}} = \tau_{\text{PID\_adjusted}} + \tau_{\text{lost}} \tag{3.130}$$

$$\tau_{\text{PID\_adjusted}} = \begin{cases} \max\left\{ -\tau_{\max}, \dfrac{K}{T_i}\displaystyle\int_0^t e(\tau)d\tau + K \cdot T_d \dfrac{de(t)}{dt} \right\} & \text{if } \left(\dfrac{K}{T_i}\displaystyle\int_0^t e(\tau)d\tau + K \cdot T_d \dfrac{de(t)}{dt} \right) < 0 \\[4mm] \min\left\{ \tau_{\max}, \dfrac{K}{T_i}\displaystyle\int_0^t e(\tau)d\tau + K \cdot T_d \dfrac{de(t)}{dt} \right\} & \text{if } \left(\dfrac{K}{T_i}\displaystyle\int_0^t e(\tau)d\tau + K \cdot T_d \dfrac{de(t)}{dt} \right) \geq 0 \end{cases} \tag{3.131}$$

$$\tau_{\text{lost}} = \left(m_v \frac{dV_{\text{veh}}}{dt} + F_{\text{load}} \right) \cdot r_{\text{wh}} \tag{3.132}$$

$$F_{\text{load}} = F_{\text{rolling}} + F_{\text{aero}} + F_{\text{grade}} \tag{3.133}$$

$$\tau_{\text{demand\_brake}} = \min\{ 0, \ \tau_{\text{demand}} \} \tag{3.134}$$

$$\tau_{\text{demand\_prop}} = \max\{ 0, \ \tau_{\text{demand}} \} \tag{3.135}$$

$$P_{\text{a\_pct}} = \frac{\tau_{\text{demand\_prop}}}{\tau_{\text{max\_prop}}(V_{\text{veh}})} \times 100 \tag{3.136}$$

$$P_{\text{brake\_pct}} = \frac{\tau_{\text{demand\_brake}}}{\tau \ \text{max\_brake}} \times 100 \tag{3.137}$$

where V_{veh} is the vehicle speed (m/s), F_{rolling}, F_{aero}, F_{grade} are the rolling resistance, the aero-drag resistance, and the grade weight forces, respectively, $\tau_{\text{demand\_prop}}$ is the propulsion torque demanded by the vehicle, $\tau_{\text{max\_prop}}$ is the maximal achievable propulsion torque at the current vehicle speed, $\tau_{\text{demand\_brake}}$ is the brake torque demanded by the vehicle, $P_{\text{a\_pct}}$ is the position of the acceleration pedal given as a percentage, and $P_{\text{brake\_pct}}$ is the brake pedal position given as a percentage.

References

Åstrom, K. and Hägglund, T. *PID Controllers: Theory, Design, and Tuning*, 2nd edition, Instrument Society of America, 1995.

Benford, H. L. and Leising, M. B. 'The Lever Analogy: A New Tool in Transmission Analysis,' SAE technical paper, 1981-810102, Society of Automotive Engineering (SAE) (http://www.sae.org), 1981.

Conlon, B. 'Comparative Analysis of Single and Combined Hybrid Electrically Variable Transmission Operating Modes,' SAE technical paper, 2005-01-1162, Society of Automotive Engineering (SAE) (http://www.sae.org), 2005.

Hall, M. and Dourra, H. 'Dynamic Analysis of Transmission Torque Utilizing the Lever Analogy,' SAE technical paper, 2009-01-1137, Society of Automotive Engineering (SAE) (http://www.sae.org), 2009.

Miller, J. M. 'Hybrid Electric Vehicle Propulsion System Architectures of the e-CVT Type,' *IEEE Transactions on Power Electronics*, 21(3), 756–767, 2006.

4

Power Electronics and Electric Motor Drives in Hybrid Electric Vehicles

A hybrid electric vehicle is a complex electrical and mechanical system. No matter what the type of system configuration, a hybrid electric vehicle has two energy flow paths – electrical and mechanical. Mechanical energy flows through a conventional powertrain, while an electric powertrain provides a path on which electrical energy can flow. In hybrid electric vehicle systems, the task of the electric motor and the power electronics is to process and control the electrical energy flow to meet the vehicle's power demands with maximal efficiency in various driving situations.

Power electronics generally combine electronic components, electrical power, and control methodology. Their performance affects the overall vehicle fuel economy and drivability, but they also take up a big portion of the entire material cost of the vehicle. Depending on the type and configuration of a hybrid electric vehicle system, the electric motor and the power electronics could make up more than 25% of the total material cost, which is almost equal to the expense of the energy storage system. Since DC electrical energy conversion is implemented by power electronic circuits, and conversion efficiency is one of the most important performance considerations in hybrid electric vehicle system design, this chapter presents the basic principles of commonly used power electronics in hybrid vehicle applications.

4.1 Basic Power Electronic Devices

Power electronic devices, or power semiconductor devices, deal with the control and conversion of electrical power from one form to another. They play an intermediary role between the energy storage system and the propulsion motor in a hybrid electric vehicle.

Hybrid Electric Vehicle System Modeling and Control, Second Edition. Wei Liu.
© 2017 John Wiley & Sons Ltd. Published 2017 by John Wiley & Sons Ltd.

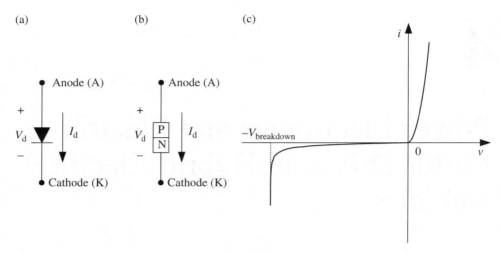

Figure 4.1   The symbol (a), structure (b), and *i-v* characteristics (c) of a diode

In order to achieve low cost and high efficiency power control and conversion in hybrid vehicle applications, power electronic devices always work in switching mode. This section reviews the basic characteristics of commonly used power electronic devices in hybrid vehicles.

4.1.1   Diodes

A diode has one *pn*-junction that makes it conduct in only one way. The circuit symbol, structure, and *i-v* characteristics of a diode are shown in Fig. 4.1(a), (b), and (c) respectively. When the potential on the anode is greater than the potential on the cathode, the state of the diode is called *forward biased*, and the diode conducts. A conducting diode has a 0.3–0.8 V forward voltage drop across the *pn*-junction depending on the junction material, the manufacturing process, and the junction temperature. When the potential on the cathode is greater than the potential on the anode, the state of the diode is called *reverse biased*. In this state, the current flow is blocked and there is only a negligible leakage current through the diode until the applied voltage reaches the reverse breakdown voltage (also called the avalanche or zener voltage).

Depending on the application requirements, a diode can be selected from the following types of power diodes:

- **Rectifier diodes:** These diodes are less expensive but have relatively long reverse recovery time. They are suitable for low-speed applications, for example, AC to DC rectifiers. The voltage and current ratings of this type of diode can go up to 5000 V and 2000 A.

- **Schottky diodes:** These diodes are ideal for high current and low voltage DC power suppliers as they have a low forward voltage drop. However, the lower blocking voltage capabilities (50–100 V) limit the applications of this type of diode.
- **Fast-recovery diodes:** At a several-kilowatt power level, these diodes can have a recovery time rating of less than several μs. They are widely used in DC–DC converters and DC–AC inverter circuits, where the fast recovery time is necessary to meet high-frequency circuit and high-frequency controllable switch requirements. The voltage and current ratings of this type of diode can be from 50 V to 3 kV and from 1 A to several hundred amperes.

All the major parameters of diodes can be found in the manufacturers' data sheets. Some principal ratings of diodes are explained below:

- **Maximal average forward current** (I_{FAV}): The maximal average current that a diode can conduct in the forward direction for a given wave at a specified temperature. Since the average current calculation is based on an AC current with a sine form, the average current rating of a diode decreases with a reduction in conduction angle.
- **Maximal RMS forward current** (I_{FRMS}): The maximal root mean square current that a diode can conduct in the forward direction for a given wave at a specified temperature.
- **Maximal working peak reverse voltage** (V_{RWM}): The maximal reverse voltage that can be connected across a diode without breakdown. The maximal working peak reverse voltage is also called the peak reverse voltage or the reverse breakdown voltage.
- **Maximal DC reverse voltage** (V_R): The voltage that can be applied in the reverse direction across the diode for a certain period of time. It is an important parameter for sizing the diode in DC–DC converter and DC–AC inverter circuit design.
- **Maximal forward surge current** (I_{FSM}): The maximal current that the diode can handle as an occasional transient at a specified temperature. This parameter indicates the capability of the diode to withstand non-repetitive surge or fault current.
- **Maximal junction temperature** (T_J): The maximal junction temperature that a diode can withstand without failure.

4.1.2 Thyristors

Thyristors are four-layer *pnpn* power semiconductor devices. These devices have three *pn*-junctions and three terminals called the anode, the cathode, and the gate. The electrical circuit symbol, *pn*-junctions, and *i-v* characteristics of thyristors are shown in Fig. 4.2(a), (b), and (c) respectively. Unlike power diodes, a thyristor has three operational modes: reverse block mode, forward blocking mode, and forward conducting mode.

When the potential on the anode is greater than the potential on the cathode, the junctions J_1 and J_3 are forward biased, while the junction J_2 is reverse biased, as a result of which only a small amount of leakage is able to flow from the anode to the cathode. In this condition, the thyristor is said to be in the forward blocking state or off state, and the leakage current is

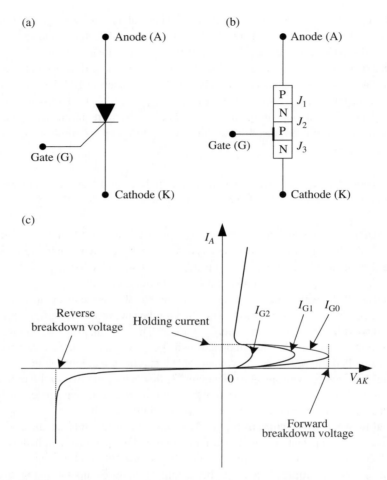

Figure 4.2 The symbol (a), structure (b), and $i\text{-}v$ characteristics (c) of a thyristor

called an off-state current. If the potential difference between the anode and cathode is increased to a certain value, resulting in the reverse-biased J_2 junction breaking, the thyristor conducts as a diode and this state is called the forward conducting state or on state. The other way to turn the thyristor from the forward blocking state to the on state is to trigger it by applying a pulse of positive gate current for a short period of time; furthermore, once the thyristor begins to conduct, the on state is latched and the gate current can be removed. It should be noted that the thyristor cannot be turned off by the gate, and it will continue to conduct like a diode until the forward current drops below the holding current, I_H or the thyristor is reverse biased. From the $i\text{-}v$ characteristic curve, it can be seen that the gate current, I_G also controls the value of the forward breakdown voltage. The bigger the gate current, I_G, the lower the forward breakdown voltage.

The principal ratings of thyristors include:

- **Maximal average forward current** (I_{AV}): The maximal average anode to cathode current value assuming that the current has a sine waveform and the conduction is 180°.
- **Maximal RMS on-state current** (I_{RMS}): The maximal root mean square current that the thyristor can handle.
- **Holding current** (I_H): The minimal value of the anode current to maintain conduction after the thyristor is latched on. If the anode current is reduced below this value, the thyristor will turn off.
- **The maximal non-repetitive (surge) peak forward current** (I_{FSM}): The maximal allowable peak anode current value that a thyristor can withstand within the given duration.
- **Peak repetitive reverse blocking voltage** (V_{RRM}): The maximal allowable peak off-state transient value applied in the reverse direction and occurring every cycle.
- **Peak repetitive forward blocking voltage** (V_{DRM}): The maximal allowable peak off-state transient voltage applied in the forward direction and occurring every cycle.
- **Non-repetitive peak reverse voltage** (V_{RSM}): The maximal allowable peak value of non-repetitive transient reverse voltage that the thyristor can withstand within the given duration.
- **Maximal peak positive gate current** (I_{GM}): The maximal DC gate current allowed to turn the thyristor on.
- **Maximal peak positive gate voltage** (V_{GM}): The maximal allowed gate-to-cathode DC voltage.
- **Critical rate of rise of on-state current** (dI/dt): The maximal rise rate of on-state current that the device will withstand after switching from an off state to an on state.
- **Critical rate of rise of off-state voltage** (dV/dt): The maximal rise rate of off-state voltage that will not cause the device to switch from an off state to an on state.
- **Maximal junction temperature** (T_J): The maximal junction temperature at which the thyristor can be continuously operated.

4.1.3   Bipolar Junction Transistors (BJTs)

A bipolar transistor is a three-layer device with two *pn*-junctions stacked opposite to each other. If a transistor has two *n*-regions and one *p*-region, it is called an *npn*-type transistor; otherwise, it is called a *pnp*-type transistor. The three terminals of a transistor are called the base, the collector, and the emitter. The electrical circuit symbols and structures of bipolar junction transistors are shown in Fig. 4.3(a) and Fig. 4.3(b). The arrow direction on the emitter of the circuit symbol distinguishes an *npn*-type transistor from a *pnp*-type transistor. The *pnp*-type transistor has an arrowhead pointing toward the base, while the *npn*-type has an arrowhead pointing away from the base. Based on the common emitter circuit of an *npn* transistor, its steady-state input characteristics and output characteristics are shown in Fig. 4.4.

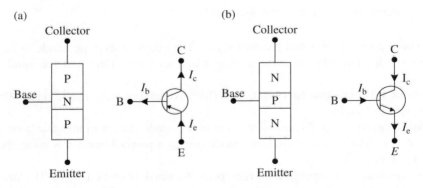

Figure 4.3   The electrical circuit symbol and *pn*-structure of a transistor. (a) PNP transistor, (b) NPN transistor

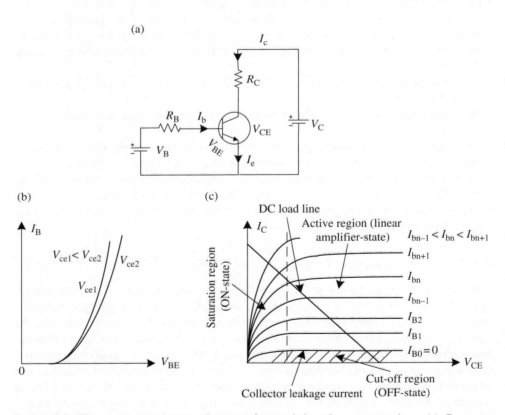

Figure 4.4   The steady-state input and output characteristics of an *npn* transistor. (a) Common-emitter circuit diagram, (b) Input characteristics, (c) Output (i-v) characteristics

4.1.3.1 The Operating Regions and Switching Characteristics

As shown in Fig. 4.4(c), a transistor has three operating regions: cutoff (OFF), saturation (ON), and active. In power applications, a transistor works in the cutoff and saturation regions as a switch. In the cutoff region, the collector–emitter voltage, V_{CE}, is equal to V_C as there is no current going through it, so the transistor acts as an opened switch in this state. In the saturation region, the base current is sufficiently high that the collector–emitter voltage is very low; therefore, the transistor acts as a closed switch in this state. In the active region, the transistor works like an amplifier to adjust the collector–emitter voltage based on the base current.

From the characteristics of a transistor, it can be seen that the base current needs to be enough to have a transistor work as a switch. Thus, the ON or OFF state of a transistor is determined by the base current, and the collector and the emitter are considered as two terminals of a switch. The values of V_C and R_C determine the operating points of a transistor from the DC load line shown in Fig. 4.4(c).

In the ON state, the collector–emitter voltage, V_{CE} is near to zero, so the collector current should be equal to $\frac{V_C - V_{CE\_sat}}{R_C}$ where V_{CE_sat} is the saturation voltage at which the voltage drop across the collector–emitter terminals of a transistor is the smallest. In other words, the collector–emitter voltage, V_{CE} is saturated when the base current is equal to or greater than the saturation current, $I_B \geq I_{B\_sat}$; at this point, the collector current is maximal and the transistor has the smallest voltage drop across the collector–emitter terminals. In the OFF state, the collector current, I_C is near to zero and the collector–emitter voltage, V_{CE} is equal to the supply voltage, V_C, which occurs when the base current $I_B = 0$.

4.1.3.2 BJT Power Losses

The BJT power losses mainly include the ON-state power loss and the turn-on/off switching loss. The ON-state power loss is determined by the saturation voltage of the power transistor, and the lost power is equal to $P_{loss\_ON} = I_C \cdot V_{CE\_sat}$. The switching frequency has a significant impact on the switching losses of a power transistor. From a power transistor point of view, the higher the switching frequency, the more power loss there is in the transistor; however, an increase in switching frequency in a power device may result in a size reduction of other power components and a power loss decrease on these components.

4.1.4 Metal Oxide Semiconductor Field Effect Transistors (MOSFETs)

As mentioned above, the ON or OFF state of a BJT is controlled by its base current; furthermore, a larger base current is needed to maintain the ON state. These characteristics make BJTs' base drive circuit design very complicated and expensive. In order to overcome these drawbacks, a new type of power electronic device, a MOSFET, was invented in the early 1970s. Unlike BJTs, MOSFETs are voltage-controlled devices and their switching speed is very high, so that they can be turned on and turned off in the order of nanoseconds.

So far, powerMOSFETs have been major power electronic components in low-power, high-frequency converters. There are two types of MOSFET: depletion and enhancement. The circuit diagram and *i-v* characteristics of an *n*-channel MOSFET are shown in Fig. 4.5(a) and Fig. 4.5(b). The transfer characteristics of the *n*-channel depletion-type and enhancement-type MOSFETs are shown in Fig. 4.6(a) and Fig. 4.6(b).

4.1.4.1　Steady-state Model

Similar to power BJTs, a MOSFET also has three operational regions: ON, OFF, and linear regions. As shown in Fig. 4.6(b), an *n*-channel enhancement-type MOSFET will be in the OFF region if $V_{GS} \leq V_T$, in the ON region if $V_{GS} > V_T$ and $V_{DS} > V_{GS} - V_T$, and in the linear region if $V_{GS} > V_T$ and $V_{DS} \leq V_{GS} - V_T$, where V_T is the threshold voltage. The steady-state equivalent circuit of an *n*-channel MOSFET is shown in Fig. 4.7, where the transconductance, g_m and output resistance, r_o are defined as:

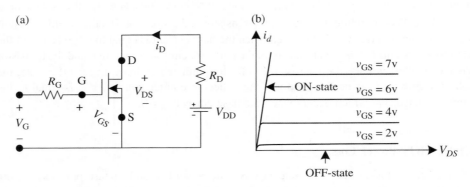

Figure 4.5　Circuit symbol and output characteristics of an *n*-channel MOSFET. (a) The n-channel MOSFET symbol and circuit diagram, (b) Output characteristics

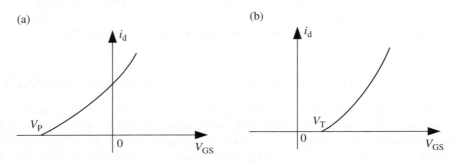

Figure 4.6　Transfer characteristics of *n*-channel depletion-type and enhancement-type MOSFETs. (a) The n-channel depletion-type MOSFET, (b) The n-channel enhancement-type MOSFET

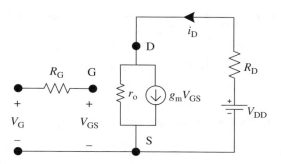

Figure 4.7   Steady-state equivalent circuit of an *n*-channel MOSFET

$$g_m = \frac{\Delta I_D}{\Delta V_{GS}}\bigg|_{V_{DS}=\text{constant}} \qquad r_0 = \frac{\Delta V_{DS}}{\Delta I_D} \qquad\qquad (4.1)$$

4.1.4.2   Switching Model

From the MOSFET transfer characteristic curves, it can be seen that an *n*-channel depletion-type MOSFET remains in the ON state even if the gate voltage, $V_{GS} = 0$, while an *n*-channel enhancement-type MOSFET remains in the OFF state at zero gate voltage. Therefore, *n*-channel enhancement-type MOSFETs are commonly used as switching devices in power electronic applications. The switching-mode equivalent circuit of an *n*-channel MOSFET is shown in Fig. 4.8, where C_{gs} and C_{gd} are the parasitic capacitances from gate to source and from gate to drain, and C_{ds} and r_{ds} are the drain to source capacitance and resistance, respectively.

4.1.5   *Insulated Gate Bipolar Transistors (IGBTs)*

In order to overcome the base-current control problem of power BJTs and to improve the large current-handling (power) capability of power MOSFETs, another power electronic device, the IGBT, was developed; this takes the advantages of both BJTs and MOSFETs. The characteristics and specification of IGBTs mean that they are predominantly used as power electronics in hybrid electric vehicle applications.

Figure 4.9 shows the circuit symbol and equivalent circuit of an IGBT, in which the three terminals are called the gate (G), the collector (C), and the emitter (E). An IGBT is a voltage-controlled device similar to a power MOSFET, and it is turned on simply by applying a positive gate voltage and turned off by removing the gate voltage. IGBTs have lower switching and conducting losses, share many attractive features of power MOSFETs, and require a

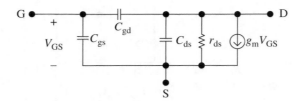

Figure 4.8 Switching-mode equivalent circuit of an *n*-channel MOSFET

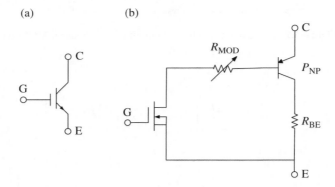

Figure 4.9 Circuit symbol and equivalent circuit of an IGBT. (a) Circuit symbol of IGBT, (b) The equivalent circuit of IGBT

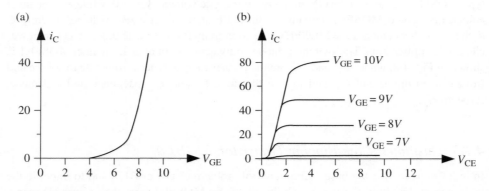

Figure 4.10 Typical transfer and *i-v* characteristics of an IGBT. (a) Transfer characteristics, (b) *i-v* characteristics

very simple control circuit. The typical transfer characteristics of i_C versus v_{GE} and *i-v* (output) characteristics of i_C versus v_{CE} of an IGBT are shown in Fig. 4.10.

IGBTs can be turned on and off in the order of microseconds, which is essentially faster than BJTs but inferior to MOSFETs. IGBT modules are available with current ratings as high as 1700 V and 1200 A.

4.2 DC–DC Converters

In hybrid electric vehicle applications, DC–DC converters are commonly used to supply electrical power to 12 V DC loads from the fluctuating high-voltage DC bus of an HEV/EV. They are also used in DC motor drives. In order to meet the fuel economy requirements of an HEV and the electric mileage requirements of a PHEV/EV, the efficiency of an automotive DC–DC converter needs to be above 95%. This section presents the basics of DC–DC converters and commonly used DC–DC converters in hybrid vehicle systems.

4.2.1 Basic Principle of a DC–DC Converter

The purpose of a DC–DC converter is to convert DC power from a voltage lever (input) to other levels (outputs). In other words, a DC–DC converter changes and controls the DC voltage magnitude; like an AC transformer, it can be used to step down or step up a DC voltage source. The goal of current converter technology is to ensure that the manufactured converters are small in size and light in weight as well as having high efficiency.

The operating principle of DC–DC converters can be illustrated by Fig. 4.11(a) and Fig. 4.11(b). To obtain the average voltage, V_o, on the load R in Fig. 4.11(a), the switch needs to be controlled at a constant frequency ($^1/_{T_s}$) shown in Fig. 4.11(b). Thus, the average output voltage, V_o, will be controlled by adjusting the switch's ON duration. The output average voltage, V_o, and the average load current, I_o, can be calculated by the following equations:

$$V_o = \frac{1}{T_s}\int_0^{T_s} v_0 dt = \frac{1}{T_s}\int_0^{T_{on}} V_{in} dt = \frac{T_{on}}{T_s} V_{in} = f_s T_{on} V_{in} = D V_{in} \tag{4.2}$$

$$I_o = \frac{V_o}{R} = \frac{D V_{in}}{R} \tag{4.3}$$

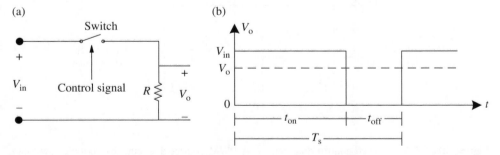

Figure 4.11 Basic principle circuit of a DC–DC converter. (a) Principle circuit diagram, (b) The switching control signal

   The above control strategy is called the pulse-width modulation (PWM) control method and is commonly used in DC–DC converters and DC–AC inverters. It is obvious that the longer the switch is ON in a given signal period, T_s, the higher the average voltage of the device output is. In PWM control, this characteristic can be described by the duty ratio, D, defined as the ratio of the ON duration to the switching time period, T_s:

$$D = \frac{t_{on}}{T_s} \tag{4.4}$$

   Sometimes, the term *duty cycle* is also used to describe the proportion of time in the ON state compared to the period of time for a given signal; furthermore, the duty cycle is expressed as a percentage, and a low duty cycle corresponds to a low voltage; for example, a 100% duty cycle means fully on.

   A PWM switch control diagram is given in Fig. 4.12, where the power electronics are switched ON and OFF following the PWM control command at a constant frequency. In the diagram, a PID controller outputs the control voltage, $V_{control}$, based on the

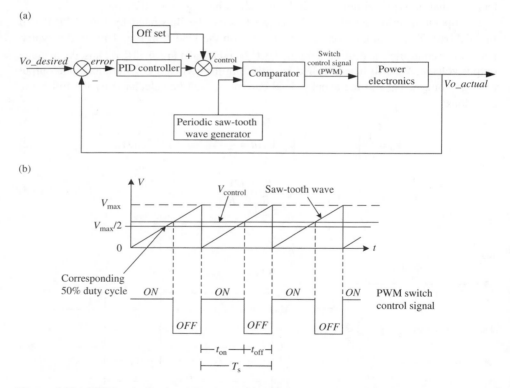

Figure 4.12   PWM switch control diagram. (a) A PWM control diagram, (b) Periodic saw-tooth waveform and PWM control signal formation

difference between the desired output voltage and the actual output voltage of the DC–DC converter, and then the control voltage, $V_{control}$, is compared with the periodic saw-tooth wave signal to generate the required PWM signal to switch the power electronics ON or OFF, as shown in Fig. 4.12(b), where the frequency of the periodic saw-tooth wave establishes the switch control frequency which is kept constant. In practice, the actual switching frequency of a DC–DC converter can be chosen from a few kHz to a few hundred kHz depending on the detailed performance requirements and the power electronic devices selected.

An important note on PWM switching control is that the PWM signal frequency has to be much faster than anything that would affect the load; otherwise, it will result in a discontinuous effect on the load. Depending on the load property, a DC–DC converter may be operated in two distinct modes – continuous current conduction and discontinuous current conduction – so the switch control signal needs to be properly designed to have the converter work well in both operational modes.

4.2.2   Step-down (Buck) Converter

A step-down converter, also called a buck converter, produces an average output voltage that is lower than the DC input voltage. The principle circuit of a step-down converter is shown in Fig. 4.13; it has a low-pass filter to avoid the output voltage fluctuating between zero and the input voltage.

During the switch ON period of time, t_{on}, the circuit can be simplified as in Fig. 4.14(a), and there are the following relationships:

$$V_{in} = v_1 + v_0; \quad v_L = L\frac{di_L}{dt};$$
$$i_L = i_C + i_R; \quad i_C = C\frac{dv_o}{dt}; \quad i_R = \frac{v_o}{R}$$

(4.5)

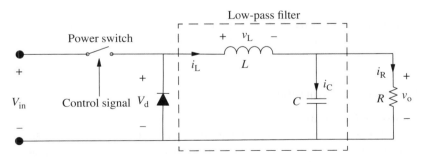

Figure 4.13   Principle circuit for step-down converters

During steady-state operation, the current waveform, i_L, repeats from one period to the next, and the current at the beginning, $I_L(0)$, must be equal to the current at the end, $I_L(T_s)$. Thus, from Eqs 4.9 and 4.10, we can obtain the following equation describing the inductor voltage waveform:

$$i_L(T_s) - i_L(0) = \frac{1}{L}\int_0^{T_s} v_L dt \cong \int_0^{t_{on}} \frac{V_{in} - V_0}{L} dt + \int_{t_{on}}^{T_s} \left(-\frac{V_0}{L}\right) dt = \frac{V_{in} - V_0}{L} t_{on} - \frac{V_0}{L}(T_s - t_{on}) = 0$$

(4.11)

That is:

$$(V_{in} - V_0)t_{on} - V_0(T_s - t_{on}) = 0 \quad \Rightarrow \quad V_{in}t_{on} - V_0 T_s = 0 \quad \Rightarrow \quad V_0 = \frac{t_{on}}{T_s} V_{in} = DV_{in} \quad (4.12)$$

If we assume that the voltage waveform across the capacitor repeats from one period to another during steady-state operation, we can analyze the voltage on the capacitor as:

$$i_C(t) = C\frac{dv_c(t)}{dt} \quad \Rightarrow \quad v_c(t) = \int_0^t i_C(t)dt \quad \Rightarrow \quad v_c(T_s) - v_c(0) = \int_0^{T_s} i_C(t)dt = 0 \quad (4.13)$$

The foregoing equations imply the following under steady-state operation:

1. The total area enclosed by the inductor voltage waveform is zero, that is, the area A during the period of time t_{on} is equal to the area B during the period of time t_{off}, shown in Fig. 4.15.
2. The total area enclosed by the capacitor current waveform is zero.
3. The output voltage is linearly controlled by the PWM duty ratio, D.

Discontinuous-conduction Mode
In some cases, the amount of load energy required is larger than that stored by the inductor, so the inductor is completely discharged. In such cases, the current through the inductor falls to zero during part of the switch OFF period, resulting in the converter operating in the discontinuous-conduction mode. Figure 4.16 shows the switch state, inductor current, and voltage waveforms for a buck converter in discontinuous-conduction mode.

During the t_{disc} period of time in discontinuous-conduction mode, the inductor current is zero and the power to the load is supplied by the filter capacitor alone. If we assume the output voltage ripple is small enough, the following equations hold for the equivalent circuit shown in Fig. 4.14:

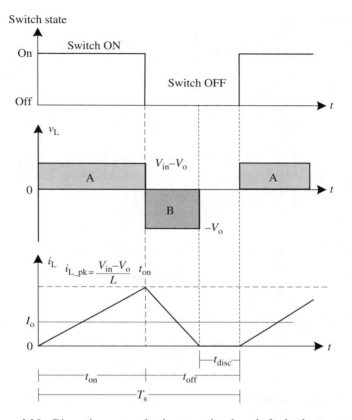

Figure 4.16 Discontinuous-conduction operational mode for buck converters

$$v_L(t) = V_{in} - v_o(t) \cong V_{in} - V_o; \quad i_C(t) = i_L(t) - \frac{v_0(t)}{R} \cong i_L(t) - \frac{V_0}{R} \quad t \in [0, t_{on})$$

$$v_L(t) = -v_o(t) \cong -V_o; \qquad i_C(t) = i_L(t) - \frac{v_0(t)}{R} \cong i_L(t) - \frac{V_0}{R} \quad t \in [t_{on}, T_s - t_{disc}) \quad (4.14)$$

$$v_L(t) = 0; \qquad\qquad i_C(t) = -\frac{v_0(t)}{R} \cong -\frac{V_0}{R} \qquad\qquad t \in [T_s - t_{disc}, T_s]$$

Since the charge and discharge must be equal during steady-state operation, we have:

$$i_L(T_s - t_{disc}) - i_L(0) = \frac{1}{L} \int_0^{T_s - t_{disc}} v_L dt \cong \int_0^{t_{on}} \frac{V_{in} - V_0}{L} dt + \int_{t_{on}}^{T_s - t_{disc}} \left(-\frac{V_0}{L}\right) dt$$

$$= \frac{V_{in} - V_0}{L} t_{on} - \frac{V_0}{L}(T_s - t_{disc} - t_{on}) = 0$$

$$(4.15)$$

Solving for V_0:

$$V_0 = \frac{V_{in}t_{on}}{T_s - t_{disc}} = \frac{V_{in}t_{on}}{t_{on} + t_{off} - t_{disc}} \tag{4.16}$$

From the capacitor's charge balance and Eq. 4.14, we have:

$$
\begin{aligned}
0 &= \int_0^{T_s} i_C \, dt \cong \int_0^{T_s - t_{disc}} \left(i_L(t) - \frac{V_0}{R} \right) dt + \int_{T_s - t_{disc}}^{T_s} \left(-\frac{V_0}{R} \right) dt \\
&= \int_0^{T_s - t_{disc}} i_L(t) \, dt + \int_0^{T_s - t_{disc}} \left(-\frac{V_0}{R} \right) dt + \int_{T_s - t_{disc}}^{T_s} \left(-\frac{V_0}{R} \right) dt \\
&= \int_0^{T_s - t_{disc}} i_L(t) \, dt + \int_0^{T_s} \left(-\frac{V_0}{R} \right) dt = \int_0^{T_s - t_{disc}} i_L(t) \, dt - \frac{V_0}{R} T_s
\end{aligned}
\tag{4.17}
$$

From Fig. 4.16, the integration in the above equation is equal to:

$$\int_0^{T_s - t_{disc}} i_L(t) \, dt = \frac{V_{in} - V_0}{2L} t_{on}(T_s - t_{disc}) \tag{4.18}$$

Hence, Eq. 4.17 may be rewritten as:

$$\frac{V_o}{R} = \frac{V_{in} - V_0}{2T_s L} t_{on}(T_s - t_{disc}) \tag{4.19}$$

Solving the system of equations in Eq. 4.16 and Eq. 4.19 for V_o gives:

$$
\begin{aligned}
\frac{V_o}{R} &= \frac{V_{in} - V_0}{2T_s L} t_{on} \frac{V_{in}}{V_0} t_{on}, \quad \text{let } K = \frac{2L}{RT_s}, \quad \Rightarrow \quad \frac{D^2}{K}\left(\frac{V_{in}}{V_o}\right)^2 - \frac{D^2}{K}\left(\frac{V_{in}}{V_o}\right) - 1 = 0 \\
&\Rightarrow \quad \left(\frac{V_{in}}{V_o}\right)^2 - \left(\frac{V_{in}}{V_o}\right) - \frac{K}{D^2} = 0
\end{aligned}
\tag{4.20}
$$

Thus, the output voltage can be achieved when the converter operates in discontinuous-current mode, as follows:

$$V_o = V_{in} \frac{2}{1 + \sqrt{1 + 4K/D^2}} \tag{4.21}$$

where D is the PWM duty ratio.

From the foregoing discussion, we can note the following:

1. A continuous- or discontinuous-conduction mode of operation depends not only on circuit parameters and input voltage, but also on the PWM control frequency, the PWM duty ratio, and the load resistance.

2. During the discontinuous-conduction mode of operation, the output voltage is not linearly controlled by the PWM duty ratio, D.

4.2.2.2 Output Voltage Ripple

Switch-mode-based DC–DC converters have the characteristic whereby the output voltage rises when the switch turns on and falls when the switch turns off. This phenomenon is called *voltage ripple*, and is mainly determined by the output filter capacitance as well as being affected by switching frequency and filter inductance. The voltage ripple calculation is an important design calculation, but this section only presents the very basic concept. For more detailed information, the interested reader should refer to the books by Mohan *et al.* (2003) and Erickson and Maksimovic (2001) listed at the end of this chapter.

From the previous discussion, we have seen that the practical output voltage of a switch-mode DC–DC converter is a rippled DC rather than a solid DC, and it can be expressed as:

$$v_o(t) = V_o + v_{\text{ripple}}(t) \tag{4.22}$$

If we assume that the load resistor takes the average component in current flow $i_L(t)$ and the ripple component goes through the filter capacitor, the charge on the capacitor can be calculated based on the capacitor's initial and final currents, $i_C(0)$ and $i_C(t_{\text{on}})$, which are obtained from Eq. 4.14 and Fig. 4.16 as follows:

$$
\begin{aligned}
i_C(0) &= i_L(0) - I_o = i_L(0) - \frac{V_o}{R} \\
i_C(t_{\text{on}}) &= i_L(t_{\text{on}}) - I_o = i_L(0) + \frac{V_{\text{in}} - V_o}{L} t_{\text{on}} - \frac{V_o}{R}
\end{aligned}
\tag{4.23}
$$

$$
\begin{aligned}
\Delta i_{\text{cpp}} &= i_c(t_{\text{on}}) - i_C(0) = i_L(0) + \frac{V_{\text{in}} - V_o}{L} t_{\text{on}} - \frac{V_o}{R} - i_L(0) + \frac{V_o}{R} = \frac{V_{\text{in}} - V_o}{L} t_{\text{on}} \\
&= \frac{1}{L}\left(\frac{V_{\text{in}}}{T_s} t_{\text{on}} T_s - \frac{V_o}{T_s} t_{\text{on}}\right) = \frac{1}{L}(V_{\text{in}} D T_s - V_o D T_s) = \frac{1}{L}(V_o - V_o D) T_s = \frac{V_o}{L}(1 - D) T_s
\end{aligned}
\tag{4.24}
$$

The charge stored in the capacitor shown in the shaded area of Fig. 4.17 is equal to the integration of the positive current flowing through it over time, that is:

$$Q = \int_0^t i_C(t) dt \Big|_{i_C(t) \geq 0} = \int_0^{T_s/2} i_C(t) dt \Big|_{i_C(t) \geq 0} = \frac{1}{2} \frac{\Delta i_{\text{cpp}}}{2} \frac{T_s}{2} = \frac{V_o}{8L}(1-D)T_s^2 \tag{4.25}$$

Thus, the output voltage ripple on the load is equal to:

$$v_{\text{ripple}}(t) = \Delta V_C = \frac{Q}{C} = \frac{V_o}{8LC}(1-D)T_s^2 \tag{4.26}$$

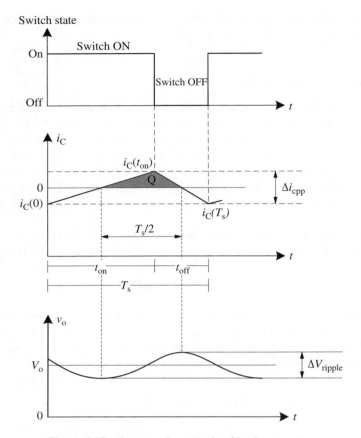

Figure 4.17   Output voltage ripple of buck converters

The voltage ripple on the high-voltage bus is an important parameter and needs to be specified as being less than a certain value in hybrid vehicle applications. It should be noted that the magnitude of the output voltage ripple can be diminished by increasing the output capacitor, the filter inductor, and the switching frequency; however, the capacitor and inductor are normally selected based on cost and physical size, and a higher switching frequency generally lowers a converter's efficiency and may also raise electromagnetic interference (EMI) concerns in hybrid vehicle system applications.

Example 4.1

Consider the DC converter shown in Fig. 4.18, which is to be used to charge an HEV battery pack. Assume that the system's input voltage, $V_{in} = 250\,\text{V}$, the battery resistance, $R_b = 0.1\,\Omega$, and the battery potential, $E_b = 190\,\text{V}$. If the average charge current is designed as $I_b = 45\text{A}$ and the PWM frequency is selected as $f_s = 20\,\text{kHz}$, determine the required inductance of the inductor that limits the charge current ripple to 10% of I_b.

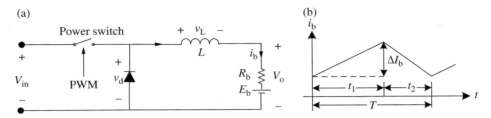

Figure 4.18 DC converter with a battery pack. (a) Schematic circuit, (b) Current waveform

Solution:

As $V_{in} = 250\,V$, $R_b = 0.1\,\Omega$, $E_b = 190\,V$, $f_s = 20\,kHz$ $\Rightarrow$ $T_s = {}^1/_{f_s} = 50\,\mu s$, and $\Delta I_b = I_b \cdot 0.1 = 4.5\,A$, the average output voltage $V_o = DV_{in} = I_b \cdot R_b + E_b = 4.5 + 190 = 194.5\,V$, and $D = \dfrac{V_o}{V_{in}} = \dfrac{194.5}{330} = 0.778$. The voltage across the inductor is described by the following equation:

$$L\frac{di}{dt} = V_{in} - R_b I_b - E_b = V_{in} - DV_{in} = V_{in}(1-D) \tag{4.27}$$

In the worst ripple case, we can assume the charge current I_b moves up and down linearly, as shown in Fig. 4.18(b), $di = \Delta I_b$, $dt = t_1 = DT = 0.778 \cdot 50 \cdot 10^{-6} = 38.9\,\mu s$. Therefore, from Eq. 4.27 we have:

$$L\frac{di}{dt} = L\frac{4.5}{38.9 \times 10^{-6}} = 250 \cdot 0.222 \quad \Rightarrow \quad L = \frac{250 \cdot 0.222 \cdot 38.9 \times 10^{-6}}{4.5} \approx 0.48\,mH$$

Thus, the required inductor needs to have an inductance of 0.48 mH.

4.2.3 Step-up (Boost) Converter

A boost converter is capable of providing the average output voltage at a level higher than the input or source voltage; therefore, the boost converter is often referred to as a *step-up converter*. The principle circuit of a boost converter is shown in Fig. 4.19, where the DC input voltage is in series with a large inductor acting as a current source. When the switch is ON, the diode is reverse biased, thus the output is isolated from the input, and the output voltage is maintained by the capacitor; meanwhile, the input supplies energy to the inductor. When the switch is OFF, the load and capacitor receive energy from the input and the inductor, so the average output voltage is higher than the input voltage. Corresponding with the switch state, the principle circuit can be split into two circuits, as shown in Fig. 4.20(a) and Fig. 4.20(b).

Since, in steady-state operation, the waveform must repeat for each cycle, the volt-second of the inductor must be balanced between the switch ON state and the switch OFF state, that is, the integral of the inductor voltage over one time period must be equal to zero in steady-state operation. In other words, the shaded area A must be equal to the shaded area B in

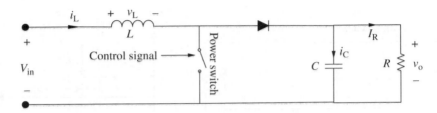

Figure 4.19   Principle circuit of boost converters

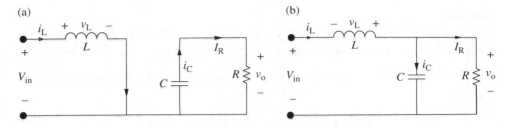

Figure 4.20   Principle circuit of boost converters corresponding with the switch being (a) ON and (b) OFF

Fig. 4.21. If we assume that the current through the inductor changes linearly, the voltage conversion ratio of a boost DC–DC converter can be obtained as follows:

$$V_{in}t_{on} + (V_{in} - V_o)t_{off} = 0 \quad \Rightarrow \quad V_{in}(t_{on} + t_{off}) = V_o t_{off} \quad \Rightarrow \quad V_{in}T_s = V_o t_{off}$$

$$\Rightarrow \quad \frac{V_o}{V_{in}} = \frac{T_s}{t_{off}} = \frac{1}{\dfrac{t_{off}}{T_s}} = \frac{1}{\dfrac{T_s - t_{on}}{T_s}} = \frac{1}{1-D} \tag{4.28}$$

The relationship between the voltage across the inductor L and the current flowing through it can be described by the following differential equation, and the current can be determined by integrating the voltage.

$$v_L(t) = L\frac{di_L(t)}{dt} \tag{4.29}$$

During the switch ON period of time, the inductor current can be determined as:

$$v_L(t) = V_{in} = L\frac{di_L(t)}{dt} \quad \Rightarrow \quad i_L(t) = \frac{\displaystyle\int_0^t V_{in}dt}{L} = i_L(0) + \frac{V_{in}}{L}t \qquad t \le t_{on} \tag{4.30}$$

where $\frac{V_{in}}{L}$ is the change slope of the inductor current when the switch is in the ON state. The inductor current during the switch OFF period of time is:

$$i_L(t) = \frac{\displaystyle\int_{t_{on}}^t (V_{in} - V_o)dt}{L} = i_L(t_{on}) + \frac{V_{in} - V_o}{L}t \qquad t_{on} < t \le T_s \tag{4.31}$$

where $\frac{V_{in} - V_o}{L}$ is the change slope of the inductor current when the switch is in the OFF state.

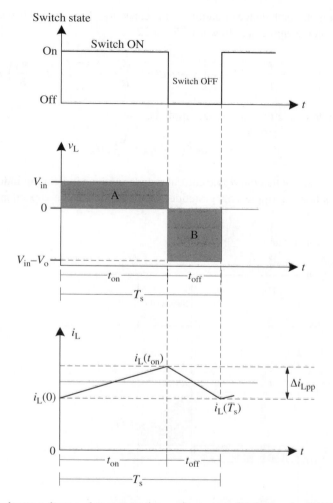

Figure 4.21 Inductor voltage and current on the continuous-conduction mode of the boost converter

Since the converter is in the steady state of operation, the current waveform, i_L, repeats from one period to the next; thus, the current at the beginning, $i_L(0)$, must be equal to the current at the end, $i_L(T_s)$. The input ripple current, Δi_{Lpp}, shown in Fig. 4.21 can be determined as:

$$\Delta i_{Lpp} = i_L(t_{on}) - i_L(0) = i_L(0) + \frac{V_{in}}{L}t_{on} - i_L(0) = \frac{V_{in}}{L}DT_s \qquad (4.32)$$

If, as in our earlier analysis of a bulk converter, we assume that the capacitor shown in Fig. 4.19 is big enough to absorb the ripple current passing through the inductor, and the load resistor only takes the DC component, the output voltage ripple can be calculated based on the following charge balance equation on the capacitor. The detailed waveforms related to the output voltage ripple are shown in Fig. 4.22.

$$\int_0^{T_s} i_C(t)dt = \int_0^{t_{on}} i_C(t)dt + \int_{t_{on}}^{T_s} i_C(t)dt = -\frac{V_0}{R}t_{on} + \int_{t_{on}}^{T_s}\left(i_L(t) - \frac{V_0}{R}\right)dt \qquad (4.33)$$

The magnitude of the output voltage ripple is:

$$\Delta V_{ripple} = \frac{Q}{C} = \frac{I_o}{C}t_{on} = \frac{V_o}{RC}DT_s \qquad (4.34)$$

Equation 4.32 shows that the ripple current reduces with an increase in inductance, while Eq. 4.34 shows that the ripple voltage reduces with an increase in capacitance.

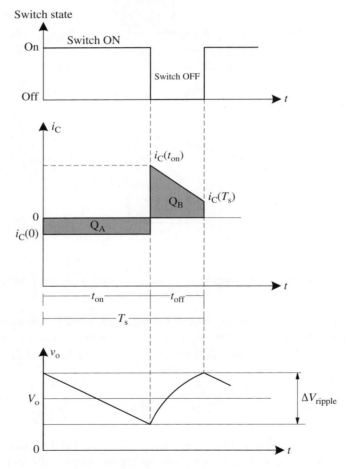

Figure 4.22 Capacitor charge and output voltage ripple of boost converters

4.2.5 DC–DC Converters Applied in Hybrid Electric Vehicle Systems

In previous sections, we introduced the basic operating principle of switch-mode DC–DC converters as well as giving details on the buck, boost, and buck-boost types of DC–DC converter. The material presented should be enough to support hybrid vehicle system modeling and performance simulation. This section briefly introduces the principles of two types of DC–DC converter that are applied practically in hybrid vehicle systems.

4.2.5.1 Isolated Buck DC–DC Converters

In an HEV/PHEV/EV system, a DC–DC converter is needed to convert the high voltage supplied by the main battery system to a lower voltage to charge a 12-volt auxiliary battery and supply electrical power to various accessories, such as headlamps, wipers, and horns. As vehicle safety regulations require that a 12 V power system has to be isolated from high voltages, the practical DC–DC converter must be an isolated buck converter. Figure 4.28 shows a principle circuit for such a converter.

This converter has two switching periods and four operational modes. During the first subinterval of the first switching period, MOSFET switches Q_1 and Q_4 are ON while Q_2 and Q_3 are OFF; during the second subinterval, all MOSFET switches are OFF. During the first subinterval of the second switching period, MOSFET switches Q_1 and Q_4 are OFF while Q_2 and Q_3 are ON; during the second subinterval, all MOSFET switches are OFF. A square-wave AC is generated at the primary winding of the transformer by turning these switches on and off. In the steady state of operation, if we neglect the ripple current passing the inductor L and magnetizing the current of the transformer, the main waveforms in the continuous-conduction mode are as shown in Fig. 4.29.

During the first subinterval of the first switching period in which Q_1 and Q_4 are ON and Q_2 and Q_3 are OFF for time t_{on}, the voltage across the primary winding of the transformer is equal to $v_p = V_{in}$, and the diode D_5 conducts. During this period of time, the voltage across the second winding of the transformer is stepped down to:

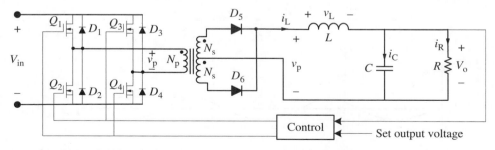

Figure 4.28 Principle circuit of a full-bridge isolated buck converter

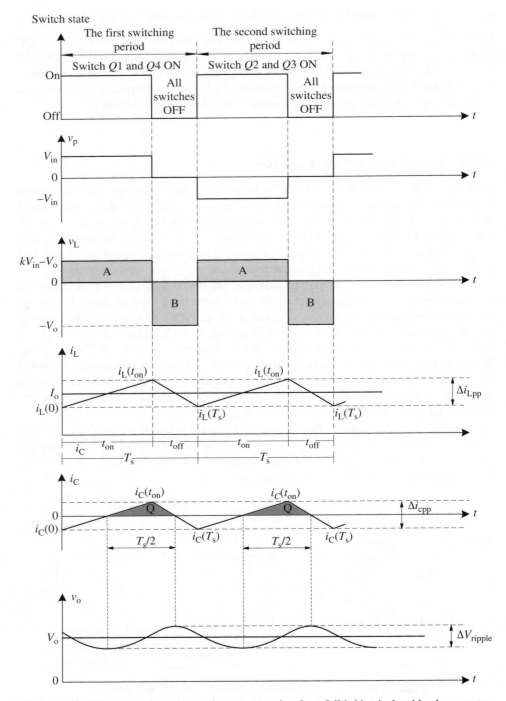

Figure 4.29 Main waveforms in steady-state operation for a full-bridge isolated buck converter

$$v_s = k v_p = \frac{N_s}{N_p} v_p \tag{4.39}$$

The voltage across the filter inductor L is given by:

$$v_L = v_s - V_o = \frac{N_s}{N_p} v_p - V_o = \frac{N_s}{N_p} V_{in} - V_o \tag{4.40}$$

During the second subinterval of the first switching period in which all switches, Q_1, Q_2, Q_3, and Q_4, are OFF for time t_{off}, the voltage across the primary winding of the transformer is equal to zero, and both diodes D_5 and D_6 conduct. During this period of time, the voltage across the filter inductor L is given by:

$$v_L = V_o \tag{4.41}$$

Thus, the current flows through the inductor L can be obtained by:

$$v_L(t) = L \frac{di_L(t)}{dt} \quad \Rightarrow \quad i_L(t) = \begin{cases} \dfrac{1}{L} \displaystyle\int_0^t \left(\dfrac{N_s}{N_p} V_{in} - V_o \right) dt = i_L(0) + \dfrac{1}{L} \left(\dfrac{N_s}{N_p} V_{in} - V_o \right) t & 0 \leq t \leq t_{on} \\[4mm] \dfrac{1}{L} \displaystyle\int_{t_{on}}^t V_o \, dt = i_L(t_{on}) - \dfrac{V_o}{L} t & t_{on} < t \leq T_s \end{cases} \tag{4.42}$$

During the first interval of the second switching period of time, switches Q_1 and Q_4 are OFF and switches Q_2 and Q_3 are ON for time t_{on}, which results in the voltage across the primary winding, v_p, being opposite to the input voltage V_{in} and the diode D_6 conducts on the second side of the transformer. The operating principles of the output filters and load are the same as for the first switching period of time shown in Fig. 4.29.

Applying the volt-second balancing principle in the steady state of operation to the inductor L, the shaded area A equals the shaded area B in Fig. 4.29, which yields the following input voltage and output voltage relationship:

$$\left(\frac{N_s}{N_p} V_{in} - V_o \right) t_{on} - V_o t_{off} = 0 \quad \Rightarrow \quad V_o(t_{on} + t_{off}) = \frac{N_s}{N_p} V_{in} t_{on}$$

$$\Rightarrow \quad V_o T_s = \frac{N_s}{N_p} V_{in} t_{on} \quad \Rightarrow \quad V_o = \frac{N_s}{N_p} V_{in} \frac{t_{on}}{T_s} = \frac{N_s}{N_p} V_{in} D \tag{4.43}$$

As, in the steady state of operation, $i_L(0) = i_L(T_s)$, the peak-to-peak current ripple of the inductor is:

$$\Delta i_{Lpp} = i_L(t_{on}) - i_L(0) = \left(\frac{N_s}{N_p} V_{in} - V_o \right) D T_s \tag{4.44}$$

If we assume, as in previous analysis, that the filter capacitor is big enough such that all ripple components can be absorbed by the capacitor and the DC component goes through the load resistor, the charge Q on the capacitor during the half period of the duty cycle, shown in Fig. 4.29, can be calculated as follows:

$$Q = \int_0^t i_C(t)dt\big|_{i_C(t)\geq 0} = \int_0^{T_s/2} i_C(t)dt\big|_{i_C(t)\geq 0} = \frac{1}{2}\frac{\Delta i_{cpp}}{2}\frac{T_s}{2} = \frac{V_o}{8L}(1-D)T_s^2 \tag{4.45}$$

where

$$\Delta i_{cpp} = i_c(t_{on}) - i_C(0) = i_L(0) + \frac{\frac{N_s}{N_p}V_{in}-V_o}{L}t_{on} - \frac{V_o}{R} - i_L(0) + \frac{V_o}{R} = \frac{\frac{N_s}{N_p}V_{in}-V_o}{L}t_{on}$$

$$= \frac{1}{L}\left(\frac{N_s}{N_p}\frac{V_{in}}{T_s}t_{on}T_s - \frac{V_o}{T_s}t_{on}T_s\right) = \frac{1}{L}\left(\frac{N_s}{N_p}V_{in}DT_s - V_oDT_s\right) = \frac{1}{L}(V_o - V_oD)T_s = \frac{V_o}{L}(1-D)T_s \tag{4.46}$$

Thus, the ripple output voltage on the load is:

$$v_{ripple}(t) = \Delta V_C = \frac{Q}{C} = \frac{V_o}{8LC}(1-D)T_s^2 \tag{4.47}$$

It should be noted that the above equation and waveforms are obtained by neglecting the transformer magnetizing current. If the transformer magnetizing current needs to be taken into account in practical analysis, the actual waveforms will deviate from the above idealized analysis.

4.2.5.2   Four-quadrant DC–DC Converters

Traction motors and other higher voltage auxiliary power units such as the electrically powered air-conditioner in a hybrid vehicle need to be powered by a shielded high-voltage DC–DC converter. Unlike the converters introduced in previous sections, this type of converter sometimes also needs to be able to transfer the DC power bidirectionally, which means that the converter can not only power electric loads, but also take the regenerative braking energy to charge the high-voltage battery. Such a converter is called a *four-quadrant DC–DC converter*. Figure 4.30 shows the principle circuit, and the operating principle is described below.

Forward power control operates in the first quadrant, where MOSFET power switches Q_1 and Q_4 are turned ON and Q_2 and Q_3 are turned OFF, which results in the DC motor being powered for forward turning. Forward regeneration operates in the fourth quadrant, and this happens when the back electromotive force (emf) E_a of the motor is greater than the input DC voltage V_{in}, which forces diodes D_1 and D_4 to conduct, and the motor, acting as a

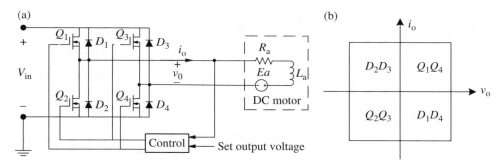

Figure 4.30   The principle circuit of a four-quadrant DC–DC converter with a DC motor load. (a) Principle circuit, (b) The quadrant

generator, returns energy to the supply (charging the battery). The third quadrant is for the reverse power operation, where the MOSFET power switches Q_2 and Q_3 are turned ON and Q_1 and Q_4 are turned OFF, thus the motor is powered for turning in the reverse direction. Reverse regeneration happens in the second quadrant, when the back emf E_a of the motor is greater than the input DC voltage V_{in}; in this case, the motor works as a generator in the reverse direction to return energy to the supply.

It should be noted that the principle circuit presented is also able to adjust the output voltage v_0 through a PWM control scheme, so that the speed or torque of the motor can be controlled. For detailed control design and analysis, the interested reader should refer to the books by Mohan *et al.* (2003) and Rashid (2008) listed at the end of the chapter.

4.3   DC–AC Inverters

In HEV/EV systems, an inverter converts the battery's DC power into AC power for the electric motors, and also converts the motor-captured AC power into DC power to charge the battery once regenerative braking energy is available; therefore, a DC–AC inverter is one of the most important parts in hybrid electric and purely electric vehicles. This section briefly introduces the operating principle of the most common DC–AC inverters.

4.3.1   Basic Concepts of DC–AC Inverters

The function of a DC–AC inverter is to change DC power into AC power with the desired voltage magnitude and frequency. In HEV/EV systems, the inverter, as the core part of the AC motor drive, is mainly there to power the traction motor, and IGBTs are normally used as power electronic switches. Sometimes, power MOSFETs are also used for higher-end systems that require higher efficiency and higher switching frequency. Since most inverters operate at switching frequencies between 5 and 20 kHz, some special measures, detailed in chapter 9, are necessary to avoid audible noise being radiated from the motor windings, while enabling higher efficiency and lower cost.

(a)   (b)

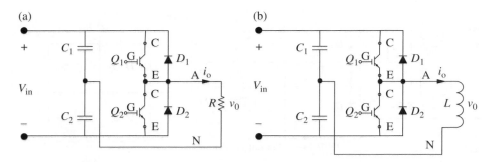

Figure 4.31 Principle circuit of DC–AC inverters. (a) Principle circuit with resistive load, (b) Principle circuit with inductive load

The basic working concept of a DC–AC inverter can be explained through the circuit shown in Fig. 4.31. For the resistive load and in the steady state of operation, IGBT power switches Q_1 and Q_2 conduct $T_s/2$ or $180°$ during the period of time T_s. That is, during the first half of the time period, $0 < t \leq T_s/2$, power switch Q_1 is turned ON and Q_2 is turned OFF, and the voltage across the load v_{AN} is equal to $V_{in}/2$; during the other half of the time period, $T_s/2 < t \leq T_s$, power switch Q_1 is turned OFF and Q_2 is turned ON, and the load voltage v_{AN} is equal to $-V_{in}/2$. The current and voltage waveforms with resistive load are shown in Fig. 4.32.

For the inductive load and in steady-state operation, each power switch Q_1 and Q_2 conducts only $T_s/4$ or $90°$, and each power diode D_1 and D_2 conducts the other $T_s/4$ or $90°$ during the period of time T_s. This is because the load current cannot immediately change even if power switch Q_1 or Q_2 is turned off, and power diodes D_1 and D_2 provide an alternative path to enable the current to continue flowing through them. The waveforms of the principle circuit with inductive load are shown in Fig. 4.33. During the first quarter ($0 < t \leq T_s/4$) of the period of time T_s, the power switch Q_1 is turned ON and Q_2 is turned OFF, so the voltage across the load v_{AN} is equal to $V_{in}/2$ and the energy is stored in the inductor through the $Q_1 \rightarrow L \rightarrow C_2$ path; during the second quarter, $T_s/4 < t \leq T_s/2$, both power switches Q_1 and Q_2 are turned OFF and diode D_2 conducts, so the stored energy returns to the DC source by the $D_2 \rightarrow L \rightarrow C_1$ path, the load voltage v_{AN} is equal to $-V_{in}/2$ in this period of time. Similarly, during the third quarter, $T_s/2 < t \leq 3T_s/4$, power switch Q_1 is turned OFF and Q_2 is turned ON, so the voltage across the load v_{AN} is equal to $-V_{in}/2$ and the energy is stored in the inductor through the $C_1 \rightarrow L \rightarrow Q_2$ path; during the fourth quarter, $3T_s/4 < t \leq T_s$, both power switches Q_1 and Q_2 are turned OFF and diode D_1 conducts to enable the stored energy to return to the DC source by the $C_2 \rightarrow L \rightarrow D_1$ path; the load voltage v_{AN} is equal to $V_{in}/2$ in this period of time.

The current passing through the inductor can be calculated based on the volt-second balance principle; thus, the inductor current during the first and second quarters is calculated as follows:

$$v_L(t) = L\frac{di_o(t)}{dt} \quad \Rightarrow \quad i_o(t) = \frac{1}{L}\int_0^t v_o dt = \frac{1}{L}\int_0^t \frac{V_{in}}{2} dt = \frac{V_{in}}{2L}t - i_o(0), \qquad 0 \leq t \leq T_s/4 \quad (4.48)$$

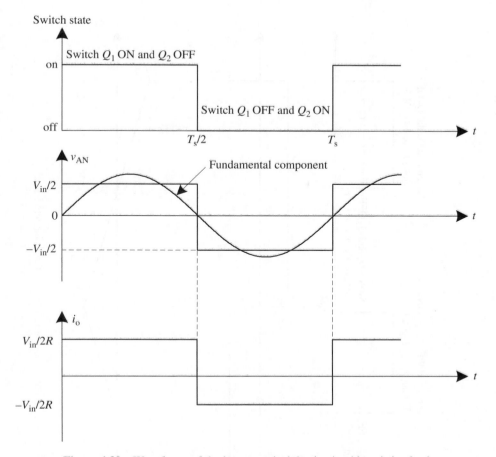

Figure 4.32   Waveforms of the inverter principle circuit with resistive load

and

$$i_o(t) = \frac{1}{L}\int_{T_s/4}^{t} v_o dt = -\frac{1}{L}\int_{T_s/4}^{t} \frac{V_{in}}{2} dt = -\frac{V_{in}}{2L}t + i_o(T_s/4), \qquad T_s/4 < t \le T_s/2 \qquad (4.49)$$

In the steady state of operation, $i_L(T_s/2) = i_L(0) = 0$, so the current at $t \le T_s/4$ can be obtained from Eq. 4.49 as:

$$i_o(T_s/2) = -\frac{V_{in}}{4L}T_s + i_o(T_s/4) = i_o(0) = 0 \;\Rightarrow\; i_o(T_s/4) = \frac{V_{in}}{4L}T_s \qquad (4.50)$$

It should be noted that the conduction time of power switches Q_1 and Q_2 for an RL load will vary from 90° to 180° depending on the detailed RL impedance.

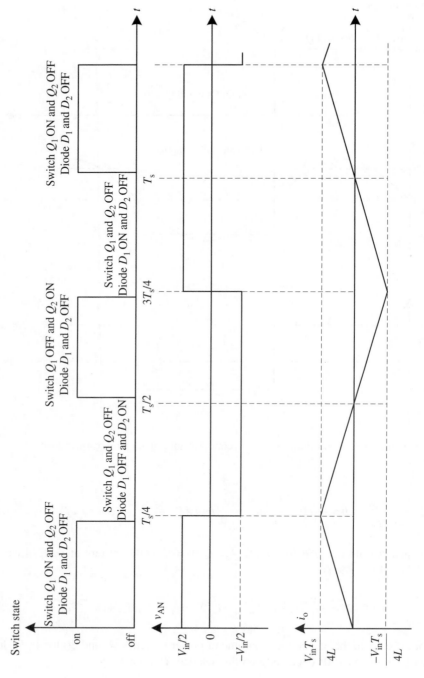

Figure 4.33 Waveforms of the inverter principle circuit with inductive load

From Fig. 4.32 and Fig. 4.33, it can be seen that the inverted AC voltage is actually a square wave rather than the expected sinusoidal waveform. In order to achieve a sinusoidal waveform, we need to expand the obtained square-wave voltage using the Fourier series as follows:

$$v_{AN}(t) = \frac{a_0}{2} + \sum_{k=1}^{\infty} [a_k \cos(k\omega t) + b_k \sin(k\omega t)] \tag{4.51}$$

where

$$a_k = \frac{2}{T_s} \int_{-\frac{T_s}{2}}^{\frac{T_s}{2}} v_{AN}(t) \cos\left(\frac{2k\pi}{T_s} t\right) dt = 0 \quad \text{as } v_{AN}(t) \text{ is an odd function over } [-T_s/2,\ T_s/2] \tag{4.52}$$

$$b_k = \frac{2}{T_s} \int_{-\frac{T_s}{2}}^{\frac{T_s}{2}} v_{AN}(t) \sin\left(\frac{2k\pi}{T_s} t\right) dt = \frac{4}{T_s} \int_0^{\frac{T_s}{2}} \frac{V_{in}}{2} \sin\left(\frac{2k\pi}{T_s} t\right) d = \frac{V_{in}}{k\pi} \int_0^{\frac{T_s}{2}} \sin\left(\frac{2k\pi}{T_s} t\right) d\left(\frac{2k\pi}{T_s} t\right)$$

$$= \frac{V_{in}}{k\pi}(-\cos(k\pi) + \cos(0)) = \frac{2V_{in}}{k\pi} \quad \text{if } k = 1,\ 3,\ 5, \cdots \tag{4.53}$$

thus, the instantaneous output voltage is expressed as:

$$v_{AN}(t) = \begin{cases} 0 & \text{if } k = 0,\ 2,\ 4, \cdots \\ \sum_{k=1}^{\infty} \left[\dfrac{2V_{in}}{k\pi} \sin(k\omega t)\right] & \text{if } k = 1,\ 3,\ 5, \cdots \end{cases} \tag{4.54}$$

where $\omega = 2\pi f$ is the angular frequency of the output voltage (rad/s). Equation 4.54 shows that only odd harmonic voltages are present and even harmonic voltages are absent, and the fundamental component is expressed as:

$$v_{AN\_1} = \frac{2V_{in}}{\pi} \sin(\omega t) = 0.637 V_{in} \sin(\omega t) \tag{4.55}$$

If filters are applied to the inverter's output to filter out all higher harmonic voltages and just allow the fundamental component to pass the filter, a sinusoidal-wave output voltage is obtained.

On the output current aspect, for the resistive load, the instantaneous output current is:

$$i_o(t) = \sum_{k=1}^{\infty} \left[\frac{2V_{in}}{k\pi R} \sin(k\omega t)\right], \quad k = 1,\ 3,\ 5, \cdots \tag{4.56}$$

for the pure inductive load, the instantaneous output current is:

$$i_o(t) = \sum_{k=1}^{\infty} \left[\frac{2V_{in}}{k^2 \pi \omega L} \sin\left(k\omega t - \frac{\pi}{2}\right) \right], \quad k = 1, 3, 5, \cdots \tag{4.57}$$

and for the RL load, the instantaneous output current can be expressed as:

$$i_o(t) = \sum_{k=1}^{\infty} \left[\frac{2V_{in}}{k\pi \sqrt{R^2 + (k\omega L)^2}} \sin(k\omega t - \theta) \right], \quad \theta = \tan^{-1}\left(\frac{k\omega L}{R}\right) \text{ for } k = 1, 3, 5, \cdots \tag{4.58}$$

4.3.2   Single-phase DC–AC Inverters

Single-phase DC–AC inverters are commonly utilized to power a single-phase AC traction motor in most light HEVs. In addition, the auxiliary power unit (APU) including the air-conditioner and cooling pumps in most HEVs/EVs is also powered by single-phase AC induction motors. The electrical power of a single-phase inverter usually flows from the DC to the AC terminal, but in some cases, reverse power flow is possible to capture regenerative energy.

The most commonly used single-phase DC–AC inverter in practical hybrid vehicle systems is the full-bridge voltage source inverter, and the principle circuit and main waveforms with inductive load are shown in Fig. 4.34(a) and Fig. 4.34(c). For the inductive load, the operating sequences of power switches and diodes are: (i) during the first quarter ($0 < t \leq T_s/4$) of the period of time T_s, power switches Q_1 and Q_4 are turned ON and switches Q_2 and Q_3 are turned OFF, so the voltage across the load v_{AB} is equal to V_{in} and the current gradually reaches the maximum and the energy is stored in the inductive load through the $Q_1 \rightarrow L \rightarrow Q_4$ path; (ii) during the second quarter, $T_s/4 < t \leq T_s/2$, all power switches are turned OFF, but power diodes D_2 and D_3 conduct, so the stored energy is able to return to the DC source and the current gradually diminishes to zero by the $D_2 \rightarrow L \rightarrow D_3$ path, the load voltage v_{AB} is equal to $-V_{in}$ in this period of time; (iii) during the third quarter, $T_s/2 < t \leq 3T_s/4$, power switches Q_2 and Q_3 are turned ON and switches Q_1 and Q_4 are turned OFF, so the voltage across the load v_{AB} is equal to $-V_{in}$ and the current gradually increases to the negative maximum and the energy is stored in the inductor; (iv) during the fourth quarter, $3T_s/4 < t \leq T_s$, all power switches are turned OFF again, but power diodes D_1 and D_4 conduct and the current gradually returns to zero from the negative maximum and the stored energy returns to the DC source, the load voltage v_{AB} is equal to V_{in} in this period of time. The detailed current passing through the inductor can be calculated based on the volt-second balance principle discussed in previous sections.

Similarly, expanding the output voltage v_{AB} using the Fourier series, we can obtain the instantaneous output voltage and current of the inverter, as follows:

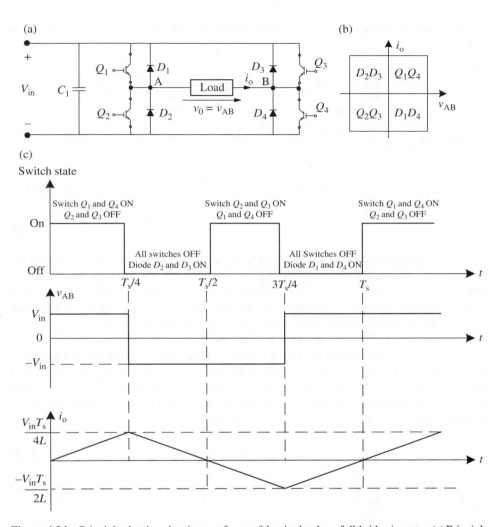

Figure 4.34 Principle circuit and main waveforms of the single-phase full-bridge inverter. (a) Principle circuit of the single phase full bridge voltage source inverter, (b) The quadrant, (c) Main waveforms

$$v_{AB}(t) = \sum_{k=1}^{\infty} \left[\frac{4V_{in}}{k\pi} \sin(k\omega t) \right] \quad \text{for } k = 1, 3, 5, \cdots \tag{4.59}$$

where $\omega = 2\pi f$ is the angular frequency of the output voltage (rad/s). It shows that the output of a single-phase full-bridge inverter does not include even harmonic voltages, and the fundamental component is:

$$v_{AN\_1} = \frac{4V_{in}}{\pi} \sin(\omega t) = 1.27 V_{in} \sin(\omega t) \tag{4.60}$$

Depending on the load property, the instantaneous output current can be expressed as follows:

$$i_o(t) = \sum_{k=1}^{\infty} \left[\frac{4V_{in}}{k\pi R} \sin(k\omega t) \right], \quad k = 1, 3, 5, \cdots \text{ for the pure resistive load} \tag{4.61}$$

$$i_o(t) = \sum_{k=1}^{\infty} \left[\frac{4V_{in}}{k^2 \pi \omega L} \sin\left(k\omega t - \frac{\pi}{2}\right) \right], \quad k = 1, 3, 5, \cdots \text{ for the pure inductive load} \tag{4.62}$$

$$i_o(t) = \sum_{k=1}^{\infty} \left[\frac{4V_{in}}{k\pi \sqrt{R^2 + (k\omega L)^2}} \sin(k\omega t - \theta) \right], \theta = \tan^{-1}\left(\frac{k\omega L}{R}\right) \text{ for } k = 1, 3, 5, \cdots \text{ for an } RL \text{ load} \tag{4.63}$$

Since the power diode pairs D_1–D_4 and D_2–D_3 are able to return the stored energy to the DC source, the above full-bridge inverter can operate bidirectionally or in four quadrants, as shown in Fig. 4.34(b). In an HEV/EV, when regenerative braking is applied to the vehicle and the back emf of the motor is greater than the input DC voltage V_{in}, the power diode pair D_1–D_4 or D_2–D_3 conducts to return the captured regenerative energy to the DC source (charging the battery), and the motor acts as a generator.

A high-voltage battery is the DC power source, and this is one of the most expensive subsystems in HEV/EV systems; therefore, how the DC–AC inverter impacts on the DC side must be well understood. Since the efficiency requirement for HEV/EV inverters is very high, the power losses on the inverter can be neglected, so the DC current ripple can be calculated based on the following instantaneous power balance:

$$v_{AB}(t)i_o(t) = V_{in}i_{in}(t) \tag{4.64}$$

For an RL load, if we only take the fundamental frequency component into account, from Eq. 4.63, we know that the DC-side current is equal to:

$$i_{in}(t) = \frac{v_{AB}(t)i_o(t)}{V_{in}} = \frac{1}{V_{in}} \frac{4V_{in}\sin(\omega t)}{\pi} \frac{4V_{in}}{\pi\sqrt{R^2 + (\omega L)^2}} \sin(\omega t - \theta)$$

$$= \frac{16}{\pi^2} \frac{V_{in}}{\sqrt{R^2 + (\omega L)^2}} \sin(\omega t)\sin(\omega t - \theta) = \frac{8}{\pi^2} \frac{V_{in}}{\sqrt{R^2 + (\omega L)^2}} (\cos(\theta) - \cos(2\omega t - \theta)) \tag{4.65}$$

where $\omega = 2\pi f_s$ is the fundamental frequency, f_s is the switching frequency, and θ is the load impedance angle at the fundamental frequency.

Equation 4.65 indicates that the current ripple frequency on the DC side doubles the frequency on the AC side, and the magnitude depends on the load property. In hybrid vehicle applications, the battery can act as a capacitor to absorb a certain degree of current ripple; however, if the ripple is too large, it could violate the battery's limits and damage the battery. The capacitors C_1 and C_2, shown in Fig. 4.31, need to be carefully designed to diminish the ripple by considering the battery's capability, but it also needs to be kept in mind that large, high-voltage capacitors are costly and require space to install.

4.3.3 Three-phase DC–AC Inverters

HEV/EV systems generally use a three-phase DC–AC inverter to power the traction motor. The principle circuit of a three-phase bridge inverter is shown in Fig. 4.35. In this circuit, each switch conducts 180° and three switches are ON at any instant of time. If we assign labels A, B, and C to the output terminals, the inverter outputs three-phase, square-wave AC power to the connected three-phase load, which could be a Δ connection or a Y connection internally, as shown on the right side of Fig. 4.35. There are eight switching states for the presented three-phase inverter circuit, from state 0, where all output terminals are clamped to the negative DC bus, to state 7, where they are all clamped to the positive bus. These eight switching states are listed in Table 4.1, and the inverted voltage waveforms are shown in Fig. 4.36.

From Fig. 4.36, it can be seen that the inverted AC voltages are square waves containing various harmonic components. As discussed in the previous section, the obtained voltages can be expanded in the Fourier series as:

$$V_{xx}(t) = \frac{a_0}{2} + \sum_{k=1}^{\infty} [a_k \cos(k\omega t) + b_k \sin(k\omega t)] \tag{4.66}$$

Since the voltage V_{AB} is symmetrical with respect to the origin when it is shifted by $\pi/6$, the even harmonic voltages are absent, that is, $a_k = 0$　$k = 0, 1, 2, \cdots$. Thus, the instantaneous voltage, V_{AB}, is equal to:

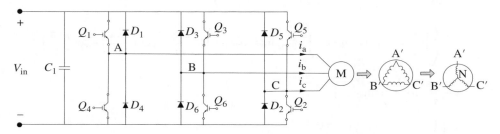

Figure 4.35　Principle circuit of three-phase bridge inverters

Table 4.1 States and output voltages of the voltage source three-phase inverter

State no.	Switch state	Terminal A state	Terminal B state	Terminal C state	Voltage V_{AB}	Voltage V_{BC}	Voltage V_{CA}
0	Q_2 Q_4 Q_6 ON Q_1 Q_3 Q_5 OFF	−	−	−	0	0	0
1	Q_1 Q_2 Q_3 ON Q_4 Q_5 Q_6 OFF	+	+	−	0	V_{in}	$-V_{in}$
2	Q_2 Q_3 Q_4 ON Q_1 Q_5 Q_6 OFF	−	+	−	$-V_{in}$	V_{in}	0
3	Q_3 Q_4 Q_5 ON Q_1 Q_2 Q_6 OFF	−	+	+	$-V_{in}$	0	V_{in}
4	Q_4 Q_5 Q_6 ON Q_1 Q_2 Q_3 OFF	−	−	+	0	$-V_{in}$	V_{in}
5	Q_5 Q_6 Q_1 ON Q_2 Q_3 Q_4 OFF	+	−	+	V_{in}	$-V_{in}$	0
6	Q_6 Q_1 Q_2 ON Q_3 Q_4 Q_5 OFF	+	−	−	V_{in}	0	$-V_{in}$
7	Q_1 Q_3 Q_5 ON Q_2 Q_4 Q_6 OFF	+	+	+	0	0	0

$$V_{AB}(t) = \sum_{k=1}^{\infty} \left[b_k \sin\left(k\left(\omega t + \frac{\pi}{6} \right) \right) \right] \quad \omega t \in [-\pi \ \ \pi] \ \ k = 1, 3, 5, \cdots \qquad (4.67)$$

where

$$b_k = \frac{1}{\pi} \int_{-\pi}^{\pi} V_{AB}(t) \sin(k\omega t) dt$$

$$= \frac{1}{\pi} \left(\int_{-\pi}^{-\frac{5\pi}{6}} 0 \sin(k\omega t) d\omega t + \int_{-\frac{5\pi}{6}}^{-\frac{\pi}{6}} V_{in} \sin(k\omega t) d\omega t + \int_{-\frac{\pi}{6}}^{\frac{\pi}{6}} 0 \sin(k\omega t) d\omega t \right.$$

$$\left. + \int_{\frac{\pi}{6}}^{\frac{5\pi}{6}} -(V_{in}) \sin(k\omega t) d\omega t + \int_{\frac{5\pi}{6}}^{\pi} 0 \sin(k\omega t) d\omega t \right)$$

$$= \frac{1}{\pi} \left(\int_{-\frac{5\pi}{6}}^{-\frac{\pi}{6}} V_{in} \sin(k\omega t) d\omega t - \int_{\frac{\pi}{6}}^{\frac{5\pi}{6}} V_{in} \sin(k\omega t) d\omega t \right) = \frac{2V_{in}}{k\pi} \left(\cos\left(\frac{5k\pi}{6} \right) - \cos\left(\frac{k\pi}{6} \right) \right)$$

$$= \frac{4V_{in}}{k\pi} \sin\left(\frac{k\pi}{2} \right) \sin\left(\frac{k\pi}{3} \right)$$

$$= \begin{cases} 0 & \text{if } k = 0, 2, 4, \cdots \\ \frac{4V_{in}}{k\pi} \sin\left(\frac{k\pi}{3} \right) & \text{if } k = 1, 3, 5, \cdots \end{cases} \quad \omega t \in [-\pi \ \ \pi] \qquad (4.68)$$

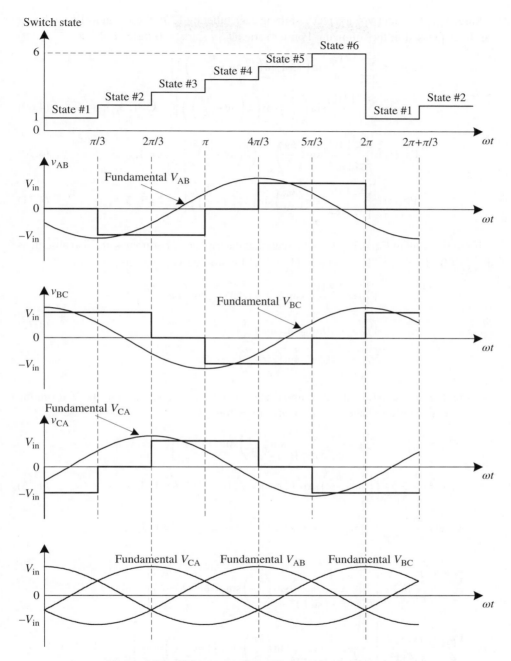

Figure 4.36   Voltage waveforms between the three terminals

Since V_{AB}, V_{BC}, and V_{CA} are phase shifting each other by 120°, these line-to-line voltages can be expressed in the following Fourier series if ωt starts at 0 instead of $-\pi$:

$$
\begin{aligned}
V_{AB}(t) &= \sum_{k=1}^{\infty} \left[\frac{4V_{in}}{k\pi} \sin\left(\frac{k\pi}{3}\right) \sin\left(k\left(\omega t + \frac{\pi}{6} - \pi\right)\right) \right] \\
&= \sum_{k=1}^{\infty} \left[\frac{4V_{in}}{k\pi} \sin\left(\frac{k\pi}{3}\right) \sin\left(k\left(\omega t - \frac{5\pi}{6}\right)\right) \right] \quad k = 1, 3, 5, \cdots \quad (4.69)
\end{aligned}
$$

$$
V_{BC}(t) = \sum_{k=1}^{\infty} \left[\frac{4V_{in}}{k\pi} \sin\left(\frac{k\pi}{3}\right) \sin\left(k\left(\omega t + \frac{\pi}{2}\right)\right) \right] \quad k = 1, 3, 5, \cdots \quad (4.70)
$$

$$
V_{CA}(t) = \sum_{k=1}^{\infty} \left[\frac{4V_{in}}{k\pi} \sin\left(\frac{k\pi}{3}\right) \sin\left(k\left(\omega t - \frac{\pi}{6}\right)\right) \right] \quad k = 1, 3, 5, \cdots \quad (4.71)
$$

From Eq. 4.69 to Eq. 4.71, it can be seen that the triple odd harmonics are also absent, as $\sin\left(\frac{k\pi}{3}\right) = 0$ if $k = 3, 6, 9, \cdots$. The fundamental line-to-line voltages are:

$$
\begin{aligned}
V_{AB} &= \frac{2\sqrt{3}V_{in}}{\pi} \sin(\omega t) = 1.103 V_{in} \sin\left(\omega t - \frac{5\pi}{6}\right) \\
V_{BC} &= 1.103 V_{in} \sin\left(\omega t + \frac{\pi}{2}\right) \\
V_{CA} &= 1.103 V_{in} \sin\left(\omega t - \frac{\pi}{6}\right)
\end{aligned} \quad (4.72)
$$

If the three-phase RL load with impedance Z is internally connected in the Y form, the instantaneous phase voltages and currents are as follows:

$$
V_{A'N} = \frac{\dot{V}_{AB}}{\sqrt{3}} \angle \left(-\frac{\pi}{6}\right) = \frac{4V_{in}}{\pi\sqrt{3}} \sum_{k=1}^{\infty} \left[\frac{1}{k} \sin\left(\frac{k\pi}{3}\right) \sin(k\omega t - \pi) \right] \quad k = 1, 3, 5, \cdots
$$

$$
V_{B'N} = \frac{4V_{in}}{\pi\sqrt{3}} \sum_{k=1}^{\infty} \left[\frac{1}{k} \sin\left(\frac{k\pi}{3}\right) \sin\left(k\left(\omega t + \frac{\pi}{3}\right)\right) \right] \quad k = 1, 3, 5, \cdots \quad (4.73)
$$

$$
V_{C'N} = \frac{4V_{in}}{\pi\sqrt{3}} \sum_{k=1}^{\infty} \left[\frac{1}{k} \sin\left(\frac{k\pi}{3}\right) \sin\left(k\left(\omega t - \frac{\pi}{3}\right)\right) \right] \quad k = 1, 3, 5, \cdots
$$

$$
i_a = \frac{V_{A'N}}{Z} = \frac{4V_{in}}{\pi\sqrt{3}} \sum_{k=1}^{\infty} \left[\frac{1}{k\sqrt{R^2 + (k\omega L)^2}} \sin\left(\frac{k\pi}{3}\right) \sin(k(\omega t - \pi) - \theta) \right] \quad k = 1, 3, 5, \cdots
$$

$$
i_b = \frac{V_{B'N}}{Z} = \frac{4V_{in}}{\pi\sqrt{3}} \sum_{k=1}^{\infty} \left[\frac{1}{k\sqrt{R^2 + (k\omega L)^2}} \sin\left(\frac{k\pi}{3}\right) \sin\left(k\left(\omega t + \frac{\pi}{3}\right) - \theta\right) \right] \quad k = 1, 3, 5, \cdots
$$

$$
i_c = \frac{V_{C'N}}{Z} = \frac{4V_{in}}{\pi\sqrt{3}} \sum_{k=1}^{\infty} \left[\frac{1}{k\sqrt{R^2 + (k\omega L)^2}} \sin\left(\frac{k\pi}{3}\right) \sin\left(k\left(\omega t - \frac{\pi}{3}\right) - \theta\right) \right] \quad k = 1, 3, 5, \cdots
$$

$$
(4.74)
$$

4.4 Electric Motor Drives

A motor drive consists of an electric motor, a power electronic converter or inverter, and a torque/speed controller with related sensors. In HEV/EV applications, there are basically three types of motor drives: a BLDC motor drive, an induction motor drive, and a switched reluctance motor drive. In all drives, the torque and speed need to be controlled, and the power DC–DC converter or DC–AC inverter plays the roles of a controller and an interface between the input power and the motor.

Unlike industrial applications, HEV/EV applications require that the motors operate well under all conditions including frequent start and stop, high rates of acceleration/deceleration, high-torque/low-speed hill climbing, and low-torque/high-speed cruising. In order to meet these drivability requirements, HEV/EV electric motors are operated in two modes: constant torque mode and constant power mode, also called the normal mode and the extended mode. Within the rated speed range, the motor exerts constant torque regardless of the speed, and once past the rated speed, the motor outputs constant power and the output torque is reduced proportionally to the speed. Depending on the vehicle system architecture, the electric motor can be used as a main mover, a peak power device, or a load-sharing device, which means that the electric motor has to be able to deliver the necessary torque for adequate acceleration during its constant torque mode before it changes to its constant power mode for steady speeds.

4.4.1 BLDC Motor and Control

The brushless DC (BLDC) motor operates in a similar way to a brushed DC motor, and has the same torque and speed characteristic curve. As the name hints, BLDC motors do not have mechanical brushes for commutation as traditional DC motors do, but they must be electronically commutated in order to produce rotational torque. Since there are no brushes to wear out and replace, BLDC motors are highly reliable and almost maintenance free. The high reliability, efficiency, and capability of providing a large amount of torque over a wide operational speed range make BLDC motors acceptable as high-performance propulsion motors in HEV/EV applications.

4.4.1.1 Basic Operating Principle of BLDC Motors

The function of electric motors is to convert electrical energy into mechanical energy to generate the mechanical torque required by the vehicle. A basic BLDC motor typically consists of three main parts: a stator, a rotor, and electronic commutation with/without position sensors. The stator in most HEV/EV BLDC motors is a three-phase stator to generate a rotating magnetic field, but the number of coils is generally replicated to reduce torque ripple. The rotor of a BLDC motor consists of an even number of permanent magnets. The number of magnetic poles in the rotor also affects the step size and torque ripple of the motor, and more poles generate smaller steps and less torque ripple. Commutation is the act of changing the motor phase current at the appropriate times to produce rotational torque, and the exact commutation timing is provided either by position sensors or by coil back electromotive force (back emf) measurements. Figure 4.37(a) shows a stator and rotor configuration of a BLDC

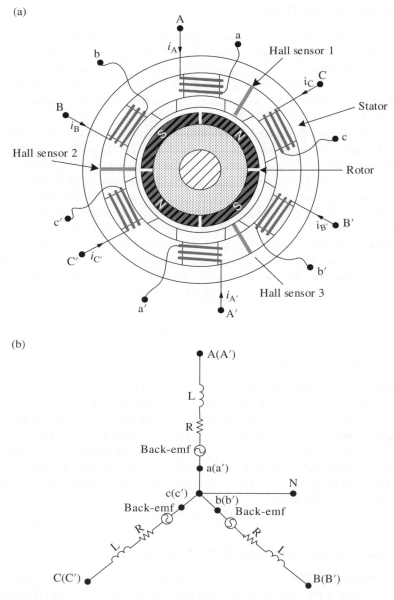

Figure 4.37 Stator and rotor diagram of a BLDC motor. (a) The configuration of a rotor with 2 pairs of poles and 3 phase stator with 6 coils, (b) The electrical diagram of three phase stator

motor. Figure 4.37(b) shows the electrical diagram of the three-phase stator that has six coils, each including coil inductance, resistance, and back electromotive force.

4.4.1.2 Torque and Rotating Field Production

Based on the fundamental physical principle, the torque developed on the motor shaft is directly proportional to the interaction of the field flux produced by the rotor magnets and the current in the stator coil. The general relationship among the developed torque, the field flux, and the current is:

$$\tau_m = K_m \psi\, i \tag{4.75}$$

where τ_m is the mechanical torque of the motor (Nm), ψ is the rotor magnetic flux (Wb), i is the stator coil current (A), and K_m is the torque constant determined by the detailed BLDC motor construction.

In addition to the developed torque, when the rotor rotates, a voltage is generated across each stator coil terminal. This voltage is proportional to the shaft velocity and tends to oppose the coil current flow, so it is called the back emf. The relationship between the back emf and the shaft velocity is:

$$E_a = K_e \psi\, \omega_m \tag{4.76}$$

where E_a denotes the back emf (V), ω_m is the shaft velocity (rad/s) of the motor, and K_e is the voltage constant of the motor. The constants K_m and K_e are numerically equal, which can be proved through the following electrical power and mechanical power balance in the steady state of operation:

$$P_e = E_a i = K_e \psi\, \omega_m i = P_m = \tau_m \omega_m = K_m \psi\, \omega_m i \quad \Rightarrow \quad K_e = K_m \tag{4.77}$$

In HEV/EV applications, the stator current i is controlled by the phase voltage, V_{xN}, and is established by the following voltage balance equation:

$$V_{xN} = E_x + R_x i_x + L_x \frac{di_x}{dt} \quad x = A,\ B,\ \text{and}\ C \tag{4.78}$$

The motor speed ω_m is built up based on the following torque balance equation:

$$\tau_m = \tau_{load} + B_m \omega_m + J_m \frac{d\omega_m}{dt} \tag{4.79}$$

where τ_m is the mechanical torque developed by the motor, J_m is the rotor inertia of the motor, B_m is the viscous frictional coefficient, and τ_{load} is the equivalent work load torque. Equations 4.75 to 4.79 form the basis of BLDC motor operation.

In order to generate a rotating field, the current flow needs to be commutated against the rotor position, which is normally detected by Hall sensors placed every 120°, as shown in Fig. 4.37(a). With these sensors, three phase commutations can be achieved so that at any time one of the three phase coils is positively energized, the second coil is negatively energized, and the third one is not energized. The six-step commutation process and the rotation of a BLDC motor are illustrated in Fig. 4.38(a) to Fig. 4.38(f) as an example, and the associated signals of the Hall sensors, the back emf, and the output torque are given in Fig. 4.39, which also shows that it needs two cycles of six commutation steps for one motor rotation for the given three-phase BLDC motor. The power switches and states of the Hall sensors in the principle circuit, shown in Fig. 4.35 and Fig. 4.37, are given in Tables 4.2 and 4.3 for a motor rotating in a clockwise and a counter-clockwise direction, respectively.

4.4.1.3　BLDC Motor Control

Since the propulsion motor in an HEV/EV needs to be operated in four quadrants, as shown in Fig. 4.40, the BLDC motor control tasks include determining the direction of rotation, maintaining the required torque, and regulating the motor's speed. The rotational direction of the BLDC motor can be easily controlled by changing the sequence of turning the power switches ON/OFF based on the Hall sensor values. As examples, Table 4.2 gives the sequence for clockwise rotation and Table 4.3 shows the counter-clockwise rotation based on the example motor and inverter circuit. However, it is challenging to control the motor's torque and regulate its speed in HEV/EV applications. This is because the vehicle is operated under various road and weather conditions and using different driving styles; also, the vehicle is accelerated and decelerated unpredictably. Such a load-varying environment demands higher fidelity and more accurate control strategies to deliver and maintain the vehicle's required torque and speed. From the control system point of view, a closed loop control is the minimum to enable the motor to respond to the driver's requests and load disturbance quickly, accurately, and robustly. This section introduces a basic PID control strategy, but there are more advanced control strategies such as adaptive control and model predictive control that are capable of meeting the strict control performance requirements for emerging HEV/EV applications.

A PID torque control diagram of an HEV/EV BLDC motor is shown in Fig. 4.41, where the input is the required torque from the vehicle controller, which is calculated based on the state of charge of the battery, the engine speed, and the position of the acceleration pedal or brake pedal. This required torque, $\tau^*_{\text{req'd}}$, is first examined if it is below the motor's torque capability by comparing it with torque–speed characteristic curves for the present operational speed of the motor, and will be passed if it is within the capability; otherwise, the torque will be saturated to the maximal torque of the motor. The processed $\tau^*_{\text{req'd}}$ is then converted to the required stator current, $I^*_{\text{req'd}}$, based on Eq. 4.75, and the current is then examined to see whether it exceeds the maximal allowable stator current to generate the desired stator current, I^*_{desired}, which is further fed to the closed PID control loop. The error $e_I(k)$ between the desired stator current I^*_{desired} and the actual measured stator current I_{actual}

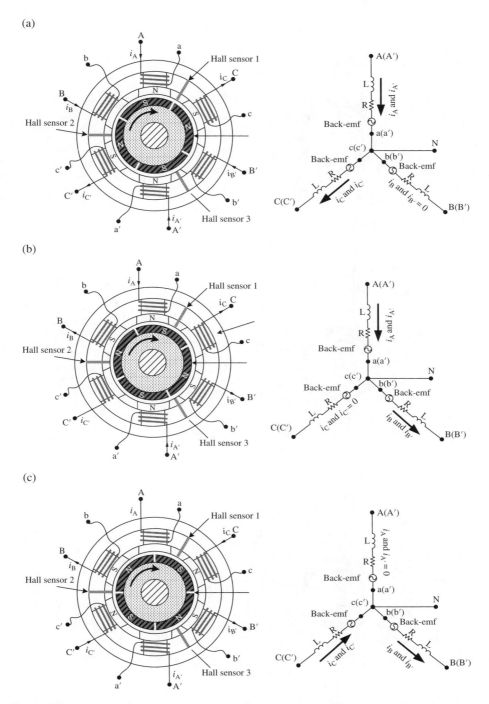

Figure 4.38 Magnetic field rotation (a) step 1; (b) step 2; (c) step 3; (d) step 4; (e) step 5; (f) step 6

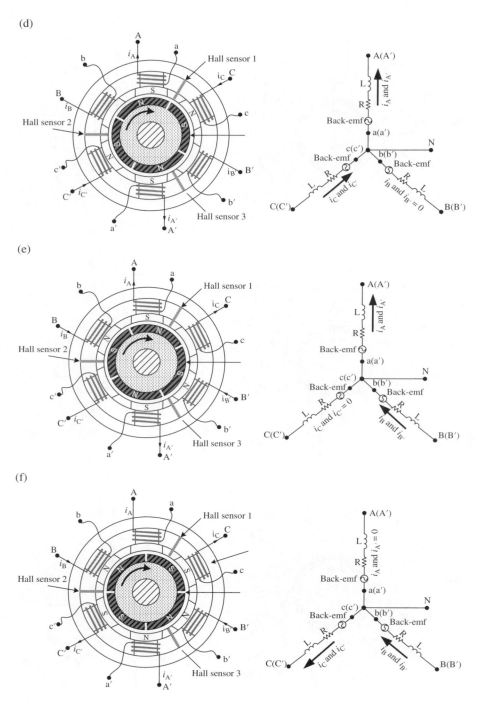

Figure 4.38 (Continued)

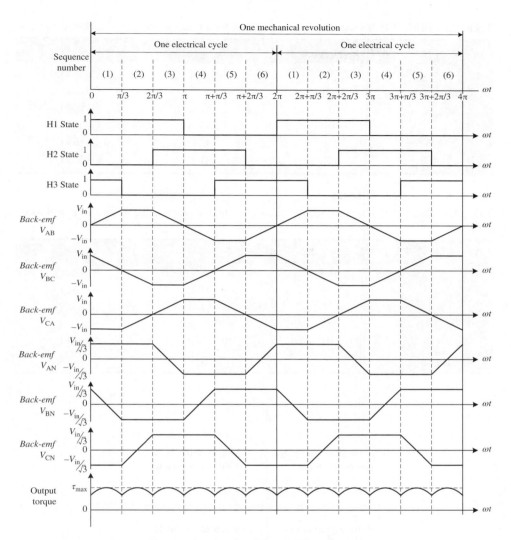

Figure 4.39 Hall sensor signal, back emf, and output torque waveforms

feeds into the following PID controller, which computes corrective action $\Delta u(k)$ to the PWM ratio, which is then sent to the three-phase inverter to produce the expected phase current for the BLDC motor. The ON/OFF sequence for the power switches is based on the actual rotor position measured by Hall sensors and the desired rotational direction of the BLDC motor.

$$\Delta u(k) = k_p(e(k) - e(k-1)) + k_i T_s e(k) + \frac{k_d}{T_s}(e(k) - 2e(k-1) + e(k-2)) \qquad (4.80)$$

Table 4.2 States of Hall sensors and power switches for a motor rotating in a clockwise direction

Sequence	Hall sensor states (H3H2H1)	Phase	ON state power switches
1	101	A–C	Q_1 and Q_2
2	001	A–B	Q_1 and Q_6
3	011	C–B	Q_5 and Q_6
4	010	C–A	Q_5 and Q_4
5	110	B–A	Q_3 and Q_4
6	100	B–C	Q_3 and Q_2

Table 4.3 States of Hall sensors and power switches for a motor rotating in a counter-clockwise direction

Sequence	Hall sensor states (H3H2H1)	Phase	ON state power switches
1	100	A–C	Q_1 and Q_2
2	110	B–C	Q_3 and Q_2
3	010	B–A	Q_3 and Q_4
4	011	C–A	Q_5 and Q_4
5	001	C–B	Q_5 and Q_6
6	101	A–B	Q_1 and Q_6

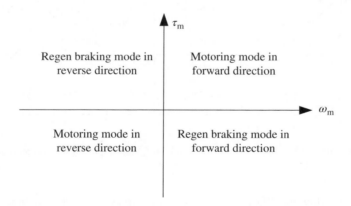

Figure 4.40 Four-quadrant operation of the electric motor in an HEV/EV

where $k_p = K$, $k_i = \frac{K}{T_i}$, and $k_d = K \cdot T_d$ are called the proportional, integral, and derivative gain, respectively, T_s is the sampling period of time. For more details on PID controller design and analysis, the interested reader should refer to the book by Åstrom and Hägglund (1995) listed at the end of the chapter.

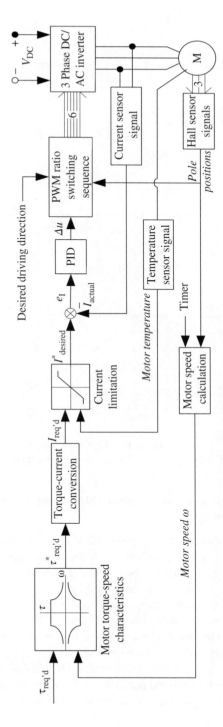

Figure 4.41 PID control diagram of a BLDC motor for HEV/EV applications

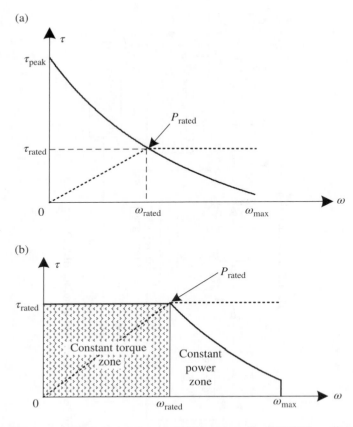

Figure 4.42 Torque–speed characteristics of BLDC motors for HEV/EV applications. (a) The principle torque-speed characteristic curve, (b) The specified torque-speed characteristic curve

4.4.1.4 BLDC Motor Torque–Speed Characteristics and Typical Technical Parameters

As is well known, one key measure to improve HEV/EV performance is using a proper propulsion motor that matches the characteristics of the battery system. From Eqs 4.75 to 4.77, we have the torque–speed characteristic curve shown in Fig. 4.42(a). In HEV/EV applications, the torque is normally specified in terms of applicable torque–speed characteristics, with a constant torque zone and a constant power zone, as shown in Fig. 4.42(b). In the constant torque zone, the output torque remains constant up to the rated speed, and once the motor speed is beyond the rated speed, the output torque starts dropping and is limited by the rated power. The maximal operating speed of a BLDC motor can be up to double the rated speed. In practical HEV/EV system design, the motor is principally sized based on the required peak torque, the RMS torque, and the operational speed range. Table 4.4 gives typical technical specification parameters for BLDC motors in HEV/EV applications.

Table 4.4   Typical technical specification parameters for BLDC motors in HEV/EV applications

Parameter	Symbol	Unit	Definition
Rated power	P_{rated}	kW	The motor delivers/regenerates this power when it is operated at the rated torque and speed
Maximal power	P_{max}	kW	The power that the motor can maximally deliver/ regenerate in a short period of time
Nominal voltage	V_{nom}	V	The rated operating terminal voltage; it normally matches the battery system terminal voltage
Rated current	I_{rated}	A	The current when the motor is operated at the rated torque
No load current	I_{no_load}	A	The current when the motor delivers zero torque, which is used to overcome frictional torque of the motor
Inductance	L	mH	The motor's stator winding inductance which can be used to calculate the electrical time constant of the motor
Resistance	R	Ω	The motor's stator winding resistance
Rated speed	N_{rated}	rpm	The rated speed of the motor beyond which the motor has to be operated in the reduced torque zone
Rated torque	τ_{rated}	Nm	The torque that the motor can continuously deliver
Efficiency	η	%	The ratio of the motor's output power over input power
Net weight	W	kg	The weight of the motor without any connections
Rotor inertia	J_m	kg.m^2	The rotor's moment of inertia
Frictional torque	T_{fr}	Nm	The lumped torque loss due to friction
Ambient temperature	T	°C or °F	The operating ambient temperature
Maximal winding temperature	T_w	°C or °F	The maximal allowed winding temperature

4.4.1.5 Sensorless BLDC Motor Control

From the foregoing discussion, we know that a BLDC motor drive generally uses Hall sensors to provide rotor position information to the controller in order to maintain synchronization. Such Hall sensors are costly and require additional wiring, which may result in lowering the entire vehicle system's reliability and durability. Since the 1980s, sensorless BLDC motor control technologies have been developed to lower the cost while meeting the performance requirements.

Sensorless BLDC motor control is based on the principle that the phase back emf has a relationship with the rotor position when the permanent magnet rotor is rotating; in other words, the rotor position can be detected by the measured phase back emf, so synchronized phase commutations can be implemented without using physical position sensors.

The key technique when implementing sensorless BLDC motor control is to detect the rotor position effectively under different working conditions based on the stator current and voltage. Various methods have been explored to improve the system performance, including the extended Kalman-filter-based state estimator, the robust state observer, and artificial intelligence-based rotor position estimation methods such as fuzzy logic and artificial neural networks.

4.4.2 AC Induction Motor and Control

An induction motor is another type of motor that is widely accepted as a propulsion motor in HEV/EV applications due to its lower cost, decreased need for maintenance, and longer lifetime. Since induction motors require an AC supply, they might seem unsuitable as the DC source in HEV/EV applications; however, nowadays, AC can easily be inverted from a DC source with modern power electronics. There are two common types of induction motor based on the rotor structure: the wound rotor and the squirrel cage rotor. Three-phase squirrel cage induction motors are normally used in lower-end HEV/EV systems due to their rigidity and construction simplicity.

4.4.2.1 Basic Principle of AC Induction Motor Operation

Regardless of the type of three-phase induction motor used, the stator is the same, and it consists of three phase windings distributed in the stator slots, displaced by 120° from each other. These stator windings produce a rotating magnetic field. A typical squirrel cage rotor consists of a series of conductive bars that are welded together at either end, forming the short-circuited windings shown in Fig. 4.43.

As we know, a balanced three-phase sinusoidal current flowing through the stator establishes a constant rotating magnetic field with synchronous speed $\omega_s = 2\omega/p$, where p is the number of magnetic poles created by the stator winding and ω is the angular frequency (rad/s) of the three-phase voltage applied to the stator. The rotating magnetic field cuts the rotor conductors at slip speed ω_{slip}, inducing a corresponding emf which causes current to flow in the short-circuited windings. The slip speed is the differential speed between the synchronous speed, ω_s, of the rotating magnetic field and the rotor spinning speed, $\omega_{\text{slip}} = \omega_s - \omega_r$. Although the slip speed, ω_{slip}, is a very important variable for the induction motor, the slip speed relative to the synchronous speed is most used in induction motor

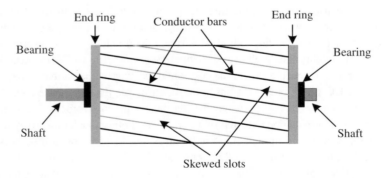

Figure 4.43 Typical structure of a squirrel cage rotor for an AC induction motor

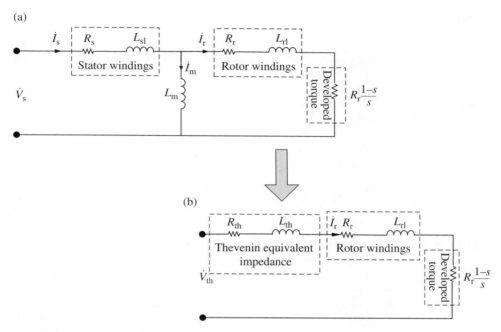

Figure 4.44 The equivalent circuit of an AC induction motor in steady-state operation. (a) The equivalent circuit (b) The Thevenin equivalent circuit

analysis; furthermore, it is called the *slip*, with symbol *s*. Technically, the slip is defined from the following equation:

$$\text{Slip} \quad s = \frac{\text{Slip speed}}{\text{Synchronous speed}} = \frac{\omega_{\text{slip}}}{\omega_{\text{s}}} = \frac{\omega_{\text{s}} - \omega_{\text{r}}}{\omega_{\text{s}}} \tag{4.81}$$

The phase equivalent circuit of an induction AC motor in steady-state operation is shown in Fig. 4.44(a), where $\dot{V}_{\text{s}}$ is the supply voltage, R_{s} and L_{sl} are the resistance and leakage inductance of the stator windings, L_{m} is the magnetic inductance, R_{r} and L_{rl} are the resistance and leakage inductance of the rotor windings, and $R_{\text{r}} \frac{1-s}{s}$ represents the developed torque.

The equivalent circuit of Fig. 4.44(a) can be simplified to Fig. 4.44(b) using Thevenin's theorem, where:

$$\dot{V}_{\text{th}} = \frac{j\omega_{\text{s}}L_{\text{m}}\dot{V}_{\text{s}}}{R_{\text{s}} + j\omega_{\text{s}}(L_{\text{sl}} + L_{\text{m}})} = \frac{\omega_{\text{s}}L_{\text{m}}\dot{V}_{\text{s}}}{\sqrt{R_{\text{s}}^2 + (\omega_{\text{s}}(L_{\text{sl}} + L_{\text{m}}))^2}} \angle \left(90° - tg^{-1}\left(\frac{\omega_{\text{s}}(L_{\text{sl}} + L_{\text{m}})}{R_{\text{s}}}\right)\right)$$

$$Z_{\text{th}} = R_{\text{th}} + j\omega_{\text{s}}L_{\text{th}} = \frac{(R_{\text{s}} + j\omega_{\text{s}}L_{\text{sl}})j\omega_{\text{s}}L_{\text{m}}}{(R_{\text{s}} + j\omega_{\text{s}}L_{\text{sl}}) + j\omega_{\text{s}}L_{\text{m}}} \tag{4.82}$$

$$= \frac{R_{\text{s}}\,\omega_{\text{s}}^2 L_{\text{m}}(L_{\text{sl}} + L_{\text{m}}) - R_{\text{s}}\,\omega_{\text{s}}^2 L_{\text{m}}L_{\text{sl}} + j\left(R_{\text{s}}^2\omega_{\text{s}}L_{\text{m}} + \omega_{\text{s}}^3 L_{\text{sl}}L_{\text{m}}(L_{\text{sl}} + L_{\text{m}})\right)}{R_{\text{s}}^2 + (\omega_{\text{s}}(L_{\text{sl}} + L_{\text{m}}))^2}$$

Thus,

$$R_{th} = \frac{R_s\,\omega_s^2\,L_m^2}{R_s^2 + (\omega_s(L_{sl}+L_m))^2} \quad \text{and} \quad L_{th} = \frac{\left(R_s^2 L_m + \omega_s^2 L_{sl} L_m (L_{sl}+L_m)\right)}{R_s^2 + (\omega_s(L_{sl}+L_m))^2} \tag{4.83}$$

Since $\omega_s L_m \gg \omega_s L_{sl}$, $\omega_s L_m \gg R_s$:

$$\dot{V}_{th} \cong \frac{\omega_s L_m \dot{V}_s}{\sqrt{R_s^2 + (\omega_s L_m)^2}} \angle \left(90° - tg^{-1}\left(\frac{\omega_s L_m}{R_s}\right)\right) \cong \dot{V}_s \tag{4.84}$$

$$R_{th} \cong \frac{R_s\,\omega_s^2\,L_m^2}{R_s^2 + (\omega_s L_m)^2} \cong R_s \quad \text{and} \quad L_{th} \cong \frac{\left(R_s^2 L_m + \omega_s^2 L_{sl} L_m^2\right)}{R_s^2 + (\omega_s L_M)^2} \cong L_{sl} \tag{4.85}$$

From the Thevenin equivalent circuit of Fig. 4.44(b), the rotor current is solved as:

$$\dot{I}_r = \frac{\dot{V}_{th}}{R_{th} + \dfrac{R_r}{s} + j\omega_s(L_{th}+L_{rl})} = \frac{\dot{V}_s}{R_s + \dfrac{R_r}{s} + j\omega_s(L_{sl}+L_{rl})}$$

$$= \frac{sV_s}{\sqrt{(sR_s+R_r)^2 + (s\omega_s(L_{sl}+L_{rl}))^2}} \angle \phi - tg^{-1}\left(\frac{\omega_s(L_{sl}+L_{rl})}{sR_s+R_r}\right) \tag{4.86}$$

Since the absorbed electrical power of the rotor is $3I_r^2 \frac{1-s}{s} R_r$, which is equal to the output mechanical power, $P_{me} = \tau_d \cdot \omega_m = \tau_e \cdot \omega_r$, the torque–speed relationship of the AC induction motor can be obtained as follows:

$$\begin{aligned}
\tau_e = \frac{P_{me}}{\omega_r} &= \frac{3\,I^{r2} R_r \dfrac{1-s}{s}}{\omega_r} = \frac{3s^2\,V_s^2}{\omega_r\left[(sR_s+R_r)^2 + (s\omega_s(L_{sl}+L_{rl}))^2\right]} R_r \frac{1-s}{s} \\[2mm]
&= \frac{3s\,V_s^2 R_r}{\dfrac{\omega_r}{1-s}\left[(sR_s+R_r)^2 + (s\omega_s(L_{sl}+L_{rl}))^2\right]} = \frac{3s\,V_s^2 R_r}{\omega_s\left[(sR_s+R_r)^2 + (s\omega_s(L_{sl}+L_{rl}))^2\right]} \\[2mm]
&= \frac{3ps\,V_s^2 R_r}{4\pi f\left[(sR_s+R_r)^2 + (s\omega_s(L_{sl}+L_{rl}))^2\right]}
\end{aligned} \tag{4.87}$$

where τ_e is the produced electromagnetic torque (Nm), ω_r is the electrical angular speed (rad/s) of the rotor, τ_d is the developed mechanical torque (Nm) at the shaft, p is the number of magnetic poles created by the stator winding, V_s is the phase stator voltage, and f is the frequency (Hz) of the input voltage.

If the motor speed is represented by n in terms of rpm (revolutions per minute), then $n = \frac{60}{2\pi}\omega_r$; thus, the motor speed, n, can be expressed as:

$$n = \frac{60}{2\pi}(1-s)\omega_s = (1-s)\frac{60}{2\pi}\frac{2\omega}{p} = (1-s)\frac{60}{2\pi}\frac{2\cdot2\pi f}{p} = (1-s)\frac{120f}{p} \qquad (4.88)$$

From Eq. 4.87, we have the typical characteristic curve of torque–speed for an induction motor shown in Fig. 4.45.

4.4.2.2 AC Induction Motor Control

The torque–speed relationship equation (Eq. 4.87) implies the following three methods that can be used to control the torque and speed of a squirrel cage induction motor:

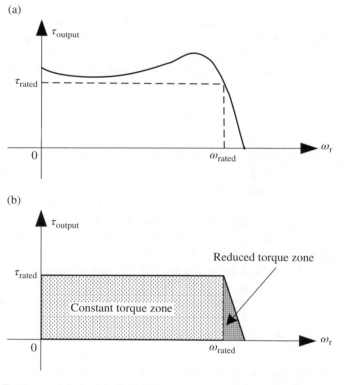

Figure 4.45 Torque–speed characteristics of an AC induction motor. (a) Nature torque-speed characteristics (b) The specified torque-speed characteristics

1. **Change the number of poles**

   Since the synchronous speed of the rotating magnetic field is determined by the number of poles of an induction motor, the speed of an induction motor can be changed by varying the number of magnetic poles by physically reorganizing the stator windings. This method is generally suitable for applications in which two constant speed settings are needed.

2. **Change the magnitude of the supplied voltage**

   From Eq. 4.87, it can be seen that the developed torque will decrease when the supplied voltage is reduced, so the motor speed will decrease. This method is normally applied to fan- or pump-type loads, as their required torque is also reduced with a reduction in speed.

3. **Combine torque and speed control by adjusting the supply voltage and frequency**

   This control method makes use of power electronic switches to achieve the desired motor torque to meet the load torque requirement. This control method, also called the vector control method, provides independent control of air-gap flux and torque, so that a squirrel cage induction motor can be driven as a DC motor. The principle of vector control and its application to the induction motor is described in the following part of this section.

From the previous discussion, we know that the induction motor is singly excited and the rotor field is induced by the stator field; in other words, compared with DC or BLDC motors, the stator and rotor magnetic fields of the induction motor are not fixed orthogonal to one another. As we know, the developed torque of a DC motor can be expressed by the following equation:

$$\tau_{DC} = K\Psi_e I_a = K_\tau I_f I_a \tag{4.89}$$

where I_f is the field current to generate the field flux linkage and I_a is the armature current to generate armature flux linkage. These two flux linkages are naturally orthogonal in a DC motor, so it is easy to control and has a fast transient response.

The idea of vector control of an induction motor is to decouple the stator current into a vector with two orthogonal components: the air-gap flux current or *d-axis*, i_{ds} and the torque current or *q-axis*, i_{qs}. The idea can be implemented by two transformations that first transform a three-phase stationary reference frame of variables (a_s–b_s–c_s) into a two-phase stationary reference frame of orthogonal variables (d_s–q_s), and then synchronously rotate the reference frame variables (D–Q) or the DC quantities.

Assume that i_{as}, i_{bs}, and i_{cs} are the instantaneous balanced three-phase stator currents:

$$i_{as} + i_{bs} + i_{cs} = 0 \tag{4.90}$$

this three-phase stationary reference frame (a, b, c) can be transformed into the two-phase stationary reference frame (d, q), shown in Fig. 4.46(a) by the following transformation:

$$
\begin{bmatrix} i_{qs} \\ i_{ds} \\ i_{0s} \end{bmatrix} = \frac{2}{3} \begin{bmatrix} \cos\theta & \cos\left(\theta-\dfrac{2\pi}{3}\right) & \cos\left(\theta+\dfrac{2\pi}{3}\right) \\ \sin\theta & \sin\left(\theta-\dfrac{2\pi}{3}\right) & \sin\left(\theta+\dfrac{2\pi}{3}\right) \\ \dfrac{1}{2} & \dfrac{1}{2} & \dfrac{1}{2} \end{bmatrix} \begin{bmatrix} i_{as} \\ i_{bs} \\ i_{cs} \end{bmatrix} \tag{4.91}
$$

where i_{0s} is called the zero-sequence component and makes the matrix invertible so that the two frames can be one-to-one transformed. This transformation can be applied to all the quantities of an induction motor including the phase currents, the voltages, and the flux linkages.

If we ignore the zero-sequence component and set $\theta = 0$, that is, the q_s-*axis* is aligned with the a_s-*axis*, the transformation relation in Eq. 4.91 simplifies to:

$$
\begin{bmatrix} i_{qs} \\ i_{ds} \end{bmatrix} = \frac{2}{3} \begin{bmatrix} \cos(0) & \cos\left(-\dfrac{2\pi}{3}\right) & \cos\left(\dfrac{2\pi}{3}\right) \\ \sin(0) & \sin\left(-\dfrac{2\pi}{3}\right) & \cos\left(\dfrac{2\pi}{3}\right) \end{bmatrix} \begin{bmatrix} i_{as} \\ i_{bs} \\ i_{cs} \end{bmatrix} = \frac{2}{3} \begin{bmatrix} 1 & -\dfrac{1}{2} & -\dfrac{1}{2} \\ 0 & \dfrac{\sqrt{3}}{2} & -\dfrac{\sqrt{3}}{2} \end{bmatrix} \begin{bmatrix} i_{as} \\ i_{bs} \\ i_{cs} \end{bmatrix} = \mathbf{C} \begin{bmatrix} i_{as} \\ i_{bs} \\ i_{cs} \end{bmatrix}
$$

$$\tag{4.92}$$

Thus, the (a, b, c) to (d, q) transformation transforms a three-phase time-varying system into a two-phase time-varying system in which two variables are orthogonal. As shown in Fig. 4.46(c), the d and q components in the two-phase stationary reference frame are still AC signals.

In vector control, all quantities must be stated in the same reference frame. Since the components i_{ds} and i_{qs} still depend on time and speed, the stationary reference frame is not suitable for the control process. In order to convert these two orthogonal AC signals to DC quantities, these components must be transformed from the stationary frame to the synchronous D–Q frame that rotates at the same speed as the angular frequency of the phase current by the following transformation:

$$
\begin{bmatrix} i_{Qs} \\ i_{Ds} \end{bmatrix} = \begin{bmatrix} \cos(\theta_f) & \sin(\theta_f) \\ -\sin(\theta_f) & \cos(\theta_f) \end{bmatrix} \begin{bmatrix} i_{qs} \\ i_{ds} \end{bmatrix} = D \begin{bmatrix} i_{qs} \\ i_{ds} \end{bmatrix}; \quad \theta_f = \omega_s t \tag{4.93}
$$

The D and Q components in Eq. 4.93 actually rotate at the synchronous speed ω_s with respect to the (d, q) axes at the angle $\theta_f = \omega_s t$, which is equivalent to mounting the two-phase (d, q) windings on the hypothetical frame rotating at the synchronous speed, resulting in the conversion of the sinusoidally varying waveforms to the DC vector shown in Fig. 4.47. After obtaining these DC quantities, each component of the DC vector can be controlled independently to generate the desired torque, which mimics the torque equation of a DC motor.

i_{Ds} = Stator direct-axis current (A)
i_{Qr} = Rotor quadrature-axis current (A)
i_{Dr} = Rotor direct-axis current (A)
Ψ_{Qs} = Stator quadrature-axis flux linkage (Wb)
Ψ_{Ds} = Stator direct-axis flux linkage (Wb)
Ψ_{Qr} = Rotor quadrature-axis flux linkage (Wb)
Ψ_{Dr} = Rotor direct-axis flux linkage (Wb)
R_s = Stator phase resistance (Ω)
R_r = Rotor phase resistance (Ω)
L_s = Stator phase inductance (H)
L_r = Rotor phase inductance (H)
L_m = Magnetizing inductance (H)
ω_s = Synchronous speed (rad/s)
ω_r = Electrical rotor speed (rad/s)
ω_m = Mechanical rotor speed (rad/s)
p = Number of poles
τ_d = Developed torque (Nm)
τ_{load} = Load torque (Nm)
τ_{loss} = Lumped loss torque (Nm)
J = Rotor inertia (kg.m^2)

The above vector or field-oriented control method was invented in the early 1970s, and it enabled an induction motor to be controlled like a separately excited DC motor. Vector control initiated a renaissance in the performance control of AC drives. With advances in modern power electronics and microelectronics, as well as the development of control techniques, vector control has been accepted as the standard control method in AC drives and has been widely applied to HEV/EV propulsion systems as well.

An implementation block diagram of the vector control method is shown in Fig. 4.48 for HEV/EV applications. In this example, once it has received the position signal for the accelerator or brake pedal, the vehicle controller calculates the torque demand based on the vehicle's speed, the DC link voltage, and the motor speed, and sends a torque request to the motor controller. The motor controller then calculates the desired stator direct-axis current, i_{Ds}^*, and the quadrature-axis current, i_{Qs}^*, in response to the corresponding command τ_d^* from the vehicle controller. These two desired current values are compared with the actual stator direct-axis current, i_{Ds}, and the quadrature-axis current, i_{Qs}. The errors between the desired and actual currents, $i_{Ds\_error} = i_{Ds}^* - i_{Ds}i_{Qs}^*$ and $i_{Qs\_error} = i_{Qs}^* - i_{Qs}$, are fed to the PID controller or an advanced controller to calculate the control requests $i_{Ds}^{req'd}$ and $i_{Qs}^{req'd}$. The motor controller makes use of two stages of inverse transformation to generate the three-phase request signals $i_{as}^{req'd}$, $i_{bs}^{req'd}$, and $i_{cs}^{req'd}$ to produce a PWM switching sequence for the three-phase inverter. At the same time, the measured rotor flux position signal, θ_r, is sent by the inverse transformation block to ensure that current $i_{Ds}^{req'd}$ has the correct alignment with the rotor flux Ψ_{Dr} and is perpendicular to current $i_{Qs}^{req'd}$. On the other hand, the inverter generates current i_a,

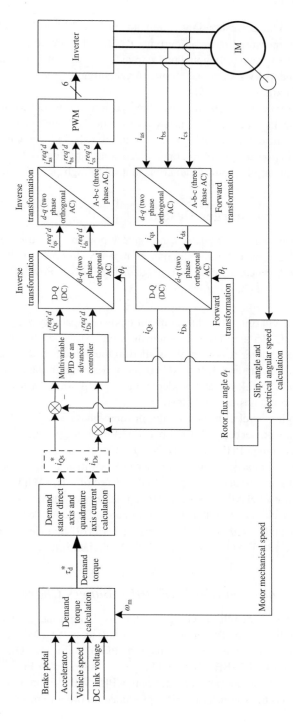

Figure 4.48 Block diagram of vector control for an AC induction motor in HEV/EV applications

i_b, and i_c in response to the corresponding command currents $i_{as}^{req'd}$, $i_{bs}^{req'd}$, and $i_{cs}^{req'd}$ from the motor controller. The measured terminal currents i_a, i_b, and i_c are converted to i_{ds} and i_{qs} through a forward three-to-two orthogonal phase transformation. They are further converted to i_{Ds} and i_{Qs} components by a stationary frame to synchronous frame transformation with the measured rotor flux position signal θ_f.

In general, the vector control technique uses the dynamic equivalent circuit of the induction motor to decouple the stator current into two perpendicular components: one generating the magnetic field and the other producing torque so they can be independently controlled as a DC drive. Progress in power electronics, microelectronics, and control techniques has substantially supported the application of induction motors to high-performance drives.

4.5 Plug-in Battery Charger Design

Plug-in hybrid electric vehicles (PHEVs) and battery-powered electric vehicles (BEVs) have been developed and have entered the market recently. In order to charge the battery of a PHEV/BEV, a high-efficiency charger is needed. The Electric Power Research Institute defined the following three charging levels:

- **Level 1:** This plug-in charging level has a charging power capability of up to 1.5 kW using a standard 120 V AC, 15 amp branch circuit. It is typically used for residential garage charging. This level charger is normally installed in the vehicle as the on-board charger and is connected to the standard 120 V AC power outlet through the provided plug-cord set. For a PHEV with 40 miles of charge-depleting range, it normally takes 8–10 hours to charge the battery to the desired SOC level.
- **Level 2:** This charging level utilizes a single- or three-phase 240 V AC, 40 amp branch circuit with a charging power capability of up to 9.6 kW. It is the preferred charging method for high electric mile range PHEVs and BEVs. This level charger can be installed in the vehicle as the on-board charger or at vehicle service facilities as an off-board charger. It generally takes 2–3 hours to achieve a 50% SOC charge of a 16 kWh battery pack.
- **Level 3:** This charging level is designed for commercial service stations, so the PHEV/BEV battery system can be charged to full in 10 to 20 minutes. The level 3 charger is the highest power level charger, and is also called a fast charger; it is an off-board charger and normally employs three-phase 480 V AC and has a 50–150 kW variable charging power capability.

4.5.1 Basic Configuration of a PHEV/BEV Battery Charger

No matter what the level of the plug-in charger is, the operating principle and the basic system configuration are the same. The entire charger consists of an EMI filter, a full-wave rectifier, in-rush protection, power factor correction (PFC), and a DC–DC converter. The general architecture of a PHEV/BEV battery charger is illustrated in Fig. 4.49.

There are many semiconductor power switches in a PHEV/BEV battery charger, and the switch-mode operation is the source of the EMI produced. In order to meet the requirements

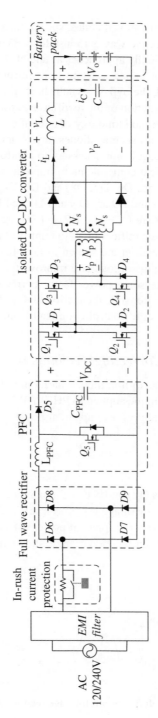

Figure 4.49 General architecture of a PHEV/BEV battery charger

of EMI regulations, standards, and national codes, special measures must be taken to limit the EMI level. The most cost-effective way is to install an EMI filter at the AC input port.

Since a discharged capacitor initially approaches a state of short-circuit when a voltage is applied to it, a pre-charge resistor is necessary to limit in-rush current during power-up. Once the pre-charge is completed, the bypass switch shorts the pre-charge resistor to eliminate undesired power loss. The function of the rectifier is to transfer sinusoidal AC voltage into DC voltage, and the most common topology of a rectifier circuit is the full-bridge rectifier. Any type of power electronics such as diodes, thyristors, or MOSFETs can perform the transformation; however, since it is not necessary to control the output voltage, diodes are the easiest and most economical components to implement the function. The rectifier shown in Fig. 4.49 comprises diodes D6, D7, D8, and D9. The operating principle is that when the input AC is positive, D6 and D9 are ON and D7 and D8 are OFF while when the input AC is negative, D6 and D9 are OFF and D7 and D8 are ON. In most cases, a filter capacitor is placed in parallel to the rectifier output to improve the quality of the DC voltage. The waveforms of a full-bridge rectifier are shown in Fig. 4.50.

In order to minimize harmonic distortions and improve the power factor, most PHEV/BEV battery chargers incorporate some power factor correction devices. Following the PFC, the DC–DC converter finally converts the rectified DC voltage to the desired DC voltage to charge the battery based on the control command from the battery system.

4.5.2 Power Factor and Correcting Techniques

Power factor correction, as a means of minimizing harmonic distortions and improving the power factor, is an important technique applied to PHEV/BEV battery chargers. The power factor is defined as the ratio of real power to apparent power as in the following equation:

$$PF = \frac{P}{S} \tag{4.104}$$

where P is the real power in watts (W) and S is the apparent power in volt-amperes (VA).

If we assume that both current and voltage waveforms are ideal sinusoidal waveforms, the three power components, real power (P), apparent power (S), and reactive power (Q), with units of volt-amperes (VA) form a right-angled triangle shown in Fig. 4.51, The relationship between them is:

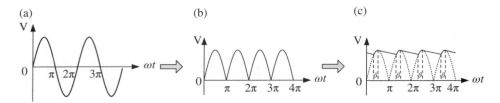

Figure 4.50 Voltage waveforms of a full-bridge rectifier with a filter capacitor

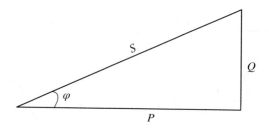

Figure 4.51   The AC power triangle

$$S^2 = P^2 + Q^2 \quad \Rightarrow \quad P = S \cdot \cos(\phi) \quad \Rightarrow \quad PF = \cos(\phi) \tag{4.105}$$

$$S = V_{RMS} I_{RMS} \quad \Rightarrow \quad P = V_{RMS} I_{RMS} \cdot \cos(\phi) \tag{4.106}$$

The φ in the equation is also the angle between the input voltage and the current, and it is defined as the phase lag/lead angle or displacement angle in the ideal sinusoidal waveform case.

If we consider the case of a PHEV/BEV charger where the input voltage stays undistorted and the current waveform is periodic non-sinusoidal, the real power and apparent power are equal to:

$$S = V_{RMS} I_{RMS} \quad \text{and} \quad P = V_{RMS} I_{1,RMS} \cdot \cos(\phi) \tag{4.107}$$

where $I_{1,RMS}$ is the RMS (root mean square) value of the fundamental frequency component of the current. As a non-sinusoidal waveform contains harmonic components in addition to the fundamental frequency, a distortion factor k_d is further defined to quantify the impact of the harmonic components on the real power:

$$k_d = \frac{I_{1,RMS}}{I_{RMS}} \tag{4.108}$$

Applying the Fourier transform to the current, the RMS value of the current can be obtained as:

$$I_{RMS} = \sqrt{I_0^2 + I_{1,RMS}^2 + I_{2,RMS}^2 + \cdots + I_{n,RMS}^2} \tag{4.109}$$

where I_0 is the DC component of the current and $I_0 = 0$ as the distorted current is still a pure AC waveform, $I_{1,RMS}$ is the RMS value of the fundamental component of the current, and $I_{2,RMS} \cdots I_{n,RMS}$ are the RMS values of the harmonic components of the current.

Figure 4.52 shows the input current waveform of a full-bridge rectifier with a filter capacitor and without a PFC corresponding to Fig. 4.50(c). Since the typical rectifier only charges the capacitor when the input voltage is greater than the voltage on the capacitor, the input current flows only near to the peak input voltage; therefore, the

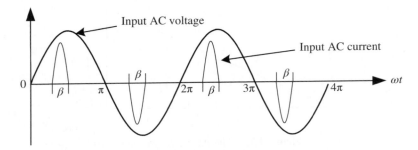

Figure 4.52   Input voltage and current of the full-bridge rectifier with filter capacitor

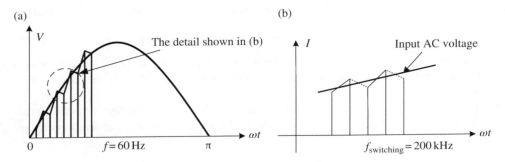

Figure 4.53   Operating principle of a PFC

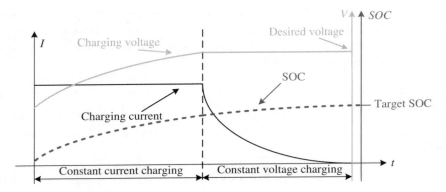

Figure 4.54   Process required to charge a PHEV/BEV battery

value of $I_{1,\mathrm{RMS}}$ is very small, which results in the actual power factor generally being less than 0.5:

$$PF = \frac{P}{S} = \frac{V_{\mathrm{RMS}}I_{1,\mathrm{RMS}}}{V_{\mathrm{RMS}}I_{\mathrm{RMS}}}\cos(\varphi) = \frac{I_{1,\mathrm{RMS}}}{I_{\mathrm{RMS}}}\cos(\varphi) < 0.5 \quad \text{normally} \tag{4.110}$$

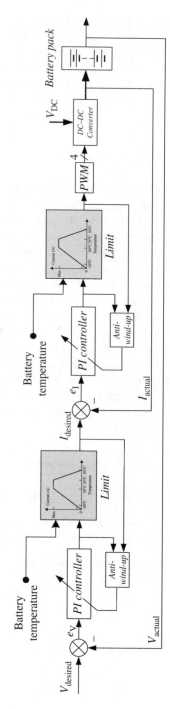

Figure 4.55 PHEV/BEV charging control scheme

There are many methods of improving the power factor; however, the most common technique used in AC plug-in chargers is to apply a PFC in the circuit shown in Fig. 4.49. The objective of inserting a PFC is to extend the current conduction angle β to 180°, while the operating principle is similar to that of the boost DC–DC converter shown in Fig. 4.19. By means of inductor L_{PFC} and switching Q5 to a higher frequency, the input AC current waveform closely tracks the rectified AC voltage shown in Fig. 4.53.

4.5.3 Controls of a Plug-in Charger

The PHEV/BEV battery system is normally charged through two stages of operation: the constant current mode and the constant voltage mode, as shown in Fig. 4.54. The switching point for the charging mode is controlled by the charging termination voltage sent by the battery system controller. In the constant current mode, the charging current is set based on the battery temperature and the AC power outlet capability, while in the constant voltage mode, the charging termination voltage is normally determined by the open-circuit voltage associated with the target battery SOC. To implement constant current and constant voltage operation, the charging control scheme normally consists of two closed loops: a voltage loop and a current loop. A common charging control scheme is shown in Fig. 4.55, where the voltage loop is positioned as an outer loop and the PI controller generates the desired current to the inner loop based on the difference between the desired charging voltage and the actual battery voltage. The PI controller cooperates with a saturation block and an anti-wind-up block to limit the generated current reference. In the current loop, the difference between the generated desired current and the actual charging current is fed to the PI controller to generate a PWM signal to turn the corresponding power switches on or off to obtain the desired battery-charging curve.

References

Åstrom, K. and Hägglund, T. *PID Controllers: Theory, Design and Tuning*, 2nd edition, The Instrument Society of America, 1995.

Erickson, R. W. and Maksimovic, D. *Fundamentals of Power Electronics*, 2nd edition, Springer, 2001.

Mohan, N., Undeland, T. P., and Robbins, W. P. *Power Electronics – Converters, Applications, and Design*, 3rd edition, John Wiley & Sons, Inc., 2003.

Rashid, M. H. *Power Electronics Circuits, Devices, and Applications*, 3rd edition, Prentice-Hall of India, New Delhi, 2008.

5

Energy Storage System Modeling and Control

5.1 Introduction

Advanced energy storage systems (ESSs) are providing new opportunities in the automotive industry. The algorithms relating to ESSs now play a central role in hybrid electric vehicle applications; for example, the SOC of the battery system is often a critical factor in hybrid vehicle performance, fuel economy, and emissions. From simple voltage, current, and temperature measurements that establish the values of operating variables, to the SOC estimation, power capability prediction, and the state of life (SOL) of the battery system, battery algorithms have permeated every aspect of a hybrid vehicle system. An ESS algorithm functionality diagram is shown in Fig. 5.1, and a short description is given below.

- **Initialization.** Initialization sets all the necessary parameters and variables to ensure that the algorithm starts properly. Battery cells may still have some dynamic functionality during the key-down period, which mainly includes Li-ion diffusion and 'self-discharge'. During the initialization, the stored previous cycle data need to be reloaded, and these include the previous SOC, the previous SOL, the estimated open-circuit voltage, and the key-off time. The controller should also check the system status to see if the level of self-discharge is too high or other fault conditions exist.
- **Battery measurement data acquisition.** The ESS controller receives cell voltages, temperatures, and pack voltage and current in real time. In addition, the measurements have to be synchronized and meet accuracy requirements.
- **SOC estimation.** Once the pack voltages, current, and temperatures are available, the ESS SOC must be estimated based on these measurements. The SOC estimation algorithm can

Hybrid Electric Vehicle System Modeling and Control, Second Edition. Wei Liu.
© 2017 John Wiley & Sons Ltd. Published 2017 by John Wiley & Sons Ltd.

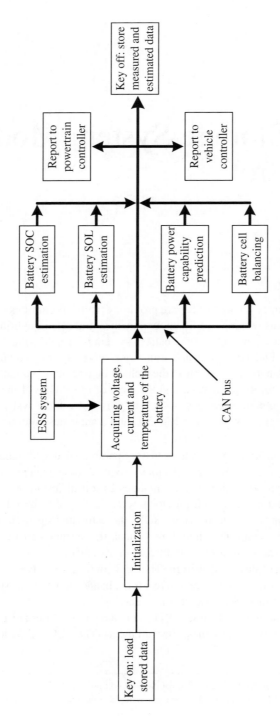

Figure 5.1 Diagram of required ESS algorithms for HEV/PHEV/EV application

be developed from many approaches for various batteries and applications. This chapter presents several SOC estimation algorithms, including current-integration-based, voltage-based, state-space estimation, and artificial-intelligence-based algorithms.

- **SOL estimation.** Since the battery's capacity and other parameters vary over its lifetime, the SOL of an ESS has to be estimated in real time to ensure vehicle safety and optimal performance. While SOC estimation techniques have been published for several types of batteries, SOL determination is less well-reported, primarily because SOL is a qualitative rather than quantitative measure of the aging of an ESS. The SOL determination should be able to provide a battery's nominal capacity over extended lengths of time. In most cases, SOL is estimated by monitoring the degradation of specific battery model parameters or by close inspection of the operational phenomena. Different SOL estimation approaches are introduced in this chapter.

- **Maximal power capability prediction.** The maximal charge/discharge power capabilities of an ESS must be known at any time, ensuring that the battery voltage and current limits will not be violated. The maximal charge/discharge power capabilities vary with SOC, temperature, and the state of health of the battery system. In order to achieve the optimal fuel economy, it is essential to predict instantaneous, short-term (e.g. two second), long-term (e.g. ten second), and continuous charge/discharge power capabilities of the battery system. In this chapter, power prediction algorithms are also addressed.

- **Cell balance.** The deviations in cell capacity, self-discharge rate, and internal resistance of an ESS will result in the cells in a series string being unbalanced, even though all cells start with the same SOC value. In addition, cell temperature differences during operation can also lead to imbalances from cell to cell. The imbalances may decrease ESS capacity, shorten battery life, or even create potential safety hazards, so it is necessary to keep all cells balanced.

Overall, from a hybrid vehicle application perspective, there exist several paramount challenges for ESS modeling and control. These are driven by the interplay of electronic technologies, the complexity of electrochemistry, and the demands of current and future applications. To articulate these challenges, this chapter introduces ESS modeling, control, and the algorithms required in hybrid electric vehicle applications.

5.2   Methods of Determining the State of Charge

In HEV/EV applications, the SOC acts as the gas gauge function analogous to the fuel tank in a vehicle, giving the driver an indication of how much energy is left in the ESS to power the vehicle before it will need recharging. The SOC of a battery can be defined as a percentage of the available amount of energy over its maximal achievable amount. As the accuracy and robustness of SOC estimation plays a crucial role in vehicle performance and safety, many SOC estimation methods have been developed. Some are particularly based on cell chemistry, but most rely on measurable variables that vary with SOC, such as battery voltage, current, and temperature. Because battery performance degrades when the battery ages, the aging factors must be taken into account if a higher-fidelity and accurate SOC estimation is required.

5.2.1 Current-based SOC Determination Method

The current-based SOC estimation method is also called the coulomb counting method. This is the most common technique for SOC calculation. If an initial SOC_i is given, the SOC at time t can be calculated based on the following current integration:

$$SOC(t) = SOC_i + \frac{\eta_{\text{bat}}(i(t), T)}{Cap(T)3600} \int_{t_0}^{t} i(t) dt \tag{5.1}$$

where $Cap(T)$ and $\eta_{\text{bat}}(i(t), T)$ are the capacity and coulombic efficiency of the battery, which are functions of battery temperature, T. An implementation diagram for this approach is shown in Fig. 5.2. In the diagram, the capacity of the battery is found by a 1D look-up table, and the coulombic efficiency is indexed by a 2D look-up table.

The current-based SOC estimation method is relatively easy and reliable as long as the following three requirements are met under operational conditions:

- The initial SOC point can be set precisely;
- The current going through the battery can be measured accurately;
- The capacity of the battery can be obtained correctly in real time.

Although this approach has been applied widely, the following three main complications arise in hybrid vehicle applications.

1. **The calculated SOC is off due to measurement noise and errors**
   Based on the nature of the algorithm, the incorrect current measurement could add up to a large error if the battery has been used for a long period of time. In addition, the coulomb counting method only depends on the external current flow and does not take account of self-discharge current. Taking a 6 Ahr Li-ion battery as an example, if the current sensor

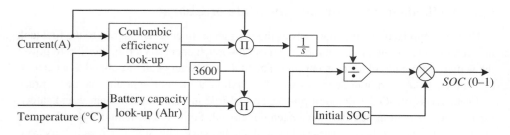

Figure 5.2 Diagram of current integration-based *SOC* estimation

(a)

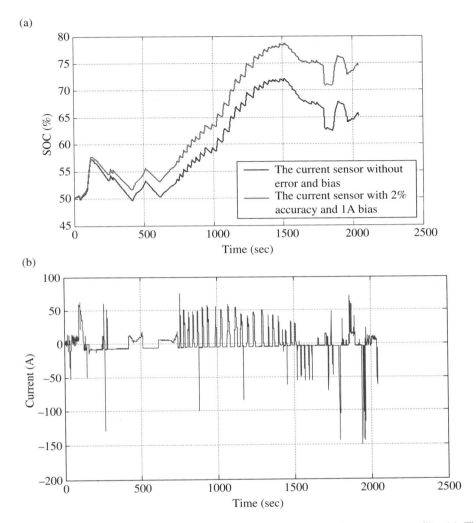

(b)

Figure 5.3 Estimation error caused by sensor accuracy for the given current profile. (a) The calculated SOC; (b) the current profile

has ±2% accuracy and a 1 A bias, over 10% SOC error will result, as shown in Fig. 5.3(a), based on the given drive cycle shown in Fig. 5.3(b).

2. **The calculated SOC is off due to power losses during cycling**

As a small portion of current has to go through the measurement circuit, a small power loss will take place in the circuit, which results in a small amount of SOC error. Taking the above example again, the measurement circuit will cause about 0.5% SOC estimation error, as shown in Fig. 5.4 at the end point of the given current profile shown in Fig. 5.3(b).

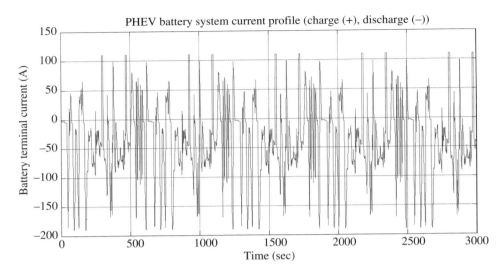

Figure 5.7   Example of a charge-depleting current profile of a PHEV battery system

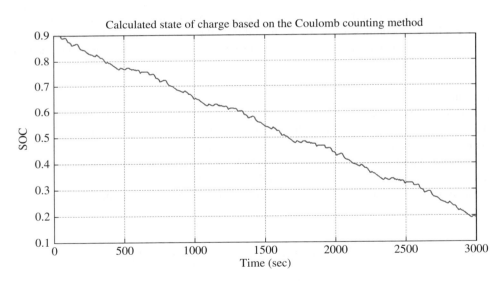

Figure 5.8   Calculated SOC based on the Coulomb counting method over the given current profile

5.2.2 Voltage-based SOC Determination Method

The voltage-based SOC estimation method determines the SOC based on the measured or calculated open-circuit voltage (OCV) of the battery. The OCV is the terminal voltage of a battery when there is no external load connected and the battery has been rested for a certain period of time. Sometimes, it is given the symbol V_{oc}. The V_{oc} of a battery can be described by the following modified Nernst equation:

$$V_{oc,cell} = V_o - \frac{R_g T}{n_e F} \ln(Q)$$

$$Q = f(SOC)$$

(5.3)

where:

F is the Faraday constant, the number of coulombs per mole of electrons: $F = 9.6485309 \times 10^4 \, C \, mol^{-1}$

n_e is the number of electrons transferred in the cell reaction

R_g is the universal gas constant: $R_g = 8.314472 \, J \, K^{-1} \, mol^{-1}$

T is the actual temperature in kelvin

V_o is the standard cell potential

Q is the reaction quotient, which is a function of SOC

Figure 5.9 shows the relationship between V_{oc} and the SOC of a Li-ion cell at a constant temperature. The technical challenge of the voltage-based SOC estimation method is how to

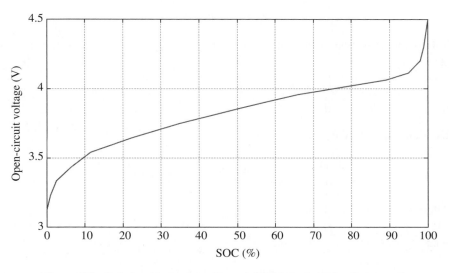

Figure 5.9 Relationship between V_{oc} and the SOC of a Li-ion battery cell

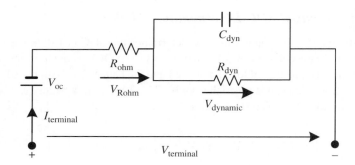

Figure 5.10   Electrical circuit equivalent model of a battery

obtain V_{oc} accurately and robustly under operational conditions whereby noise is present alongside voltage, current, and temperature measurements.

   In practical hybrid electric vehicle system applications, the battery system can be empirically modeled by an *RC* equivalent circuit shown in Fig. 5.10, and the V_{oc} in the model is estimated through a recursive estimation technique introduced in Appendix A and based on the measured battery terminal voltage, current, and temperature. From the *RC* circuit model, the following equations hold at the given SOC and temperature:

$$V_{terminal} = V_{oc} + V_{Rohm} + V_{dynamic} = V_{oc} + R_{ohm} \cdot I_{terminal} + V_{dynamic} \tag{5.4}$$

$$C_{dyn}\frac{dV_{dynamic}}{dt} + \frac{V_{dynamic}}{R_{dyn}} = I_{terminal}$$

$$\text{i.e.} \quad \frac{dV_{dynamic}}{dt} + \frac{V_{dynamic}}{R_{dyn}C_{dyn}} = \frac{I_{terminal}}{C_{dyn}} \tag{5.5}$$

Taking the derivative of Eq. 5.4 and assuming that V_{oc} is constant in a short period of time, we get:

$$\frac{dV_{terminal}}{dt} = \frac{dV_{oc}}{dt} + \frac{dV_{Rohm}}{dt} + \frac{dV_{dynamic}}{dt} = R_{ohm} \cdot \frac{dI_{terminal}}{dt} + \frac{dV_{dynamic}}{dt} \tag{5.6}$$

Dividing Eq. 5.4 by $R_{dyn}C_{dyn}$,

$$\frac{V_{terminal}}{R_{dyn}C_{dyn}} = \frac{V_{oc}}{R_{dyn}C_{dyn}} + \frac{R_{ohm}}{R_{dyn}C_{dyn}} \cdot I_{terminal} + \frac{V_{dynamic}}{R_{dyn}C_{dyn}} \tag{5.7}$$

Adding Eqs 5.6 and 5.7, we have:

$$\frac{dV_{terminal}}{dt} + \frac{V_{terminal}}{R_{dyn}C_{dyn}} = R_{ohm} \cdot \frac{dI_{terminal}}{dt} + \frac{dV_{dynamic}}{dt} + \frac{V_{oc}}{R_{dyn}C_{dyn}} + \frac{R_{ohm}}{R_{dyn}C_{dyn}} \cdot I_{terminal} + \frac{V_{dynamic}}{R_{dyn}C_{dyn}}$$

$$\tag{5.8}$$

Substituting Eq. 5.5 into 5.8, we get:

$$
\begin{aligned}
\frac{dV_{\text{terminal}}}{dt} + \frac{V_{\text{terminal}}}{R_{\text{dyn}}C_{\text{dyn}}} &= R_{\text{ohm}} \cdot \frac{dI_{\text{terminal}}}{dt} + \frac{I_{\text{terminal}}}{C_{\text{dyn}}} + \frac{V_{\text{oc}}}{R_{\text{dyn}}C_{\text{dyn}}} + \frac{R_{\text{ohm}}}{R_{\text{dyn}}C_{\text{dyn}}} \cdot I_{\text{terminal}} \\
&= R_{\text{ohm}} \cdot \frac{dI_{\text{terminal}}}{dt} + \left(\frac{R_{\text{dyn}} + R_{\text{ohm}}}{R_{\text{dyn}}C_{\text{dyn}}} \right) I_{\text{terminal}} + \frac{V_{\text{oc}}}{R_{\text{dyn}}C_{\text{dyn}}}
\end{aligned}
\tag{5.9}
$$

Discretizing Eq. 5.9, the following difference equation can be obtained:

$$
\begin{aligned}
\frac{V_{\text{terminal}}(k) - V_{\text{terminal}}(k-1)}{\Delta T} + \frac{V_{\text{terminal}}(k-1)}{R_{\text{dyn}}C_{\text{dyn}}} &= R_{\text{ohm}} \cdot \frac{I_{\text{terminal}}(k) - I_{\text{terminal}}(k-1)}{\Delta T} \\
&+ \left(\frac{R_{\text{dyn}} + R_{\text{ohm}}}{R_{\text{dyn}}C_{\text{dyn}}} \right) I_{\text{terminal}}(k-1) + \frac{V_{\text{oc}}}{R_{\text{dyn}}C_{\text{dyn}}} \\
\text{i.e.} \quad V_{\text{terminal}}(k) &= \left(1 - \frac{\Delta T}{R_{\text{dyn}}C_{\text{dyn}}} \right) \cdot V_{\text{terminal}}(k-1) + R_{\text{ohm}} I_{\text{terminal}}(k) \\
&+ \left(\frac{\left(R_{\text{dyn}} + R_{\text{ohm}}\right)\Delta T}{R_{\text{dyn}}C_{\text{dyn}}} - R_{\text{ohm}} \right) \cdot I_{\text{terminal}}(k-1) + \frac{\Delta T \cdot V_{\text{oc}}}{R_{\text{dyn}}C_{\text{dyn}}}
\end{aligned}
\tag{5.10}
$$

where ΔT is the sampling time. The above equation can be further expressed as:

$$
V_{\text{terminal}}(k) = \alpha_1 V_{\text{terminal}}(k-1) + \alpha_2 I_{\text{terminal}}(k) + \alpha_3 I_{\text{terminal}}(k-1) + \alpha_4 \tag{5.11}
$$

where

$$
\alpha_1 = \left(1 - \frac{\Delta T}{R_{\text{dyn}}C_{\text{dyn}}} \right), \alpha_2 = R_{\text{ohm}}, \alpha_3 = \frac{\left(R_{\text{dyn}} + R_{\text{ohm}}\right)\Delta T}{R_{\text{dyn}}C_{\text{dyn}}} - R_{\text{ohm}}, \quad \text{and} \quad \alpha_4 = \frac{\Delta T \cdot V_{\text{oc}}}{R_{\text{dyn}}C_{\text{dyn}}}
$$

The parameters α_1, α_2, α_3, and α_4 can be estimated in real time based on the measured battery terminal voltage V_{terminal} and the current I_{terminal} through a recursive estimation method. Once these parameters are estimated, the electrical circuit model parameters can be obtained by solving the following equations:

$$
\begin{cases}
R_{\text{ohm}} = \alpha_2 \\
V_{\text{oc}} = \dfrac{\alpha_4}{1 - \alpha_1} \\
R_{\text{dyn}} = \dfrac{\alpha_3 + \alpha_1 \cdot \alpha_2}{1 - \alpha_1} \\
C_{\text{dyn}} = \dfrac{\Delta T}{\alpha_3 + \alpha_1 \cdot \alpha_2}
\end{cases}
\tag{5.12}
$$

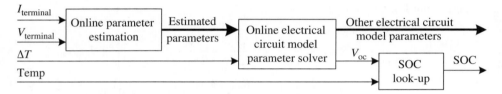

Figure 5.11   Diagram of the voltage-based SOC estimation method

Based on the obtained open-circuit voltage V_{oc}, the battery's SOC can be determined from the pre-established $V_{oc}-SOC$ look-up table.

Since the dynamics of a battery system are fundamentally nonlinear, resulting from an electrochemical reaction, there generally exist mismatches between actual behaviors and model outputs. It would be very challenging to obtain an accurate and robust V_{oc} under all operational conditions. However, due to its relative simplicity, the voltage-based SOC approach has been applied to practical vehicles (Verbrugge and Tate, 2004). The diagram of the voltage-based SOC estimation method is shown in Fig. 5.11.

Example 5.2
An HEV battery system consists of 32 4.5 Ahr Li-ion cells connected in series. Given the usage profiles shown in Fig. 5.12 and the V_{oc} versus SOC look-up chart given in Table 5.1, determine the battery's SOC using the voltage-based approach.

Solution:
Based on the model (Eq. 5.11), we have the following estimation equations:

$$
\begin{aligned}
V_{\text{terminal}}(k) &= \alpha_1 V_{\text{terminal}}(k-1) + \alpha_2 I_{\text{terminal}}(k) + \alpha_3 I_{\text{terminal}}(k-1) + \alpha_4 \\
&= \left[V_{\text{terminal}}(k-1) \;\; I_{\text{terminal}}(k) \;\; I_{\text{terminal}}(k-1) \;\; 1 \right] \boldsymbol{\theta}
\end{aligned}
\tag{5.13}
$$

where V_{terminal} and I_{terminal} are the actual measured battery terminal voltage and current, $\boldsymbol{\theta} = \begin{bmatrix} \alpha_1 & \alpha_2 & \alpha_3 & \alpha_4 \end{bmatrix}^{\mathrm{T}}$ is the parameter vector to be estimated. The following recursive least squares can be used to obtain these parameters:

$$
\hat{\boldsymbol{\theta}}(k) = \hat{\boldsymbol{\theta}}(k-1) + \mathbf{K}(k) \left[V_{\text{terminal}}(k) - \left[V_{\text{terminal}}(k-1) \;\; I_{\text{terminal}}(k) \;\; I_{\text{terminal}}(k-1) \;\; 1 \right] \hat{\boldsymbol{\theta}}(k-1) \right]
\tag{5.14}
$$

$$
\mathbf{K}(k) = \frac{\mathbf{P}(k-1) \left[V_{\text{terminal}}(k-1) \;\; I_{\text{terminal}}(k) \;\; I_{\text{terminal}}(k-1) \;\; 1 \right]^{\mathrm{T}}}{\lambda + \left[V_{\text{terminal}}(k-1) \;\; I_{\text{terminal}}(k) \;\; I_{\text{terminal}}(k-1) \;\; 1 \right] \mathbf{P}(k-1) \left[V_{\text{terminal}}(k-1) \;\; I_{\text{terminal}}(k) \;\; I_{\text{terminal}}(k-1) \;\; 1 \right]^{\mathrm{T}}}
\tag{5.15}
$$

$$
\mathbf{P}(k) = \frac{1}{\lambda} \left[\mathbf{I} - \mathbf{K}(k) \left[V_{\text{terminal}}(k-1) \;\; I_{\text{terminal}}(k) \;\; I_{\text{terminal}}(k-1) \;\; 1 \right] \right] \mathbf{P}(k-1)
\tag{5.16}
$$

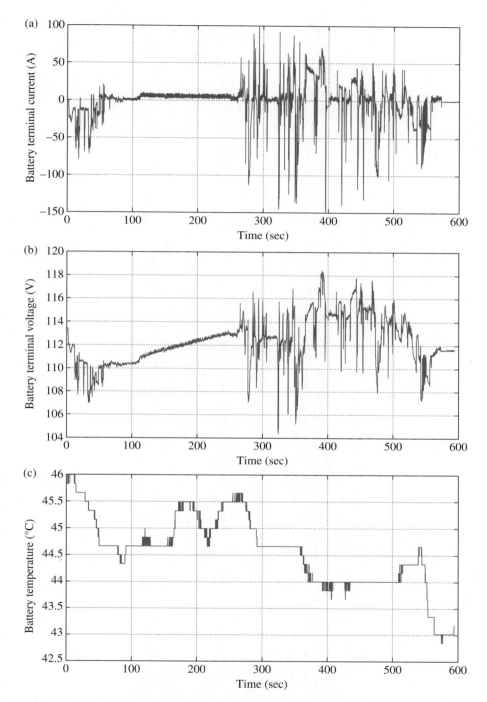

Figure 5.12 The battery usage profiles. (a) The battery's current (charge: +, discharge: −); (b) the battery's terminal voltage; (c) the battery's temperature

Table 5.1 The SOC (%) look-up chart indexed by V_{oc} (V) and temperature (°C) of a system with 32 4.5 Ahr Li-ion batteries

	88.6	95.0	101.1	106.6	111.4	113.3	115.8	117.8	119.7	121.6	123.5	125.4	127.4	129.0	130.6
−30	0	10	20	30	40	44	50	55	60	65	71	78	86	93	100
−20	0	10	20	30	40	44	50	55	60	65	71	78	86	93	100
−10	0	10	20	30	40	44	50	55	60	65	71	78	86	93	100
0	0	10	20	30	40	44	50	55	60	65	71	78	86	93	100
10	0	10	20	30	40	44	50	55	60	65	71	78	86	93	100
20	0	10	20	30	40	44	50	55	60	65	71	78	86	93	100
30	0	10	20	30	40	44	50	55	60	65	71	78	86	93	100
40	0	10	21	31	41	45	51	56	61	67	72	79	86	93	100
50	0	10	21	31	41	45	51	56	61	67	72	79	86	93	100

where λ is the calibratable forgetting factor which is less than 1. The parameters are estimated from the following recursive process:

$$\alpha_{1'}(k-1), \alpha_{2'}(k-1), \alpha_{3'}(k-1), \alpha_{4'}(k-1), \mathbf{P}(k-1) \xrightarrow{V_{\text{terminal}}(k), \ I_{\text{terminal}}(k)}$$

$$\mathbf{K}(k), \mathbf{P}(k) \xrightarrow{V_{\text{terminal}}(k), \ I_{\text{terminal}}(k)} \alpha_{1'}(k), \alpha_{2'}(k), \alpha_{3'}(k), \alpha_{4'}(k) \tag{5.17}$$

Once parameters $\boldsymbol{\theta} = [\alpha_1 \ \alpha_2 \ \alpha_3 \ \alpha_4]^{\mathrm{T}}$ have been obtained, the electrical model parameters $R_{\text{ohm}}, V_{\text{oc}}, R_{\text{dyn}}, C_{\text{dyn}}$ can be determined by Eq. 5.12, and the SOC can be looked up using the measured temperature and estimated V_{oc}.

Since all model parameters are calculated in real time, the power capability of the battery system can also be calculated based on these parameters. The estimated V_{oc} and SOC over the given cycle are shown in Fig. 5.13.

5.2.3   Extended Kalman-filter-based SOC Determination Method

As described in Appendix A, a Kalman filter provides a method of estimating the states of a linear dynamic system, while an extended Kalman filter (EKF) can be used to estimate the states of a nonlinear system. From a system point of view, an HEV battery system is a nonlinear dynamic system. Once the state-space equation is properly established and the SOC is defined as a state, the EKF technique can be applied to obtain the SOC. If we assume that a battery is modeled by the two RC pair electrical circuit model shown in Fig. 5.14, where one RC pair represents the fast dynamic response behavior such as the charge transfer process and the other RC pair models the slow dynamic response behavior such as the solid diffusion process, the detailed relationships between variables can be described by the following equations:

$$\text{SOC}: \quad SOC(t) = \int_{t_i}^{t} \frac{\eta_{\text{bat}}\Delta T}{Cap} I_{\text{terminal}} dt \quad \Rightarrow \quad \frac{dSOC}{dt} = \frac{\eta_{\text{bat}}\Delta T}{Cap} I_{\text{terminal}}$$

$$V_{\text{diff}}: \quad C_{\text{diff}}\frac{dV_{\text{diff}}}{dt} + \frac{V_{\text{diff}}}{R_{\text{diff}}} = I_{\text{terminal}} \quad \Rightarrow \quad \frac{dV_{\text{diff}}}{dt} = -\frac{V_{\text{diff}}}{C_{\text{diff}}R_{\text{diff}}} + \frac{I_{\text{terminal}}}{C_{\text{diff}}} \tag{5.18}$$

$$V_{\text{dl}}: \quad C_{\text{dl}}\frac{dV_{\text{dl}}}{dt} + \frac{V_{\text{dl}}}{R_{\text{ct}}} = I_{\text{terminal}} \quad \Rightarrow \quad \frac{dV_{\text{dl}}}{dt} = -\frac{V_{\text{dl}}}{C_{\text{dl}}R_{\text{ct}}} + \frac{I_{\text{terminal}}}{C_{\text{dl}}}$$

$$V_{\text{terminal}}: \quad V_{\text{terminal}} = V_{\text{oc}} + V_{\text{diff}} + V_{\text{dl}} + V_{\text{ohm}} = f(SOC) + V_{\text{diff}} + V_{\text{dl}} + R_{\text{ohm}}I_{\text{terminal}}$$

Discretizing Eq. 5.18 and defining the states as $x_1 = SOC$, $x_2 = V_{\text{diff}}, x_3 = V_{\text{dl}}$, the following state-space equations of the battery can be obtained:

$$\text{State equation}: \quad \begin{bmatrix} x_1(k+1) \\ x_2(k+1) \\ x_3(k+1) \end{bmatrix} = \begin{bmatrix} 1 & 0 & 0 \\ 0 & 1-\dfrac{\Delta T}{R_{\text{diff}}C_{\text{diff}}} & 0 \\ 0 & 0 & 1-\dfrac{\Delta T}{R_{\text{ct}}C_{\text{dl}}} \end{bmatrix} \begin{bmatrix} x_1(k) \\ x_2(k) \\ x_3(k) \end{bmatrix} + \begin{bmatrix} \dfrac{\eta_{\text{bat}}\Delta T}{Cap} \\ \dfrac{\Delta T}{C_{\text{diff}}} \\ \dfrac{\Delta T}{C_{\text{dl}}} \end{bmatrix} I_{\text{terminal}}(k)$$

$$\text{Output equation}: \quad V_{\text{terminal}}(k) = f(x_1(k)) + x_2(k) + x_3(k) + R_{\text{ohm}}I_{\text{terminal}}(k)$$

$$\tag{5.19}$$

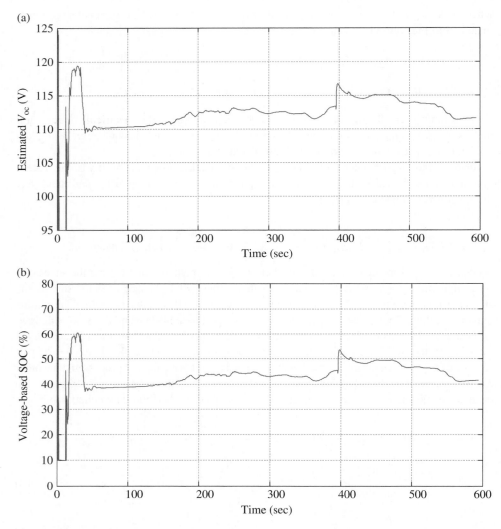

Figure 5.13   Estimated V_{oc} and SOC over the given profiles. (a) The estimated open-circuit voltage of the battery pack; (b) the estimated battery SOC using the voltage-based approach

where η_{bat} is the coulombic efficiency, $V_{terminal}(k)$ and $I_{terminal}(k)$ are the terminal voltage and current of the battery system, ΔT is the sampling time, Cap is the nominal capacity, C_{diff} is the diffusion capacitance, R_{diff} is the diffusion resistance, C_{dl} is the double layer capacitance, R_{ct} is the charge transfer resistance, R_{ohm} is the ohmic resistance, and $f(SOC)$ is a monotonous converting function from SOC to V_{oc} based on the Nernst equation (Eq. 5.3).

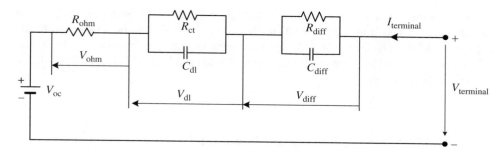

Figure 5.14   Two RC pair electrical circuit equivalent model of a battery

If the model parameters C_{diff}, R_{diff}, C_{dl}, R_{ct}, R_{ohm} are unknown and need to be estimated with the states simultaneously, we can define them as new states as follows:

$$
\begin{aligned}
x_4 &= C_{\text{diff}}(k+1) = C_{\text{diff}}(k) \\
x_5 &= R_{\text{diff}}(k+1) = R_{\text{diff}}(k) \\
x_6 &= C_{\text{dl}}(k+1) = C_{\text{dl}}(k) \\
x_7 &= R_{\text{ct}}(k+1) = R_{\text{ct}}(k) \\
x_8 &= R_{\text{ohm}}(k+1) = R_{\text{ohm}}(k)
\end{aligned}
\tag{5.20}
$$

Combining Eqs 5.19 and 5.20, the following augmented nonlinear state-space model is obtained:

State equations :

$$
\begin{bmatrix}
x_1(k+1) \\
x_2(k+1) \\
x_3(k+1) \\
x_4(k+1) \\
x_5(k+1) \\
x_6(k+1) \\
x_7(k+1) \\
x_8(k+1)
\end{bmatrix}
=
\begin{bmatrix}
1 & 0 & 0 & 0 & 0 & 0 & 0 & 0 \\
0 & 1-\dfrac{\Delta T}{x_4 x_5} & 0 & 0 & 0 & 0 & 0 & 0 \\
0 & 0 & 1-\dfrac{\Delta T}{x_6 x_7} & 0 & 0 & 0 & 0 & 0 \\
0 & 0 & 0 & 1 & 0 & 0 & 0 & 0 \\
0 & 0 & 0 & 0 & 1 & 0 & 0 & 0 \\
0 & 0 & 0 & 0 & 0 & 1 & 0 & 0 \\
0 & 0 & 0 & 0 & 0 & 0 & 1 & 0 \\
0 & 0 & 0 & 0 & 0 & 0 & 0 & 0
\end{bmatrix}
\begin{bmatrix}
x_1(k) \\
x_2(k) \\
x_3(k) \\
x_4(k) \\
x_5(k) \\
x_6(k) \\
x_7(k) \\
x_8(k)
\end{bmatrix}
+
\begin{bmatrix}
\dfrac{\eta_i \Delta T}{Cap} \\
\dfrac{\Delta T}{x_4} \\
\dfrac{\Delta T}{x_6} \\
0 \\
0 \\
0 \\
0 \\
0
\end{bmatrix}
I_{\text{terminal}}(k)
$$

Output equations : $V_{\text{terminal}}(k) = f(x_1(k)) + x_2(k) + x_3(k) + x_8(k) I_{\text{terminal}}(k)$

$$\tag{5.21}$$

Based on the measured battery terminal voltage and current, the states in Eq. 5.21 can be estimated in real time by the EKF method. For detailed examples of applying the EKF technique to estimate the SOC for hybrid vehicle systems, the interested reader should refer to the papers of Plett (2004a, 2004b, 2004c, 2004d) and Barbarisi *et al.* (2006) listed at the end of this chapter.

5.2.4 *SOC Determination Method Based on Transient Response Characteristics (TRCs)*

In recent years, advances in battery technology have led to substantial changes in the battery design for HEV/EV applications. Some batteries, such as Li-on iron phosphate ($LiFePO_4$), possess a nearly constant V_{oc} across most of the range. Figure 5.15 shows the relationship between the open-circuit voltage and SOC of an $LiFePO_4$ cell at 25 °C. Compared with the curve of a traditional Li-ion cell shown in Fig. 5.9, the curve of V_{oc} versus SOC for the $LiFePO_4$ cell is much flatter. Thus, for this type of battery, it will be very difficult to estimate SOC using the V_{oc} approach.

The method introduced in this section calculates the battery's SOC based on the characteristics of transient time response. The battery model, shown in Fig. 5.16, consists of a potential with hysteresis, ohmic resistance, and a linear dynamic subsystem. The transient portion of the time response is the part of the system response that goes to zero as time becomes large. Based on the linear system theory, the transient response characteristics of a system are related to the locations of system poles. The closer the system pole is to the imaginary axis, the faster the system responds to the input. Figure 5.17

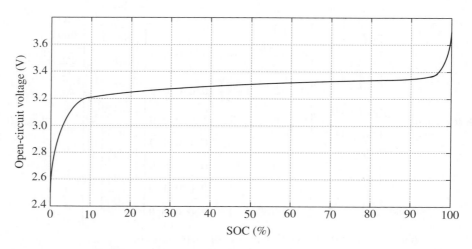

Figure 5.15 V_{oc} vs. SOC curve of an $LiFePO_4$ cell at 25 °C

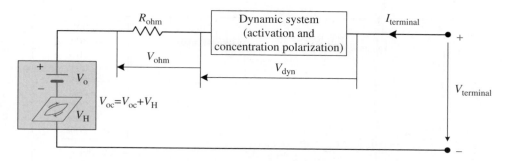

Figure 5.16   Generalized electrical circuit equivalent model of a battery system

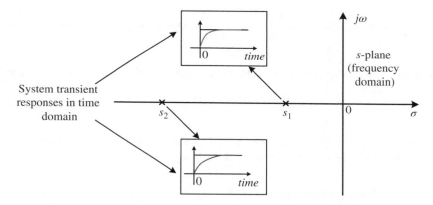

Figure 5.17   Relationship between a system's transient response and pole location in the system

illustrates how the transient time response varies with pole location as an example of a first-order system.

On the other hand, test data show that the transient response of the battery terminal voltage to a given load current varies with the battery's SOC. That is, under the same operational conditions, the transient voltage response of a battery to its current is influenced by the SOC of the battery. Figure 5.18 shows an example of how the terminal voltage of an LiFePO$_4$ battery responds to a 100 A step discharge current at different levels of SOC at 25 °C. If the transient performance of a battery system can be characterized under the operational conditions of the vehicle, the SOC of the battery can be determined. The method introduced in this section identifies the locations of the system poles in real time based on the measured terminal current and voltage. Therefore, the transient response characterization approach to determining the SOC of a battery works exclusively on the dynamic operational conditions of the battery system.

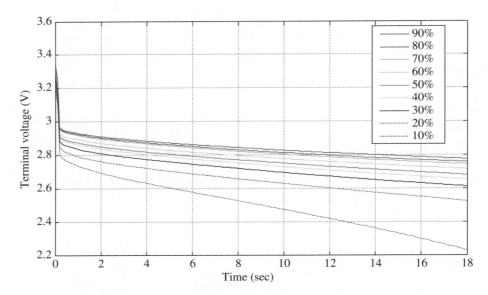

Figure 5.18   Voltage responses of an LiFePO4 battery cell to a 100 A step discharge current at 25 °C and different SOCs

The procedures for determining the SOC of the battery system based on transient response characteristics to a given load current are as follows:

1. Establish a linear system equation describing the dynamics of the battery system, for example, a proper order difference equation such as:

$$V(k) = a_1 \cdot V(k-1) + a_2(t) \cdot v(k-2) + \cdots + a_n(k-n) \cdot V(k-n)$$
$$+ b_0 \cdot I(k) + b_1 \cdot I(k-1) + \cdots + b_m \cdot I(k-m)$$

(5.22)

2. Estimate the parameters of the system equation based on an online estimation algorithm such as a recursive least squares algorithm.
3. After obtaining the parameters, the system given in Eq. 5.22 can be expressed by a z-transfer function:

$$\frac{V(z)}{I(z)} = \frac{\hat{b}_1 z^{-1} + \hat{b}_2 z^{-2} + \cdots + \hat{b}_{m_0} z^{-m_0}}{1 + \hat{a}_1 z^{-1} + \hat{a}_2 z^{-2} + \cdots + \hat{a}_{n_0} z^{-n_0}}$$

(5.23)

where $V(z)$ is the output of the system (voltage) and $I(z)$ is the input of the system (terminal current).

4. Rewrite the z-transfer function in Eq. 5.23 in pole/zero form:

$$\frac{V(z)}{I(z)} = \frac{k(z+z_1)(z+z_2)\cdots(z+z_{m_0})}{(z+p_1)(z+p_2)\cdots(z+p_{n_0})} \tag{5.24}$$

where z_i and p_j are the ith zero and the jth pole of the system.

5. Determine the dominant pole location of the battery system in the s-plane under the present operational conditions.

6. Find out the SOC from the pre-established *SOC-pole-location* look-up table based on the dominant pole location and the present operational temperature.

5.2.5 Fuzzy-logic-based SOC Determination Method

Fuzzy logic is a logic form derived from fuzzy set theory to work out a precise solution based on approximate data. Fuzzy logic can be used to model a complex system through a vague expression abstracted from the knowledge and experience of an expert. This knowledge expression can have subjective concepts such as a long or short time, a fast or slow speed. This section briefly introduces how to use fuzzy logic techniques to determine the SOC of a battery system.

The diagram of the fuzzy-logic-based battery SOC determination algorithm is shown in Fig. 5.19, in which battery dynamic behaviors are described by fuzzy logic. The SOC of a battery system under the present operational conditions can be determined by the following procedures.

1. Reason the dynamic voltage V_{dyn} based on the measured terminal voltage, $V_{\mathrm{terminal}}(k)$, the current, $I_{\mathrm{terminal}}(k)$, derivatives of the voltage and current $\dot{V}_{\mathrm{terminal}}(k)$ and $\dot{I}_{\mathrm{terminal}}(k)$, the temperature $T(k)$, and the calculated $SOC(k-1)$ in the previous time step.

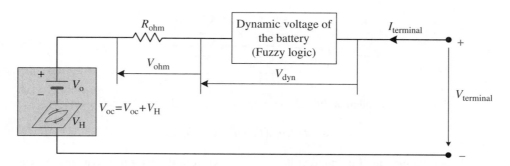

Figure 5.19 Battery system model with fuzzy logic

Depending on the type of battery, the dynamic voltage V_{dyn} can be reasoned as follows:

i. Establish input universe $\mathbf{I}$, $\dot{\mathbf{I}}$, $\mathbf{V}$, and $\mathbf{T}$ based on the designed operational conditions as

$$\mathbf{I}=[I_1 \ \ I_2 \ \ \cdots \ \ I_n], \dot{\mathbf{I}}=\left[\dot{I}_1 \ \ \dot{I}_2 \ \ \cdots \ \ \dot{I}_n\right], \mathbf{V}=[V_1 \ \ V_2 \ \ \cdots \ \ V_n], \mathbf{T}=[T_1 \ \ T_2 \ \ \cdots \ \ T_n]$$

where $\mathbf{I}$ is the measured current, $\mathbf{V}$ is the measured terminal voltage, $\mathbf{T}$ is the measured temperature, and $\dot{\mathbf{I}}$ is the derivative of the current.

ii. Set up fuzzy sets $\mathbf{A}_i$, $\mathbf{B}_i$, $\mathbf{C}_i$, $\mathbf{D}_i$ and membership functions $\mu_{A_i}(I)$, $\mu_{B_i}(\dot{I})$, $\mu_{C_i}(V)$, $\mu_{D_i}(T)$ and their values which are represented as follows

$$\mathbf{A}_i=\mu_{A_i}(I_1)/I_1+\mu_{A_2}(I_2)/I_2+\cdots+\mu_{A_n}(I_n)/I_n$$

$$\mathbf{B}_i=\mu_{B_i}(\dot{I}_1)/\dot{I}_1+\mu_{B_2}(\dot{I}_2)/\dot{I}_2+\cdots+\mu_{B_n}(\dot{I}_n)/\dot{I}_n$$

$$\mathbf{C}_i=\mu_{C_i}(V_1)/V_1+\mu_{C_2}(V_2)/V_2+\cdots+\mu_{C_n}(V_n)/V_n$$

$$\mathbf{D}_i=\mu_{D_i}(T_1)/T_1+\mu_{D_2}(T_2)/T_2+\cdots+\mu_{D_n}(D_n)/T_n$$

$$i=1, \ 2, \cdots r$$

iii. Establish rules based on the test data from vehicles or labs; these rules represent the dynamic behavior of the battery, for example:

Rule 1 : $IF \, I(t)$ is $\mathbf{A}_1, \dot{I}(t)$ is $B_1, V(t)$ is C_1, and $T(t)$ is D_1; THEN $V_{\text{dyn}}(t)$ is E_1;

Rule 2 : $IF \, I(t)$ is $\mathbf{A}_2, \dot{I}(t)$ is $B_2, V(t)$ is C_2, and $T(t)$ is D_2; THEN $V_{\text{dyn}}(t)$ is E_2;

    $\cdots$ $\cdots$ $\cdots$ $\cdots$ $\cdots$ $\cdots$ $\cdots$

Rule r : $IF \, I(t)$ is $\mathbf{A}_r, \dot{I}(t)$ is $B_r, V(t)$ is C_r, and $T(t)$ is D_r; THEN $V_{\text{dyn}}(t)$ is E_r;

iv. Give the conclusion of the reasoning by:

$$V_{\text{dyn}}(t)=\frac{\displaystyle\sum_{i=1}^{r}w_i V_{\text{dyn}}^i}{\displaystyle\sum_{i=1}^{r}w_i}=\frac{\displaystyle\sum_{i=1}^{r}w_i E_i}{\displaystyle\sum_{i=1}^{r}w_i} \tag{5.25}$$

where w_i is the adaptability of the premise part of rule i and is calculated by:

$$w_i=\mu_{A_i}(I) \wedge \mu_{B_i}(\dot{I}) \wedge \mu_{C_i}(V) \wedge \mu_{D_i}(T) \tag{5.26}$$

2. Subtract the measured terminal voltage $V_{\text{terminal}}(k)$ from the reasoned dynamic voltage $V_{\text{dyn}}(k)$ using fuzzy logic; that is, $\tilde{V}(k)=V_{\text{terminal}}(k)-V_{\text{dyn}}$.

3. Construct the regression equation $\tilde{V}(k) = R_{\text{ohm}}I_{\text{terminal}}(k) + V_{\text{oc}}$.
4. Estimate the open-circuit voltage $\hat{V}_{\text{oc}}$ and the ohmic resistance $\hat{R}_{\text{ohm}}$ through a recursive least squares estimation method.
5. Use a predetermined look-up table to find out $SOC(k)$ from the estimated $\hat{V}_{\text{oc}}$.

5.2.6 Combination of SOCs Estimated Through Different Approaches

The engineering practices in hybrid electric vehicle systems have proven that it is very challenging to determine the SOC of a battery system accurately using a single approach. In order to meet the robustness requirement of the SOC determination, the final reported SOC of an HEV battery system should be a combination of different approaches. This section presents a method of combining SOCs based on the sensitivity of individually calculated SOCs. Taking the three approaches presented above as examples, a combination can be implemented through the diagram shown in Fig. 5.20.

In the figure, the coulomb-counting-based, voltage-based, and transient-response-based approaches are used to calculate individual SOC_I, SOC_v, and SOC_{Trans}. The comprehensive information processing unit is used to handle these SOCs in order to generate a robust SOC based on the following equation:

$$SOC = \alpha_I SOC_I + \alpha_v SOC_v + \alpha_T SOC_{\text{Trans}} \tag{5.27}$$

where α_I, α_v, α_T are the weight factors which are calculated in real time based on the sensitivities of SOC_I to the accuracy of the current measurement, SOC_v to open-circuit voltage V_{oc}, and SOC_{Trans} to the system pole p_i. The calculations are as follows:

$$\alpha_I' = \frac{\partial SOC_I}{\partial I_{\text{accuracy}}}, \alpha_v' = \frac{\partial SOC_v}{\partial \hat{V}_{\text{oc}}} \text{ and } \alpha_T' = \frac{\partial SOC_{\text{Trans}}}{\partial p_i} \tag{5.28}$$

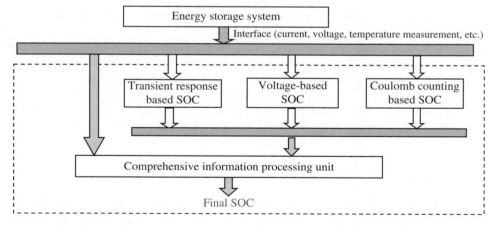

Figure 5.20   Diagram of combined SOCs calculated using different approaches

where p_i is the calculated dominant pole in real time. The raw weight factors α'_I, α'_v, and α'_T are then normalized as $\alpha_I + \alpha_v + \alpha_T = 1$ and $0 \leq \alpha_I \leq 1$, $0 \leq \alpha_v \leq 1$, and $0 \leq \alpha_T \leq 1$ from α'_I, α'_v, and α'_T.

5.2.7 Further Discussion on SOC Calculations in Hybrid Electric Vehicle Applications

Determining the SOC of an energy storage system under operational conditions is a crucial and challenging real-time computation in hybrid and battery-powered electric vehicles. The accuracy of the SOC estimation not only affects the overall fuel economy, but also impacts the drivability and safety of the vehicle. However, due to the limits of manufacturing cost and practical operational conditions and environments, technically, it is extremely difficult to achieve the same level of accuracy as for an SOC determined in a battery lab.

In the previous section, several SOC determination methods were introduced for hybrid electric vehicle system applications. This section, provides the following comments on these SOC determination approaches.

5.2.7.1 Comment 1

The coulomb-counting-based SOC determination method is the easiest and most robust method; however, in order to apply it to hybrid vehicle systems, it must be combined with other methods for the following three reasons:

1. The current sensor on a practical vehicle system is not as accurate as that used in labs due to manufacturing cost considerations, which results in large deviations in SOC when the vehicle has been driven over a long period of time. Therefore, the calculated SOC must be corrected or reset periodically by other methods over a period of driving time.
2. The initial SOC at each key cycle needs to be determined properly by other methods, and it is normally set via the measured open-circuit voltage at the key-up moment.
3. The battery aging factor must be taken into account; as a battery ages, its capacity diminishes, and each battery system applied in a hybrid vehicle system normally has a different fading mode. Thus, a capacity estimation method is needed to determine the SOC directly based on the coulomb counting method.

5.2.7.2 Comment 2

The voltage-based SOC determination method is the primary SOC correction technique in hybrid electric vehicle systems. This is because the *OCV–SOC* relationship of a battery is determined by the properties of the battery materials, which theoretically should be relatively stable over the service period of time; therefore, once the open-circuit voltage is obtained at the given temperature, the SOC of the battery can be found by looking up the *OCV–SOC* curve.

It is not difficult to obtain the open-circuit voltage (V_{oc}) when the battery system is at rest (load current equals zero), for example, at the moment of key-start; however, it is very challenging to estimate V_{oc} under load conditions. In order to obtain the correct V_{oc}, a good but relatively simple battery model is required to describe the behaviors of a battery. The equivalent electrical circuit model and state-space model mentioned in previous sections are good attempts to achieve this goal; however, so far there is no satisfactory solution achieved from this approach due to the complexity of the dynamic relationship between the terminal voltage and the current of a battery.

The battery terminal voltage is made up of the potential (V_{oc}), the internal resistance voltage drop, activation polarization, and concentration polarization for a given load current. The internal resistance of a battery is the equivalent resistance across two terminals of a battery, which is the sum of the resistance of each component of a battery. The internal resistance voltage drop is also referred to as *ohmic* or *resistance polarization*, which is proportional to the load current. The activation polarization is the over-potential due to charge transfer kinetics of the electrochemical reactions. The concentration polarization is a result of the formation of a solid diffusion layer adjacent to the electrode surface. Although the internal resistance increases with the age of a battery, it is relatively stable and is little affected by load current; however, not only do the characteristics of activation and concentration polarizations change as a battery ages, but their magnitudes also vary nonlinearly with load current. It is the activation and concentration polarizations that cause difficulty in estimating the V_{oc} of a battery under operational conditions. The relationship between the terminal voltage and the terminal current can be illustrated briefly by Fig. 5.21.

Figure 5.21(a) shows that the voltage drop on the internal resistance increases approximately linearly with load current, but activation polarization and concentration polarization apparently increase nonlinearly with the current. Thus, successfully modeling activation polarization and concentration polarization of a battery exclusively based on the terminal voltage and current of the battery is the key to solving the SOC estimation problem in hybrid vehicle applications. Figure 5.21(c) demonstrates the V_{oc}, V_{ohm}, and the combined polarization of activation and concentration of a Li-ion cell responding to the charge and discharge current pulse shown in Fig. 5.21(b).

Due to the complexity of modeling activation polarization and concentration polarization of a battery using an exact mathematical expression, artificial intelligence techniques have been applied to set these polarizations under present operational conditions, and fuzzy logic is the most used method. The biggest advantage of this approach is its robustness, but it does have the following drawbacks:

1. The set rule base is hardly adjustable as the battery ages.
2. It significantly relies on experts' knowledge of the battery.
3. It is very difficult to achieve an accurate SOC estimation.

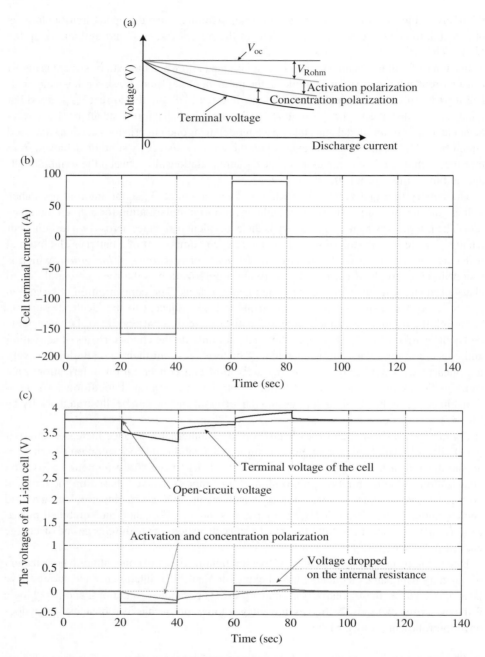

Figure 5.21 The relationship between terminal voltage and terminal current. (a) Polarization of a battery under discharge current; (b) applied charge/discharge current at cell terminals; (c) voltage responses to the applied charge/discharge current

5.2.7.3 Comment 3

As discussed, once we have obtained the V_{oc} under operational conditions, the SOC can be found; however, this is exclusively suitable to batteries in which the gradient of V_{oc} with respect to the SOC is sufficiently high, such as the characteristics shown in Fig. 5.9. For some advanced batteries such as an LiFePO$_4$ cell, even though the V_{oc} can be estimated accurately, the SOC still cannot be determined due to the super flat characteristics of V_{oc} versus the SOC.

Another big challenge when estimating the SOC using the voltage-based approach is the shifting of the OCV–SOC relationship of the Li-ion battery with age due to the impurity of the battery's materials, lithium metal plating, material fatigue, and uneven capacity degradation across the cell structure. Figure 5.22 shows one-hour rest OCV–SOC relationships over the calendar life of an NMC-based Li-ion battery at 35 °C and 80% SOC storage condition. Figure 5.23 shows the shifting of the one-hour rest OCV–SOC relationship over 80% depth of discharge (DoD) cycle life at 35 °C. The OCV–SOC shifting characteristics of Li-ion batteries often result in the battery SOC being overestimated, and the overestimation can have a significant impact on vehicle performance, drivability, or even safety. A common scenario is that a PHEV may not start in a very cold winter if the aged battery has not been plug-in charged, while the worst-case scenario would be that of a BEV with an aged battery stalling on the highway if the driver attempts to use the full electric range. One feasible solution to overcoming the challenge of OCV–SOC relationship shifting is to iterate the

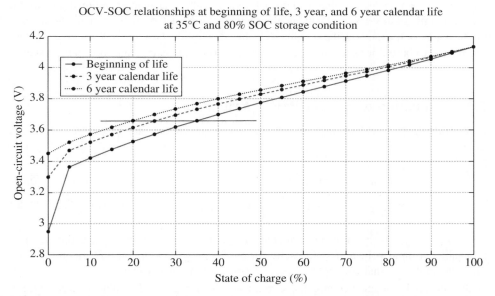

Figure 5.22 OCV–SOC relationship shifting over the calendar life of a Li-ion battery at 35 °C and 80% SOC storage condition

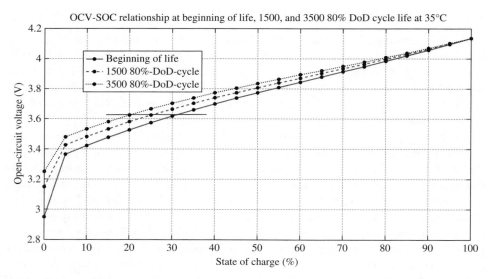

Figure 5.23   OCV–SOC relationship shifting over 80% DoD cycle life of a Li-ion battery at 35 °C

battery's Ahr capacity estimation and V_{oc} estimation in real time so that the OCV–SOC relationship can be adapted online.

The TRC approach determines the SOC of a battery based on changes in the transient response of a battery. Since this method only works under dynamic operational conditions, it is very difficult to verify and validate the accuracy of the determination; in addition, based on the nature of the method, the stability, robustness, and fidelity of the SOC estimation algorithm need to be further studied before being applied to a practical hybrid vehicle system.

5.3   Estimation of Battery Power Availability

In hybrid electric vehicle applications, the available charge and discharge power are defined as:

$$P_{\text{max\_chg}}(t) = \min\left\{ I_{V_{\max}}(t) \cdot V_{\max},\ I_{\text{max\_chg}} \cdot V_{I_{\text{max\_chg}}}(t) \right\} \qquad (5.29)$$

$$P_{\text{max\_dischg}}(t) = \min\left\{ I_{V_{\min}}(t) \cdot V_{\min},\ I_{\text{max\_dischg}} \cdot V_{I_{\text{max\_dischg}}}(t) \right\} \qquad (5.30)$$

where $P_{\text{max\_chg}}(t)$ is the maximal available charge power of a battery, $V_{\max}$ is the maximal allowable terminal voltage, $I_{V_{\max}}(t)$ is the current keeping the terminal voltage on the maximal allowable terminal voltage, $I_{\text{max\_chg}}$ is the maximal allowable charge current, $V_{I_{\text{max\_chg}}}(t)$ is the terminal voltage keeping the current on the maximal allowable charge current, $P_{\text{max\_dischg}}(t)$ is the available maximal discharge power of a battery, $V_{\min}$ is the minimal allowable terminal voltage, $I_{V_{\min}}(t)$ is the current keeping the battery's terminal voltage on the minimal allowable terminal voltage, $I_{\max\_dischg}$ is the maximal allowable discharge current, and $V_{I_{\max\_dischg}}(t)$ is the terminal voltage keeping the current on the maximal allowable discharge current.

The power availability of a battery system decays with time. Actual examples of the maximal charge and discharge power availability of a Li-ion battery at 25 °C are given by Fig. 5.24 and Fig. 5.25. In the hybrid vehicle system, the energy management expects to

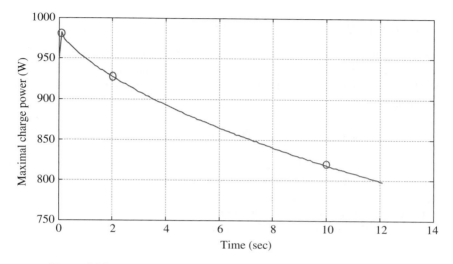

Figure 5.24   Maximal charge power availability of a Li-ion cell at 25 °C

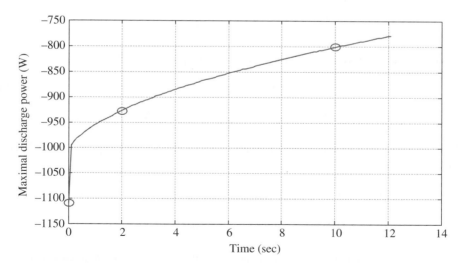

Figure 5.25   Maximal discharge power availability of a Li-ion cell at 25 °C

5.3.3 Power Availability Estimation Based on the Electrical Circuit Equivalent Model

Since the electrical circuit equivalent model shown in Fig. 5.10 can describe the terminal behavior of a battery, the power availabilities of the battery can also be calculated from the model.

In order to obtain the power availability calculation formula, we rewrite the terminal voltage equation (Eq. 5.8) of the one RC pair electrical circuit equivalent model below:

$$\frac{dV_{\text{terminal}}}{dt} + \frac{V_{\text{terminal}}}{R_{\text{dyn}}C_{\text{dyn}}} = R_{\text{ohm}} \cdot \frac{dI_{\text{terminal}}}{dt} + \left(\frac{R_{\text{dyn}} + R_{\text{ohm}}}{R_{\text{dyn}}C_{\text{dyn}}}\right) I_{\text{terminal}} + \frac{V_{\text{oc}}}{R_{\text{dyn}}C_{\text{dyn}}} \tag{5.44}$$

If we apply the maximal allowable constant current, $I_{\text{max\_limit}}$, to the battery, the terminal voltage response is described as:

$$\frac{dV_{\text{terminal}}}{dt} + \frac{V_{\text{terminal}}}{R_{\text{dyn}}C_{\text{dyn}}} = \left(\frac{R_{\text{dyn}} + R_{\text{ohm}}}{R_{\text{dyn}}C_{\text{dyn}}}\right) I_{\text{max\_limit}} + \frac{V_{\text{oc}}}{R_{\text{dyn}}C_{\text{dyn}}} \tag{5.45}$$

The solution of Eq. 5.45 is:

$$V_{\text{terminal}}(t) = R_{\text{dyn}}I_{\text{max\_limit}}\left(1 - e^{\left(-\frac{t}{R_{\text{dyn}}C_{\text{dyn}}}\right)}\right) + R_{\text{ohm}}I_{\text{max\_limit}} + V_{\text{oc}} \tag{5.46}$$

Hence, the maximal constant current-based charge and discharge power availabilities in the period Δt are:

$$P_{\text{max\_chg\_IB}}(\Delta t) = R_{\text{dyn}}I_{\text{max\_chg\_limit}}^2\left(1 - e^{\left(-\frac{\Delta t}{R_{\text{dyn}}C_{\text{dyn}}}\right)}\right) + R_{\text{ohm}}I_{\text{max\_chg\_limit}}^2$$

$$+ V_{\text{oc}}I_{\text{max\_chg\_limit}}$$

$$P_{\text{max\_dischg\_IB}}(\Delta t) = V_{\text{oc}}I_{\text{max\_dischg\_limit}} - R_{\text{ohm}}I_{\text{max\_dischg\_limit}}^2 \tag{5.47}$$

$$- R_{\text{dyn}}I_{\text{max\_dischg\_limit}}^2\left(1 - e^{\left(-\frac{\Delta t}{R_{\text{dyn}}C_{\text{dyn}}}\right)}\right)$$

If we apply the maximal allowable constant voltage, $V_{\text{max\_limit}}$, to the battery, the terminal current response is described as:

$$\frac{dI_{\text{terminal}}}{dt} + \left(\frac{R_{\text{dyn}} + R_{\text{ohm}}}{R_{\text{dyn}}C_{\text{dyn}}R_{\text{ohm}}}\right) I_{\text{terminal}} = \frac{(V_{\text{max\_limit}} - V_{\text{oc}})}{R_{\text{ohm}}R_{\text{dyn}}C_{\text{dyn}}} \tag{5.48}$$

The solution of Eq. 5.48 is:

$$I_{\text{terminal}}(t) = \frac{V_{\text{max\_limit}} - V_{\text{oc}}}{R_{\text{dyn}} + R_{\text{ohm}}} + \frac{(V_{\text{max\_limit}} - V_{\text{oc}})R_{\text{dyn}}}{R_{\text{ohm}}(R_{\text{dyn}} + R_{\text{ohm}})} e^{\left(-\frac{R_{\text{dyn}} + R_{\text{ohm}}}{R_{\text{dyn}} C_{\text{dyn}} R_{\text{ohm}}} t\right)} \tag{5.49}$$

Then, the maximal constant voltage-based charge and discharge power availabilities in the period Δt are:

$$P_{\text{max\_chg\_VB}}(\Delta t) = V_{\text{max\_chg\_limit}} I_{\text{terminal}} = \frac{V_{\text{max\_chg\_limit}} - V_{\text{oc}}}{R_{\text{dyn}} + R_{\text{ohm}}} V_{\text{max\_chg\_limit}}$$

$$+ \frac{(V_{\text{max\_chg\_limit}} - V_{\text{oc}})R_{\text{dyn}}}{R_{\text{ohm}}(R_{\text{dyn}} + R_{\text{ohm}})} V_{\text{max\_chg\_limit}} e^{\left(-\frac{R_{\text{dyn}} + R_{\text{ohm}}}{R_{\text{dyn}} C_{\text{dyn}} R_{\text{ohm}}} \Delta t\right)}$$

$$P_{\text{max\_dischg\_VB}}(\Delta t) = V_{\text{max\_dischg\_limit}} I_{\text{terminal}} = \frac{V_{\text{max\_dischg\_limit}} - V_{\text{oc}}}{R_{\text{dyn}} + R_{\text{ohm}}} V_{\text{max\_dischg\_limit}}$$

$$+ \frac{(V_{\text{max\_dischg\_limit}} - V_{\text{oc}})R_{\text{dyn}}}{R_{\text{ohm}}(R_{\text{dyn}} + R_{\text{ohm}})} V_{\text{max\_dischg\_limit}} e^{\left(-\frac{R_{\text{dyn}} + R_{\text{ohm}}}{R_{\text{dyn}} C_{\text{dyn}} R_{\text{ohm}}} \Delta t\right)}$$

$$\tag{5.50}$$

The maximal charge and discharge power availabilities of a battery in the period Δt should be the minimum of the maximal constant current-based charge and discharge power availabilities and the maximal constant voltage-based charge and discharge power availabilities, that is:

$$P_{\text{max\_chg}}(\Delta t) = \min\left\{ P_{\text{max\_chg\_IB}}(\Delta t), \ P_{\text{max\_chg\_VB}}(\Delta t) \right\}$$

$$P_{\text{max\_dischg}}(\Delta t) = \min\left\{ P_{\text{max\_dischg\_IB}}(\Delta t), \ P_{\text{max\_dischg\_VB}}(\Delta t) \right\} \tag{5.51}$$

Example 5.3

A 5.3 Ahr Li-ion battery cell has the model parameters shown in Table 5.2 based on an HPPC test at 50% SOC and 25 °C, and the operational limits are listed in Table 5.3. Determine the two-second and ten-second charge/discharge power availabilities based on the introduced methods. (Assume that the charge efficiency is 0.99 and the discharge efficiency equals 1.)

In Table 5.2, $R_{\text{ohm-c}}$, $R_{\text{ohm-d}}$, $R_{\text{ct-c}}$, $R_{\text{ct-d}}$, $C_{\text{dl-c}}$, and $C_{\text{dl-d}}$ are the parameters of the one-pair electrical circuit equivalent model, V_{oc} is the open-circuit voltage, $R_{\text{int-2s-c}}$ and $R_{\text{int-2s-d}}$ are the two-second internal resistances, $R_{\text{int-10s-c}}$ and $R_{\text{int-10s-d}}$ are the ten-second resistances, and $dV_{\text{oc}}/dSOC$ is the slope of V_{oc} versus SOC at 50% SOC and 25 °C.

Table 5.2 Extracted model parameters based on HPPC test at 50% SOC and 25 °C for a 5.3 Ahr Li-ion cell

Parameters	R_{ohm_c}	R_{ohm_d}	R_{ct_c}	R_{ct_d}	C_{dl_c}	C_{dl_d}
Value	0.002225	0.00245	0.002775	0.003063	3732	4147
Parameters	V_{oc}	$R_{int\_2s\_c}$	$R_{int\_2s\_d}$	$R_{int\_10s\_c}$	$R_{int\_10s\_d}$	$dV_{oc}/dSOC$
Value	3.75	0.00274	0.00284	0.00393	0.0041	1.1669

Table 5.3 Operational limits of the 5.3 Ahr Li-ion battery cell at 50% SOC and 25 °C

Limit	Inst. chg voltage (V)	2s chg voltage (V)	10s chg voltage (V)	Inst. dischg voltage (V)	2s dischg voltage (V)	10s dischg voltage (V)
Value	4.3	4.25	4.25	2.5	2.5	2.5
Limit	Inst. chg current (A)	2s chg current (A)	10s chg current (A)	Inst. dischg current (A)	2s dischg current (A)	10s dischg current (A)
Value	130	120	100	220	200	180

Solution:

1. By the PNGV HPPC power availability estimation method

Based on Eq. 5.32, the current driven by the maximal two-second charge voltage is:

$$I_{max\_chg}(2) = \frac{V_{max} - V_{oc}(SOC)}{R_{int}^{chg}} = \frac{4.25 - 3.75}{0.00274} = 182.48(A)$$

Since this is greater than the two-second charge current (120 A), it is necessary to calculate the terminal voltage driven by the maximal two-second charge current as:

$$V_{max\_chg}(2) = V_{oc}(SOC) + R_{int}^{chg} I_{max\_chg\_limit} = 3.75 + 0.00274 \cdot 120 = 4.08(V)$$

Therefore, the maximal two-second charge power is $P_{max\_chg}(2) = 4.08 \cdot 120 = 490$ (W)

Similarly, the current driven by the maximal ten-second charge voltage is:

$$I_{max\_chg}(10) = \frac{V_{max} - V_{oc}(SOC)}{R_{int}^{chg}} = \frac{4.25 - 3.75}{0.00393} = 127.23(A)$$

while the terminal voltage driven by the maximal ten-second charge current is:

$$V_{max\_chg}(10) = V_{oc}(SOC) + R_{int}^{chg} I_{max\_chg\_limit} = 3.75 + 0.00393 \cdot 100 = 4.14(V)$$

Thus, the maximal ten-second charge power is $P_{\max\_chg}(10) = 4.14 \cdot 100 = 414$ (W)

Check the current driven by the minimal two-second discharge voltage as:

$$I_{\max\_dischg}(2) = \frac{V_{oc}(SOC) - V_{\max\_2s}}{R_{int}^{dischg\_2s}} = \frac{3.75 - 2.5}{0.00284} = 440.14(A)$$

which is greater than the two-second discharge current (200 A); it is also necessary to calculate the terminal voltage driven by the maximal two-second discharge current as:

$$V_{\max\_dischg}(2) = V_{oc}(SOC) - R_{int\_2s}^{dischg} I_{\max\_dischg\_limit} = 3.75 - 0.00284 \cdot 200 = 3.18(V)$$

Thus, the maximal two-second discharge power is $P_{\max\_dischg}(2) = 3.18 \cdot 200 = 636$ (W)

In the same way, the current driven by the minimal ten-second discharge voltage is:

$$I_{\max\_dischg}(10) = \frac{V_{oc}(SOC) - V_{\max\_10s}}{R_{int\_10s}^{dischg}} = \frac{3.75 - 2.5}{0.0041} = 304.88(A)$$

and the terminal voltage driven by the maximal ten-second discharge current is:

$$V_{\max\_dischg}(10) = V_{oc}(SOC) - R_{int\_10s}^{dischg} I_{\max\_chg\_limit} = 3.75 - 0.0041 \cdot 180 = 3.01(V)$$

and the maximal ten-second discharge power is $P_{\max\_dischg}(10) = 3.01 \cdot 180 = 542$ (W)

2. **By the revised PNGV HPPC power availability estimation method**

Based on Eq. 5.40, the current driven by the maximal two-second charge voltage is:

$$I_{\max\_chg}(2) = \frac{V_{\max} - V_{oc}(SOC)}{\dfrac{\partial V_{oc}}{\partial SOC}\dfrac{\Delta t \eta_{bat}}{3600 Cap} + R_{int\_2s}^{chg}} = \frac{4.25 - 3.75}{\dfrac{1.1669 \cdot 2 \cdot 0.99}{3600 \cdot 5.3} + 0.00274}$$

$$= \frac{0.5}{0.002861} = 174.8(A)$$

As this is greater than the two-second charge current 120 A, we need to calculate the terminal voltage driven by the maximal two-second charge current as:

$$V_{\max\_chg}(2) = V_{oc}(SOC) + \left(\frac{\partial V_{oc}}{\partial SOC}\frac{\Delta t \eta_{bat}}{3600 Cap} + R_{int\_2s}^{chg} \right) I_{\max\_chg\_limit}$$

$$= 3.75 + 0.002861 \cdot 120 = 4.09(V)$$

Therefore, the maximal two-second charge power is $P_{\max\_chg}(2) = 4.09 \cdot 120 = 491$ (W)

Similarly, the current driven by the maximal ten-second charge voltage is:

$$I_{\max\_chg}(10) = \frac{V_{\max} - V_{oc}(SOC)}{\dfrac{\partial V_{oc}}{\partial SOC}\dfrac{\Delta t\eta_{bat}}{3600Cap} + R_{int\_10}^{chg}} = \frac{4.25 - 3.75}{0.00405} = 123(A)$$

but the terminal voltage driven by the maximal ten-second charge current is:

$$V_{\max\_chg}(10) = V_{oc}(SOC) + \left(\frac{\partial V_{oc}}{\partial SOC}\frac{\Delta t\eta_{bat}}{3600Cap} + R_{int\_10}^{chg}\right)I_{\max\_chg\_limit}$$

$$= 3.75 + 0.00405 \cdot 100 = 4.16(V)$$

Therefore, the maximal ten-second charge power is $P_{\max\_chg}(10) = 4.16 \cdot 100 = 416$ (W)

To calculate the maximal two-second discharge power, we need to calculate the current driven by the minimal two-second voltage as:

$$I_{\max\_dischg}(2) = \frac{V_{oc}(SOC) - V_{\max\_2s}}{R_{int\_2s}^{dischg} + \dfrac{\partial V_{oc}}{\partial SOC}\dfrac{\Delta t\eta_{bat}}{3600Cap}} = \frac{3.75 - 2.5}{0.00296} = 422(A)$$

Since this is greater than the two-second discharge current (200 A), it is necessary to calculate the terminal voltage driven by the maximal two-second discharge current as:

$$V_{\max\_dischg}(2) = V_{oc}(SOC) - \left(R_{int\_2s}^{dischg} + \frac{\partial V_{oc}}{\partial SOC}\frac{\Delta t\eta_{bat}}{3600Cap}\right)I_{\max\_dischg\_limit}$$

$$= 3.75 - 0.00345 \cdot 200 = 3.06(V)$$

Thus, the maximal two-second discharge power is $P_{\max\_dischg}(2) = 3.06 \cdot 200 = 612$ (W)

The current driven by the minimal ten-second discharge voltage is:

$$I_{\max\_dischg}(10) = \frac{V_{oc}(SOC) - V_{\max\_10s}}{\dfrac{\partial V_{oc}}{\partial SOC}\dfrac{\Delta t\eta_{bat}}{3600Cap} + R_{int\_10s}^{dischg}} = \frac{3.75 - 2.5}{0.00471} = 265(A)$$

and the terminal voltage driven by the maximal ten-second discharge current is:

$$V_{\max\_\mathrm{dischg}}(10) = V_{\mathrm{oc}}(SOC) - \left(\frac{\partial V_{\mathrm{oc}}}{\partial SOC} \frac{\Delta t \eta_{\mathrm{bat}}}{3600 Cap} + R_{\mathrm{int\_10s}}^{\mathrm{dischg}} \right) I_{\max\_\mathrm{chg\_limit}}$$

$$= 3.75 - 0.00471 \cdot 180 = 2.90(\mathrm{V})$$

So, the maximal ten-second discharge power is $P_{\max\_\mathrm{dischg}}(10) = 2.9022 \cdot 180 = 522$ (W)

3. **By the one *RC* pair electrical circuit model**
   Based on Eq.5.46, the terminal voltage driven by the maximal two-second charge current is:

$$V_{\mathrm{terminal}}(2) = R_{\mathrm{dyn}} I_{\max\_\mathrm{limit}} \left(1 - e^{\left(-\frac{t}{R_{\mathrm{dyn}} C_{\mathrm{dyn}}} \right)} \right) + R_{\mathrm{ohm}} I_{\max\_\mathrm{limit}} + V_{\mathrm{oc}}$$

$$= 0.002775 \cdot 120 \left(1 - e^{-\frac{2}{0.002775 \cdot 3732}} \right) + 0.00225 \cdot 120 + 3.75$$

$$= 0.002775 \cdot 120 \cdot 0.176 + 0.00225 \cdot 120 + 3.75 = 4.08(\mathrm{V})$$

and the current driven by the maximal two-second charge voltage is:

$$I_{\mathrm{terminal}}(2) = \frac{V_{\max\_\mathrm{limit}} - V_{\mathrm{oc}}}{R_{\mathrm{dyn}} + R_{\mathrm{ohm}}} + \frac{(V_{\max\_\mathrm{limit}} - V_{\mathrm{oc}}) R_{\mathrm{dyn}}}{R_{\mathrm{ohm}} (R_{\mathrm{dyn}} + R_{\mathrm{ohm}})} e^{\left(-\frac{R_{\mathrm{dyn}} + R_{\mathrm{ohm}}}{R_{\mathrm{dyn}} C_{\mathrm{dyn}} R_{\mathrm{ohm}}} t \right)}$$

$$= \frac{4.25 - 3.75}{(0.00225 + 0.002775)}$$

$$+ \frac{4.25 - 3.75}{(0.00225 + 0.002775) 0.00225} e^{\left(-\frac{(0.00225 + 0.002775)}{0.00225 \cdot 0.002775 \cdot 3732} 2 \right)}$$

$$= 99.5 + 122.7 \cdot 0.65 = 179(\mathrm{A})$$

Therefore, the maximal two-second charge power is $P_{\max\_\mathrm{chg}}(2) = 4.08 \cdot 120 = 489$ (W)

The terminal voltage driven by the maximal ten-second charge current is:

$$V_{\mathrm{terminal}}(10) = 0.002775 \cdot 100 \left(1 - e^{-\frac{10}{0.002775 \cdot 3732}} \right) + 0.00225 \cdot 100 + 3.75 = 4.15(\mathrm{V})$$

and the current driven by the maximal ten-second charge voltage is:

$$I_{\text{terminal}}(10) = \frac{4.25 - 3.75}{(0.00225 + 0.002775)}$$

$$+ \frac{4.25 - 3.75}{(0.00225 + 0.002775) \cdot 0.00225} e^{\left(-\frac{(0.00225 + 0.002775)}{0.00225 \cdot 0.002775 \cdot 3732} 10 \right)}$$

$$= 99.5 + 122.7 \cdot 0.116 = 114 (\text{A})$$

Therefore, the maximal ten-second charge power is $P_{\text{max\_chg}}(10) = 4.15 \cdot 100 = 415$ (W)

In the same way, we can calculate the maximal two-second and ten-second discharge power as:

$$P_{\text{max\_dischg}}(2) = \left(V_{\text{oc}} - R_{\text{ohm}} I_{\text{max\_limit}} - R_{\text{dyn}} I_{\text{max\_limit}} \left(1 - e^{\left(-\frac{2}{R_{\text{dyn}} C_{\text{dyn}}} \right)} \right) \right) \cdot I_{\text{max\_limit}}$$

$$= (3.75 - 0.003063 \cdot 200 - 0.002775 \cdot 200 \cdot 0.146) \cdot 200 = 611 (\text{W})$$

$$P_{\text{max\_dischg}}(10) = \left(V_{\text{oc}} - R_{\text{ohm}} I_{\text{max\_limit}} - R_{\text{dyn}} I_{\text{max\_limit}} \left(1 - e^{\left(-\frac{10}{R_{\text{dyn}} C_{\text{dyn}}} \right)} \right) \right) \cdot I_{\text{max\_limit}}$$

$$= (3.75 - 0.003063 \cdot 180 - 0.002775 \cdot 180 \cdot 0.545) \cdot 180 = 524 (\text{W})$$

The calculated power availabilities from the three different methods are summarized in Table 5.4. The calculated values are different but very close. The difference between these power values is due to the extracted parameter deviation from real test data as well as the difference in modeling principles.

Table 5.4 Calculated power availabilities using different methods of the example cell at 50% SOC and 25 °C

Unit: (W)	2s charge power	10s charge power	2s discharge power	10s discharge power
PNGV	490	414	636	542
Revised PNGV	491	416	612	522
1 RC elec. circuit model	489	415	611	524

5.4   Battery Life Prediction

The battery system is one of the core subsystems in a hybrid vehicle. If the battery fails, the vehicle's performance will be reduced and a catastrophic failure may even result. Effectively monitoring the state of health of the battery system significantly improves the reliability of an HEV/EV. This section presents battery life prediction methods for hybrid vehicle applications.

5.4.1   Aging Behavior and Mechanism

Aging of batteries is becoming a bigger and bigger concern because hybrid electric vehicle systems require batteries to have a very long lifetime, sometimes even exceeding ten years. As an example, Fig. 5.26 illustrates the capacity fade obtained after 5000 cycles at 80% depth of discharge (DoD), and Fig. 5.27 shows the increase in internal resistance during certain storage days.

Battery aging, resulting in an increase in the internal resistance, power capability fading, and energy capacity decay, originates from multiple and complex mechanisms. Many researchers (Broussely *et al.*, 2001, 2005; Sarre *et al.*, 2004; Meissner and Richter, 2005; Vetter *et al.*, 2005) have explored the aging mechanism and believe that battery aging is mainly attributable to the following three reasons:

1. Degradation of active material reactions with the electrolyte at electrode interfaces;
2. Self-degradation of the active material structure on cycling;
3. Aging of non-active components.

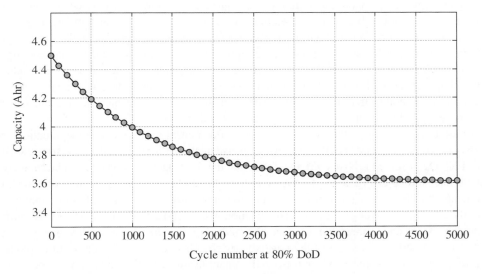

Figure 5.26   Capacity fade with cycle number (80% DoD)

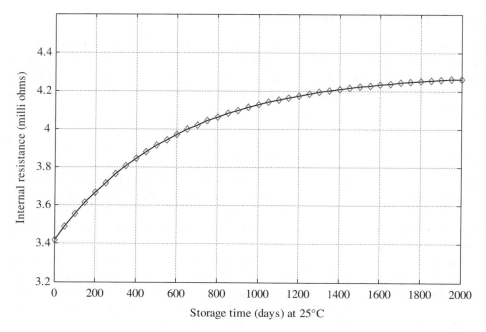

Figure 5.27   Internal resistance increase with storage time at 30 °C

In HEV/EV applications, there are two distinguishable aging situations: storage and cycling. The storage and cycling conditions have a significant impact on battery lifetime and performance; for example, either a very high or a very low SOC may deteriorate performance and shorten battery life depending on the cell chemistry. At high temperatures, this decay is accelerated, but low temperatures, especially during charging, can also have negative impacts. Further studies show that aging in storage conditions is due to side reactions resulting from thermodynamic instability of the materials present, and cycling adds kinetically induced effects to the degradation of the active material structure due to volume variations or concentration gradients. Although these two aging mechanisms are often considered additive, they sometimes interact with each other.

Furthermore, aging in the cycling condition often comes from reversible degradation of active materials, as most HEV Li-ion batteries use materials for the positive electrode, such as LiCoO2 or mixed LiNixCoyMzO2, for which the degradation is reversible. The capacity fade rate, shown in Fig. 5.26, decreases with cycle number, and this is mainly due to stabilization with time of utilization. This phenomenon is in agreement with a continuation of the passivation layer growth and consolidation of the solid electrolyte interface (SEI) producing a more and more stable interface, thus reducing the corrosion rate. However, during the storage period, the aging of the positive electrode happens mainly as a result of oxidization. Temperature and SOC are the two dominant factors affecting the oxidizing process, and this results in an increase in internal resistance and power capability decay, as shown in

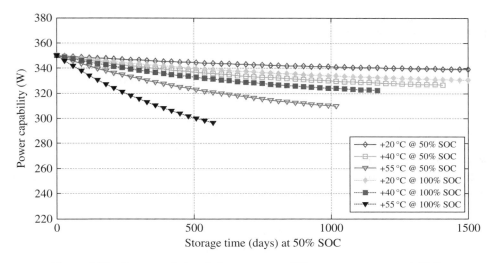

Figure 5.28   Power evolution during storage at different temperatures and SOCs

Fig. 5.27 and Fig. 5.28. In addition to an increase in internal resistance and power decay, high temperature and SOC in the storage condition may induce CO_2 gas evolution, which results in an increase in internal pressure. Gas evolution may be a life-limiting factor when batteries are maintained and stored at high temperatures.

5.4.2   Definition of the State of Life

State of life (SOL) is a figure of merit when describing the degree of degradation of a battery in HEV applications. The SOL is an essential variable to guarantee full functionality of electrically powered components and ensure that the battery operates in the best state. SOL can be defined in several ways, depending on specific properties and individual degradation. In addition, a battery may be unable to fulfill one specification, but may still be ready to achieve another.

If the power availability decay is designed as the health indicator of a battery, the SOL can be defined as:

$$\text{SOL}_{\text{Power\_based}} = \frac{P_{\text{ACT}}(SOC, T) - P_{\text{EOL}}(SOC, T)}{P_{\text{BOL}}(SOC, T) - P_{\text{EOL}}(SOC, T)} \times 100\% \tag{5.52}$$

where $P_{\text{BOL}}(SOC, T)$ is the power availability at the beginning of life, $P_{\text{EOL}}(SOC, T)$ is the defined power availability at the end of life, and $P_{\text{ACT}}(SOC, T)$ is the actual power availability under the given (SOC, T) conditions.

where $P_{BOL}(SOC, T)$ is the battery power availability at the beginning of life, $P_{EOL}(SOC, T)$ is the designed battery power availability at the end of life, and $G(P)$ is the decline in the power availability under the given storage conditions.

If the aging indicator is set as the internal resistance, the SOL of a battery can be determined based on the storage time from the following equation:

$$SOL_{R\_based} = \frac{R_{EOL} - R_{ACT}}{R_{EOL} - R_{BOL}} \times 100\% = \frac{R_{EOL} - R_{BOL} - G(R)}{\Delta R_{max}} \times 100\% = \frac{\Delta R_{max} - G(R)}{\Delta R_{max}} \times 100\%$$

$$= \left(1 - \frac{A(R) \cdot \exp\left(-\dfrac{E_a}{R_g(T + 273.15)} \right) \cdot t_R^{f(SOC)}}{\Delta R_{max}} \right) \times 100\%$$

(5.60)

where R_{BOL} is the battery's internal resistance at the beginning of life, R_{EOL} is the designed battery internal resistance at the end of life, and $G(R)$ is the increment of the internal resistance during the period of time, t, under the given storage conditions.

If the aging indicator is set as the Ahr capacity, the SOL of a battery can be determined based on the storage time from the following equation:

$$SOL_{Cap\_based} = \frac{Cap_{ACT}(T) - Cap_{EOL}(T)}{Cap_{BOL}(T) - Cap_{EOL}(T)} \times 100\% = \frac{Cap_{BOL} - G(Cap) - Cap_{EOL}(T)}{Cap_{BOL}(T) - Cap_{EOL}(T)} \times 100\%$$

$$= \frac{\Delta Cap_{max} - G(Cap)}{\Delta Cap_{max}} \times 100\%$$

$$= \left(1 - \frac{A(Cap) \cdot \exp\left(-\dfrac{E_a}{R_g(T + 273.15)} \right) \cdot t_{Cap}^{f(SOC)}}{\Delta Cap_{max}} \right) \times 100\%$$

(5.61)

where Cap_{BOL} is the battery's Ahr capacity at the beginning of life, Cap_{EOL} is the designed battery Ahr capacity at the end of life, and $G(Cap)$ is the decline in the Ahr capacity under the given storage conditions.

Example 5.4

According to the design, a 5.3 Ahr Li-ion HEV battery system reaches the end of its life when the internal resistance increases by 80% or the capacity drops to 40% of the value at the beginning of life. Based on actual tests, Eq. 5.55 is identified as follows:

1. The resistance increase follows the formula

$$\Delta R(t) = 0.125 \cdot \exp\left(-\frac{1638}{8.315 \cdot (T + 273.15)} \right) \cdot t_{month}^{\left(0.7706 \cdot SOC^2 - 0.8238 \cdot SOC + 0.729 \right)}$$

(5.62)

2. The capacity decay follows the formula

$$\Delta Cap(t) = -0.078125 \cdot \exp\left(-\frac{1725}{7.615 \cdot (T + 273.15)}\right) \cdot t_{\text{month}}^{(0.7706 \cdot SOC^2 - 0.8238 \cdot SOC + 0.729)}$$

(5.63)

where the calendar life time, t, is in months, the temperature, T, is in °C, and $0 \le SOC \le 1$. If the battery system is stored at 10 °C with 50% SOC, or at 25 °C with 70% SOC, determine the respective expected battery life.

Solution:

1. **When the battery is stored at 10 °C and 50% SOC**

   If an internal resistance increase of 80% is defined as the end of life, based on Eq. 5.62, the service time is calculated as follows:

   $$\Delta R(t) = 0.125 \cdot \exp\left(-\frac{1638}{8.315 \cdot (T + 273.15)}\right) \cdot t_{\text{month}}^{(0.7706 \cdot SOC^2 - 0.8238 \cdot SOC + 0.729)} = 0.8$$

   $$\Rightarrow t_{\text{month}}^{(0.7706 \cdot 0.5^2 - 0.8238 \cdot 0.5 + 0.729)} = \frac{0.8}{0.125} \cdot \exp\left(\frac{1638}{8.315 \cdot (10 + 273.15)}\right)$$

   $$\Rightarrow t^{0.50975} = 6.4 e^{0.6957}$$

   $$\Rightarrow t_R = (12.833)^{1/0.50975} = 149 \text{(month)}$$

   If an Ahr capacity drop to 40% is defined as the end of life, based on Eq. 5.63, the service time is calculated as:

   $$\Delta Cap(t) = -0.078125 \cdot \exp\left(-\frac{1725}{7.615 \cdot (T + 273.15)}\right) \cdot t_{\text{month}}^{(0.7706 \cdot SOC^2 - 0.8238 \cdot SOC + 0.729)}$$

   $$= -0.4$$

   $$\Rightarrow t_{\text{month}}^{(0.7706 \cdot 0.5^2 - 0.8238 \cdot 0.5 + 0.729)} = \frac{-0.4}{-0.078125} \cdot \exp\left(\frac{1725}{7.615 \cdot (10 + 273.15)}\right)$$

   $$\Rightarrow t^{0.50975} = 11.395$$

   $$\Rightarrow t_{Cap} = (11.395)^{1/0.50975} = 118 \text{(month)}$$

   Therefore, if the battery system is stored at 10 °C and with 50% SOC, the designed service life is $\min\{t_R \quad t_{Cap}\} = 118$ months.

2. **When the battery is stored at 25 °C and 70% SOC**

   In the same way, if an internal resistance increase of 80% is defined as the end of life, based on Eq. 5.62, the service time is calculated as follows:

$$\Delta R(t) = 0.125 \cdot \exp\left(-\frac{1638}{8.315 \cdot (T + 273.15)}\right) \cdot t_{month}^{\left(0.7706 \cdot SOC^2 - 0.8238 \cdot SOC + 0.729\right)}$$

$$= 0.8$$

$$\Rightarrow t_{month}^{\left(0.7706 \cdot 0.7^2 - 0.8238 \cdot 0.7 + 0.729\right)} = \frac{0.8}{0.125} \cdot \exp\left(\frac{1638}{8.315 \cdot (25 + 273.15)}\right)$$

$$\Rightarrow t^{0.529934} = 6.4 e^{0.6607}$$

$$\Rightarrow t_R = (12.392)^{1/0.529934} = 115 \,(\text{month})$$

If an Ahr capacity drop to 40% is defined as the end of life, based on Eq. 5.63, the service time is calculated as follows:

$$\Delta Cap(t) = -0.078125 \cdot \exp\left(-\frac{1725}{7.615 \cdot (T + 273.15)}\right) \cdot t_{month}^{\left(0.7706 \cdot SOC^2 - 0.8238 \cdot SOC + 0.729\right)}$$

$$= -0.4$$

$$\Rightarrow t_{month}^{\left(0.7706 \cdot 0.5^2 - 0.8238 \cdot 0.5 + 0.729\right)} = \frac{-0.4}{-0.078125} \cdot \exp\left(\frac{1725}{7.615 \cdot (25 + 273.15)}\right)$$

$$\Rightarrow t^{0.529934} = 5.12 e^{0.7598} = 10.9455$$

$$\Rightarrow t_{Cap} = (10.9455)^{1/0.529934} = 91 \,(\text{month})$$

Thus, if the battery system is stored at 25 °C and with 70% SOC, the designed service life is $\min\{t_R \quad t_{Cap}\} = 91$ months.

5.4.4   SOL Determination under Cycling Conditions

Compared with those under storage conditions, the aging processes under cycling conditions are even more complicated. The aging in cycling adds kinetically induced effects, such as volume variations or concentration gradients. The capacity decrease process and power fading do not originate from one single cause, but from a number of processes and their interactions. In this section, we discuss how battery temperature, cycle SOC depth, Ahr throughput, duty period, and the intensity of cycling affect the life of the battery in HEV/EV applications.

5.4.4.1   Offline Lifetime Determination under Cycling Conditions

The offline lifetime calculation needs to take various cycling conditions into account; based on the calculations, hybrid vehicle system engineers are able to size the energy storage system, set system operational points, determine the warranty period, evaluate system performance, and calculate the overall system cost. As discussed above, the lifetime of a battery under cycling conditions is generally expressed with the following functional form:

$$life\_time = f\left(e^{\alpha T}, \quad SOC_{sp}, \quad \Delta SOC, \quad ATP_{nor} \quad I_{nor}^2\right) \qquad (5.64)$$

where T is the averaged operational temperature (°C), SOC_{sp} is the operational SOC setpoint, ΔSOC is the designed SOC swing range (operational window), ATP_{nor} is the normalized ampere-hour throughput, and I_{nor}^2 is the normalized current square based on the battery capacity. ATP_{nor} and I_{nor}^2 are related to the designed annual mileage and cycling intensity.

Based on Taylor expansion theorem, the model in Eq. 5.64 can be approximated (second order) as:

$$
\begin{aligned}
life\_time \doteq {}& a_0 + a_1 T + a_2 SOC_{sp} + a_3 \Delta SOC + a_4 ATP_{nor} + a_5 I_{nor}^2 \\
& + a_6 T^2 + a_7 SOC_{sp}^2 + a_8 (\Delta SOC)^2 + a_9 (ATP_{nor})^2 + a_{10} (I_{nor}^2)^2 \\
& + a_{11} T \cdot SOC + a_{12} T \cdot \Delta SOC + a_{13} T \cdot ATP_{nor} + a_{14} T \cdot I_{nor}^2 \\
& + a_{15} SOC \cdot \Delta SOC + a_{16} SOC \cdot ATP_{nor} + a_{17} SOC \cdot I_{nor}^2 \\
& + a_{18} \Delta SOC \cdot ATP_{nor} + a_{19} \Delta SOC \cdot I_{nor}^2 + a_{20} ATP_{nor} \cdot I_{nor}^2
\end{aligned}
\tag{5.65}
$$

where the coefficients a_i $i = 0, 1, \cdots, 20$ are determined by test data.

Since this model is an empirical statistical model, the sample size has to be big enough to meet the accuracy and fidelity requirements, and the raw data are normally acquired from practical vehicles owned by customers rather than from labs.

5.4.4.2   Online SOL Determination under Cycling Conditions

Online SOL Determination Based on the Estimated Internal Resistance in Real Time
As defined in Section 5.4.2, if the internal resistance of a battery can be estimated in real time, the SOL can be determined in real time by the following equation:

$$
SOL_{R\_based} = \frac{R_{EOL} - R_{ACT}}{R_{EOL} - R_{BOL}} \times 100\% = \frac{R_{EOL} - \hat{R}_{real\_time}}{R_{EOL} - R_{BOL}} \times 100\%
\tag{5.66}
$$

where R_{BOL} is the BOL resistance, R_{EOL} is the defined EOL resistance, and $\hat{R}_{real\_time}$ is the estimated resistance in real time.

As presented in Section 5.2.2, the battery terminal voltage can be expressed as:

$$
V_{terminal} = V_{oc} + V_{Rohm} + V_{dynamic} = V_{oc} + R_{ohm} \cdot I_{terminal} + V_{dynamic}
\tag{5.67}
$$

If the battery is modeled by the one RC pair electrical circuit shown in Fig. 5.10, the terminal voltage can be described by the following difference equation at time k:

$$
V_{terminal}(k) = \alpha_1 V_{terminal}(k-1) + \alpha_2 I_{terminal}(k) + \alpha_3 I_{terminal}(k-1) + \alpha_4
\tag{5.68}
$$

where

$$\alpha_1 = \left(1 - \frac{\Delta T}{R_{\text{dyn}} C_{\text{dyn}}}\right), \alpha_2 = R_{\text{ohm}}, \alpha_3 = \frac{\left(R_{\text{dyn}} + R_{\text{ohm}}\right)\Delta T}{R_{\text{dyn}} C_{\text{dyn}}} - R_{\text{ohm}}, \text{ and } \alpha_4 = \frac{\Delta T \cdot V_{\text{oc}}}{R_{\text{dyn}} C_{\text{dyn}}} \quad (5.69)$$

The following relationships hold between the parameters of the difference equation (Eq. 5.68) and the physical parameters of the battery model:

$$\begin{cases} R_{\text{ohm}} = \alpha_2 \\ V_{\text{oc}} = \dfrac{\alpha_4}{1 - \alpha_1} \\ R_{\text{dyn}} = \dfrac{\alpha_3 + \alpha_1 \cdot \alpha_2}{1 - \alpha_1} \\ C_{\text{dyn}} = \dfrac{\Delta T}{\alpha_3 + \alpha_1 \cdot \alpha_2} \end{cases} \quad (5.70)$$

The parameters α_1, α_2, α_3, and α_4 in Eq. 5.68 can be estimated in real time based on the measured battery terminal voltage, V_{terminal}, and current I_{terminal} through a recursive estimation method. Thus, the ohmic resistance can be obtained in real time. In practice, a moving average filter is normally used to further smooth the estimated resistance $\hat{R}_{\text{ohm}}(k)$ over the drive cycle as:

$$\bar{R}_{\text{ohm}} = \frac{\displaystyle\sum_{i=1}^{N} \hat{R}_{\text{ohm}}(i)}{N} \quad (5.71)$$

After $\bar{R}_{\text{ohm}}$ has been obtained, the SOL can be determined by Eq. 5.66 in real time.

Example 5.5

A PHEV uses a 45 Ahr Li-ion battery system which consists of 96 cells connected in series. Given the battery usage profile shown in Fig. 5.29(a), the measured terminal voltage shown in Fig. 5.29(b), and the current shown in Fig. 5.29(c), the estimated average internal resistance over this profile is $\bar{R}_{\text{ohm}} = 0.3725\,\Omega$, and the corresponding dynamic curve of the estimation is shown in Fig. 5.29(d). If the internal resistance of the battery system is $R_{\text{BOL}} = 0.2933\,\Omega$ at the beginning of life and the end of life of the battery system is defined as the resistance increasing by 90% of the value at the beginning of life, determine the state of life of the battery.

Solution:

By comparison with the $R_{\text{BOL}} = 0.2933\,\Omega$ at the beginning of life, the resistance increases by about 27%; therefore, the state of life of the battery is about 70% based on Eq. 5.66:

$$\text{SOL} = \frac{R_{\text{EOL}} - R_{\text{ACT}}}{R_{\text{EOL}} - R_{\text{BOL}}} \times 100\% = \frac{0.2933 \cdot 1.9 - 0.3725}{0.2933 \cdot 1.9 - 0.2933} \times 100\% = 70\%$$

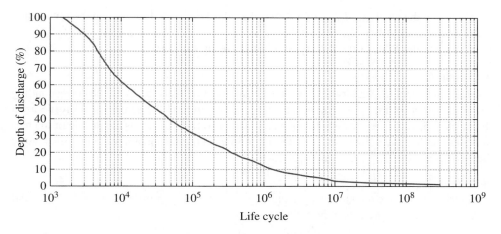

Figure 5.30   Life cycles vs. depth of discharge

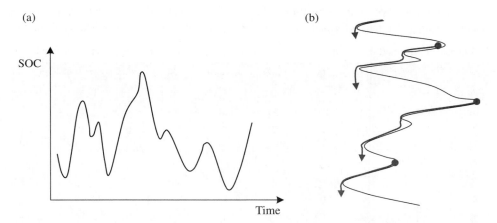

Figure 5.31   Rainflow cycles corresponding to a given drive cycle. (a) The SOC swing diagram over a drive cycle; (b) the corresponding rainflow over the drive cycle

2. Establish a battery $S–N$ curve, as shown in Fig. 5.31(a) for each drive cycle.
3. Calculate the actual performed cycle number, n_i, on the vehicle at different DoDs and create a histogram of cyclic stress in real time. Normally, it is difficult to determine the DoD of the battery when the vehicle is in operation, and it is necessary to have a good algorithm to decide where a cycle starts and where it completes. The rainflow algorithm or rainflow cycle counting is the most popular method for stress and fatigue analysis, and it can also be applied to determine the DoD of a battery during operation (Sarre *et al.*, 2004; Vetter *et al.*, 2005). The principle of the rainflow counting algorithm is that the cycle ends when rain is running from one roof and meets rain from a lower roof, as illustrated in Fig. 5.31(b).

4. Calculate the used life, L_{used}, of a battery based on the cycle life, N_i, and the obtained n_i in step 3 using Miner's rule (Eq. 5.79).
5. Calculate the SOL of the battery in real time using Eq. 5.80.

Example 5.7

The S–N test data for a Li-ion PHEV battery are given in Table 5.5, and the corresponding curve is shown in Fig. 5.30. Based on the same battery and the same usage profile as in Example 5.5, calculate the performed cycle number over the given usage profile of the battery, and update the used cycle life after the profile is completed, assuming that the battery has experienced $n_i(S_i)$ $1 \leq i \leq 100$ cycles.

Solution:

From the given usage profile, the cycle number over the usage profile can be calculated, and the used cycle life of the battery is updated as:

$$L_{used\_update} = \sum_{i=1}^{100} \frac{n_i + n_{new}(i)}{N_i}$$

where N_i is the life cycle number in Table 5.5, n_i is the historical performed cycle number of the battery, and $n_{new}(i)$ is the performed cycle number over the given usage profile listed in Table 5.6.

Table 5.6 Performed cycle number for Example 5.7 over the usage profile given in Example 5.5

DoD	No. of cycles	DoD	No. of cycles	DoD	No. of cycles	DoD	No. of cycles	DoD	No. of cycles
100	0	80	0	60	0	40	0	20	0
99	0	79	0	59	0	39	0	19	0
98	0	78	0	58	0	38	0	18	0
97	0	77	0	57	0	37	0	17	0
96	0	76	0	56	0	36	0	16	1
95	0	75	0	55	1	35	0	15	1
94	0	74	0	54	0	34	0	14	1
93	0	73	0	53	0	33	0	13	0
92	0	72	0	52	0	32	0	12	0
91	0	71	0	51	0	31	0	11	0
90	0	70	0	50	0	30	0	10	2
89	0	69	0	49	0	29	0	9	0
88	0	68	0	48	0	28	0	8	0
87	0	67	0	47	0	27	0	7	0
86	0	66	0	46	0	26	0	6	0
85	0	65	0	45	0	25	0	5	6
84	0	64	0	44	0	24	0	4	8
83	0	63	0	43	0	23	0	3	15
82	0	62	0	42	0	22	0	2	21
81	0	61	0	41	0	21	0	1	56

5.4.5   Lithium Metal Plating Issue and Symptoms in Li-ion Batteries

Lithium plating, also called lithium deposition, is a mechanism of Li-ion battery degradation and failure. In general, the lithium metal is deposited on the surface of the anode in circumstances where the transport rate of Li^+ ions to the negative electrode exceeds the rate that Li^+ can be inserted into the graphite. Since the Li^+ insertion rate is much slower at low temperature, lithium plating often occurs when the battery is used in operating conditions involving high charge rates and low temperatures, but it may also depend on other factors such as the SOC and the detailed chemistry of the battery.

Lithium plating may also result in changes in a battery's characteristics, such as OCV–SOC relationships, internal resistance, and charge transfer and diffusion behaviors. A change in battery characteristics would have a significant impact on the accuracy of the battery's SOC estimation, power capability prediction, and capacity estimation in the vehicle. For example, Fig. 5.32 shows that lithium plating results in an upper voltage plateau and the width of the plateau is proportional to the amount of metallic lithium (Ratnakumar and Smart, 2010).

Lithium plating normally results in the following symptoms, and therefore these can be used as indicators to help detect and avoid plating:

1. A higher discharge voltage than normal is shown, especially when operating at low temperatures.

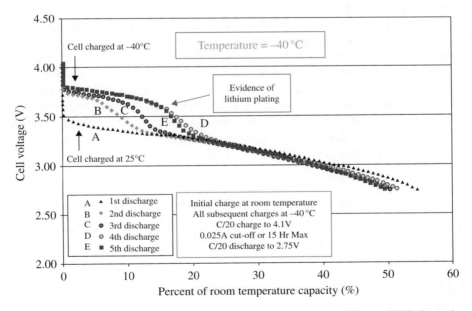

Figure 5.32   Comparsion of the discharge curves of a Li-ion battery at C/20 at −40 °C. Reproduced from Ratnakumar and Smart, 'Lithium plating behavior in Lithium-ion Cells,' *ECS Transactions*, **25**(36), 241–252, 2010 with permission from The Electrochemical Society

2. An unexpectedly high charge voltage over potential is shown.
3. The cell is generating more heat than normal.
4. There is a rapid increase in cell resistance.

5.5   Cell Balancing

An HEV/EV battery system consists of several cells which are connected in series and parallel. This combination enables the battery pack to meet both the voltage and power requirements of the hybrid vehicle based on the available battery cells. However, over a period of operation, the SOCs of different cells in a battery system will be different to each other, which will limit the capacity of the entire battery system to that of the weakest cell. The task of cell balancing is to equalize the SOCs of different cells such that the performance of the battery system is maintained.

5.5.1   SOC Balancing

In a battery system, the cells are considered balanced when: $SOC_1 = SOC_2 = \cdots = SOC_n$ where n is the number of cells in the series string of the battery system and SOC_i is the SOC of the ith individual cell. However, cell imbalance may occur over a long period of time. Since V_{oc} is a good measure of the SOC, cells can be balanced based on the measured cell V_{oc}.

The principle of balancing cells in the series string is to provide differential amounts of current on every cycle. For example, in a three-cell series stack where the capacities are $Cap_1 > Cap_2 = Cap_3$, the only way to balance the cells of this stack is to apply additional discharging current to the cell with higher capacity (Cap_1) or to bypass some charge current from the weaker cell. In this case, if the cells are not balanced, the total stack capacity diminishes because both charge and discharge shut off when the cell with the lowest capacity reaches its cut-off voltage. It is obvious that the cell with the lowest capacity reaches cut-off while the other cells still have capacity remaining, which results in the cell degrading faster than the others, and thus it will actually accelerate the capacity loss over many charge/discharge cycles unless cell balancing is used.

5.5.2   Hardware Implementation of Balancing

Typically, cell balancing is implemented by one of the following two electronic circuit configurations: current bypass or charge/discharge redistribution.

5.5.2.1   Current Bypass Hardware Implementation

Current bypass is a low-cost and practical cell-balancing method used widely in HEV/EV applications. A current bypass cell-balancing circuit is composed of a power transistor (switch) and a current limiting resistor which connect to each series cell unit in a battery pack, shown in Fig. 5.33. Generally, the power electronic switch is a field effect transistor (FET) or metal oxide semiconductor field effect transistor (MOSFET). During the charging operation, if the power transistor is turned ON, part of the charging current will divert through the resistor so the cell will be charged at a slower rate than the other cells in the pack. During discharging, if the power transistor is turned ON, additional current has to pass through the resistor, resulting in the effective load on the cell increasing, so the cell discharges faster than the rest of the cells in the pack.

The bypass current is typically designed to be between 10 and 200 mA, depending on the thermal capacity of the electronic circuit of the battery system and the vehicle's operational needs for a given battery system.

5.5.2.2   Charge Redistribution Hardware Implementation

The disadvantage of the current bypass approach is that the energy of the bypassed charge is wasted. Since every Ahr is precious in terms of improving the overall fuel economy of an HEV/EV, it is significant to have a cell-balancing approach that is able to transfer energy from the 'high' cells to the 'low' rather than burning it by using a bypass resistor. There are two techniques capable of implementing this type of cell balancing – one uses a capacitor-based charge shuttle and the other an inductor-based converter – but the trade-off for accomplishing this is higher cost due to the increased size and complexity.

The principle circuit for the capacitor-based charge shuttle approach is shown in Fig. 5.34. To redistribute the energy between the cells, the shuttle capacitor is first connected to the higher voltage cell to get it charged, and then switched to the lower voltage cell to charge it. For example, if we assume that cell #1 is the highest voltage cell and cell #3 is the lowest voltage cell, to balance cells #1 and #3, switch #1 and switch #2 are first closed and shuttle

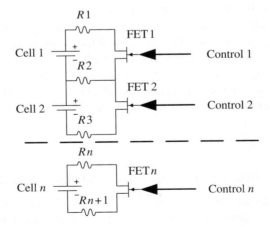

Figure 5.33   Diagram of a current bypass cell-balancing circuit

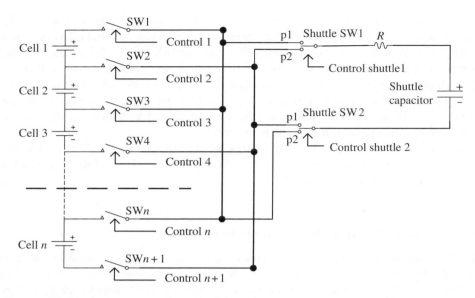

Figure 5.34 Diagram of capacitor-based charge shuttle cell-balancing circuit

switch #1 and shuttle switch #2 are toggled to the position p_1 to charge the shuttle capacitor; then, switch #1 and switch #2 are opened and switch #3 and switch #4 are closed, but both shuttle switch #1 and shuttle switch #2 are kept in the position p_1 to charge cell #3.

Engineering design problems concerning capacitor-based shuttle cell balancing implementation include sizing the current limitation resistor, R, and calculating the balancing efficiency. Because the transfer rate of this method is proportional to voltage differences, it becomes inefficient near the end of charge/discharge action.

The other cell-balancing implementation capable of transferring energy from the 'high' cells to the 'low' uses inductive-based conversion technology. Figure 5.35 shows the principle circuit for this approach, where the control principle is similar to the capacitor-based shuttle cell-balancing circuit. The cell with higher voltage is first connected to the buck-boost converter to charge the inductor, then the converter connects to the lower voltage cell to charge it. This cell-balancing method is free from the disadvantage that a small voltage difference between cells decreases the balancing rate in the capacitor-based charge shuttle method.

5.5.2.3 Engineering Design Considerations on Cell-balancing Implementation

When designing a cell-balancing electrical circuit and developing a cell-balancing algorithm, an engineer has to determine the balancing current flow rate and maximal allowable balancing time based on a given cell capacity. If there are multiple cells that are connected in

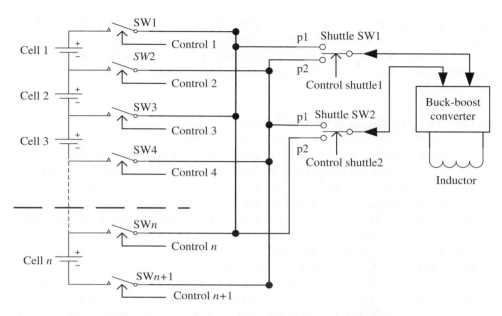

Figure 5.35 Diagram of an inductive-converter-based cell-balancing circuit

parallel, the required maximal balancing current to balance $x\%$ SOC deviation in a period of time t can be calculated by the following formula:

$$I_{\text{balancing}} = \frac{num_{\text{parallel}} \times Cap_{\text{cell}} \times x\%}{t} \tag{5.81}$$

where $I_{\text{balancing}}$ is the maximal balancing current, num_{parallel} is the number of cells connected in parallel, and Cap_{cell} is the cell capacity (Ahr).

The main components in the cell-balancing electrical circuit that need to be sized are the power transistor to switch the balancing circuit ON/OFF and the current-limiting resistor, while the main engineering consideration is the heat capacity of the circuit. If the balancing current is designed too high, power dissipation may be considerable or even exceed the heat capacity, resulting in the failure of the entire battery system. If the balancing current is designed too low, it may take too long, resulting in ineffective balancing.

5.5.3 Cell-balancing Control Algorithms and Evaluation

No matter which type of hardware is implemented, the control algorithm plays a central role in cell balancing. The related FETs/MOSFETs or other power electronic switches are actuated by control instructions. The simplest algorithm is based on the voltage difference among cells. Taking the current bypass circuit as an example, the bypass resistor is engaged if the SOC difference exceeds a predefined threshold. The cell-balancing algorithm determines which switches actuate, when to start to balance, and when to complete the balancing action. Based on the most used current-bypass-type cell-balancing circuit in HEV/EV applications, the following steps are generally taken to develop an SOC cell-balancing algorithm.

1. **Determine the SOC of the individual cell**

   Since an HEV/EV battery system may consist of tens to over one hundred cells, the common method of finding the SOC of each cell is to directly measure the V_{oc} of the cell which is at rest and obtain the SOC from an SOC–V_{oc} look-up table. The obtained SOC information is then used to decide which cells need to be balanced.

2. **Set up a balancing strategy based on the specific design**

   Based on the overall system design, the design engineer decides whether the balancing will be performed during an offline charging period, a complete rest period, or during the normal operational period of the vehicle.

3. **Direct balancing time determination algorithm**

   Once the V_{oc} of each cell has been obtained, it is easy to calculate the SOC difference among cells. Since HEV/EV application requires that multiple cells are balanced simultaneously, the individual (*i*th) cell SOC_i needs to be compared with the average cell SOC_{ave} of the battery system. If the *i*th cell, SOC_i, is higher than the average, SOC_{ave}, the *i*th cell needs to be balanced, and its balancing time can be calculated based on the following equation as an example:

$$t_i^{\text{balancing}} = \frac{\Delta SOC_i \times Cap_{ave}}{I_{\text{balancing}} \times 100} = \frac{(SOC_i - SOC_{ave}) \times Cap_{ave}}{I_{\text{balancing}} \times 100} \qquad (5.82)$$

where $t_i^{\text{balancing}}$ is the balancing time (h) for the *i*th cell, SOC_i is the *i*th cell's SOC expressed as a percentage, SOC_{ave} is the battery pack or average SOC, Cap_{ave} is the average cell capacity (Ahr) of the battery system, and $I_{\text{balancing}}$ is the balancing current determined by the cell-balancing circuit.

Example 5.8

A PHEV battery system consists of 96 cells connected in series with a cell capacity of 45 Ahr, and a balancing current of 0.2 A. The cell-balancing circuit uses current bypass, and balancing is actuated only when the vehicle is in the key-on state. If the vehicle operates for an hour per day, the average initial SOC_{ave} of the battery system is 80%, but the *i*th cell SOC_i is equal to 85%, determine the balancing time to achieve an SOC difference among cells within 1%.

Solution:

Based on Eq. 5.82, the calculated balancing time $t_i^{\text{balancing}}$ is $\dfrac{5 \times 45}{0.2 \times 100} = 11.25(\text{h})$ at the first key-on moment. Based on the given driving scenario, the cell is balanced back to about 0.44% SOC each day, and it will take ten days to balance the *i*th cell within the threshold.

each cell. Through the interface block (input selection), the cell model shown in Figure 5.38(b) picks up the necessary input signals including the cell current, the cell's initial SOC, the coolant temperature (air temperature), the cell-balancing command from the balancing algorithm, the cell-balancing resistance, and the operational state of the battery system. The cell-balancing algorithm in Fig. 5.38(a) outputs a balancing command to turn on/off the balancing switches. The required inputs of the cell-balancing algorithm are passed from the battery system model through the algorithm interface block (inputs signal selection).

The SOC of an individual cell, determined based on the current integration, is affected by its parasitic load and the variances in coulombic efficiency and capacity of the cell. An individual cell's terminal voltage, calculated by the cell electrical sub-model, is affected by its input current, SOC, and temperature. The temperature of the individual cell is determined by the thermal sub-model, the inputs of which are the temperature of the coolant and the heat power generated by the cell, while the model parameters include the battery cell-specific heat, the cell heat transfer coefficient, the cell mass, and the cell's initial temperature. Deviations in model parameters will result in differences in cell terminal voltages, temperature, and SOC in the battery system.

Before carrying out cell-balancing performance evaluation using the model, it is necessary to identify the various factors that can influence cell imbalance within the battery system and set up an analyzing plan systemically varying these parameters on a cell-by-cell basis to affect the rate of imbalance.

Two study examples based on a battery system which consists of 32 individual Li-ion cell modules each with a 5.3 Ahr nominal capacity are given in the following paragraphs. The study objective is to evaluate the cell-balancing algorithm and understand how the battery system responds to the cell-balancing command over time under different usage scenarios. A simulated vehicle operation profile is shown in Table 5.7. This one-week profile is repeated for 15 weeks to understand the battery system response to balancing over a longer period of time than just one balancing event, drive cycle, or vehicle idle period. The ½-hour battery power profile shown in Fig. 5.39 is configured three times through a US06 drive schedule for HEV applications. The corresponding vehicle drive profile has a 38.67 km route with an average speed of 77.33 km/h.

Table 5.7 Vehicle operation profile

Day	Drive 1	Park 1	Drive 2	Park 2
Monday–Friday	½ hour	8 hours	½ hour	15 hours
Saturday	0 hours	24 hours	0 hours	0 hours
Sunday	0 hours	24 hours	0 hours	0 hours

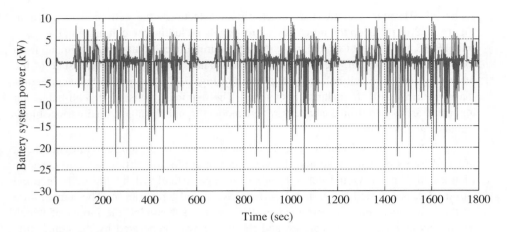

Figure 5.39   The battery usage profile repeating the US06 drive schedule three times

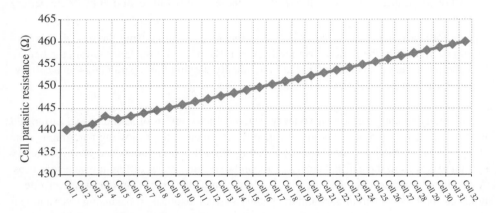

Figure 5.40   Cell-to-cell parasitic load in a 32-cell pack

Example 5.9   Battery system balancing ability versus cell parasitic load variation

This example is to study the ability of the battery system and the balancing algorithm to handle deviations in parasitic load from cell to cell. If we assume that there is no cell-to-cell capacity variation, parasitic load variation is the main cause of cell imbalance in the battery system. In the presented battery system model, the cell parasitic load includes cell resistance, cell self-discharge resistance, and battery system controller resistances in the vehicle key-on and key-off states. Figure 5.40 shows the set parasitic load at each cell during vehicle key-on. Figure 5.41 shows each cell set initial SOC before vehicle key-on, where the minimal to maximal cell SOC difference is forced to 10% and the cell-to-cell SOC standard deviation is 3%. The objective of the simulation is to evaluate the ability of the cell-balancing system to

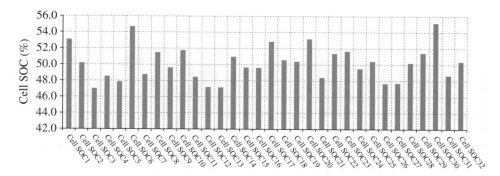

Figure 5.41   Cell SOC variation at the start point of the evaluation period

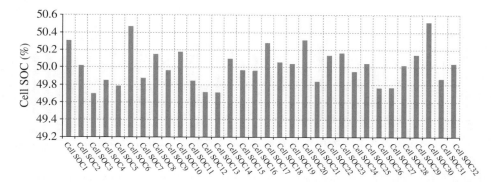

Figure 5.42   Cell SOC variation after 15 weeks of vehicle operation with cell-balancing hardware and algorithm active

bring the battery cells into balance. The simulation results show that once the cells are balanced, the system is able to keep all cells within a 0.5% SOC difference over the course of the 20-week operation, as shown in Fig. 5.42.

Example 5.10   Battery system balancing ability versus cell capacity variation

Example 5.9 focused on cell-to-cell SOC imbalance in the case where cell-to-cell capacity variation was set to zero; however, the cell capacity variation of an HEV battery system is an important element that affects battery system performance and it should be understood for hybrid vehicle operation. Depending on the actual situation, there are a few different ways to handle capacity variation in a cell-balancing system. If the cell-to-cell capacity variation is small enough and the system is operated in a small SOC band, it may be ignored and cells balanced only for SOC variation. A more effective way to handle capacity variation is to perform balancing based on the detailed information of the capacity variation using a closed-loop balancing strategy such as the method introduced in the previous section.

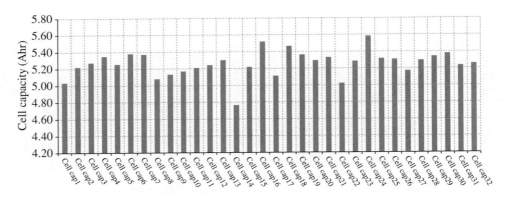

Figure 5.43   Capacity variation of the battery pack

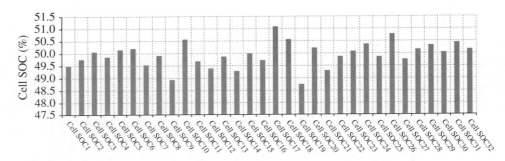

Figure 5.44   Cell SOC variation after 15 weeks of vehicle operation under the given conditions

Figure 5.43 shows the capacity variation of the battery system for the case study that was overlaid on the parasitic load variation shown in Fig. 5.40 to evaluate the ability of the battery system to handle capacity variation. The capacity variation is generated based on the 5.3 Ahr nominal cell capacity with ±5% deviation and 3% standard deviation.

The initial cell SOCs of the battery system are set as in Example 5.8 (shown in Fig. 5.41), and the cell capacities have ±5% deviation, shown in Fig. 5.43. The simulation shows that the cell-balancing unit and algorithm are able to bring the deviation in cell SOC to ±1.5% from ±5% at the start point and keep the SOC deviation of the battery system within 1.5% after the vehicle has been operated in a certain period of time. Figure 5.44 shows the SOCs of the battery system after 15 weeks of operation.

5.6   Estimation of Cell Core Temperature

5.6.1   Introduction

Battery temperature is a critical factor affecting battery life and performance. In order for the battery system to operate in the best state, it is necessary to know the battery cell temperature

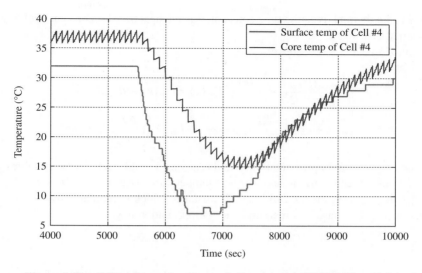

Figure 5.45   Cell surface and core temperatures over an HEV driving cycle

accurately. In HEV/EV applications, the available temperatures are the surface temperatures of the battery cells, which are measured by temperature sensors installed on the cells' skin; however, a cell's performance is actually determined by the core temperature. There may be a large difference between the cell surface temperature and the core temperature, especially in an air-cooled battery system. Figure 5.45 shows that the actual difference can be more than 10 °C, which significantly affects battery life and performance. For example, a battery system can be used for over 12 years if it operates at 25 °C while it only lasts for 7 years if it operates at 35 °C, and the battery system can deliver 14 kW of power at 0 °C but can only deliver 8 kW of power at −10 °C. This section presents a technical solution to estimate the core temperature of battery cells in HEV/EV applications.

5.6.2   Core Temperature Estimation of an Air-cooled, Cylinder-type HEV Battery

In this section, a core temperature estimation algorithm is introduced for an air-cooled, cylinder-type HEV battery system. The algorithm inputs include measured cell surface temperature, current, voltage, and vehicle off time.

5.6.2.1   Thermal Dynamic Model

Based on the physical principle, the following thermal dynamic model can be used to estimate the core temperature of the cell:

$$\tau \frac{dT_{\text{est}}(t)}{dt} + T_{\text{est}}(t) = T_{\text{sur}}(t) + R_{\text{int}}\alpha \cdot I^2(t) \tag{5.87}$$

where τ is the thermal dynamic time constant, α is the heat transfer constant, and R_{int} is the cell internal resistance.

Discretizing Eq. 5.87, we have:

$$\tau \frac{T_{\text{est}}(k+1) - T_{\text{est}}(k)}{\Delta T} + T_{\text{est}}(k) = T_{\text{sur}}(k) + \alpha R_{\text{int}} I^2(k) \tag{5.88}$$

$$T_{\text{est}}(k+1) = \frac{\tau - \Delta T}{\tau} T_{\text{est}}(k) + \frac{\Delta T}{\tau} T_{\text{sur}}(k) + \frac{\Delta T}{\tau} \alpha R_{\text{int}} I^2(k) \tag{5.89}$$

For an engineering implementation, τ and α are set as calibratable look-up tables which are functions of cell current and surface temperature; that is, $\tau = f(I_{\text{cell}}, T_{\text{cell\_sur}})$ and $\alpha = g(I_{\text{cell}}, T_{\text{cell\_sur}})$. R_{int} is the cell internal resistance which is estimated by recursive least squares in real time based on measured cell current and voltage.

5.6.2.2 Electrical Circuit Model and the Estimation of Internal Resistance

In order to estimate core temperature, a two-parameter battery electrical circuit model, shown in Fig. 5.46, is used to represent the battery's terminal behavior, and the model parameters are estimated in real time based on the measured terminal current, I_{terminal}, and voltage, V_{terminal}, through the following technique.

From the circuit equation, we have:

$$V_{\text{terminal}}(k) = V_{\text{oc}}(k) - R_{\text{int}} I_{\text{terminal}}(k) \tag{5.90}$$

where V_{terminal} and I_{terminal} are the measured cell terminal voltage and current, R_{int} and V_{oc} are the internal resistance and the open-circuit voltage of the cell, which are estimated in real time based on the following recursive LS method:

$$\begin{bmatrix} R_{\text{int}}(k+1) \\ V_{\text{oc}}(k+1) \end{bmatrix} = \begin{bmatrix} R_{\text{int}}(k) \\ V_{\text{oc}}(k) \end{bmatrix} + \begin{bmatrix} K_1(k+1)(V_{\text{terminal}}(k+1) - R_{\text{int}}(k) \cdot I_{\text{terminal}}(k+1) - V_{\text{oc}}(k)) \\ K_2(k+1)(V_{\text{terminal}}(k+1) - R_{\text{int}}(k) \cdot I_{\text{terminal}}(k+1) - V_{\text{oc}}(k)) \end{bmatrix} \tag{5.91}$$

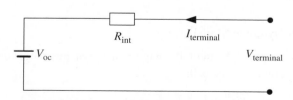

Figure 5.46 Two-parameter battery electrical circuit equivalent model

$$\begin{bmatrix} K_1(k+1) \\ K_2(k+1) \end{bmatrix} = \frac{\begin{bmatrix} p_{11}(k) & p_{12}(k) \\ p_{21}(k) & p_{22}(k) \end{bmatrix} \begin{bmatrix} I_{\text{terminal}}(k+1) \\ 1 \end{bmatrix}}{\lambda + \begin{bmatrix} I_{\text{terminal}}(k+1) & 1 \end{bmatrix} \begin{bmatrix} p_{11}(k) & p_{12}(k) \\ p_{21}(k) & p_{22}(k) \end{bmatrix} \begin{bmatrix} I_{\text{terminal}}(k+1) \\ 1 \end{bmatrix}} \tag{5.92}$$

$$\begin{bmatrix} p_{11}(k+1) & p_{12}(k+1) \\ p_{21}(k+1) & p_{22}(k+1) \end{bmatrix} = \frac{1}{\lambda^2} \begin{bmatrix} 1-K_1(k+1)I_{\text{terminal}}(k+1) & -K_1(k+1) \\ -K_2(k+1)I_{\text{terminal}}(k+1) & 1-K_2(k+1) \end{bmatrix} \begin{bmatrix} p_{11}(k) & p_{12}(k) \\ p_{21}(k) & p_{22}(k) \end{bmatrix}$$
$$\tag{5.93}$$

The parameters are estimated from the following recursive process:

$$R_{\text{int}}(k), V_{\text{oc}}(k), P(k) \xrightarrow{V_{\text{terminal}}(k+1),\ I_{\text{terminal}}(k+1)} K(k+1), P(k+1) \xrightarrow{R_{\text{int}}(k), V_{\text{oc}}(k)} R_{\text{int}}(k+1), V_{\text{oc}}(k+1) \tag{5.94}$$

The advantage of this approach is that it deals properly with battery aging, as R_{int} is the main heat source of internal temperature and increases with aging. This approach finds the actual R_{int} regardless of how the battery ages.

5.6.2.3 Initialization of Core Temperature Estimation

The equations discussed above are used to describe the thermal behaviors when the battery is being utilized. Since all measured variables are not available during the key-off period, the core temperature needs to be initialized properly at the moment of key-up.

Thermal dynamics during the vehicle key-off period can be illustrated by Fig. 5.47. If the battery temperature is higher than the ambient temperature, the battery is undergoing

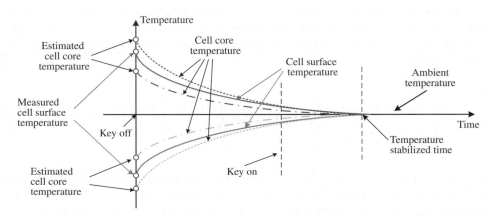

Figure 5.47 Cooling down or heating up process for battery temperature during the vehicle key-off period

a cooling down process; if the battery temperature is lower than the ambient temperature, the battery is being heated up naturally. Furthermore, let us take the cooling down process as an example to describe cell core temperature variation during the key-off period. If the cell core temperature is higher than the surface temperature at the key-down moment, the core temperature (dark gray dotted line) gradually converges to the surface temperature (dark gray solid line) with key-off time, and both temperatures will converge to the ambient temperature if the key-off time is long enough (longer than the stabilized time). Correspondingly, if the cell core temperature is lower than the surface temperature at the key-down moment, the core temperature (dark gray dash-dotted line) also gradually converges to the surface temperature (dark gray solid line), and both temperatures will converge to the ambient temperature with key-off time. If the battery is in the heating up process (pale gray lines), the core temperature converges to the surface temperature and both temperatures further converge to the ambient temperature in the same way as for the cooling down process.

Based on the thermal process of the battery system during the key-off period described above, the cell thermal dynamics can be described by the following equations:

$$
\begin{cases}
\tau_{T_{\text{sur}}} \dfrac{d(T_{\text{sur}}(t_{\text{off}}))}{dt_{\text{off}}} + T_{\text{sur}}(t_{\text{off}}) = T_{\text{ambient}} & \text{if } t_{\text{off}} < t_{\text{stable}} \\
T_{\text{sur}}(t_{\text{off}}) = T_{\text{ambient}} & \text{if } t_{\text{off}} \geq t_{\text{stable}}
\end{cases}
\tag{5.95}
$$

Solving the differential equation, we obtain the solution:

$$
\begin{aligned}
T_{\text{sur}}(t_{\text{off}}) &= T_{\text{sur}}(0)\exp(-t_{\text{off}}/\tau_{T_{\text{sur}}}) + T_{\text{ambient}}(1-\exp(-t_{\text{off}}/\tau_{T_{\text{sur}}})) \\
&= T_{\text{ambient}} + (T_{\text{sur}}(0)-T_{\text{ambient}})\exp(-t_{\text{off}}/\tau_{T_{\text{sur}}})
\end{aligned}
\tag{5.96}
$$

From the above analysis, the core temperature of the battery cell at the key-up moment can be calculated based on the measured surface temperatures at the key-down and key-up moments and the estimated core temperature at the key-down moment and key-off duration of the vehicle, as follows:

$$
\begin{cases}
T_{\text{est}}(t_{\text{off}}) = (T_{\text{est}}(0)-T_{\text{sur}}(0))\left(1 - \dfrac{t_{\text{off}}}{t_{\text{stable}}}\right) + T_{\text{sur}}(t_{\text{off}}) & \text{if } t_{\text{off}} < t_{\text{stable}} \\
T_{\text{est}}(t_{\text{off}}) = T_{\text{sur}}(t_{\text{off}}) & \text{if } t_{\text{off}} \geq t_{\text{stable}}
\end{cases}
\tag{5.97}
$$

where $T_{\text{est}}(0)$ is the estimated core temperature at the key-down moment, $T_{\text{sur}}(0)$ is the measured surface temperature at the key-down moment, t_{off} is the time during which the vehicle is in key-off, t_{stable} is the time when the core temperature and the surface temperature become the same and both converge to the ambient temperature; t_{stable} is calibrated based on tests or experiments.

5.7 Battery System Efficiency

In HEV/EV applications, battery system efficiency has a significant influence on fuel economy and electric mileage. In order to maximize fuel economy, it is necessary to know the battery system's efficiency. The actual efficiency tests should be performed on the detailed vehicle configuration and a realistic battery system usage scenario. For example, for HEV applications, high C-rate test profiles should be used, while for EV applications, a relatively low C-rate is preferable. This section introduces a general definition and testing procedure to obtain the efficiency.

The efficiency of a battery system is defined as the ratio of energy out and energy in during a round charge/discharge trip subject to charge balance (Ahr) during the cycle; in other words, the SOC at the end needs to be exactly the same as the SOC at the beginning. Thus, the battery system efficiency can be calculated as:

$$\eta_{bat}(T,\ SOC) = \frac{\int_{t_i}^{t_f} V_{terminal}(t) I_{dischg}(t) dt}{\int_{t_i}^{t_f} V_{terminal}(t) I_{chg}(t) dt} \times 100\% \tag{5.98}$$

where η_{bat} is the battery system efficiency at the given temperature and SOC, t_i is the start time, t_f is the end time, $V_{terminal}(t)$ is the cell terminal voltage, $I_{chg}(t)$ is the charge current, and $I_{dischg}(t)$ is the discharge current.

In order to obtain the battery efficiency table, it is necessary to perform efficiency tests at different temperature and SOC setpoints, and the SOC will swing about 5% depending on the detailed applications. A test profile example is shown in Fig. 5.48 at 25 °C and 50% SOC for a 6.0 Ahr Li-ion HEV battery system. Depending on the battery cell chemistry and the

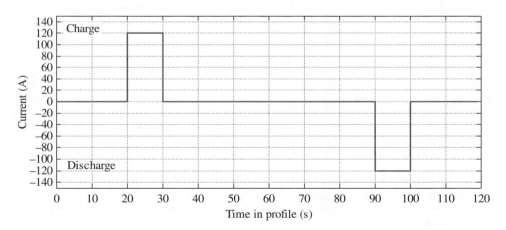

Figure 5.48 Example of a battery system efficiency test profile at 25 °C and 50% SOC

Table 5.8 Efficiencies of a Li-ion battery system for HEV application

Temperature (°C)	−40	−30	−20	−10	0	10	25	35	45	50
10% SOC	0.75	0.80	0.85	0.87	0.88	0.89	0.89	0.89	0.90	0.90
20% SOC	0.76	0.81	0.86	0.88	0.89	0.91	0.90	0.90	0.91	0.91
30% SOC	0.79	0.82	0.87	0.89	0.91	0.91	0.91	0.91	0.92	0.92
40% SOC	0.80	0.83	0.88	0.89	0.91	0.92	0.92	0.92	0.93	0.92
50% SOC	0.80	0.84	0.88	0.89	0.91	0.92	0.92	0.93	0.94	0.93
60% SOC	0.80	0.84	0.87	0.89	0.91	0.92	0.92	0.93	0.94	0.92
70% SOC	0.79	0.83	0.86	0.89	0.91	0.99	0.92	0.92	0.93	0.92
80% SOC	0.78	0.82	0.85	0.88	0.90	0.90	0.92	0.92	0.92	0.90
90% SOC	0.77	0.81	0.85	0.88	0.89	0.90	0.91	0.91	0.90	0.89

system configuration, the efficiency of an HEV/PHEV battery system is normally between 75% and 98% in the operable temperature and SOC range. The efficiency table of the example battery system is given in Table 5.8.

References

Barbarisi, O., Vasca, F., and Gliemo, L. 'State of Charge Kalman filter estimator for automotive batteries,' *Control Engineering Practice*, **14**, 267–275, 2006.

Broussely, M., Herreyre, S., Biensan, P. *et al.*, 'Aging Mechanism in Li ion Cells and Calendar Life Predictions,' *Journal of Power Sources*, **97–98**, 13–21, 2001.

Broussely, M., Biensan, P., Bonhomme, F. *et al.*, 'Main Aging Mechanisms in Li-ion Batteries,' *Journal of Power Sources*, **146**, 90–96, 2005.

Meissner, E. and Richter, G. 'The challenge to the automotive battery industry: the battery has to become an increasingly integrated component within the vehicle electric power system,' *Journal of Power Sources*, **144**, 438–460, 2005.

Palmgren, A. and Kugellagern, D. *Zeitschrift des Vereins Deutscher Ingenieure*, **68**, 339–341, 1924.

Plett, G. L. 'Extended Kalman filtering for battery management systems of LiPB-based HEV battery packs – part 1,' *Journal of Power Sources*, **134**, 252–261, 2004a.

Plett, G. L. 'Extended Kalman filtering for battery management systems of LiPB-based HEV battery packs – part 2,' *Journal of Power Sources*, **134**, 262–276, 2004b.

Plett, G. L. 'Extended Kalman filtering for battery management systems of LiPB-based HEV battery packs – part 3,' *Journal of Power Sources*, **134**, 277–292, 2004c.

Plett, G. L. 'High-Performance Battery-Pack Power Estimation Using a Dynamic Cell Model,' *IEEE Transactions On Vehicular Technology*, **53**(5), 1586–1593, 2004d.

Ratnakumar, B. V. and Smart, M. C. 'Lithium plating behavior in lithium-ion cells,' *Journal of the Electrochemical Society Transactions*, **25**, 241–252, 2010.

Sarre, G., Blanchard, P., and Broussel, M. 'Aging of Lithium-ion Batteries,' *Journal of Power Sources*, **127**, 65–71, 2004.

Verbrugge, M. and Tate, E. 'Adaptive state of charge algorithm for nickel metal hydride batteries including hysteresis phenomena,' *Journal of Power Sources*, **126**, 236–249, 2004.

Vetter, J., Novak., P., Wagner, M. R. *et al.*, 'Ageing Mechanisms in Lithium-ion Batteries,' *Journal of Power Sources*, **147**, 269–281, 2005.

6

Energy Management Strategies for Hybrid Electric Vehicles

6.1 Introduction

As stated in previous chapters, hybrid electric vehicles, using a combination of an internal combustion engine (ICE) and electric motor(s), have become a viable alternative to conventional internal-combustion-engine-based vehicles (Powers and Nicastri, 1999). The overall performance of an HEV with respect to fuel economy and emissions reduction depends on the efficiency of the individual components and good coordination of these components. In other words, the energy management strategy (also called the power control strategy) in an HEV plays a very important role in the improvement of fuel economy and the reduction of emissions.

Although there are a variety of HEV configurations, achieving maximal fuel economy, minimal emissions, and the lowest system cost are the key goals of HEV energy management strategies. In addition, the following problems are often taken into account in the development of HEV energy management strategies:

- **Optimizing the engine operational points/region:** The operational points of the engine are set on the optimal points of the torque–speed plane, based on engine-mapping data of fuel economy and emissions, and there is often a compromise between fuel economy and emissions.
- **Minimizing engine dynamics:** The operating speed of an engine is regulated in such a way that any fast fluctuations are avoided, hence minimizing the engine dynamics.
- **Minimizing engine idle time:** The engine's idle time needs to be minimized in order to improve fuel economy and decrease emissions.

Hybrid Electric Vehicle System Modeling and Control, Second Edition. Wei Liu.
© 2017 John Wiley & Sons Ltd. Published 2017 by John Wiley & Sons Ltd.

- **Optimizing engine ON–OFF times:** The engine ON and OFF times are optimized based on the driver's habits, road and weather conditions, and traffic situations by taking advantage of the dual power sources of HEVs.
- **Maximally capturing regenerative energy:** The SOC level of the battery system is optimized in order to maximally capture free regenerative energy based on the driver's habits, road and weather conditions, and traffic situations.
- **Optimizing the SOC operational window of the battery system:** The SOC operational window and swing rates are optimized in order to find a compromise between the battery's life and fuel economy of an HEV.
- **Optimizing operating points/region of the electric motor set:** The motor set has a preferred operating region on the torque–speed plane in which the overall efficiency of the system remains optimal.
- **Following zero emissions policy:** In certain areas such as tunnels and workshops, an HEV may need to be operated in the purely electric mode. The driving mode of some HEVs is controlled manually and/or automatically.

Having said all of the above, the most important and challenging task of the energy management strategy is to distribute the vehicle's power demand between the ICE and the electric motor optimally in real time (Rahman, Butler, and Ehsani, 1999). In order to achieve such a goal, various attempts have been made to develop optimal energy management strategies. For example, Johnson *et al.* presented a control strategy which optimizes vehicle dynamic operating conditions in real time to improve both fuel economy and emissions (Johnson, Wipke, and Rausen, 2000), Kheir *et al.* developed fuzzy-logic-based energy management strategies for parallel HEVs (Schouten, Salman, and Kheir, 2003; Kheir, Salman, and Schouten, 2004), while Sciarretta *et al.* proposed a cost-function-based real-time energy management strategy (Sciarretta, Back, and Guzzella, 2004).

We discuss several practical and advanced energy management strategies of an HEV in this chapter. Section 6.2 describes a general, rule-based energy management strategy. Section 6.3 presents the implementation of a fuzzy-logic-based HEV energy management strategy. A dynamic programming approach and implementation are proposed in Section 6.4. The optimization results and comparisons are given in Section 6.5.

6.2   Rule-based Energy Management Strategy

A rule-based energy management strategy is one of the most commonly used strategies in light to mild HEVs, especially in the early development stage. The fundamental rules of this type of energy management strategy include:

Rule 1: With a low power demand and low vehicle speed state, use the electric motor only.
Rule 2: Use both the electric motor and the ICE in high power demand states.
Rule 3: On a highway or in stable driving conditions, use the ICE only.

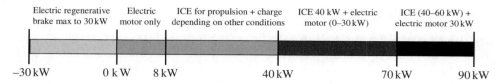

Figure 6.1   Power split for a vehicle's power demand based on a rule-based energy strategy

Rule 4: Use the ICE for driving the vehicle and/or powering the electric motor to charge the battery depending on the SOC level of the battery system:

a. If the SOC is too high, do not charge the battery;
b. If the SOC is too low, charge the battery as much as possible.

Rule 5: Maximally use regenerative braking.

Rule 6: Optimize overall efficiency of the entire vehicle system by adjusting the output power of the electric motor:

a. Use electric motor power if this action will result in the operational points of the ICE being shifted to a higher efficiency region;
b. Charge the battery if the electric motor speed is in the optimal range;
c. Keep the SOC of the battery system within the 0.5–0.7 range and have a battery with the most operational efficiency and favorable battery life;
d. Charge the battery when the vehicle demand power is low.

An example of a power split resulting from the rule-based energy management strategy to a parallel HEV is shown in Fig. 6.1. From the strategy, it can be found that if the vehicle's power demand is less than 8 kW, only the electric motor is used to drive the vehicle; if the vehicle's power demand is between 8 and 40 kW, the ICE mainly powers the vehicle, and whether the battery is charged or not depends on the other inputs such as the SOC of the ESS; if the vehicle's power demand is between 40 and 70 kW, the ICE outputs a constant 40 kW of power, and the electric motor generates additional mechanical power to meet the drivability requirements; and if the vehicle's power demand is between 70 and 90 kW, the electric motor outputs its maximum of 30 kW of power and the ICE outputs additional power to meet the power requirements of the vehicle.

6.3   Fuzzy-logic-based Energy Management Strategy

Due to the complexity of a hybrid electric vehicle system, conventional design methods that rely on an exact mathematical model are limited. Fuzzy control, where fuzzy set theory is applied, is one of the most active and fruitful areas. It is feasible and advantageous to employ fuzzy logic control techniques to design the energy management strategy for a hybrid vehicle system. In this section, a fuzzy logic control method is briefly introduced, and then a fuzzy-logic-based energy management strategy is presented, including a detailed design example.

6.3.1 *Fuzzy Logic Control*

Fuzzy logic control theory mainly includes fuzzy set theory and fuzzy logic. Fuzzy set theory is an extension of conventional set theory, and fuzzy logic is an extension of conventional logic (Tanaka, 1996). Some basic concepts of fuzzy sets and logic are introduced below.

6.3.1.1 Fuzzy Sets and Membership Functions

Fuzzy sets were proposed to deal with vague words and expressions. Lotfi A. Zadeh first introduced the concept of fuzzy sets, whose elements have degrees of membership (Zadeh, 1965). In mathematics, a fuzzy set A on the universe X is a set defined by a membership function, μ_A, representing a mapping:

$$\mu_A : \; X \rightarrow \{0 \;\; 1\} \tag{6.1}$$

where the value of $\mu_A(x)$ in the fuzzy set A is called the *membership value*, which represents the degree of x belonging to the fuzzy set A. The membership value indicated by Eq. 6.1 can be an arbitrary real value between 0 and 1. The closer the value of μ_A to 1, the higher the grade of membership of the element x in fuzzy set A. For example, if $\mu_A = 1$, the element x completely belongs to the fuzzy set A, while if $\mu_A = 0$, the element x does not at all belong to the fuzzy set A. In order to illustrate the features of fuzzy sets, the following examples are given based on the speed of a vehicle and compared with conventional sets which are exactly defined and are called *crisp sets*.

Example 6.1

In this example, Fig. 6.2 and Fig. 6.3 show the characteristic functions of conventional sets and membership functions of fuzzy sets for 'low', 'medium', and 'high' speeds described by crisp sets and fuzzy sets.

Suppose that three vehicle speeds, A, B, and C, are given as:

$$A: \; 30\text{km/h}; \; B: \; 55\text{km/h}; \; C: \; 95\,\text{km/h}$$

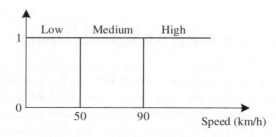

Figure 6.2 Crisp set of vehicle speed

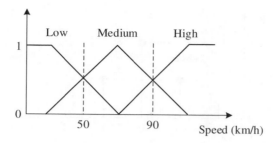

Figure 6.3   Fuzzy set of vehicle speed

Table 6.1   The values of the characteristic function in crisp sets and the membership function in fuzzy sets

Cases	Speed	Characteristic function value			Membership function value		
		Low	Medium	High	Low	Medium	High
A	30	1	0	0	8/9	1/9	0
B	55	0	1	0	1/3	2/3	0
C	95	0	0	1	0	2/3	1/3

Comparing the vehicle speeds defined in Fig. 6.2 and Fig. 6.3, we can obtain the values of the characteristic function in crisp sets and the membership function in fuzzy sets shown in Table 6.1. The characteristic function values in a conventional set indicate that A belongs to the 'Low' set, B belongs to the 'Medium' set, and C belongs to the 'High' set. The values of the membership function in the fuzzy set indicate that A belongs to the 'Low' set with a grade of 8/9 and to the 'Medium' set with a grade of 1/9, and A does not belong to the 'High' set; B belongs to the 'Low' set with a grade of 1/3 and to the 'Medium' set with a grade of 2/3; and C belongs to the 'Medium' set with a grade of 2/3 and to the 'High' set with a grade of 1/3. In reality, if we are to express these speeds linguistically, the expressions may be along these lines:

$$A: \text{ Higher low; } B: \text{ Relatively medium; } C: \text{ Lower high}$$

6.3.1.2   Mathematical Expression of Fuzzy Sets and Membership Functions

The detailed values of a membership function in the defined fuzzy set can be calculated based on the following definition:

Definition: Given a universe $X = \{x_1 \ \ x_2 \ \ \cdots \ \ x_n\}$, a fuzzy set A on X can be defined as follows:

$$A = \mu_A(x_1)/x_1 + \mu_A(x_2)/x_2 + \cdots + \mu_A(x_n)/x_n = \sum_{i=1}^{n} \mu_A(x_i)/x_i \qquad (6.2)$$

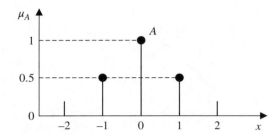

Figure 6.4   Triangular fuzzy set expression

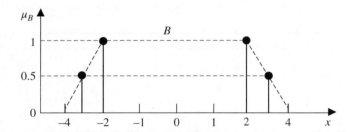

Figure 6.5   Trapezoidal fuzzy set expression

The symbol '/' in Eq. 6.2 is called a separator. An element of the universe is written on the right of the separator and the membership value of the fuzzy set A is written on the left.

Example 6.2   The finite expression of triangular fuzzy sets
If the universe X is given as $X = \{-2 \ -1 \ 0 \ 1 \ 2\}$, then the expression of the triangular fuzzy set A is $A = 0.5/-1 + 1.0/0 + 0.5/1$.

Example 6.3   The finite expression of trapezoidal fuzzy sets
If the universe X is given as $X = \{-5 \ -4 \ -3 \ -2 \ -1 \ 0 \ 1 \ 2 \ 3 \ 4 \ 5\}$, then the expression of the trapezoidal fuzzy set B is $B = 0.5/-3 + 1/-2 + 1/-1 + 1/0 + 1/1 + 1/2 + 0.5/3$.
These examples are shown in Fig. 6.4 and Fig. 6.5, respectively.

6.3.1.3   Properties, Equality, and Inclusion of Fuzzy Sets

If we assume that A, B, and C are fuzzy sets on the universe X, they have the following properties:

- Idempotent law: $A \cup A = A$, $A \cap A = A$
- Commutative law: $A \cup B = B \cup A$, $A \cap B = B \cap A$
- Associative law: $(A \cup B) \cup C = A \cup (B \cup C)$, $(A \cap B) \cap C = A \cap (B \cap C)$

- Distributive law: $A \cup (B \cap C) = (A \cup B) \cap (A \cup C)$, $A \cap (B \cup C) = (A \cap B) \cup (A \cap C)$
- Double negation law: $A = \bar{\bar{A}}$
- De Morgan's law: $\overline{A \cup B} = \bar{A} \cap \bar{B}$; $\overline{A \cap B} = \bar{A} \cup \bar{B}$
- Equality: $A = B \iff \mu_A(x) = \mu_B(x)$, $\forall x \in X$
- Inclusion: $A \subset B \iff \mu_A(x) \leq \mu_B(x)$, $\forall x \in X$

6.3.1.4 Fuzzy Relations

Fuzzy relations are an extension of relations in conventional set theory. In addition to being applied in practical engineering systems, fuzzy relations are also widely applied in many other aspects of life, including psychology, medicine, economics, and sociology. A definition of an n-ary fuzzy relation is given by Kazuo Tanaka as follows (Tanaka, 1996).

Definition: An n-ary fuzzy relation, R, is defined on the Cartesian product of $X_1 \times \cdots \times X_n$ of sets $X_1, \cdots, X_n$ such as:

$$R = \int_{X_1 \times \cdots x X_n} \mu_R(x_1, x_2, \cdots, x_n) / (x_1, x_2, \cdots, x_n) \tag{6.3}$$

where μ_R is the membership function of R given as:

$$\mu_R : X_1 \times \cdots \times X_n \rightarrow [0 \ 1] \tag{6.4}$$

In practice, the most used fuzzy relation is the binary fuzzy relation, and this is expressed as follows.

If we take the universes X and Y:

$$X = \{x_1 \ x_2 \ \cdots \ x_n\}, \ Y = \{y_1 \ y_2 \ \cdots \ y_m\} \tag{6.5}$$

the binary fuzzy relation of X and Y is expressed as:

$$R = \begin{array}{c} \\ x_1 \\ x_2 \\ \vdots \\ x_n \end{array} \begin{array}{cccc} y_1 \qquad\ y_2 \qquad\ \cdots \ \ y_{m-1} \qquad\ y_m \\ \begin{bmatrix} \mu_R(x_1,y_1) & \mu_R(x_1,y_2) & \cdots & \mu_R(x_1,y_{m-1}) & \mu_R(x_1,y_m) \\ \mu_R(x_2,y_1) & \mu_R(x_2,y_2) & \cdots & \mu_R(x_2,y_{m-1}) & \mu_R(x_2,y_m) \\ \vdots & \vdots & \vdots & \vdots & \vdots \\ \mu_R(x_n,y_1) & \mu_R(x_n,y_2) & \cdots & \mu_R(x_n,y_{m-1}) & \mu_R(x_n,y_m) \end{bmatrix} \end{array} \tag{6.6}$$

Example 6.4   Fuzzy relations

Let us assume that son x_1 and daughter x_2 are the elements of set X and father y_1 and mother y_2 are the elements of set Y; that is, $X = \{x_1 \ x_2\}$, $Y = \{y_1 \ y_2\}$. Then, the fuzzy relation 'similar' can be expressed as:

$$\begin{array}{c} \quad\quad y_1 \quad y_2 \\ \text{'similar'} \ = \ \begin{array}{c} x_1 \\ x_2 \end{array} \begin{bmatrix} 0.8 & 0.3 \\ 0.3 & 0.6 \end{bmatrix} \end{array}$$

In this fuzzy relation, 'similar', the degree of resemblance between son x_1 and father y_1 is 0.8; the degree of resemblance between son x_1 and mother y_2 is 0.3; the degree of resemblance between daughter x_2 and father y_1 is 0.3; and the degree of resemblance between daughter x_2 and mother y_2 is 0.6. This relation indicates that the daughter is more similar to the mother than the son, but the son is much more similar to the father than the daughter.

6.3.1.5   Fuzzy Reasoning

The inference rules of fuzzy reasoning are expressed in IF–THEN format, and are called fuzzy IF–THEN rules. The fuzzy reasoning methods can be classified as direct methods and indirect methods. Most popular reasoning methods are direct methods, and these include the Mamdani method, the Takagi–Sugeno method, and other simplified methods (Mamdani, 1974; Takagi and Sugeno, 1985). Direct fuzzy reasoning methods use the following inference rule format:

$$\text{IF} \quad x \text{ is } A \quad \text{and } y \text{ is } B \quad \text{THEN } z \text{ is } C \tag{6.7}$$

where A, B, and C are fuzzy sets. In the IF–THEN rule, the term following IF is called the *premise*, and the term following THEN is called the *consequence*. Therefore, variables x and y are premise variables and the variable z is a consequence variable.

Example 6.5   The fuzzy logic of automatic driving based on the relative speed and distance from other cars

- Rule 1: IF   the distance between cars is short
           AND   the vehicle speed is low
           THEN   maintain the speed (hold the current acceleration pedal position)
- Rule 2: IF   the distance between cars is short and two cars are closing
           AND   the vehicle speed is high
           THEN   reduce the speed (step on the brake pedal)
- Rule 3: IF   the distance between cars is long
           AND   the vehicle speed is low
           THEN   increase the speed (step on the acceleration pedal)

- Rule 4: IF   the distance between cars is long
  - AND   the vehicle speed is high
  - THEN   maintain the speed (hold the current acceleration pedal position)

6.3.1.6   Conversion of IF–THEN Rules to Fuzzy Relations

Let us consider a fuzzy reasoning with two positions in Eq. 6.7. This can be simplified to:

$$A \text{ and } B \rightarrow C \tag{6.8}$$

It should be noted that Eq. 6.8 cannot be described by the preceding relation $A \cap B \rightarrow C$ as A and B are subsets of different sets X and Y. Zadeh and Mamdani gave the following conversion formulas, respectively (Zadeh, 1965; Mamdani, 1974):

Zadeh's formula:

$$R = A \text{ and } B \rightarrow C \quad \Rightarrow \quad \mu_R(x,y,z) = 1 \wedge (1 - (\mu_A(x) \wedge \mu_B(y)) + \mu_c(z)) \tag{6.9}$$

Mamdani's formula:

$$R = A \text{ and } B \rightarrow C \quad \Rightarrow \quad \mu_R(x,y,z) = \mu_A(x) \wedge \mu_B(y) \wedge \mu_c(z) \tag{6.10}$$

where $\mu_R(x, y, z)$, $\mu_A(x)$, $\mu_B(y)$, $\mu_c(z)$ are membership values.

Example 6.6   Composition of fuzzy relations
If we assume that there are the following fuzzy IF–THEN rules:

Rule 1: IF x is A_1 and y is B_1 THEN z is C_1
Rule 2: IF x is A_2 and y is B_2 THEN z is C_2

where $X = \{x_1 \ x_2 \ x_3\}$ and $A_1, A_2 \subset X$, $Y = \{y_1 \ y_2 \ y_3\}$ and $B_1, B_2 \subset Y$, $Z = \{z_1 \ z_2 \ z_3\}$ and $C_1, C_2 \subset Z$; the fuzzy sets A_1, A_2, B_1, B_2, C_1, C_2 are given as follows:

$$A_1 = 1.0/x_1 + 0.6/x_2$$

$$A_2 = 0.8/x_2 + 1.0/x_3$$

$$B_1 = 1.0/y_1 + 0.5/y_2$$

$$B_2 = 0.2/y_2 + 0.9/y_3$$

$$C_1 = 1.0/z_1 + 0.6/z_2 + 0.1/z_3$$

$$C_2 = 0.2/z_1 + 0.8/z_2 + 0.9/z_3$$

then, Rule 1 and Rule 2 can be converted into a fuzzy relation by Mamdani's method as:

$$A_1 \text{ and } B_1 = A_1 \times B_1 = \begin{bmatrix} 1.0 \\ 0.6 \\ 0 \end{bmatrix} \times \begin{bmatrix} 1.0 & 0.5 & 0 \end{bmatrix} = \begin{bmatrix} 1.0 & 0.5 & 0 \\ 0.6 & 0.5 & 0 \\ 0 & 0 & 0 \end{bmatrix}$$

Thus,

$$R_1 = (A_1 \text{ and } B_1) \rightarrow C_1 = (A_1 \times B_1) \wedge C_1 = \begin{bmatrix} 1.0 & 0.5 & 0 \\ 0.6 & 0.5 & 0 \\ 0 & 0 & 0 \end{bmatrix} \wedge \begin{bmatrix} 1.0 & 0.6 & 0.1 \end{bmatrix}$$

That is:

$$R_1 = \begin{bmatrix} 1.0 & 0.5 & 0 \\ 0.6 & 0.5 & 0 \\ 0 & 0 & 0 \end{bmatrix} \begin{bmatrix} 0.6 & 0.5 & 0 \\ 0.6 & 0.5 & 0 \\ 0 & 0 & 0 \end{bmatrix} \begin{bmatrix} 0.1 & 0.1 & 0 \\ 0.1 & 0.1 & 0 \\ 0 & 0 & 0 \end{bmatrix}$$

In the same way, we have:

$$A_2 \text{ and } B_2 = A_2 \times B_2 = \begin{bmatrix} 0 \\ 0.8 \\ 1.0 \end{bmatrix} \times \begin{bmatrix} 0 & 0.2 & 0.9 \end{bmatrix} = \begin{bmatrix} 0 & 0 & 0 \\ 0 & 0.2 & 0.8 \\ 0 & 0.2 & 0.9 \end{bmatrix}$$

$$R_2 = (A_2 \text{ and } B_2) \rightarrow C_2 = (A_2 \times B_2) \wedge C_2 = \begin{bmatrix} 0 & 0 & 0 \\ 0 & 0.2 & 0.8 \\ 0 & 0.2 & 0.9 \end{bmatrix} \wedge \begin{bmatrix} 0.2 & 0.8 & 0.9 \end{bmatrix}$$

That is:

$$R_2 = \begin{bmatrix} 0 & 0 & 0 \\ 0 & 0.2 & 0.2 \\ 0 & 0.2 & 0.2 \end{bmatrix} \begin{bmatrix} 0 & 0 & 0 \\ 0 & 0.2 & 0.8 \\ 0 & 0.2 & 0.8 \end{bmatrix} \begin{bmatrix} 0 & 0 & 0 \\ 0 & 0.2 & 0.8 \\ 0 & 0.2 & 0.9 \end{bmatrix}$$

Thus, the overall fuzzy relation $R = R_1 \cup R_2$ is

$$R = \begin{bmatrix} 1.0 & 0.5 & 0 \\ 0.6 & 0.5 & 0.2 \\ 0 & 0.2 & 0.2 \end{bmatrix} \begin{bmatrix} 0.6 & 0.5 & 0 \\ 0.6 & 0.5 & 0.8 \\ 0 & 0.2 & 0.8 \end{bmatrix} \begin{bmatrix} 0.1 & 0.1 & 0 \\ 0.1 & 0.2 & 0.8 \\ 0 & 0.2 & 0.9 \end{bmatrix}$$

6.3.2 Fuzzy-logic-based HEV Energy Management Strategy

In the recent development of hybrid vehicle systems, fuzzy logic has been used to control the energy flow of an HEV based on the following principles:

- The vehicle's required power has to be met at all times during operation;
- The driver's inputs from the brake and acceleration pedals must be satisfied consistently;
- The SOC of the battery system must be maintained in a certain window at all times;
- The overall system efficiency must be maximized during operation.

Fuzzy-logic-based HEV energy management strategies can be designed using the following procedures.

6.3.2.1 Establish Fuzzy Rules

Based on the configuration of an HEV and the engineering design objectives, fuzzy rules should be established (an example is shown in Table 6.2, where ω_{veh} is the vehicle speed, P_{veh} is the vehicle's power demand, P_{mot} is the required power from the electric motor

Table 6.2 Fuzzy logic rules of an energy management strategy for an HEV

No.	Rule
1	If $SOC \geq SOC_{max}$, ω_{veh} = low, and P_{veh} = negative high, then P_{mot} = zero
2	If $SOC \geq SOC_{max}$, ω_{veh} = low, and P_{veh} = negative low, then P_{mot} = zero
3	If $SOC \geq SOC_{max}$, ω_{veh} = low, and P_{veh} = zero, then P_{mot} = zero
4	If $SOC \geq SOC_{max}$, ω_{veh} = low, and P_{veh} = positive low, then P_{mot} = positive low
5	If $SOC \geq SOC_{max}$, ω_{veh} = low, and P_{veh} = positive high, then P_{mot} = positive high
6	*If* $SOC \geq SOC_{max}$, ω_{veh} = medium, and P_{veh} = negative high, then P_{mot} = zero
7	If $SOC \geq SOC_{max}$, ω_{veh} = medium, and P_{veh} = negative low, then P_{mot} = zero
8	If $SOC \geq SOC_{max}$, ω_{veh} = medium, and P_{veh} = zero, then P_{mot} = zero
9	If $SOC \geq SOC_{max}$, ω_{veh} = medium, and P_{veh} = positive low, then P_{mot} = positive low
10	If $SOC \geq SOC_{max}$, ω_{veh} = medium, and P_{veh} = positive high, then P_{mot} = positive high
11	If $SOC \geq SOC_{max}$, ω_{veh} = high, and P_{veh} = negative high, then P_{mot} = zero
12	If $SOC \geq SOC_{max}$, ω_{veh} = high, and P_{veh} = negative low, then P_{mot} = zero
13	If $SOC \geq SOC_{max}$, ω_{veh} = high, and P_{veh} = zero, then P_{mot} = zero
14	If $SOC \geq SOC_{max}$, ω_{veh} = high, and P_{veh} = positive low, then P_{mot} = positive low
15	If $SOC \geq SOC_{max}$, ω_{veh} = high, and P_{veh} = positive high, then P_{mot} = positive high
16	If $SOC_{min} \leq SOC \leq SOC_{max}$, ω_{veh} = low, and P_{veh} = negative high, then P_{mot} = negative high
17	If $SOC_{min} \leq SOC \leq SOC_{max}$, ω_{veh} = low, and P_{veh} = negative low, then P_{mot} = negative low
18	If $SOC_{min} \leq SOC \leq SOC_{max}$, ω_{veh} = low, and P_{veh} = zero, then P_{mot} = zero
19	If $SOC_{min} \leq SOC \leq SOC_{max}$, ω_{veh} = low, and P_{veh} = positive low, then P_{mot} = positive low

(*continued overleaf*)

Table 6.2 (*continued*)

No.	Rule
20	If $SOC_{min} \leq SOC \leq SOC_{max}$, $\omega_{veh} =$ low, and $P_{veh} =$ positive high, then $P_{mot} =$ positive high
21	If $SOC_{min} \leq SOC \leq SOC_{max}$, $\omega_{veh} =$ medium, and $P_{veh} =$ negative high, then $P_{mot} =$ negative high
22	If $SOC_{min} \leq SOC \leq SOC_{max}$, $\omega_{veh} =$ medium, and $P_{veh} =$ negative low, then $P_{mot} =$ negative low
23	If $SOC_{min} \leq SOC \leq SOC_{max}$, $\omega_{veh} =$ medium, and $P_{veh} =$ zero, then $P_{mot} =$ zero
24	If $SOC_{min} \leq SOC \leq SOC_{max}$, $\omega_{veh} =$ medium, and $P_{veh} =$ positive low, then $P_{mot} =$ positive low
25	If $SOC_{min} \leq SOC \leq SOC_{max}$, $\omega_{veh} =$ medium, and $P_{veh} =$ positive high, then $P_{mot} =$ positive low
26	If $SOC_{min} \leq SOC \leq SOC_{max}$, $\omega_{veh} =$ high, and $P_{veh} =$ negative high, then $P_{mot} =$ negative high
27	If $SOC_{min} \leq SOC \leq SOC_{max}$, $\omega_{veh} =$ high, and $P_{veh} =$ negative low, then $P_{mot} =$ negative low
28	If $SOC_{min} \leq SOC \leq SOC_{max}$, $\omega_{veh} =$ high, and $P_{veh} =$ zero, then $P_{mot} =$ zero
29	If $SOC_{min} \leq SOC \leq SOC_{max}$, $\omega_{veh} =$ high, and $P_{veh} =$ positive low, then $P_{mot} =$ positive low
30	If $SOC_{min} \leq SOC \leq SOC_{max}$, $\omega_{veh} =$ high, and $P_{veh} =$ positive high, then $P_{mot} =$ zero
31	If $SOC \leq SOC_{min}$, $\omega_{veh} =$ low, and $P_{veh} =$ negative high, then $P_{mot} =$ negative high
32	If $SOC \leq SOC_{min}$, $\omega_{veh} =$ low, and $P_{veh} =$ negative low, then $P_{mot} =$ negative low
33	If $SOC \leq SOC_{min}$, $\omega_{veh} =$ low, and $P_{veh} =$ zero, then $P_{mot} =$ negative low
34	If $SOC \leq SOC_{min}$, $\omega_{veh} =$ low, and $P_{veh} =$ positive low, then $P_{mot} =$ negative low
35	If $SOC \leq SOC_{min}$, $\omega_{veh} =$ low, and $P_{veh} =$ positive high, then $P_{mot} =$ zero
36	If $SOC \leq SOC_{min}$, $\omega_{veh} =$ medium, and $P_{veh} =$ negative high, then $P_{mot} =$ negative low
37	If $SOC \leq SOC_{min}$, $\omega_{veh} =$ medium, and $P_{veh} =$ negative low, then $P_{mot} =$ negative low
38	If $SOC \leq SOC_{min}$, $\omega_{veh} =$ medium, and $P_{veh} =$ zero, then $P_{mot} =$ negative low
39	If $SOC \leq SOC_{min}$, $\omega_{veh} =$ medium, and $P_{veh} =$ positive low, then $P_{mot} =$ negative low
40	If $SOC \leq SOC_{min}$, $\omega_{veh} =$ medium, and $P_{veh} =$ positive high, then $P_{mot} =$ zero
41	If $SOC \leq SOC_{min}$, $\omega_{veh} =$ high, and $P_{veh} =$ negative high, then $P_{mot} =$ negative high
42	If $SOC \leq SOC_{min}$, $\omega_{veh} =$ high, and $P_{veh} =$ negative low, then $P_{mot} =$ negative low
43	If $SOC \leq SOC_{min}$, $\omega_{veh} =$ high, and $P_{veh} =$ zero, then $P_{mot} =$ negative low
44	If $SOC \leq SOC_{min}$, $\omega_{veh} =$ high, and $P_{veh} =$ positive low, then $P_{mot} =$ negative low
45	If $SOC \leq SOC_{min}$, $\omega_{veh} =$ high, and $P_{veh} =$ positive high, then $P_{mot} =$ zero

(positive is power, negative is regenerative), SOC_{min} is the allowable minimal SOC of the battery system, and SOC_{max} is the allowable maximal SOC of the battery system).

6.3.2.2 Formularize Fuzzy Relations from Fuzzy IF–THEN Rules

Based on the configuration and specification of the hybrid vehicle system, if the fuzzy sets are defined by Eqs 6.11 to 6.14 below, the fuzzy IF–THEN rules in Table 6.2 can be written in the form shown in Table 6.3.

$$\text{Battery SOC}: X = [x_1 \quad x_2 \quad x_3] = [40 \quad 50 \quad 60](\%) \tag{6.11}$$

$$\text{Vehicle speed}: Y = [y_1 \quad y_2 \quad y_3] = [20 \quad 50 \quad 80]\,(\text{km/h}) \tag{6.12}$$

Table 6.3 Formularized fuzzy logic rules

No.	Rule
1	If x is A_3, y is B_1, and z is C_1, then s is D_3
2	If x is A_3, y is B_1, and z is C_2, then s is D_3
3	If x is A_3, y is B_1, and z is C_3, then s is D_3
4	If x is A_3, y is B_1, and z is C_4, then s is D_4
5	If x is A_3, y is B_1, and z is C_5, then s is D_5
6	If x is A_3, y is B_2, and z is C_1, then s is D_3
7	If x is A_3, y is B_2, and z is C_2, then s is D_3
8	If x is A_3, y is B_2, and z is C_3, then s is D_3
9	If x is A_3, y is B_2, and z is C_4, then s is D_4
10	If x is A_3, y is B_2, and z is C_5, then s is D_5
11	If x is A_3, y is B_3, and z is C_1, then s is D_3
12	If x is A_3, y is B_3, and z is C_2, then s is D_3
13	If x is A_3, y is B_3, and z is C_3, then s is D_3
14	If x is A_3, y is B_3, and z is C_4, then s is D_4
15	If x is A_3, y is B_3, and z is C_5, then s is D_5
16	If x is A_2, y is B_1, and z is C_1, then s is D_1
17	If x is A_2, y is B_1, and z is C_2, then s is D_2
18	If x is A_2, y is B_1, and z is C_3, then s is D_3
19	If x is A_2, y is B_1, and z is C_4, then s is D_4
20	If x is A_2, y is B_1, and z is C_5, then s is D_5
21	If x is A_2, y is B_2, and z is C_1, then s is D_1
22	If x is A_2, y is B_2, and z is C_2, then s is D_2
23	If x is A_2, y is B_2, and z is C_3, then s is D_3
24	If x is A_2, y is B_2, and z is C_4, then s is D_4
25	If x is A_2, y is B_2, and z is C_5, then s is D_4
26	If x is A_2, y is B_3, and z is C_1, then s is D_1
27	If x is A_2, y is B_3, and z is C_2, then s is D_2
28	If x is A_2, y is B_3, and z is C_3, then s is D_3
29	If x is A_2, y is B_3, and z is C_4, then s is D_4
30	If x is A_2, y is B_3, and z is C_5, then s is D_5
31	If x is A_1, y is B_1, and z is C_1, then s is D_1
32	If x is A_1, y is B_1, and z is C_2, then s is D_2
33	If x is A_1, y is B_1, and z is C_3, then s is D_2
34	If x is A_1, y is B_1, and z is C_4, then s is D_2
35	If x is A_1, y is B_1, and z is C_5, then s is D_3
36	If x is A_1, y is B_2, and z is C_1, then s is D_1
37	If x is A_1, y is B_2, and z is C_2, then s is D_2
38	If x is A_1, y is B_2, and z is C_3, then s is D_2
39	If x is A_1, y is B_2, and z is C_4, then s is D_2
40	If x is A_1, y is B_2, and z is C_5, then s is D_3
41	If x is A_1, y is B_3, and z is C_1, then s is D_1
42	If x is A_1, y is B_3, and z is C_2, then s is D_2
43	If x is A_1, y is B_3, and z is C_3, then s is D_2
44	If x is A_1, y is B_3, and z is C_4, then s is D_2
45	If x is A_1, y is B_3, and z is C_5, then s is D_3

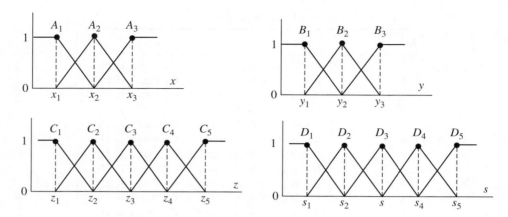

Figure 6.6 Fuzzy sets for HEV energy management

$$\text{Vehicle power demand}: Z = \begin{bmatrix} z_1 & z_2 & z_3 & z_4 & z_5 \end{bmatrix} = \begin{bmatrix} -50 & -10 & 0 & 20 & 90 \end{bmatrix} \text{(kW)} \qquad (6.13)$$

$$\text{Electric motor power}: S = \begin{bmatrix} s_1 & s_2 & s_3 & s_4 & s_5 \end{bmatrix} = \begin{bmatrix} -30 & -10 & 0 & 10 & 30 \end{bmatrix} \text{(kW)} \qquad (6.14)$$

The following fuzzy sets are defined and shown in Fig. 6.6:

$$A_1 = \begin{bmatrix} 1.0 & 0 & 0 \end{bmatrix}; \qquad A_2 = \begin{bmatrix} 0 & 1.0 & 0 \end{bmatrix}; \qquad A_3 = \begin{bmatrix} 0 & 0 & 1.0 \end{bmatrix};$$
$$B_1 = \begin{bmatrix} 1.0 & 0 & 0 \end{bmatrix}; \qquad B_2 = \begin{bmatrix} 0 & 1.0 & 0 \end{bmatrix}; \qquad B_3 = \begin{bmatrix} 0 & 0 & 1.0 \end{bmatrix};$$
$$C_1 = \begin{bmatrix} 1.0 & 0 & 0 & 0 & 0 \end{bmatrix}; \quad C_2 = \begin{bmatrix} 0 & 1.0 & 0 & 0 & 0 \end{bmatrix}; \quad C_3 = \begin{bmatrix} 0 & 0 & 1.0 & 0 & 0 \end{bmatrix};$$
$$C_4 = \begin{bmatrix} 0 & 0 & 0 & 1.0 & 0 \end{bmatrix}; \quad C_5 = \begin{bmatrix} 0 & 0 & 0 & 0 & 1.0 \end{bmatrix};$$
$$D_1 = \begin{bmatrix} 1.0 & 0 & 0 & 0 & 0 \end{bmatrix}; \quad D_2 = \begin{bmatrix} 0 & 1.0 & 0 & 0 & 0 \end{bmatrix}; \quad D_3 = \begin{bmatrix} 0 & 0 & 1.0 & 0 & 0 \end{bmatrix};$$
$$D_4 = \begin{bmatrix} 0 & 0 & 0 & 1.0 & 0 \end{bmatrix}; \quad D_5 = \begin{bmatrix} 0 & 0 & 0 & 0 & 1.0 \end{bmatrix}$$

Mamdani's conversion method can be used to establish fuzzy relations from the fuzzy IF–THEN rules described above. Because there are four elements, x, y, z, and s, the conversion formula is:

$$\mu_R\left(x_i, y_j, z_k, s_m\right) = \mu_A\left(x_i\right) \wedge \mu_B\left(y_j\right) \wedge \mu_C\left(z_k\right) \wedge \mu_D\left(s_m\right) \qquad (6.15)$$

where x is the SOC, y is the vehicle speed (km/h), z is the vehicle power demand (kW), and s is the determined motor power (positive is propulsion power and negative is brake power), i, $j = 1$, 2, 3 and k, $m = 1$, 2, 3, 4, 5.

Since the given fuzzy relation is four-dimensional, the rules contain three propositions in the premises. The three elements of the fuzzy array are calculated by Mamdani's formula as:

$$\mu_{R_1}(x_1, y_1, z_1, s_1) = \mu_{A_3}(x_1) \wedge \mu_{B_1}(y_1) \wedge \mu_{C_1}(z_1) \wedge \mu_{D_3}(s_1) = 0 \wedge 1.0 \wedge 0 \wedge 0 = 0 \tag{6.16}$$

$$\mu_{R_1}(x_1, y_1, z_1, s_3) = \mu_{A_3}(x_1) \wedge \mu_{B_1}(y_1) \wedge \mu_{C_1}(z_1) \wedge \mu_{D_3}(s_3) = 0 \wedge 1.0 \wedge 0 \wedge 1.0 = 0 \tag{6.17}$$

$$\mu_{R_1}(x_3, y_1, z_2, s_3) = \mu_{A_3}(x_3) \wedge \mu_{B_1}(y_1) \wedge \mu_{C_1}(z_2) \wedge \mu_{D_3}(s_3) = 1.0 \wedge 1.0 \wedge 1.0 \wedge 1.0 = 1.0 \tag{6.18}$$

By the conversion formula (Eq. 6.15), the fuzzy relations from R_1 to R_{45} can be obtained as follows:

$$R_{2-44} = \quad \cdots \quad \cdots \quad \cdots \quad \cdots \quad \cdots \quad \cdots \quad \cdots \quad \cdots \quad \cdots$$

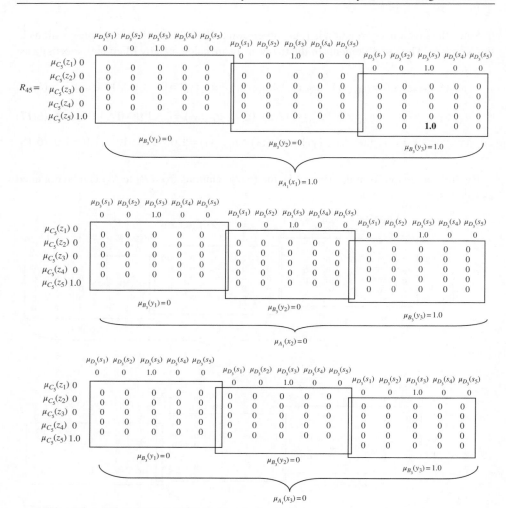

The overall fuzzy relation R is given by:

$$R = R_1 \cup R_2 \cup \cdots \cup R_{45} = \overset{45}{\underset{i=1}{\cup}} R_i \tag{6.19}$$

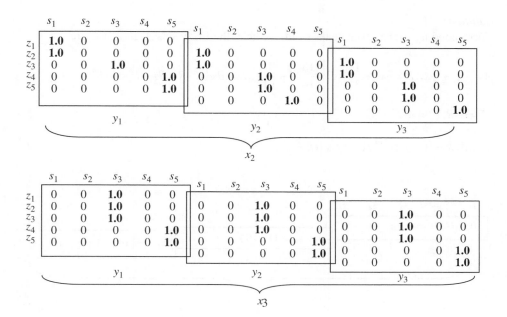

6.3.2.3 Reasoning by Fuzzy Relations

Once fuzzy IF–THEN rules and fuzzy relations have been established, the next step is to reason the control output using fuzzy relations; that is, to command the electric motor and the ICE to meet the vehicle's power demands. In the example given above, there are three inputs, x, y, z, and one output, s. For the input of fuzzy set A on X, fuzzy set B on Y, and fuzzy set C on Z, the output fuzzy set D on S can be obtained as:

$$D = (A \text{ and } B \text{ and } C) \circ R = A \circ B \circ C \circ R \tag{6.20}$$

where $\circ$ represents the composition.

When there are three variables in the premise, fuzzy relation R has four terms (relations between x, y, z, and s). In the above example, a three-stage composition process for reasoning is employed, as in Eq. 6.20.

Example 6.7

If the states of an HEV are assumed to be:

- Battery SOC = 56% (minimal $SOC_{min} = 40\%$, maximal $SOC_{max} = 60\%$)
- Vehicle speed = 30 km/h
- Vehicle's power demand = 25 kW

determine the required power from the electric motor and the ICE based on the fuzzy rules of energy management introduced above.

Solution:

Corresponding to the given states of the vehicle, the fuzzy sets A, B, and C are:

$$A = \frac{2}{5}x_2 + \frac{3}{5}x_3; B = \frac{2}{3}y_1 + \frac{1}{3}y_2; C = \frac{13}{14}z_4 + \frac{1}{14}z_5; D = B \circ A \circ C \circ R = B \circ A \circ T' = B \circ T'';$$

$$T' = C \circ R; T'' = A \circ T'$$

$$T' = C \circ R = \begin{bmatrix} 0 & 0 & 0 & {}^{13}/_{14} & {}^{1}/_{14} \end{bmatrix} \circ R =$$

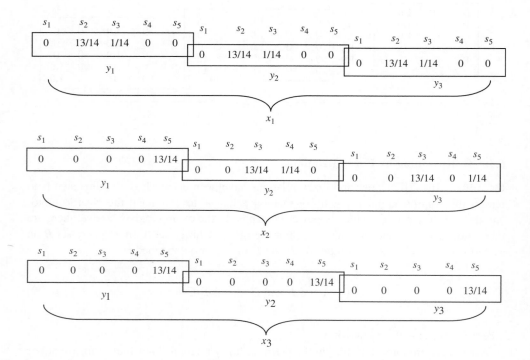

$$T'' = A \circ T' = \begin{bmatrix} 0 & 2/5 & 3/5 \end{bmatrix} \circ T' = \begin{array}{c} \\ y_1 \\ \\ y_2 \\ \\ y_3 \\ \\ \end{array} \begin{bmatrix} 0 & 0 & 0 & 0 & {}^{3}/_{5} \\ & & & & \\ 0 & 0 & {}^{2}/_{5} & {}^{1}/_{14} & {}^{3}/_{5} \\ & & & & \\ 0 & 0 & {}^{2}/_{5} & 0 & {}^{3}/_{5} \end{bmatrix} \begin{array}{c} s_1\ s_2\ s_3\ s_4\ s_5 \end{array}$$

Then the output of fuzzy reasoning D is:

$$D = B \circ T'' = [2/3 \quad 1/3 \quad 0] \circ \begin{bmatrix} 0 & 0 & 0 & 0 & 3/5 \\ 0 & 0 & 2/5 & 1/14 & 3/5 \\ 0 & 0 & 2/5 & 0 & 3/5 \end{bmatrix} = [0 \quad 0 \quad 1/3 \quad 1/14 \quad 3/5]$$

Through defuzzification, the definite value of the required power from the electric motor is 18.626 kW; therefore, the required power from the ICE equals $25 - 18.626 = 6.374$kW to meet the drivability requirements of the vehicle.

6.3.2.4 Diagram of Fuzzy-logic-based HEV Energy Management Strategy

The overall fuzzy-logic-based energy management strategy diagram is shown in Fig. 6.7. This strategy first employs fuzzy logic methodology to calculate the required power from the electric motor based on the vehicle's speed and power demand as well as the SOC of the battery system, then feeds this to the fuzzy logic algorithm to determine the power required from the motor for the final decision block to decide how much power the electric motor should provide and how much power the ICE should output to meet the vehicle's power requirements. The final decision is to adjust the electric motor output power calculated by fuzzy logic to make the ICE work at the predetermined operational points. For example, in the circumstances described in Example 6.6, the motor power (18.626 kW) calculated by fuzzy logic is adjusted to 18 kW with an ICE output of 7 kW to meet the 25 kW power requirement of the vehicle.

6.4 Determination of the Optimal ICE Operational Points of Hybrid Electric Vehicles

Since the working state of the ICE directly affects the fuel economy and emissions of an HEV, it is necessary to determine the optimal operational points specifically based on the hybrid electric vehicle's driving characteristics. This section introduces a two-step optimization method for determining the optimal operational points for an HEV based on engine-mapping data and the power requirements of practical drive cycles.

6.4.1 Mathematical Description of the Problem

Selecting the operational points (OPs) for the ICE in an HEV is an important calibration task. Conventionally, they are determined by experienced engineers from torque versus speed or power versus speed mapping data. Fundamentally, determining OPs is a trade-off between emissions and the fuel consumption rate, and the optimal problem can be formalized as follows.

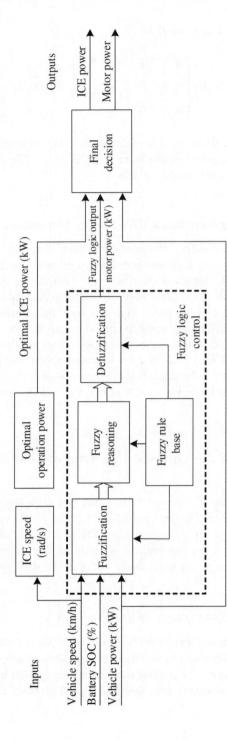

Figure 6.7 Fuzzy-logic-based HEV energy management strategy

Find the optimal operational points of the ICE in an HEV:

$$OP^* \in \text{possible } OPs \qquad (6.21)$$

Minimize the following two objective functions:

$$J_1 \text{ and } J_2 \; = \min \qquad (6.22)$$

The objective function J_1, objective function J_2, and the constraints are defined as:

$$\text{Objective function one}: \quad J_1 = a_1*fuel + a_2*\text{NOx} + a_3*\text{CO} + a_4*\text{HC} + a_5*\text{PM} \qquad (6.23)$$

$$\text{Objective function two}: \quad J_2 = b_1*J_{P1} + b_2*J_{P2} + \ldots + b_n*J_{Pn} \qquad (6.24)$$

where $a_1 - a_5$ are the weight factors corresponding to the relative importance between fuel economy and emissions, *fuel* is the fuel consumption, NOx is nitrogen oxides, CO is carbon monoxide, HC is hydrocarbon, PM is particle matter, $b_1 - b_n$ are the weight factors corresponding to the frequency of the power level at which the ICE operates, J_{P1} is the value of objective function J_1 at which the ICE works at zero power output (idle state), J_{Pn} is the value of objective function J_1 at which the ICE works at the nth power level.

$$\text{Constraint}: OP_1 \leq OP_2 \leq \; \cdots \; \leq OP_n \qquad (6.25)$$

where OP_1–OP_n are the operational speeds of an ICE corresponding to the required power/torque points; for example, OP_1 is the operational speed of the ICE at zero output power (idle state) and OP_n is the operational speed at which the ICE outputs its maximal power. This constraint represents the fact that a higher power requirement needs a higher operational speed.

6.4.2 Procedures of Optimal Operational Point Determination

The two-step optimization algorithm shown in Fig. 6.8 can be used to solve the above optimization problem. The strategy first uses the golden section search algorithm to obtain the local optimal operational point at each specific operational power level of the ICE, which minimizes the objective function J_1 (Eq. 6.23) without considering objective function J_2 (Eq. 6.24) or the constraint (Eq. 6.25); then, dynamic programming is employed to obtain the global minimum of the optimization problem which minimizes objective function J_2 subject to the practical operational constraints.

The detailed procedures are:

1. Determine how many operational points need to be calculated based on the operational specification of the ICE.
2. Calculate the values of objective function J_1 reflecting the different operational speeds at the given operational power level.

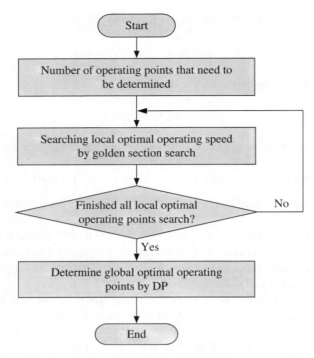

Figure 6.8 Algorithmic diagram to find the optimal points for the ICE in an HEV

3. Find the optimal operational speed which minimizes objective function J_1 (Eq. 6.23) at each given power point by an optimization method such as the golden search method described in the next section.
4. Find the global optimal operational points which minimize the objective function J_2 (Eq. 6.24) considering the constraints (Eq. 6.25) by the dynamic programming methods described in the following section.

6.4.3 Golden Section Searching Method

A golden section search is a commonly used method to solve a one-dimensional optimization problem. If we assume that the distance between two ends (say, a and c) is L, the optimal middle point, b, for searching is the fractional distance 0.618 from one end and 0.382 from the other end. As this fraction was called the golden ratio in ancient Greece, the searching method based on this section is called the *golden section search* (Rao, 1996). The golden section searching process can be illustrated as in Fig. 6.9. The minimum is originally bracketed by points 1, *, and 2. Based on the principle of the golden section search method, the objective function value is evaluated at the point 3, which is the fractional distance 0.382 from point 2 and 0.618 from point 1, then point 2 is replaced by point 3; then the objective

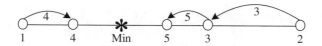

Figure 6.9   Golden section search

function value is evaluated at point 4, which is the fractional distance 0.382 from point 1 and 0.618 from point 3, and then point 1 will be replaced by point 4; then at 5, which replaces 3. After several steps, as shown in Fig. 6.9, the minimum is bracketed in a smaller and smaller range; for example, after three steps, the minimum will be bracketed between points 4 and 5.

6.4.4   Finding the Optimal Operational Points

The search process for the operational points of an ICE can be stated in the following form: once the local optimization results for the objective function J_1 have been obtained based on a search method such as the golden search, and the objective function J_2 and the constraint equation have been given, the optimal operational points of the ICE of an HEV can be found and they are the points that make the sum of values of objective function J_1 from power level 1 to n minimal. The mathematical expression is:

$$Find: OP^* = \{ op_1^* \ op_2^* \ \cdots \ op_n^* \}^{\mathrm{T}} \ \text{makes} \ \sum_{i=1}^{n} b_i J_{Pi} = \min \tag{6.26}$$

$$Subject\ to: op_1 \le op_2 \le \cdots \le op_n \tag{6.27}$$

where op_i is the ICE operational speed corresponding to the operational power level i, b_i is the weight for the given operational power level, and J_{Pi} is the value of objective function J_1 corresponding to the operational speed at the given operational power level.

It is obvious that the above problem is a multi-stage decision problem, which means that the operational speed has to be determined at the given operational power level of the ICE, as shown in Fig. 6.10. In other words, the objective function J_i needs to be minimized in order to achieve the optimal fuel economy and emissions. The dynamic programming technique is well suited to solving such problems (Bellman, 1957).

6.4.5   Example of the Optimal Determination

This example is based on a series HEV with the following system parameters:

- ICE: 50 kW maximal power diesel engine; the speed–torque and speed–power curves are shown in Fig. 6.11, and the mapped fuel consumption and emission data are shown in Table 6.4

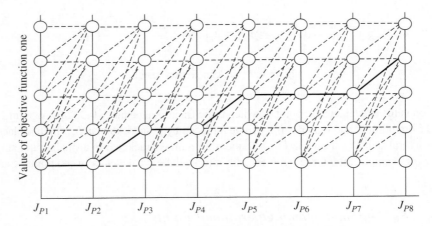

Figure 6.10 Process of optimizing operational points

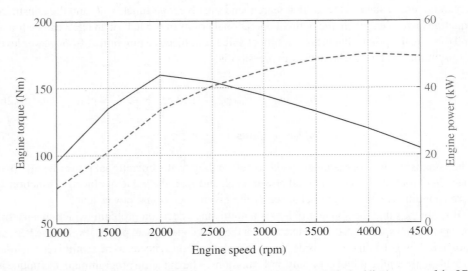

Figure 6.11 Power versus speed (solid line) and torque versus speed (dashed line) curves of the ICE

- Generator: 45 kW rated maximal power, maximal usable power 40 kW; the speed–torque and speed–power curves are shown in Fig. 6.12
- Electric motor: 67 kW rated maximal power; the speed–torque and speed–power curves are shown in Fig. 6.13
- Gear reduction ratio: 3.05:1
- DC–DC converter: 1.4 kW
- Battery system: 8.5 Ahr NiMH battery with 70% initial SOC

Table 6.4 Corresponding engine-mapping data for optimizing the operational points

	Speed [rpm]	3400	3200	3000	2800	2600	2400	2200	2000	1800	1600
No load	Fuel [kg/h]								0.64	0.64	0.32
	HC [ppm]								283.2	399.2	336
	CO [ppm]								700	880	720
	NOx [ppm]								10.4	4.98	9.12
5 kW	Fuel [kg/h]						1.28	1.28	1.28	0.96	0.96
	HC [ppm]						616	440	392	590.4	896
	CO [ppm]						216	160	160	212.8	296
	NOx [ppm]						30	31.2	30.4	24	24
10 kW	Fuel [kg/h]	4.08	3.76	3.44	2.8	2.8	2.48	2.48	2.16	1.84	1.84
	HC [ppm]	25.96	22.68	22.68	22.84	22.84	22.68	1.00	1.00	1.00	0.20
	CO [ppm]	23.7	24.5	23.7	12.5	10.9	5.72	1.82	1.3	0.58	0.01
	NOx [ppm]	112	104.8	100.8	66.4	64	59.2	52	45.6	52	56
15 kW	Fuel [kg/h]	4.96	4.4	4.08	3.76	3.44	3.12	3.12	2.8	2.48	2.48
	HC [ppm]	25.98	25.98	25.98	25.98	22.66	19.40	19.40	17.80	16.20	16.20
	CO [ppm]	20.5	18.9	18.1	12.34	6.5	4.676	2.33	1.54	1.06	1.06
	NOx [ppm]	138.4	134.4	132	79.2	74.4	73.6	76	104	100	104
20 kW	Fuel [kg/h]	5.76	5.28	4.96	4.72	4.24	4.08	4.08	3.76	3.76	
	HC [ppm]	25.98	25.98	23.00	22.68	19.40	19.40	19.40	16.20	16.20	
	CO [ppm]	14.4	14	13.2	9.12	5.216	3.36	2.336	0.776	0.48	
	NOx [ppm]	225.6	216	204	120	120	113.6	104	132	156	
25 kW	Fuel [kg/h]	6.56	6.24	5.92	5.6	5.12	4.96	5.12	4.96	4.72	
	HC [ppm]	22.60	22.20	19.40	19.40	19.40	16.20	13.00	16.20	16.20	
	CO [ppm]	16	12.8	11.32	7.36	4.64	3.36	2.59	3.36	4	
	NOx [ppm]	288	272	272	176	196	176	144	160	212	
30 kW	Fuel [kg/h]	7.52	7.28	6.96	6.56	6.24	6.24	5.92			
	HC [ppm]	19.40	16.20	16.20	19.40	13.00	13.00	16.20			
	CO [ppm]	15.2	14.4	13.2	7.6	6.26	5.6	5.74			
	NOx [ppm]	328	328	248	148	152	120	140			
35 kW	Fuel [kg/h]	8.4	8.24	8	7.68	7.52					
	HC [ppm]	16.20	13.00	13.00	13.00	13.00					
	CO [ppm]	20.8	17.6	16.16	9.2	8.88					
	NOx [ppm]	400	392	392	200	216					

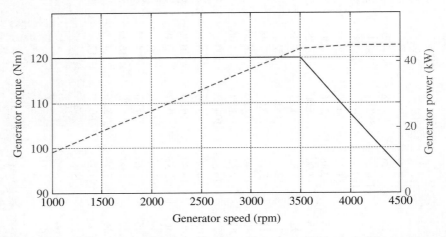

Figure 6.12   Torque versus speed (solid line) and power versus speed (dashed line) curves of the generator

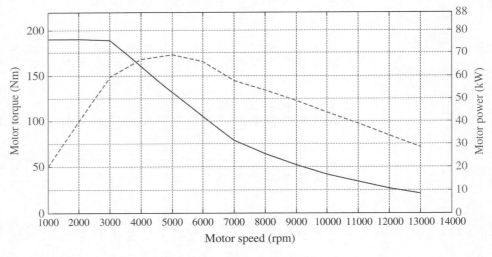

Figure 6.13   Torque versus speed (solid line) and power versus speed (dashed line) curves of the electric motor

- Final drive ratio: 4.1:1
- Vehicle weight: 2400 kg
- Vehicle Cd: 0.46
- Wheel radius: 0.335 m
- Scheduled road grade is 0%

The optimal operational points are determined from the following procedures based on the method introduced in previous sections. For more detailed descriptions of the optimal ICE operational point determination method, the interested reader should refer to the patent by Liu and Bouchon (2008a).

1. Determine the operational speed range and speed increment at various power levels from idle speed to the maximal speed; for example, the operational range and speed increment of the engine is [900 1000 ... 3200 3400] (rpm).
2. The calculated objective function J_1 values are listed in Table 6.5 based on the mapping data (Table 6.4) at the allowable operational speeds for the given operational power level.
3. The optimal values at the given operational power levels are marked in bold italic in Table 6.5.
4. The global optimal operational points are then determined by minimizing objective function J_2 subject to the constraint (Eq. 6.25), and the dynamic programming method is used to minimize the objective function.
5. The determined operational points are [900, 1500, 1500, 1700, 1800, 1800, 2100, 2600] (rpm), corresponding to the given operational power levels [0, 5, 10, 15, 20, 25, 30, 35] (kW), and the optimized operational point curve is shown in Fig. 6.14.

6.4.6   Performance Evaluation

In order to illustrate the effect of the optimized operational points, this section presents two performance studies on a parallel HEV based on the EPA75 drive cycle shown in Fig. 6.15. The required ICE power over the cycle, shown in Fig. 6.16, is commanded by the energy management algorithm of the HEV. The manually set ICE operational points are [1600 1700 2000 2200 2200 2200 2400 2600] (rpm), corresponding to the operational power levels [0 5 10 15 20 25 30 35] (kW) shown in Fig. 6.17.

6.4.6.1   Comparison Study One

If the weight factors of the objective function J_1 are set as [10 1 1 1], as shown in Eq. 6.28 below, and the weight factors of the objective function J_2 are set as [1 1 1 5 10 5 1 1], as shown in Eq. 6.29, the computed optimal operational points are [900 1500 1500 1700 1800 1800 2100 2600] (rpm), corresponding to the given operational power levels based on the engine-mapping data shown in Table 6.4. The optimized ICE operational setpoints are shown in Fig. 6.18. The objective function J_1 is to minimize fuel usage over the drive cycle by putting a bigger penalty factor on the fuel usage term; while the objective function J_2 is to minimize global fuel consumption over the cycle by putting a bigger penalty factor on the 25 kW power level at which the ICE mostly operates.

Table 6.5 Values of objective function one (J_1) at each power requirement point

Speed[rpm]	3400	3200	3000	2800	2600	2400	2200	2000	1800	1600
No load J_{P1} value						1.0813		1.1414	1.4518	***1.1354***
5 kW J_{P2} value							0.832	***0.7785***	1.007	1.4284
10 kW J_{P3} value	0.633	0.5851	0.5446	0.4079	0.402	0.3553	0.3151	0.2708	***0.2483***	0.2538
15 kW J_{P4} value	0.765	0.6992	0.6615	0.5314	0.4805	0.4409	0.4425	0.4556	***0.4137***	0.4206
20 kW J_{P5} value	0.9924	0.9263	0.8687	0.6944	0.6375	0.6081	***0.5904***	0.6009	0.6421	
25 kW J_{P6} value	1.18	1.1157	1.0784	0.8756	0.8582	0.8024	***0.7592***	0.7748	0.8407	
30 kW J_{P7} value	1.3429	1.314	1.1418	0.9257	0.8915	***0.8355***	0.8409			
35 kW J_{P8} value	1.5598	1.5227	1.4966	***1.1248***	1.1357					

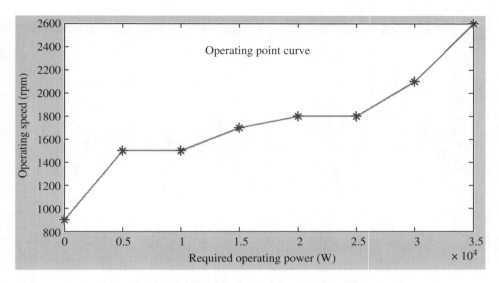

Figure 6.14 Determined optimal operational points when the objective functions one and two are set as $J_1 = 10.0*fuel + 1.0*NOx + 1.0*CO + 1.0*HC$ and $J_2 = 1.0*J_{P1} + 1.0*J_{P2} + 1.0*J_{P3} + 5.0*J_{P4} + 10.0*J_{P5} + 5.0*J_{P6} + 1.0*J_{P7} + 1.0*J_{P8}$

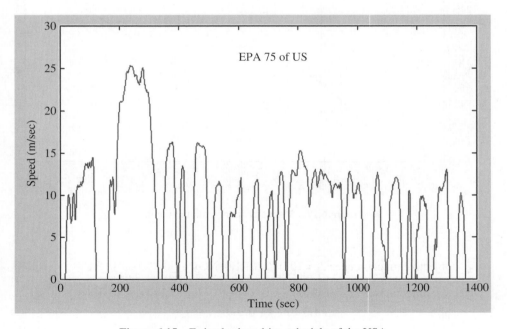

Figure 6.15 Federal urban drive schedule of the USA

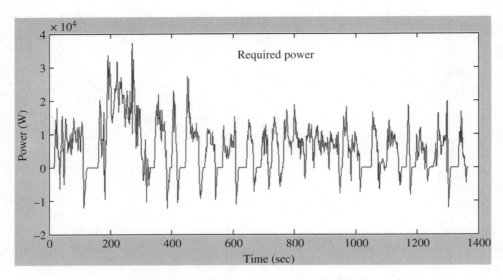

Figure 6.16   Required ICE power over the EPA75 drive cycle

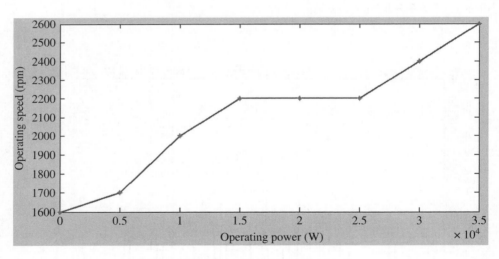

Figure 6.17   Manually set operational points

Objective function one : $J_1 = 10.0*fuel + 1.0*\text{NOx} + 1.0*\text{CO} + 1.0*\text{HC}$ (6.28)

Objective function two : $J_2 = 1.0*J_{P1} + 1.0*J_{P2} + 1.0*J_{P3} + 5.0*J_{P4} + 10.0*J_{P5} + 5.0*J_{P6}$
$$+ 1.0*J_{P7} + 1.0*J_{P8}$$

(6.29)

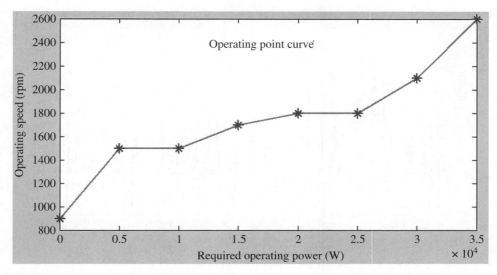

Figure 6.18 Optimized operational point curve using $J_1 = 10.0*fuel + 1.0*NOx + 1.0*CO + 1.0*HC$ and $J_2 = 1.0*J_{P1} + 1.0*J_{P2} + 1.0*J_{P3} + 5.0*J_{P4} + 10.0*J_{P5} + 5.0*J_{P6} + 1.0*J_{P7} + 1.0*J_{P8}$

The average results achieved over the drive cycle are as follows:

1. When the ICE operates at the manually set operational points, the average fuel consumption is 2.65 (kg/h), the average NOx emissions are 66.65 (ppm), the average CO emissions are 85.4 (ppm), and the average HC emissions are 112.3 (ppm).
2. When the ICE operates at the computed operational points optimized by the given objective functions, the average fuel consumption is 2.22 (kg/h), the average NOx emissions are 85.74 (ppm), the average CO emissions are 46.3 (ppm), and the average HC emissions are 84.1 (ppm).

The comparison shows that if the ICE operates at the optimal operational points over an EPA75 cycle, the fuel economy is improved by approximately 16.2%, CO and HC emissions are reduced by about 45.8% and 25.1% respectively, at the cost of a 28.7% increase in NOx emissions. The detailed comparison curves on fuel consumption and emissions are shown in Figs 6.19 to 6.22.

6.4.6.2 Comparison Study Two

Unlike Study one, here we put bigger penalty factors on the fuel consumption and NOx emissions of the objective function J_1 and bigger penalty factors on the 20 kW, 25 kW, and 30 kW operational power levels of the objective function J_2, so that the optimized

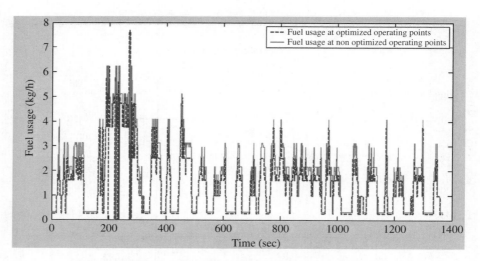

Figure 6.19   Fuel consumption comparison over an EPA drive cycle

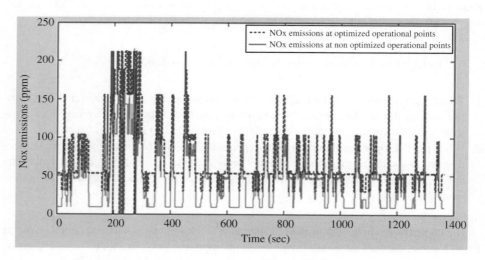

Figure 6.20   NOx emissions comparison over an EPA drive cycle

operational points are $[1700\ \ 1700\ \ 1900\ \ 2300\ \ 2300\ \ 2300\ \ 2400\ \ 2600]$ (rpm), the curve of which is shown in Fig. 6.23. The detailed objective functions are:

Objective function one: $J_1 = 10.0*fuel + 10.0*\text{NOx} + 1.0*\text{CO} + 1.0*\text{HC}$ (6.30)

Objective function two: $J_2 = 1.0*J_{P1} + 1.0*J_{P2} + 1.0*J_{P3} + 5.0*J_{P4} + 10.0*J_{P5} + 5.0*J_{P6}$
$+ 1.0*J_{P7} + 1.0*J_{P8}$

(6.31)

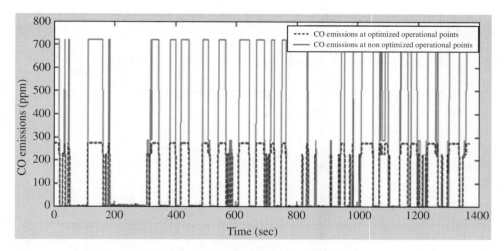

Figure 6.21 CO emissions comparison over an EPA drive cycle

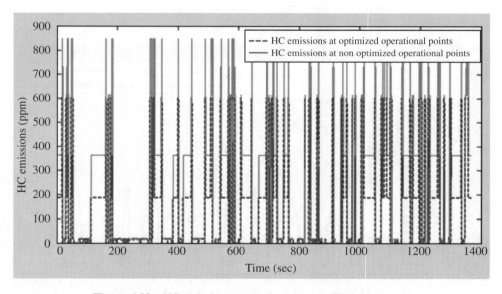

Figure 6.22 HC emissions comparison over an EPA drive cycle

The study results show that the average fuel consumption is 2.61 (kg/h), the average NOx emissions are 66.24 (ppm), the average CO emissions are 103.2 (ppm), and the average HC emissions are 120.2 (ppm) when the ICE operates at the optimized operational points given by the above objective functions. Compared with the case where the ICE operates at the manually set operational points, the optimal operational points improve the fuel economy

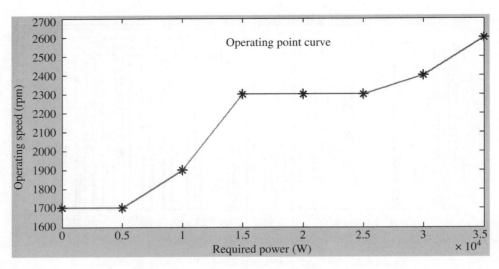

Figure 6.23   The optimized operational points when $J_1 = 10.0*fuel + 10.0*NOx + 1.0*CO + 1.0*HC$ and $J_2 = 1.0*J_{P1} + 1.0*J_{P2} + 1.0*J_{P3} + 5.0*J_{P4} + 10.0*J_{P5} + 5.0*J_{P6} + 1.0*J_{P7} + 1.0*J_{P8}$

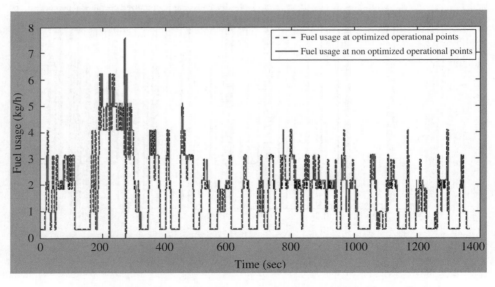

Figure 6.24   Fuel consumption comparison over an EPA drive cycle

by approximately 1.5% and reduce NOx emissions by about 0.6%; these are traded off by increasing CO and HC emissions by about 20.8% and 7%, respectively. The detailed comparison curves on fuel consumption and emissions are shown in Figs 6.24 to 6.27.

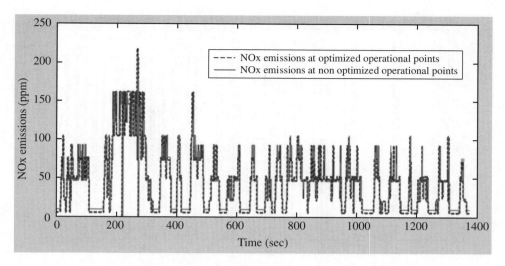

Figure 6.25 NOx emissions comparison over an EPA drive cycle

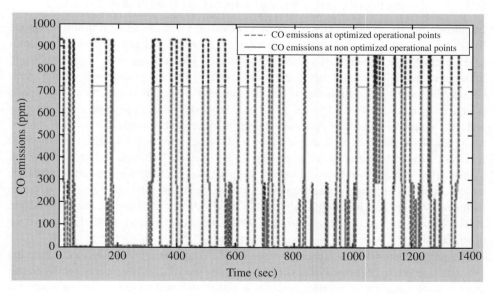

Figure 6.26 CO emissions comparison over an EPA drive cycle

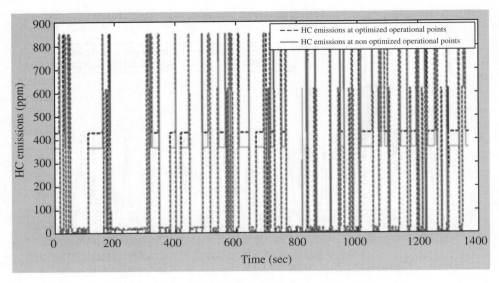

Figure 6.27   HC emissions comparison over an EPA drive cycle

6.5   Cost-function-based Optimal Energy Management Strategy

Engineering practices have proven that hybrid electric vehicles can significantly improve fuel economy and lower greenhouse gas emissions. A crucial task in hybrid electric propulsion system design is to optimally distribute a vehicle's power demand between the ICE and the electric motor. The quality of this decision directly affects the overall performance of the vehicle, including drivability, fuel economy, and emissions. Therefore, most automobile companies, including GM, Toyota, and Ford, have developed many new technologies and filed a massive number of patents on energy management strategies for hybrid vehicles. Many scientists and engineers have also proposed various approaches to achieve the optimal fuel economy and emissions (Powers and Nicastri, 1999; Rahman, Butler, and Ehsani, 1999). These include real-time optimization algorithms for fuel economy and emissions, various intelligent energy-management methods, and cost-function-based strategies (Johnson, Wipke, and Rausen, 2000; Piccolo *et al.*, 2001; Schouten, Salman, and Kheir, 2003; Won *et al.*, 2003; Fussey *et al.*, 2004; Cox and Bertness, 2005; Liu and Bouchon, 2008b).

This section outlines a cost-function-based HEV energy management strategy. A detailed example is given to illustrate how to configure the objective function and adapt the equivalent fuel consumption of the electric motor in real time, based on the amount of recovered energy from regenerative braking. This section also outlines how to take the constraints of the power requirement and battery life into account in a practical hybrid electric vehicle system design.

6.5.1 Mathematical Description of Cost-function-based Optimal Energy Management

6.5.1.1 Definition of the Cost Function

The cost function to optimize the energy flow of a hybrid electric vehicle can be defined as:

$$J = Cost_{engine} + Cost_{motor} + Cost_{battery\_life} + Cost_{energy\_balance}$$
$$= C_{fuel} \cdot Power_{eng} + C_{electric} \cdot Power_{mot} + C_{bat\_life} \cdot Power_{mot} + C_{energy\_balance} \cdot Power_{mot}$$
$$= C_{fuel} \cdot Power_{eng} + \left(C_{electric} + C_{battery\_life} + C_{energy\_balance} \right) \cdot Power_{mot}$$

$$(6.32)$$

In Eq. 6.32, C_{fuel} is the weight factor for the fuel cost, which can be calibrated based on a fuel consumption map of the engine, such as that shown in Table 6.4, and the unit is grams per kW power per sampling time of engine control system $(g/(Kw \cdot T_{eng}))$

$$C_{fuel} = f \left(Temperature_{eng}, Torque_{eng}, Speed_{eng} \right) \quad (6.33)$$

$C_{electric}$ is the weight factor for the electrical energy cost, which is normalized and equalized to the fuel cost of the engine. The normalization is based on the best fuel rate of the engine at the current operational speed and efficiencies of the electric motor and battery system, as follows:

$$C_{electric} = \frac{Min(C_{fuel})}{\eta_{mot} \cdot \eta_{bat}} \quad (6.34)$$

where η_{bat} is the efficiency of the battery system, whereby $\eta_{bat} = \eta_{discharge}$ if $Power_{mot} \leq 0$ and $\eta_{bat} = \eta_{charge}$ if $Power_{mot} > 0$ in the cost function (Eq. 6.32), η_{mot} is the efficiency of the motor/inverter system under the current operational conditions.

C_{bat_life} is the weight factor for the cost of battery life loss, which is a function of the battery's SOC, temperature, and electrical power:

$$C_{bat\_life} = f(SOC, T, P_{mot}) \quad (6.35)$$

$C_{energy\_balance}$ is the weight factor for the cost involving the battery's SOC being out of alignment with the desired operational setpoint $SOC_{desired}$, which is a function of the battery SOC, temperature, and electrical power:

$$C_{energy\_balance} = f(SOC, T, P_{mot}) \quad (6.36)$$

All cost weight factors in the cost function (Eq. 6.32) are normalized with the real fuel consumption rate of the engine under the given operational conditions.

6.5.1.2 Required Power

In a parallel hybrid electric vehicle system, the vehicle's power requirement is always equal to the sum of the engine's output power and the motor's output power. The engine always outputs positive power, but the motor's power can be positive (propulsion) or negative (regenerative braking). Therefore, the following expressions hold:

$$Power_{req\_veh} = Power_{engine} + Power_{motor} \tag{6.37}$$

$$Power_{engine} \geq 0 \tag{6.38}$$

6.5.1.3 Operational Constraints

In order to assign the vehicle's power demands to the ICE and electric motor optimally, it is necessary to consider the physical limitations of each component in a practical hybrid vehicle. The limitations can be defined as constraints when performing optimization.

- **Constraints from the energy storage system**
  Maximal charging power: The maximal allowable charging power availability of the ESS is a function of temperature, SOC, and SOH under the given operational conditions. The following equation can be used to describe this constraint:

$$Power_{max\_chg\_bat} = f(SOC, Temp, SOH) \tag{6.39}$$

where $Power_{max\_chg\_bat}$ is the maximal charging power availability of the ESS in the next control period.
  Maximal discharging power: The maximal allowable discharging power availability of the ESS is also a function of temperature, SOC, and SOH. The following equation can be used to describe this constraint:

$$Power_{max\_dischg\_bat} = f(SOC, Temp, SOH) \tag{6.40}$$

where $Power_{max\_dischg\_bat}$ is the maximal discharging power availability of the ESS in the next control period.

- **Constraints from the electric motor**
  Maximal propulsion power: The maximal propulsion power of the electric motor is a function of the temperature, speed, and torque of the motor under the given operational conditions. The following equation can be used to describe the constraint:

$$Power_{max\_prop\_mot} = f(Temp, Speed, Torque) \tag{6.41}$$

where $Power_{max\_prop\_mot}$ is the propulsive power constraint of the electric motor in the next control period.

Maximal regenerative power: The maximal regenerative power of the motor is also related to the temperature, speed, and torque of the motor under the given operational conditions. It can be expressed by the following equation:

$$Power_{max\_regen\_mot} = f(Temp,\ Speed,\ Torque) \tag{6.42}$$

where $Power_{max\_regen\_mot}$ is the regenerative power constraint of the electric motor for the next control period.

- **Constraints from the engine**

Maximal propulsion power: The maximal propulsion power of the engine is a function of its speed, which can be described by:

$$Power_{max\_eng} = f(Speed) \tag{6.43}$$

6.5.1.4 Description of the Optimization Problem

The optimization objective is to find the optimal power distribution to the ICE and electric motor of an HEV such that the vehicle has the best fuel economy under the current operational conditions. The optimization problem can be stated mathematically as follows:

Objective function:

$$J = C_{fuel} \cdot Power_{eng} + \left(C_{ele} + C_{bat\_life} + C_{energy\_balance}\right) \cdot Power_{mot} \tag{6.44}$$

Subject to:

$$Power_{eng} + Power_{mot} = Power_{demand\_veh} \tag{6.45}$$

Under the constraints:

1. If $Power_{demand\_veh} \geq 0$

$$\begin{aligned} Power_{mot} &\leq \min\left(Power_{max\_dischg\_bat},\ Power_{max\_prop\_mot}\right) \\ Power_{eng} &\leq Power_{max\_eng} \end{aligned} \tag{6.46}$$

2. If $Power_{demand\_veh} < 0$

$$Power_{mot} > \max\left(Power_{max\_chg\_bat},\ Power_{max\_regen\_mot}\right) \tag{6.47}$$

6.5.2   An Example of Optimization Implementation

The given optimization example is based on the following hybrid vehicle system configuration, parameters, and operational conditions:

- Hybrid vehicle system configuration: parallel
- Control period of time: 10 ms
- ICE: 50 kW maximal power diesel engine; the mapped fuel consumption and emissions data are listed in Table 6.4 and modified to milligrams per kW/10ms from the original grams per kWh
- Electric motor: 30 kW rated maximal power
- Vehicle operating speed: 110 km/h
- Vehicle's power demand: 40 kW
- SOC setpoint of the battery system: 60%
- SOC operational window of the battery system: 40–80%
- Battery age: beginning of life

The real-time power split optimization algorithm diagram is shown in Fig. 6.28. The detailed optimization results are listed in Table 6.6. The best power combination under the given operational conditions is that the engine outputs 25 kW power and the electric motor outputs 15 kW power.

6.6   Optimal Energy Management Strategy Incorporated with Cycle Pattern Recognition

The main drawback of the aforementioned energy management strategies is that the decision is based only on the current vehicle operating conditions without considering the driving cycle or the driver's driving style; therefore, the optimal performance cannot be guaranteed after a longer period of operation. In this section, an energy management strategy based on driving situation awareness is briefly introduced, and this is incorporated with driving cycle/style pattern recognition algorithms. The dynamic programming (DP) optimization algorithm makes the optimal power split decision between the engine and electric motor based on the present vehicle operational speed and the torque demand, the present battery system operational condition, the vehicle's operational environment conditions, and the predicted future driving power profile output from the pattern recognition algorithm unit. The strategy diagram is shown in Fig. 6.29.

6.6.1   Driving Cycle/Style Pattern Recognition Algorithm

The entire pattern recognition algorithm diagram, shown in Fig. 6.30, consists of roadway recognition, driving cycle pattern recognition, and driving style recognition algorithms; the

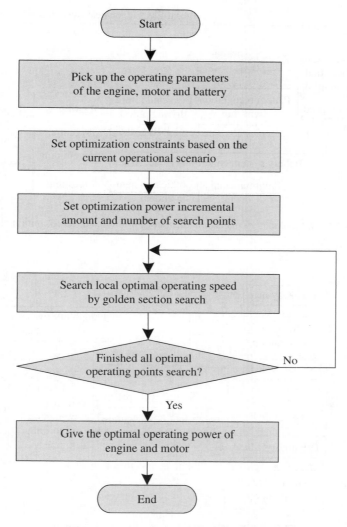

Figure 6.28 A real-time power split optimization algorithm

Table 6.6 Evaluated results in the period of 10 ms

No.	Total cost	Engine power (W)	Motor power (W)	Veh req'd power	Engine speed	Motor cost	Engine cost	Battery life cost
1	0.0623	50022	−10022	40000	2000	−0.004377	0.06675	0.001
2	0.0611	24444	15556	40000	2000	0.014077	0.044572	0.0001
3	0.0593	29986	10014	40000	2000	0.0087503	0.050573	0.001
4	0.0592	24882	15118	40000	2000	0.013563	0.045047	0.0001
5	0.0803	64972	−14972	50000	2500	−0.006092	0.086316	0.001
6	0.0793	38628	11372	50000	2500	0.010609	0.0661	0.0001
7	0.0769	35020	14980	50000	2500	0.01480	0.062068	0.001
8	0.0768	36856	13144	50000	2500	0.01247	0.06412	0.0001

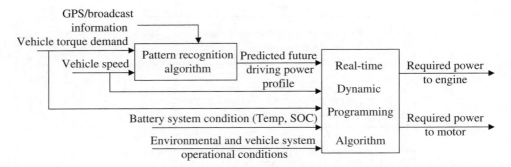

Figure 6.29 Optimal energy management strategy with a pattern recognition algorithm

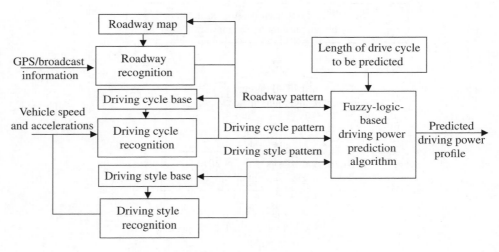

Figure 6.30 Entire pattern recognition algorithm

recognized patterns are further fed to the fuzzy-logic-based driving power prediction algorithm to generate the future driving power profile in the given prediction horizon period of time.

6.6.1.1 Roadway Recognition (RWR)

The mission of the roadway recognition algorithm is to identify the current roadway conditions combined with the traffic congestion situation based on GPS or broadcasting information and stored street maps. The roadway map is updated to the previous roadway recognition.

6.6.1.2 Driving Cycle Recognition (DCR)

The driving cycle recognition algorithm works out the current driving pattern based on the vehicle's speed and acceleration and on stored cycle data. The DCR's mission is to extract the key statistical features of the driving pattern and characteristic parameters. The acquired information will be used by the fuzzy-logic-based power prediction algorithm to predict the vehicle's power demand in a certain period of time ahead. DCR is the most important part of this approach. There are a few mature methods for extracting the pattern from an actual driving cycle, but there is no consensus among researchers as to how to define the characteristic parameters over a driving cycle (Langari and Won, 2003). Ericsson defined and grouped a set of parameters representing a typical HEV driving cycle and presented a method of extracting characteristic parameters from an actual driving cycle (Ericsson, 2001).

6.6.1.3 Driving Style Recognition (DSR)

Driving style has a strong influence on fuel economy and emissions. There are three types of driving style: *light*, *mild*, and *aggressive*. The driving style recognition algorithm is designed to provide driving style information to the power prediction algorithm based on current and historical data to predict the vehicle's power demand in a certain period of time ahead.

6.6.1.4 Fuzzy-logic-based Vehicle Power Demand Prediction

The predicted power profile will be produced by the fuzzy-logic-based algorithm. The inputs are the recognized driving roadway, driving style, driving cycle, and the desired time length of the predicted power profile horizon. The output is the vehicle's required power profile for the prediction horizon. The detailed fuzzy rule base can be developed from various practical driving scenarios. Based on the methods introduced in previous sections, fuzzy relations can be set up and fuzzy reasoning can be carried out.

6.6.2 Determination of the Optimal Energy Distribution

Once the predicted power profile is obtained, the dynamic programming (DP) method can be employed to assign the vehicle's power demand to the engine and the electric motor with maximal fuel economy globally. The optimality principle, illustrated in Fig. 6.31, can be used to find the exact command power/torque to the engine and electric motor at each time point in the predicted horizon by minimizing the given cost function over the whole predicted vehicle power profile. The given cost function may have the following form as an example:

$$J = \sum_{k=0}^{N} \left[(\textit{fuel}(k) + \alpha_1 \mathrm{NOx} + \alpha_2 \mathrm{CO} + \alpha_3 \mathrm{PM}(k)) + \beta \left((SOC(N) - SOC(0))^2 \right) \right] \qquad (6.48)$$

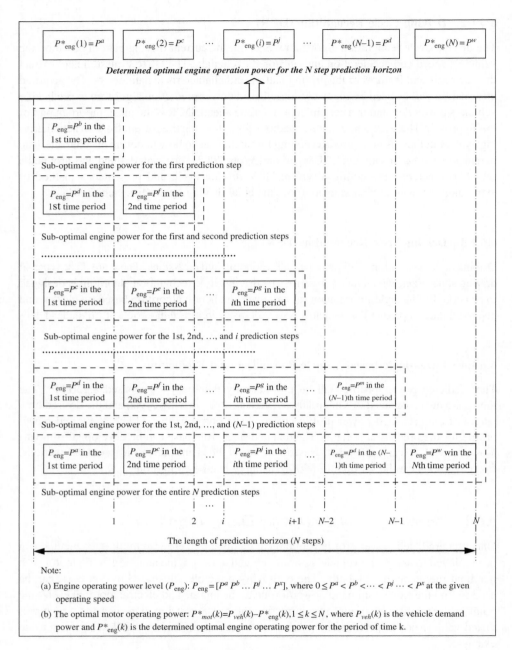

Figure 6.31 Optimality principle

where k is the index of the time step of the power profile prediction horizon, N is the length of the prediction horizon, and α_x and β are the weight factors for the emissions and battery charge deviation. Weight factor β is to ensure that the control strategy will not result in significant charge depletion.

References

Bellman, R. E. *Dynamic Programming*, Princeton University Press, Princeton, NJ, 1957.

Cox, M. and Bertness, K. *Energy Management System for Automotive Vehicle*, US patent, No. 20050024061, February 3, 2005.

Ericsson, E. 'Independent driving pattern factors and their influence on fuel use and exhaust emission factor,' *Transportation Research Part D*, **6**(5), pp. 325–345, 2001.

Fussey, P., Porter, B., Wheals, J., and Goodfellow, C. '*Hybrid Power Sources Distribution Management*,' US patent, No. 20040074682, April 22, 2004.

Johnson, V. H., Wipke, K. B., and Rausen, D. J. 'HEV Control Strategy for Real-Time Optimization of Fuel Economy and Emissions,' SAE paper 2000-01-1543.

Kheir, N. A., Salman, M. A., and Schouten, N. J. 'Emissions and Fuel Economy Trade-off for Hybrid Vehicles Using Fuzzy Logic,' *Mathematics and Computers in Simulation*, **66**, 155–172, 2004.

Langari, R. and Won, J. S. 'Intelligent Energy Management for Hybrid Vehicles via Drive Cycle Pattern Analysis and Fuzzy Logic Torque Distribution,' *Proceedings of the 2003 IEEE International Symposium on Intelligent Control*, pp. 223–228, Houston, Texas, October 5–8, 2003.

Liu, W. and Bouchon, N. 'Method, Apparatus, Signals, and Media for Selecting Operating Conditions of A Genset,' Patent Application Publication, United States, Publication No. US2008/0122228A1, May 29, 2008a.

Liu, W. and Bouchon, N. 'Method, Apparatus, Signals, and Media for Managing Power in a Hybrid Vehicle,' Patent Application Publication, United States, Publication No. US2008/0059013 A1, March 6, 2008b.

Mamdani, E. H. 'Application of Fuzzy Algorithms for Control of a Simple Dynamic Plant,' *Proceedings of IEEE*, **121**(12), 1585–1588, 1974.

Piccolo, A., Ippolito, L., zo Galdi, V., and Vaccaro, A. 'Optimisation of energy flow management in hybrid electric vehicles via genetic algorithms,' *Proceedings of the 2001 IEEE/ASME International Conference on Advanced Intelligent Mechatronics*, pt. 1, pp. 433–439.

Powers, W. F. and Nicastri, P. P. 'Automotive vehicle control challenges in the twenty-first century,' 14th World Congress of IFAC, Beijing, 1999.

Rahman, Z., Butler, K. L., and Ehsani, M. 'A Study of Design Issues on Electrically Peaking Hybrid Electric Vehicles for Diverse Urban Driving Patterns,' Society of Automotive Engineers, SAE 1999-01-1151.

Rao, S. S. *Engineering Optimization – Theory and Practice*, 3rd edition, New Age International Ltd., New Delhi, 1996.

Schouten, N. J., Salman, M. A., and Kheir, N. A. 'Energy Management Strategies for Parallel Hybrid Vehicles Using Fuzzy Logic,' *Control Engineering Practice*, **11**, 171–177, 2003.

Sciarretta, A., Back, M., and Guzzella, L. 'Optimal Control of Parallel Hybrid Electric Vehicles,' *IEEE Transactions On Control Systems Technology*, **12**, 352–363, 2004.

Takagi, T. and Sugeno, M. 'Fuzzy Identification of Systems and its Applications to Modeling and Control,' *IEEE Transactions on Systems, Man and Cybernetics*, **15**(1), 116–132, 1985.

Tanaka, K. *An Introduction to Fuzzy Logic for Practical Applications*, Springer-Verlag New York, LLC, 1996.

Won, J. S. and Langari, R. 'Intelligent energy management agent for a parallel hybrid vehicle,' *American Control Conference*, pt. 3, pp. 2560–2565, 2003.

Zadeh, L. A. 'Fuzzy sets,' *Information and Control*, **8**(3), 338–353, 1965.

7

Other Hybrid Electric Vehicle Control Problems

7.1 Basics of Internal Combustion Engine Control

The primary function of engine control is to adjust the engine torque generated to meet the torque required by the driver, and at the same time to satisfy the requirements regarding emissions, fuel economy, driving comfort, and safety. For a spark-ignition (SI) engine, the torque is generated through a combustion process and is determined by air mass, fuel mass, and ignition time, as shown in Fig. 7.1. For a diesel engine, the torque is also generated by a combustion process, but it is a function of injected fuel mass and injection time, as shown in Fig. 7.2. In order to achieve these objectives, all variables that influence torque generation are controlled by the electronic/engine control unit (ECU). This section will briefly describe the general schematic diagrams of SI and diesel engine control. The detailed engine controls are very sophisticated and are beyond the scope of this book. The interested reader should refer to the book by Guzzella and Onder (2010) listed at the end of this chapter.

7.1.1 SI Engine Control

The SI engine control system is complex and consists of many subsystems. From a control point of view, it is a multi-input, multi-output (MIMO) interconnected control system where the fuel delivery and injection system, the air–fuel mixing system, and the cooling and warming up system are the main subsystems. The most important control functions include starting and post-starting enrichment control, warm-up enrichment control, acceleration enrichment control, full-throttle enrichment control, overrun fuel cut-off control,

Hybrid Electric Vehicle System Modeling and Control, Second Edition. Wei Liu.
© 2017 John Wiley & Sons Ltd. Published 2017 by John Wiley & Sons Ltd.

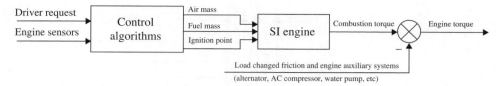

Figure 7.1   The torque generation process for an SI engine

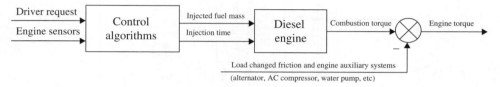

Figure 7.2   The torque generation process for a diesel engine

engine speed limit control, closed-loop engine idle-speed control, and lambda control. A diagram of the overall SI engine control system is presented in Fig. 7.3. The measured variables of the SI engine include the acceleration pedal position, brake pedal position, engine speed, coolant temperature, intake air flow, oxygen (lambda O_2 sensor), throttle position (idle and overrun, wide-open throttle), and cylinder pressure.

7.1.2   Diesel Engine Control

The objective of diesel engine control is to adjust the injected fuel mass and time to make the generated torque and emissions meet drivability and environmental requirements. The control system mainly includes minimal and maximal speed limit control, an idle speed governor, engine temperature control, and fuel mass injection control. A diagram of the overall diesel engine control system is presented in Fig. 7.4. The measured variables of a diesel engine include the engine speed, coolant temperature, intake air flow, oxygen level in the emissions, position of the acceleration pedal and the brake pedal, and cylinder pressure.

7.2   Engine Torque Fluctuation Dumping Control Through the Electric Motor

Due to the complexity of internal combustion, the generated torque presents a certain degree of fluctuation when it responds to throttle maneuvers, reflecting the power demands of the driver. As a result, the resonance vibration in drivelines is excited and transferred to passenger compartments through the suspension system and the engine mount (Abe *et al.*, 1989).

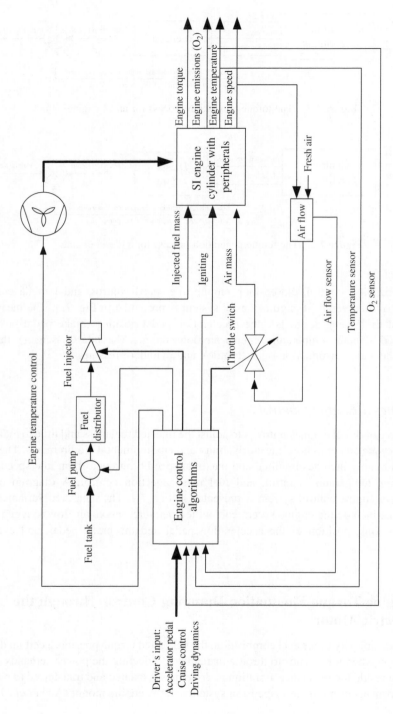

Figure 7.3 Schematic diagram of SI engine control

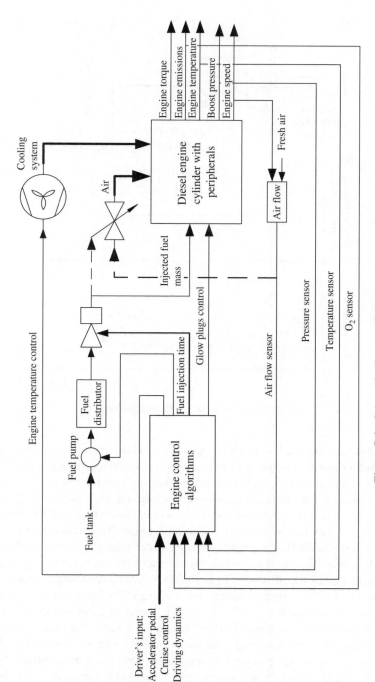

Figure 7.4 Schematic diagram of diesel engine control

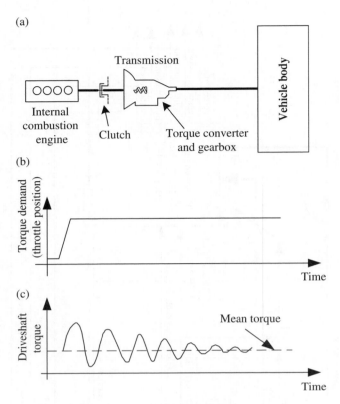

Figure 7.5   Torque vacillation process

A vibration mechanism model is shown in Fig. 7.5(a), and this consists of the engine, transmission including clutch, a torque converter and gearbox, and the vehicle body. When the acceleration pedal is depressed, fuel mass and air flow are immediately increased to create combustion which provides transient torque to the powertrain, as shown in Fig. 7.5(b); then, resonance is generated due to the mass and stiffness of the powertrain and the vehicle body, as shown in Fig. 7.5(c). In conventional powertrains, torque fluctuations are attenuated by the damper system installed in the torque converter, which generally results in a 5–15% efficiency loss; however, the hybrid powertrain provides an additional opportunity to attenuate the torque fluctuation by a well-controlled electric motor, as this can generate either positive torque (power) or negative torque (regenerative braking). The basic principle of dumping engine transient fluctuation is to enable the electric motor to generate torque opposite in phase to the engine torque, as shown in Fig. 7.6, and the control system diagram is shown in Fig. 7.7. In practice, implementing such ideas is a very intricate process and there are many challenges to controlling it; several approaches have been investigated (Nakajima *et al.*, 2000; Kim *et al.*, 2009). In taking advantage of the dual power sources of hybrid electric vehicles, it is necessary to apply advanced control strategies to achieve silence, comfort,

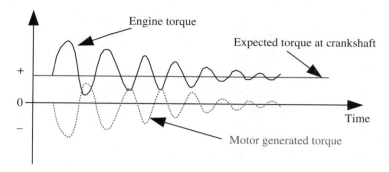

Figure 7.6 Concept of dumping engine torque fluctuation by the motor in an HEV

and unprecedented performance. In the following section, a sliding-mode-based control strategy is introduced to attenuate torque fluctuation at the driveshaft.

7.2.1 Sliding Mode Control

Sliding mode control is a nonlinear control method which alters the dynamic characteristics of a nonlinear system by a high-frequency switching control action. The designed switching control law is to drive the system state trajectory onto a predefined surface in the state space and to maintain the state trajectory on this surface subsequently. The surface is sometimes also called a sliding surface, switching hyperplane, or a sliding manifold. Under the switch control action, the system state trajectory will go to one side of the surface from an initial state, and once the state trajectory reaches the other side of the surface, the switched control will force the state trajectory back to the surface (Utkin, 1992). Therefore, the surface actually determines how to regulate the control action in this control method, and robustness is its advantage, because the controlled system will naturally slide along the surface and rest on it in finite time. Sliding mode control can be described mathematically as follows.

Consider the following nonlinear system:

$$\dot{\mathbf{x}} = \mathbf{A}(x) + \mathbf{B}(x)u \tag{7.1}$$

where: $\mathbf{x} = \begin{pmatrix} x_1 & x_2 & \cdots & x_n \end{pmatrix}^{\mathrm{T}}$ is a state vector, $\mathbf{A}(x)$ and $\mathbf{B}(x)$ are the appropriate nonlinear vector functions, and u is a scalar control variable.

Sliding control is a discontinuous control with the following form:

$$u = \begin{cases} u^+ & S = \mathbf{C}^{\mathrm{T}}\mathbf{x} > 0 \\ u^- & S = \mathbf{C}^{\mathrm{T}}\mathbf{x} < 0 \end{cases} \tag{7.2}$$

where $u^+ \neq u^-$ are control laws corresponding to the state trajectory being on different sides of hyperplane S, which is a linear switching hyperplane in Eq. 7.2, $\mathbf{C}^{\mathrm{T}} = \begin{pmatrix} c_1 & c_2 & \cdots & c_n \end{pmatrix}$ is a constant vector.

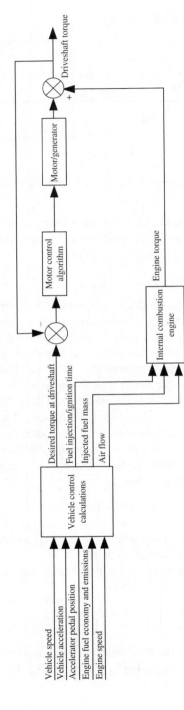

Figure 7.7 Diagram of electric motor dumping engine torque fluctuation

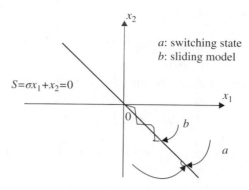

Figure 7.8   Sliding mode control and state trajectory

Take a two-phase variable case, for instance, as shown in Fig. 7.8. In this example, if we assume $S \geq 0$ when $t = 0$, the system state vector, $\mathbf{x}$, will reach the plane $S = 0$ in a finite time, t, under the control action u^+; the control u^+ forces the system state $\mathbf{x}$ to pass through the switching plane and enter the domain of $S < 0$, and the control suddenly changes from u^+ to u^-, which forces the state to pass through the switching plane again in the opposite direction and enter the domain of $S > 0$ once more; thus, a sliding mode is formed.

Therefore, the state $\mathbf{x}$ will be restricted on the switching plane $S = 0$ by the sliding control u, and the condition to form a sliding mode is:

$$\begin{cases} \lim\limits_{s \to 0^+} \dot{S} < 0 \\ \lim\limits_{s \to 0^-} \dot{S} > 0 \end{cases} \tag{7.3}$$

That is, $S\dot{S} < 0$, which forces the system motion toward the hyperplane $S = 0$.

It is apparent that the sliding mode will be formed if the following equation is held by a control action:

$$\dot{S} = -\varepsilon \operatorname{sgn}(S) - kS \tag{7.4}$$

where $\operatorname{sgn}(S)$ is the sign function of S, the term $-\varepsilon \operatorname{sgn}(S)$ generates a variable control structure to form a sliding mode at the desired hyperplane, S, and the term kS makes the trajectory's coverage exponential.

As an example, a sliding control can be designed through the following procedures:

1. Based on the desired system performance, the sliding surface can be defined as:

$$\dot{S} = \mathbf{C}^T \dot{\mathbf{x}} = \mathbf{C}^T (\mathbf{A}\mathbf{x} + \mathbf{B}u) = \mathbf{C}^T \mathbf{A}\mathbf{x} + \mathbf{C}^T \mathbf{B}u \tag{7.5}$$

2. From Eq. 7.5, the control u is derived:

$$u = \left[-\mathbf{C}^{\mathrm{T}}\mathbf{B}\right]^{-1}\left[\mathbf{C}^{\mathrm{T}}\mathbf{Ax} - \dot{S}\right] \tag{7.6}$$

3. Substituting Eq. 7.4 into Eq. 7.6, the final control yields:

$$u = -\left[\mathbf{C}^{\mathrm{T}}\mathbf{B}\right]^{-1}\left[\mathbf{C}^{\mathrm{T}}\mathbf{Ax} + \varepsilon\mathrm{sgn}(S) + kS\right] \tag{7.7}$$

where ε and k are the given positive constants which should not be sensitive to the system's operational setpoint.

7.2.2   Engine Torque Fluctuation Dumping Control Based on the Sliding Mode Control Method

Since the sliding mode control strategy is very robust in terms of system disturbances and plant parameter variations, it has been applied widely to control DC motors, permanent magnet synchronous motors, and induction motors. In addition, it is also used in DC–DC converters, DC–AC inverters, and other power electronic devices (Utkin *et al.*, 1999). Based on the nature of engine transient torque dumping control, the sliding control strategy is a good control candidate and can be implemented as follows.

Assume that the dynamics of the engine driveshaft are described by the following state-space equation:

$$\begin{bmatrix} \dot{x}_1 \\ \dot{x}_2 \end{bmatrix} = \begin{bmatrix} f_1(x_1,x_2) \\ f_2(x_1,x_2) \end{bmatrix} + \begin{bmatrix} b_1 \\ b_2 \end{bmatrix} u \tag{7.8}$$

where x_1 is the torque on the engine driveshaft produced by combustion, x_2 is the speed of the engine driveshaft, and u is the anti-torque generated either by the electric motor or the vehicle, or both.

Based on the characteristics of torque versus speed of an engine, the sliding surface can be set as a simple surface expressed by Eq. 7.9 or Eq. 7.10 and shown in Fig. 7.9(a) and Fig. 7.9(b), respectively.

$$S = \sigma x_1 + x_2, \quad \sigma > 0 \tag{7.9}$$

$$S = f(x_1, x_2), \quad \frac{dx_1}{dx_2} > 0 \tag{7.10}$$

As an example, if Eq. 7.9 is selected as the sliding surface, the control law can be designed to make the sliding mode occur under the following equation:

$$\dot{S} = -\varepsilon\mathrm{sgn}(S) - kS \quad 0 < \varepsilon, \quad 0 < k \tag{7.11}$$

where ε is a positive constant that defines the boundary of the sliding mode zone, and k is also a positive constant that determines the sliding frequency to the plane $S = 0$. Figure 7.10 shows the performance of the example sliding mode. At $t = 0$, the states move to the sliding surface under the control action and will ideally remain on the plane (shown as the thick gray arrow); however, due to system inertia, limited control action, action delay, or other imperfections, the system states will pass through the sliding plane, leading to chatter in the sliding zone within a range $[-\rho\ \ \rho]$ depending on the actual system characteristics and control ability.

If we assume that the driveshaft dynamics are described by Eq. 7.8 and the sliding surface and sliding mode behavior are defined by Eqs 7.9 and 7.11, the following sliding mode control law can be obtained to dump engine torque fluctuation on the driveshaft by the electric motor in an HEV.

From Eqs 7.8 and 7.9, we have:

$$\dot{S} = \sigma\dot{x}_1 + \dot{x}_2 = \sigma f_1(x_1,x_2) + \sigma b_1 u + f_2(x_1,x_2) + b_2 u$$
$$= \sigma f_1(x_1,x_2) + f_2(x_1,x_2) + (\sigma b_1 + b_2)u$$

(7.12)

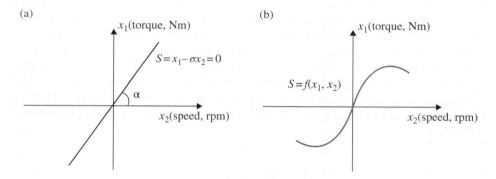

Figure 7.9   The sliding surface

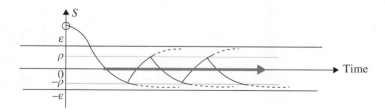

Figure 7.10   The performance of sliding modes

Substituting Eq. 7.11 into Eq. 7.12, yields the following equation:

$$\begin{aligned} \dot{S} &= \sigma f_1(x_1, x_2) + f_2(x_1, x_2) + (\sigma b_1 + b_2)u \\ &= -\varepsilon \operatorname{sgn}(S) - kS \\ &= -\varepsilon \operatorname{sgn}(\sigma x_1 + x_2) - k(\sigma x_1 + x_2) \end{aligned} \qquad (7.13)$$

By equating the right-hand sides of Eqs 7.12 and 7.13, the sliding mode control law can be derived as:

$$u = -\frac{1}{\sigma b_1 + b_2}[\sigma f_1(x_1, x_2) + f_2(x_1, x_2) + \varepsilon \operatorname{sgn}(\sigma x_1 + x_2) + k(\sigma x_1 + x_2)] \qquad (7.14)$$

7.3   High-voltage Bus Spike Control

Hybrid electric vehicles and purely electric vehicles have a high-voltage bus connected to the battery system, a DC–DC converter, a DC–AC inverter for low-voltage components, and a propulsion motor, as shown in Fig. 7.11. When the vehicle is accelerated or regeneratively braked, the DC–DC converter or DC–AC inverter will generate spikes which frequently affect battery limits, especially during low-temperature operation, as shown in Fig. 7.12.

Since these spikes are very harmful to battery life, some hybrid electric vehicles are equipped with an overvoltage protection device, shown as the dashed line box on the right-hand side of Fig. 7.11. If this is the case, the vehicle controller needs to send a control signal to turn the device on or off to attenuate the spikes and protect the battery and other components on the high-voltage bus. The general control diagram is illustrated in Fig.7.13. This section introduces three control strategies to control the overvoltage protection unit (OVPU).

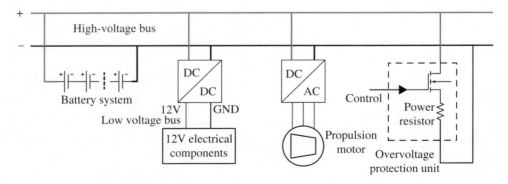

Figure 7.11   High-voltage bus and the connected loads

(a)

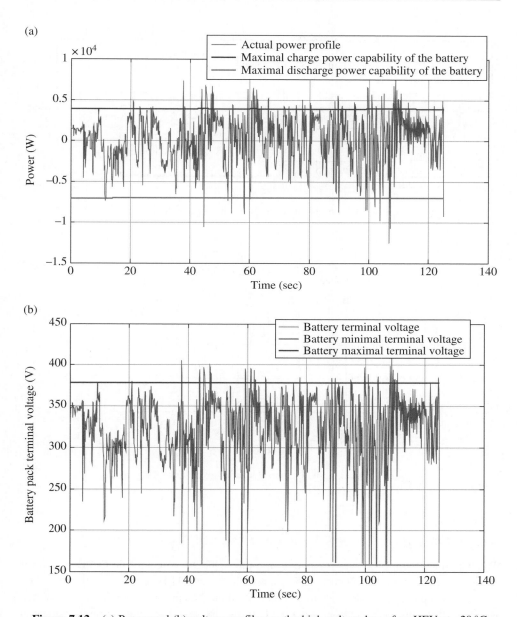

(b)

Figure 7.12 (a) Power and (b) voltage profiles on the high-voltage bus of an HEV at −20 °C

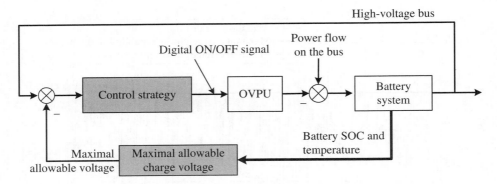

Figure 7.13   Architecture of overvoltage protection control

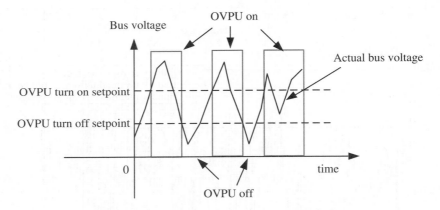

Figure 7.14   Bang-bang control signal turning ON/OFF setpoint

7.3.1   Bang-Bang Control Strategy of Overvoltage Protection

The bang-bang control to prevent an over-the-limit DC bus voltage is an ON/OFF control strategy which determines the duration for which the OVPU is on and off, based on the actual bus voltage. The control principle is that when the charging (regenerative) voltage is greater than the upper setpoint, the OVPU turns ON, and when the charging voltage is below the lower setpoint, the OVPU turns OFF, as shown in Fig. 7.14.

The control strategy and OVPU ON/OFF setpoint can be described as:

$$u = \begin{cases} 1 & \text{if bus voltage} \geq \text{OVPU turn ON point} \\ 0 & \text{if bus voltage} \leq \text{OVPU turn OFF point} \end{cases} \qquad (7.15)$$

OVPU turn ON setpoint = battery maximal acceptable charge voltage $-\delta_1$

OVPU turn OFF setpoint = battery maximal acceptable charge voltage $-\delta_2$ (7.16)

$$\delta_2 > \delta_1 \geq 0$$

where u is the signal to switch the OVPU ON or OFF, and δ_1 and δ_2 are calibratable variables. Obviously, bang-bang control is not optimal, but it is effective and has a low implementation cost.

7.3.2 PID-based ON/OFF Control Strategy for Overvoltage Protection

The OVPU can be controlled based on the following PID control strategy:

$$u(t) = \begin{cases} 1 & \text{if } K\left(e(t) + \dfrac{1}{T_i}\displaystyle\int_0^t e(\tau)d\tau + T_d\dfrac{de(t)}{dt}\right) \geq 0 \\[4mm] 0 & \text{if } K\left(e(t) + \dfrac{1}{T_i}\displaystyle\int_0^t e(\tau)d\tau + T_d\dfrac{de(t)}{dt}\right) < 0 \end{cases} \tag{7.17}$$

where $e(t) = V_{\text{setpoint}}(t) - V_{\text{actual}}(t)$ is the control error, $u(t)$ is the signal to command the OVPU to switch ON or OFF, V_{setpoint} is the maximal bus voltage, V_{actual} is the actual bus voltage, and K, T_i, T_d are the parameters of the PID controller.

7.3.3 Fuzzy-logic-based ON/OFF Control Strategy for Overvoltage Protection

A fuzzy-logic-based ON/OFF control strategy can also be developed to control the OVPU. The control inputs are the error between the maximal bus voltage setpoint and the actual bus voltage and the derivative of the error. The output is the signal that commands the OVPU to switch ON or OFF. A diagram of a fuzzy-logic-based OVPU ON/OFF control strategy is presented in Fig. 7.15. Examples of the control design procedures are as follows:

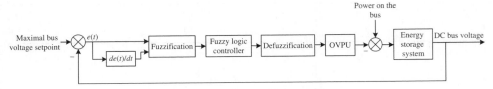

Figure 7.15 Diagram of overvoltage protection fuzzy logic control

1. Set up fuzzy sets of the error, $e = V_{setpoint} - V_{actual}$, and its derivative, $\dot{e}$, based on the value range. In this example, they are defined as:

$$e = [0 \ -1 \ -2 \ -3 \ -4 \ -5 \ -6 \ -7 \ -8 \ -9 \ -10] \text{ and}$$
$$\dot{e} = [-6 \ -5 \ -4 \ -3 \ -2 \ -1 \ 0 \ 1 \ 2 \ 3 \ 4 \ 5 \ 6]$$

2. Set up a membership table for error e

	0	−1	−2	−3	−4	−5	−6	−7	−8	−9	−10
Z	1	0.6	0.3	0.2	0.1	0	0	0	0	0	0
NS	0.1	0.7	1	0.9	0.8	0.6	0.2	0	0	0	0
NM	0	0	0.1	0.3	0.4	0.7	0.8	1	0.8	0.6	0
NB	0	0	0	0	0	0	0.1	0.4	0.7	0.9	1

3. Set up a membership table for $\dot{e}$

μ $\dot{e}$ FS	−6	−5	−4	−3	−2	−1	0	1	2	3	4	5	6
NB	1	0.5	0.1	0	0	0	0	0	0	0	0	0	0
NM	0.1	0.7	1	0.7	0.3	0	0	0	0	0	0	0	0
NS	0	0	0.1	0.6	1	0.8	0.3	0	0	0	0	0	0
Z	0	0	0	0	0	0.3	0.7	1	0.7	0.3	0	0	0
PS	0	0	0	0	0	0.1	0.2	0.5	0.7	1	0.3	0	0
PM	0	0	0	0	0	0	0	0	0.1	0.4	0.8	1	0.8
PB	0	0	0	0	0	0	0	0	0	0	0.4	0.8	1

4. Set up a membership table for control output u

μ u FS	0	1	2	3	4
OFF	1.0	0.7	0.4	0.1	0
ON	0	0	0.1	0.7	1.0

5. Set up a fuzzy rule table from IF–THEN rules

rules \ $\dot{e}$	NB	NM	NS	Z	PS	PM	PB
e							
Z	OFF	OFF	OFF	OFF	ON	ON	ON
NS	OFF	OFF	ON	ON	ON	ON	ON
NM	OFF	ON	ON	ON	ON	ON	ON
NB	ON	ON	ON	ON	ON	ON	ON

6. Set up a fuzzy relationship based on the maximum/minimum principle from the tables established above.
7. Perform fuzzy reasoning for fuzzy control action.

uf \ $\dot{e}$	−6	−5	−4	−3	−2	−1	0	1	2	3	4	5	6
e													
0	1	1	1	1	1	1	2	2	3	3	4	4	4
−1	1	1	1	2	3	3	3	3	3	3	4	4	4
−2	1	1	1	2	3	3	3	3	3	3	4	4	4
−3	1	2	2	2	3	4	4	4	4	4	4	4	4
−4	1	2	2	2	3	4	4	4	4	4	4	4	4
−5	1	3	3	3	3	4	4	4	4	4	4	4	4
−6	1	3	3	3	3	4	4	4	4	4	4	4	4
−7	2	3	4	4	4	4	4	4	4	4	4	4	4
−8	2	3	4	4	4	4	4	4	4	4	4	4	4
−9	3	3	4	4	4	4	4	4	4	4	4	4	4
−10	4	4	4	4	4	4	4	4	4	4	4	4	4

8. Execute defuzzification to obtain the OVPU ON/OFF control.

u \ $\dot{e}$	−6	−5	−4	−3	−2	−1	0	1	2	3	4	5	6
e													
0	0	0	0	0	0	0	1	1	1	1	1	1	1
−1	0	0	0	1	1	1	1	1	1	1	1	1	1
−2	0	0	0	1	1	1	1	1	1	1	1	1	1
−3	0	1	1	1	1	1	1	1	1	1	1	1	1
−4	0	1	1	1	1	1	1	1	1	1	1	1	1
−5	0	1	1	1	1	1	1	1	1	1	1	1	1

−6	0	1	1	1	1	1	1	1	1	1	1	1
−7	1	1	1	1	1	1	1	1	1	1	1	1
−8	1	1	1	1	1	1	1	1	1	1	1	1
−9	1	1	1	1	1	1	1	1	1	1	1	1
−10	1	1	1	1	1	1	1	1	1	1	1	1

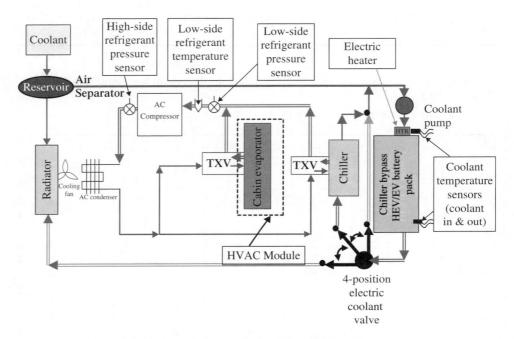

Figure 7.16   Thermal control diagram for an HEV/EV battery system

7.4   Thermal Control of an HEV Battery System

Since a battery's thermal system performance has a significant impact on the vehicle's fuel economy and battery life, vehicle manufacturers have been making great efforts to improve the performance. In general, the HEV cooling/heating system is quite complex and consists of several independent cooling loops such as an engine cooling loop, a transmission cooling loop, a motor and DC–AC inverter cooling loop, a battery pack cooling loop, and an accessory power cooling loop. Some of these are liquid cooled, others are air cooled, and each cooling loop has its own temperature setpoint. The engine cooling loop normally has the highest temperature setpoint, but the cooling loop of the battery system is the most sophisticated and the lowest temperature loop in an HEV – normally between 25 and 35 °C. A diagram of an HEV/EV battery system cooling loop is presented in Fig. 7.16.

From Fig. 7.16, the battery system can be cooled down by a chiller and/or a radiator/cooling fan set or heated up by an electric heater. The coolant pump and four-position electric coolant valve provide additional measures to control the battery temperature more efficiently. If the battery temperature is slightly lower than the setpoint, the coolant goes through a short cycling path by closing the channels to the radiator and the chiller from the four-position valve. If the battery temperature is much lower than the setpoint in the deep of winter, the heater is ON and the channels to the radiator and chiller are closed by the four-position valve. If the battery temperature is slightly higher than the setpoint, the coolant goes only through the radiator channel, and the cooling fan could be turned ON or OFF depending on the battery temperature. If the battery is very hot in the summer, the chiller will be turned ON in addition to the radiator channel. From a control point of view, the battery cooling subsystem is a multivariable system, and the electric load (battery current) can be considered a process disturbance. The simplified thermal loop diagram is shown in Fig. 7.17, and the battery system thermal behaviors can be described by the following mathematical equations:

$$
\begin{aligned}
c_{\mathrm{ESS}}M_{\mathrm{ESS}}\frac{dT_{\mathrm{ESS}}}{dt} &= P_{\mathrm{ESS}} - \dot{Q}_{\mathrm{exchg}} = I^2 R_{\mathrm{ESS}} - h(\dot{m}_{\mathrm{c}})A_{\mathrm{H/C}}(T_{\mathrm{ESS}} - T_{\mathrm{c}}) \\
&= -h(\dot{m}_{\mathrm{c}})A_{\mathrm{H/C}}T_{\mathrm{ESS}} + h(\dot{m}_{\mathrm{c}})A_{\mathrm{H/C}}T_{\mathrm{c}} + I^2 R_{\mathrm{ESS}} \\
c_{\mathrm{c}}M_{\mathrm{c}}\frac{dT_{\mathrm{c}}}{dt} &= \dot{Q}_{\mathrm{H/C}} + \dot{Q}_{\mathrm{exchg}} = \eta_{\mathrm{H/C}}P_{\mathrm{Htr/chiller}} + h(\dot{m}_{\mathrm{c}})A_{\mathrm{H/C}}(T_{\mathrm{ESS}} - T_{\mathrm{c}}) \\
&= -h(\dot{m}_{\mathrm{c}})A_{\mathrm{H/C}}T_{\mathrm{c}} + h(\dot{m}_{\mathrm{c}})A_{\mathrm{H/C}}T_{\mathrm{ESS}} + \eta_{\mathrm{H/C}}P_{\mathrm{Htr/chiller}}
\end{aligned}
\tag{7.18}
$$

That is:

$$
\begin{aligned}
\frac{dT_{\mathrm{ESS}}}{dt} &= \frac{-h(\dot{m}_{\mathrm{c}})A_{\mathrm{H/C}}}{c_{\mathrm{ESS}}M_{\mathrm{ESS}}}T_{\mathrm{ESS}} + \frac{h(\dot{m}_{\mathrm{c}})A_{\mathrm{H/C}}}{c_{\mathrm{ESS}}M_{\mathrm{ESS}}}T_{\mathrm{c}} + \frac{R_{\mathrm{ESS}}}{c_{\mathrm{ESS}}M_{\mathrm{ESS}}}I^2 \\
\frac{dT_{\mathrm{c}}}{dt} &= \frac{-h(\dot{m}_{\mathrm{c}})A_{\mathrm{H/C}}}{c_{\mathrm{c}}M_{\mathrm{c}}}T_{\mathrm{c}} + \frac{h(\dot{m}_{\mathrm{c}})A_{\mathrm{H/C}}}{c_{\mathrm{c}}M_{\mathrm{c}}}T_{\mathrm{ESS}} + \frac{\eta_{\mathrm{H/C}}}{c_{\mathrm{c}}M_{\mathrm{c}}}P_{\mathrm{Htr/chiller}}
\end{aligned}
\tag{7.19}
$$

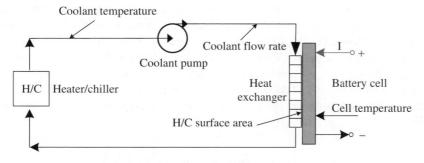

Figure 7.17   Simplified thermal loop diagram

or

$$
\begin{bmatrix} \dot{T}_{ESS} \\ \dot{T}_c \end{bmatrix} = \begin{bmatrix} a_{11}(\dot{m}_c) & a_{12}(\dot{m}_c) \\ a_{21}(\dot{m}_c) & a_{22}(\dot{m}_c) \end{bmatrix} \begin{bmatrix} T_{ESS} \\ T_c \end{bmatrix} + \begin{bmatrix} 0 \\ b_2 \end{bmatrix} P_{Htr/chiller} + \begin{bmatrix} c_1 \\ 0 \end{bmatrix} I^2
$$

$$
\eta_{H/C} = \begin{cases} \eta_h \\ \eta_c \end{cases} \quad \text{and} \quad P_{Htr/Chiller} = \begin{cases} P_h \ \text{ if the battery system needs to be heated up} \\ P_c \ \text{ if the battery system needs to be cooled down} \end{cases} \tag{7.20}
$$

$$
\dot{m}_c = k \cdot P_{pump}
$$

where I is the battery terminal current (A), R_{ESS} is the internal resistance (Ω) of the battery system, T_{ESS} is the battery cell temperature (K), T_c is the coolant temperature (K), C_{ESS} is the specific heat (J/kg.K) of the battery system, C_c is the specific heat (J/kg.K) of the coolant, h is the heat transfer coefficient (W/m^2.K), $A_{H/C}$ is the heating/cooling surface area (m^2) between the battery pack and the heating/cooling channel, M_{ESS} is the mass (kg) of the battery system, M_c is the mass of the total coolant (kg), $\dot{Q}_{H/C}$ is the energy transfer rate (W) of the heater or chiller, $\eta_{H/C}$ is the efficiency of the heater or chiller, $\dot{m}_c$ is the mass flow rate (kg/s) of the coolant, P_{pump} is the power (W) of the heating/cooling system pump, and k is the transfer coefficient of pump power to the mass flow rate.

7.4.1 Combined PID Feedback with Feedforward Battery Thermal System Control Strategy

In addition to feedback control, feedforward control can significantly improve the system performance if major disturbances can be measured before they affect the process output. Figure 7.18 shows the principle of a PID feedback with feedforward control strategy for battery thermal system control. The power of the heater/chiller is the input variable and the battery temperature is the output, while the battery load, the cooling fan, the coolant flow path and flow rate (the power of the coolant pump) are considered major disturbances. In this control system, the cooling fan speed, the system operational mode (heating or cooling), the flow path (positions of electrical valve), and the flow rate (power of the coolant pump) are logically set based on the battery temperature, the ambient temperature, and the target temperature (setpoint) of the battery system. The feedforward variables include the cooling fan speed, the coolant flow rate, and the heat generated ($I^2 R_{ESS}$) by the battery system itself. The relationships among these variables are established by look-up tables which are different in heating mode and cooling mode. Since the heat $I^2 R_{ESS}$ generated by the battery system itself has a significant impact on the battery temperature, it is directly and independently fed forward to the process control.

In this control strategy, the feedback control action is calculated by the following PID equation:

$$
u_{PID}(t) = K \left(e(t) + \frac{1}{T_i} \int_0^t e(\tau) d\tau + T_d \frac{de(t)}{dt} \right) \tag{7.21}
$$

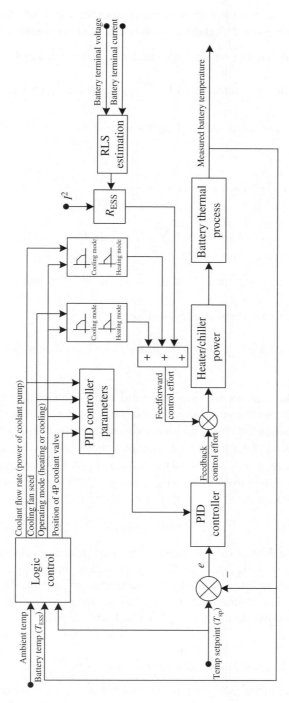

Figure 7.18 Combined PID and feedforward control diagram of battery thermal control

where $e(t) = T_{sp}(t) - T_{ESS}(t)$ is the error between the set temperature and the actual temperature, $u_{PID}(t)$ is the calculated PID control action which is a sum of three terms: the P term $(Ke(t))$ which is proportional to the error, the I term $\left(K\dfrac{1}{T_i}\displaystyle\int_0^t e(\tau)d\tau\right)$ which is proportional to the integral of the error, and the D term $\left(K \cdot T_d \dfrac{de(t)}{dt}\right)$ which is proportional to the derivative of the error.

The discrete form can be obtained if approximating the integral and the derivative terms, using:

$$\int_0^t e(\tau)d\tau \approx T_s \sum_{i=0}^k e(i) \qquad \frac{de(t)}{dt} \approx \frac{e(k)-e(k-1)}{T_s} \tag{7.22}$$

$$u_{PID}(k) = k_p e(k) + k_i \cdot T_s \sum_{i=0}^k e(i) + \frac{k_d}{T_s}(e(k)-e(k-1)) \tag{7.23}$$

where T_s is the sampling period of time, k is the discrete step at time t, $k_p = K$ $k_i = \frac{K}{T_i}$, $k_d = K \cdot T_d$

The feedforward control action is calculated as:

$$u_{ffd}(t) = u_{cfan}(t) + u_{cf}(t) + u_{heat}(t) = lookup(fan\ spd) + lookup(coolant\ flow\ rate) + I^2 \hat{R}_{ESS} \tag{7.24}$$

where u_{cfan} is the feedforward control action associated with the cooling fan which is calculated through a look-up table for the given operational mode, u_{cf} is the feedforward control action associated with the coolant flow rate which is calculated through a look-up table for the given operational mode, $I^2 \hat{R}_{ESS}$ is the feedforward control action related to the heat generated by the battery system itself, and $\hat{R}_{ESS}$ is the online estimated ESS internal resistance through a recursive least squares algorithm based on the measured ESS terminal voltage and current.

The actual control action is:

$$u(t) = u_{PID}(t) + u_{ffd}(t) \tag{7.25}$$

The parameters of the feedback PID controller are tuned offline but changed in real time, based on the outputs of logic control, reflecting the battery thermal system's operational mode and conditions. For PID control, detailed parameter tuning methods, and a recursive least squares estimation algorithm, the interested reader should refer to Åstrom and Hägglund (1995), Goodwin and Payne (1977), and Ljung (1987).

7.4.2 Optimal Battery Thermal Control Strategy

Since the performance of the battery thermal control has an impact on the overall fuel economy of an HEV, there is a need to trade-off control accuracy and cooling system power

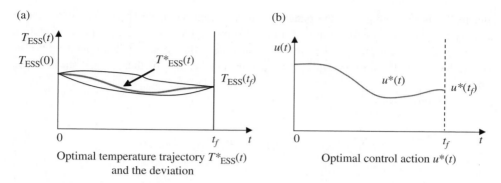

(a)

Optimal temperature trajectory $T^*_{ESS}(t)$
and the deviation

(b)

Optimal control action $u^*(t)$

Figure 7.19 Optimal control $u^*(t)$ and the change of state $T^*_{ESS}(t)$ between the initial, $T_{ESS}(0)$, and the final, $T_{ESS}(t_f)$, set value

consumption using an optimal control strategy, which means that the control, $u^*(t)$, denotes the control function, $u(t)$ giving the minimal objective function, J, in the required time, t_f, once J has been set. If we assume that the optimal control, $u^*(t)$, does exist and has the form shown in Fig. 7.19, this $u^*(t)$ minimizes the objective function J, and the control path between two end points, $T_{ESS}(0)$ and $T_{ESS}(t_f)$, is labeled $T^*_{ESS}(t)$. In practice, the optimal control is normally appended to the PID control, and once the battery temperature reaches the setpoint, the control strategy will be switched to the feedback PID with feedforward control described above to maintain the battery system at the target temperature.

If we assume that the thermal process is described by the following state-space equation, the optimal control can be determined as follows:

$$\begin{bmatrix} T_{ESS}(k+1) \\ \dot{T}_c(k+1) \end{bmatrix} = \begin{bmatrix} a_{11}(\dot{m}_c) & a_{12}(\dot{m}_c) \\ a_{21}(\dot{m}_c) & a_{22}(\dot{m}_c) \end{bmatrix} \begin{bmatrix} T_{ESS}(k) \\ T(k)_c \end{bmatrix} + \begin{bmatrix} 0 \\ b_2 \end{bmatrix} u(k) + \begin{bmatrix} R_{ESS} \\ 0 \end{bmatrix} I^2(k)$$

$$\dot{m}_c = k \cdot P_{pump}(k), \text{ i.e., } a_{11} = lookup\_a(P_{pump}(k)), \ a_{12} = lookup\_b(P_{pump}(k))$$

$$a_{21} = lookup\_c(P_{pump}(k)), \ a_{22} = lookup\_d(P_{pump}(k))$$

(7.26)

where I is the battery system terminal current (A), R_{ESS} is the internal resistance (Ω) of the battery system, T_{ESS} is the cell temperature (K), T_c is the coolant temperature (K), a_{ij} is a component of the state matrix which is a function of the operating power of the coolant pump and is implemented by look-up tables, b_2 is a component of the control vector which is a constant for a given cooling system, and u is the control variable which is the physical power of the heater/chiller.

7.4.2.1 Define the Objective Function

The objective function for optimal thermal system control can be defined as:

$$J = \int_0^{t_f} \left[\left(T_{sp} - T_{ESS}(t) \right)^2 + w_1 P_{pump}^2 + w_2 u^2 \right] dt$$

(7.27)

where T_{sp} is the setpoint of the battery system thermal control, $T_{ESS}(t)$ is the measured battery temperature, t_f is the required time to complete the control, P_{pump} is the operating power of the coolant pump which results in a coolant flow rate, u is the operating power of the heater/chiller, and w_1 and w_2 are the weight factors for the control action P_{pump} and u.

The discrete form is:

$$J = \sum_{i=1}^{N} \left[\left(T_{sp} - T_{ESS}(i) \right)^2 + w_1 P_{pump}^2(i) + w_2 u^2(i) \right] \Delta T \tag{7.28}$$

where ΔT is the control time interval and N is the required number of steps to reach the control target.

7.4.2.2 Set the Control Constraints

The constraints of the battery system thermal optimal control include:

- Coolant flow rate range, that is, the operating power range of the coolant pump:

$$P_{min} \leq P_{pump} \leq P_{max} \tag{7.29}$$

where P_{min} is the minimal operating power of the coolant pump to maintain the required minimal flow rate, P_{max} is the maximal allowable operational power of the coolant pump to generate the maximal flow rate.
- The operating power range of the heater/chiller:

$$0 \leq u \leq P_{H/C\_max} \tag{7.30}$$

where P_{H/C_max} is the maximal heating or cooling power of the heater/chiller.

7.4.2.3 Set the Initial Condition and End Point Value

$$T_c(0) = T_{c\_init}, \quad T_{ESS}(0) = T_{ESS\_init}, \quad T_c(N) = T_{c\_sp}, \quad T_{ESS}(N) = T_{ESS\_sp} \tag{7.31}$$

where T_{c_init} is the initial coolant temperature, T_{ESS_init} is the initial battery temperature, T_{c_sp} is the target coolant temperature (setpoint), and T_{ESS_sp} is the target battery temperature (setpoint).

The battery system of an HEV generally has the following setpoints and constraints to ensure maximal efficiency and state of health:

- Target operational temperature: 25–35 °C;
- Maximal coolant pressure: 40 kPa (5.8 psi);
- Thermal gradient within the battery system: ≤3 °C.

7.4.2.4 Determine the Control Law

Two methods may be used to solve this problem: one uses Bellman's dynamic programming and the other uses Pontryagin's minimum principle (Pontryagin, 1962). Pontryagin's minimum principle provides a necessary condition for optimal control rather than a direct computation of the control itself, but the detailed application depends on the type of problem posed. The fundamental theory of the dynamic programming method is the principle of optimality; that is, 'an optimal policy has the property that whatever the initial state and the initial decisions are, the remaining decisions must constitute an optimal policy with regard to the state resulting from the first decision' (Bellman, 1962). The dynamic programming technique is a favorable method to find out the optimal control law for battery system thermal control, and the interested reader should refer to Lewis and Syrmos's (1995) book to get a detailed solution for an actual battery system.

7.5 HEV/EV Traction Motor Control

Since an electric motor is the sole power source in a series HEV and an EV, the quality of the traction motor control significantly affects the vehicle's drivability and safety. In addition, compared with a conventional powertrain, the electric powertrain of an HEV/EV has superior dynamic behaviors such as a higher stalling torque and a faster transient response. In this section, two key control problems are briefly addressed.

7.5.1 Traction Torque Control

During standing starts, acceleration, or braking, the amount of force that can be transferred to the road from the wheels depends on the traction availability between the tires and the road surface. The adhesion and slip curves for acceleration and braking are shown in Fig. 7.20. The electric motor output torque during acceleration and regenerative (braking) is limited by the amount of slip, allowing the response to remain within the stable range. From the curves, it can be seen that any rise in slip up to a certain point is accompanied by a corresponding increase in available adhesion, but beyond that point, further increases in slip take the curves through the maxima and into the unstable range, resulting in a reduction in adhesion. Under regenerative braking, this will result in the wheel locking within a

Coefficient of friction under acceleration/braking $\mu_{A,B}$
torque under acceleration/braking $\lambda_{A,B}$, lateral-force
coefficient μ_s

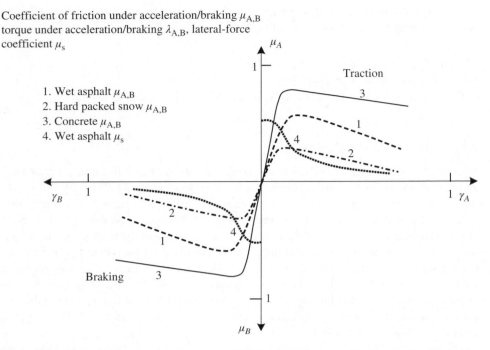

1. Wet asphalt $\mu_{A,B}$
2. Hard packed snow $\mu_{A,B}$
3. Concrete $\mu_{A,B}$
4. Wet asphalt μ_s

Figure 7.20 Adhesion/slip curves. Reproduced from *BOSCH Automotive Handbook*, 5th edition with permission from Robert Bosch GmbH

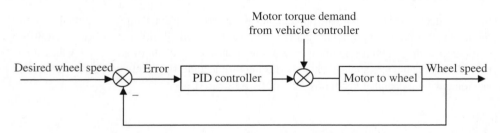

Figure 7.21 Traction motor torque control in a hybrid vehicle

short period of time. Under acceleration, driving wheels will start to spin more and more if the traction torque exceeds the maximal adhesion point.

The PID feedback with feedforward control strategy, shown in Fig. 7.21, can be used to control the traction force, which provides closed-loop control of torque at the drive wheels. In this control diagram, the vehicle speed is compared with the wheel speed and the difference is fed to the PID controller; its output is then added to the electric motor torque demanded by the driver, so that the torque at the drive wheels is adjusted. In order to further improve traction motor control performance, a more sophisticated control strategy may be needed, and the interested reader may refer to Appendix B for advanced control methods.

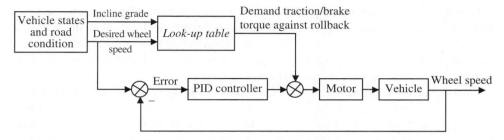

Figure 7.22   HEV/EV anti-rollback

7.5.2   Anti-rollback Control

Rollback occurs during the transition between the release of the brake pedal and application of the accelerator on an incline. If this is not done in harmony, the vehicle will roll down the incline. For a conventional vehicle with an automatic transmission, the anti-rollback feature can be implemented easily by a transmission control (Tamimi *et al.*, 2005). Since there is no such transmission in a series HEV or EV, anti-rollback control is a necessary function for handling situations when the vehicle is parked on an incline. This function is also necessary to support the stop–start feature in the hybrid electric powertrain by holding the vehicle on an incline and preventing undesired motion. Although rollback can be prevented by sending an appropriate control signal to the electrical or hydraulic brake control module, the most efficient and economical way is to apply traction/regenerative torque to prevent rollback in an HEV/EV.

A control diagram for implementing an anti-rollback function is given in Fig. 7.22 for an HEV/EV. Based on the throttle torque request from the driver, the grade of the incline, and other vehicle states such as the key state and the vehicle position (down or uphill), the vehicle controller sets a desired speed for the wheels. The desired wheel speed and the incline grade are fed to a calibratable look-up table to set a traction/brake torque, and, simultaneously, a PID controller outputs a torque based on the difference between the desired wheel speed and the actual wheel speed. The sum of the set traction/brake torque and the PID output traction/brake torque is the anti-rollback torque demand for the traction motor to generate.

7.6   Active Suspension Control in HEV/EV Systems

The purpose of a vehicle's suspension system is to provide driving comfort and operating safety, through damping vertical oscillations of the vehicle and improving vehicle body and wheel/tire transient dynamics. Traditional suspension systems mainly consist of steel springs and shock absorbers to transmit and smooth all forces between the body and the road. The primary function of a spring is to carry vehicle mass and isolate the vehicle body from road irregularities, while a shock absorber is to damp out the up–down motions of a vehicle on its springs. Such a suspension system is also called a *passive suspension system*. Generally speaking, softer dampers provide a more pleasurable drive, while the vehicle's stability is better using stiffer ones. Therefore, driving comfort and operating safety are two conflicting criteria, and the design engineer normally determines the damping properties by implementing some kind of trade-off between them.

On the other hand, suspension systems in which the damping properties can be adjusted to some extent are called *active suspension systems*, and these can be further classified into semi-active and active suspension systems. Compared with passive suspension systems, the common feature of semi-active suspension systems is that the damping characteristics of the spring and/or shock absorber can adapt to the actual demands, while active suspension systems are equipped with separate actuators that can exert an extra force input to the suspension system. Figure 7.23 illustrates principle diagrams for passive, semi-active, and active suspension systems. Conventional active suspension systems are implemented through high-speed hydraulic, hydro-pneumatic, or pneumatic systems, and the vehicle body oscillations are compensated by adjusting the hydraulic fluid or compressed air supply. However, due to the complexity of the control, the cost of implementation, and the high and quick peak power demands, so far only a few high-end conventional vehicles on the market are equipped with such active suspension systems.

Advances in electronics and control technologies have led to the development of new sensors and actuators and advanced control strategies, as a result of which, the dynamic behavior of a suspension system can be controlled by adjusting the properties of the springs and dampers. So far, the active suspension system in an HEV/EV has emerged as an active research and development area to improve driving comfort, operating safety, and fuel economy. Unlike the active suspension systems applied in conventional vehicles, the active suspension systems in an HEV/EV are controlled by means of electric actuators, and they are further divided into two major categories: rotary-motor-based systems and linear-motor-based systems, as shown in Fig. 7.24. The advantages of electric active suspension systems over conventional hydraulic/pneumatic active suspension systems are not only the superior dynamic performance, but also the regenerative capability. Using electric active suspension systems, the kinetic energy of the vehicle body vibration can be converted into electrical energy to charge the battery system, rather than wasting it as heat energy as happens in conventional suspension systems. This results in an extended mileage for an electric vehicle or improved fuel economy for a hybrid electric vehicle. Since linear-motor-based active suspension systems are still in the development stage, this section only introduces the controls for rotary-motor-based active suspension systems.

7.6.1   Suspension System Model of a Quarter Car

The free body diagram of a suspension system for a quarter car is shown in Fig. 7.25, where the following notation is used:

z_0 – Road level displacement
z_1 – Wheel/tire displacement

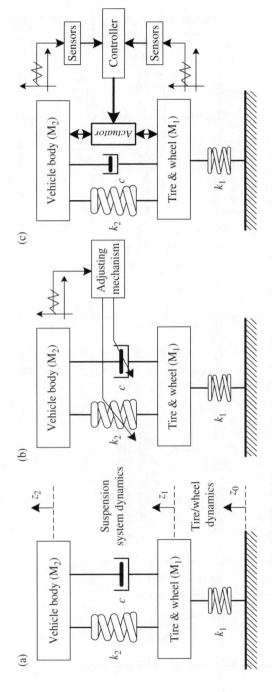

Figure 7.23 Diagrams showing (a) passive, (b) semi-active, and (c) active suspension systems

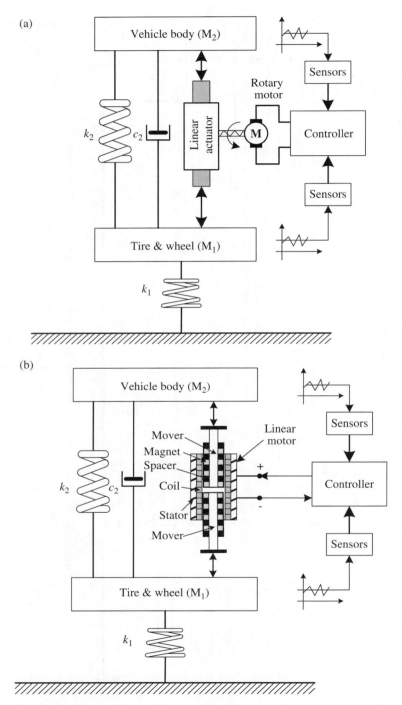

Figure 7.24   Diagrams of (a) rotary motor-based and (b) linear motor-based electric active suspension systems

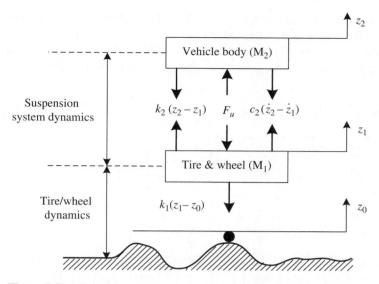

Figure 7.25　Free body diagram of the suspension system of a quarter car

z_2 – Vehicle body displacement
m_1 – Mass of wheel assembly
m_2 – Mass of quarter vehicle body
k_1 – Stiffness of tire
k_2 – Stiffness of passive spring
c_2 – Damping coefficient of the passive shock absorber
F_u – Force generated by the actuator
$F_{actuator\_max}$ – Maximal output force of the actuator
$K_{actuator}$ – Gain of the actuator from input voltage to output force
$T_{actuator}$ – Time constant of the actuator from input voltage to output force

According to Fig. 7.25 and based on Newton's second law, the dynamics of a suspension system can be described by the following set of differential equations:

$$m_2\ddot{z}_2 = -c_2(\dot{z}_2 - \dot{z}_1) - k_2(z_2 - z_1) + F_u$$
$$m_1\ddot{z}_1 = c_2(\dot{z}_2 - \dot{z}_1) + k_2(z_2 - z_1) - k_1(z_1 - z_0) - F_u$$

(7.32)

Reorganizing Eq. 7.32, we get:

$$\ddot{z}_2 = -\frac{c_2}{m_2}\dot{z}_2 - \frac{k_2}{m_2}z_2 + \frac{c_2}{m_2}\dot{z}_1 + \frac{k_2}{m_2}z_1 + \frac{F_u}{m_2}$$
$$\ddot{z}_1 = \frac{c_2}{m_1}\dot{z}_2 + \frac{k_2}{m_1}z_2 - \frac{c_2}{m_1}\dot{z}_1 - \frac{k_2 + k_1}{m_1}z_1 + \frac{k_1}{m_1}z_0 - \frac{F_u}{m_1}$$

(7.33)

If we define the disturbance input as the road level disturbance $v = z_0$, the control input as the actuator force $u = F_u$, the state variables as $x_1 = z_1$, $x_2 = \dot{z}_1$, $x_3 = z_2$, and $x_4 = \dot{z}_2$, and the outputs as the wheel/tire displacement and the vehicle body displacement, that is, $y_1 = z_1$ and $y_2 = z_2$, then we have the following state-space equation for the active suspension system:

$$
\begin{bmatrix} \dot{x}_1 \\ \dot{x}_2 \\ \dot{x}_3 \\ \dot{x}_4 \end{bmatrix} = \begin{bmatrix} 0 & 1 & 0 & 0 \\ -\dfrac{k_2+k_1}{m_1} & -\dfrac{c_2}{m_1} & \dfrac{k_2}{m_1} & \dfrac{c_2}{m_1} \\ 0 & 0 & 0 & 1 \\ \dfrac{k_2}{m_2} & \dfrac{c_2}{m_2} & -\dfrac{k_2}{m_2} & -\dfrac{c_2}{m_2} \end{bmatrix} \begin{bmatrix} x_1 \\ x_2 \\ x_3 \\ x_4 \end{bmatrix} + \begin{bmatrix} 0 \\ -\dfrac{1}{m_1} \\ 0 \\ \dfrac{1}{m_2} \end{bmatrix} u + \begin{bmatrix} 0 \\ \dfrac{k_1}{m_1} \\ 0 \\ 0 \end{bmatrix} v
$$

$$
= \begin{bmatrix} 0 & 1 & 0 & 0 \\ -a_{21} & -a_{22} & a_{23} & a_{24} \\ 0 & 0 & 0 & 1 \\ a_{41} & a_{42} & -a_{43} & -a_{44} \end{bmatrix} \begin{bmatrix} x_1 \\ x_2 \\ x_3 \\ x_4 \end{bmatrix} + \begin{bmatrix} 0 \\ -b_2 \\ 0 \\ b_4 \end{bmatrix} u + \begin{bmatrix} 0 \\ \gamma_2 \\ 0 \\ 0 \end{bmatrix} v \qquad \begin{bmatrix} y_1 \\ y_2 \end{bmatrix} = \begin{bmatrix} 1 & 0 & 0 & 0 \\ 0 & 0 & 1 & 0 \end{bmatrix} \begin{bmatrix} x_1 \\ x_2 \\ x_3 \\ x_4 \end{bmatrix}
$$

$$(7.34)$$

The transfer function between the control input voltage and the output force of the actuator can be assumed as:

$$
G_{\text{actuator}}(s) = \frac{F_{\text{actuator}}(s)}{V_{\text{actuator}}(s)} = \frac{K_{\text{actuator}}}{T_{\text{actuator}}s + 1} \qquad (7.35)
$$

7.6.2 Active Suspension System Control

The output force of an electric actuator can be controlled to combat road disturbance, and the control objective is to minimize the vehicle body displacement; that is, the output variable, y_2. This section presents PID and model predictive control strategies for the active suspension system based on the following parameters of the suspension system and the vehicle body:

$$
F_{\text{actuator\_max}} = \pm 2800 \, (\text{N})
$$

$$
K_{\text{actuator}} = 500
$$

$$
T_{\text{actuator}} = 0.12 \, (\text{ms})
$$

$$
m_1 = 45 \, (\text{kg})
$$

$$m_2 = 320\,(\text{kg})$$

$$k_1 = 150000\,(\text{N/m})$$

$$k_2 = 45000\,(\text{N/m})$$

$$c_2 = 3000\,(\text{Ns/m})$$

From the above system parameters, the system model (Eq. 7.34) is detailed as:

$$
\begin{bmatrix} \dot{x}_1 \\ \dot{x}_2 \\ \dot{x}_3 \\ \dot{x}_4 \end{bmatrix} = \begin{bmatrix} 0 & 1 & 0 & 0 \\ -4333 & -67 & 1000 & 67 \\ 0 & 0 & 0 & 1 \\ 140 & 9.375 & -140 & -9.375 \end{bmatrix} \begin{bmatrix} x_1 \\ x_2 \\ x_3 \\ x_4 \end{bmatrix} + \begin{bmatrix} 0 \\ -0.022 \\ 0 \\ 0.003125 \end{bmatrix} u + \begin{bmatrix} 0 \\ 3333 \\ 0 \\ 0 \end{bmatrix} v
$$

$$
\begin{bmatrix} y_1 \\ y_2 \end{bmatrix} = \begin{bmatrix} 1 & 0 & 0 & 0 \\ 0 & 0 & 1 & 0 \end{bmatrix} \begin{bmatrix} x_1 \\ x_2 \\ x_3 \\ x_4 \end{bmatrix}
$$

(7.36)

Furthermore, the following transfer functions can be obtained:

1. From active control input to wheel displacement:

$$
\frac{Y_1(s)}{U(s)} = G_{11}(s) = \frac{-0.02222s^2 + 7.648 \times 10^{-17}s + 4.722 \times 10^{-16}}{s^4 + 76.04s^3 + 4474s^2 + 31250s + 468800}
$$

(7.37)

2. From road disturbance to wheel displacement:

$$
\frac{Y_1(s)}{V(s)} = G_{12}(s) = \frac{3333s^2 + 31250s + 468700}{s^4 + 76.04s^3 + 4474s^2 + 31250s + 468800}
$$

(7.38)

3. From active control input to vehicle body displacement:

$$
\frac{Y_2(s)}{U(s)} = G_{21}(s) = \frac{0.003125s^2 + 3.164 \times 10^{-16}s + 10.42}{s^4 + 76.04s^3 + 4474s^2 + 31250s + 468800}
$$

(7.39)

4. From road disturbance to vehicle body displacement:

$$
\frac{Y_2(s)}{V(s)} = G_{22}(s) = \frac{31250s + 468700}{s^4 + 76.04s^3 + 4474s^2 + 31250s + 468800}
$$

(7.40)

5. The transfer function of the actuator is:

$$\frac{F_{\text{actuator\_output}}(s)}{V_{\text{actuator\_input}}(s)} = G_{\text{actuator}}(s) = \frac{500}{0.00012s + 1} \tag{7.41}$$

7.6.2.1 PID Control Strategy

A diagram of a PID-based active suspension control system is presented in Fig. 7.26, where the output of PID control is the manipulated variable of the actuator, and, in turn, the actuator outputs the associated force to the suspension system. The road-level displacement is considered as the measured disturbance input. Output limits of the PID controller are $-5(\text{V}) \le u_{\text{actuator\_input}} \le 5(\text{V})$ and the corresponding actuator output limits are $-2800(\text{N}) \le u_{\text{actuator\_output}} \le 2800(\text{N})$. Since the control objective is to minimize the displacement of the vehicle body, the tuned PID controller is actually set as a PI controller. The gain, K, and the integral time constant, T_i, are set as 50 and 0.4, respectively.

The following simulation example shows that when exerting a road disturbance signal, shown in Fig. 7.27, oscillations of the vehicle body are significantly attenuated by the active suspension control. Figures 7.28 and 7.29 present the vehicle body and tire/wheel responses of the suspension system with/without active control action. Figures 7.30 and 7.31 show the output force and consumed/regenerative electrical power of the actuator based on the PID control strategy, respectively.

7.6.2.2 Model Predictive Control Strategy

The diagram of a model predictive active suspension control system is shown in Fig. 7.32. The inputs for the model predictive controller are the reference setpoint and the measured road disturbance. Vehicle body displacement is one of the measured system outputs, and it is fed back to the input. The output of the model predictive controller is the input of the actuator, and the actuator generates the corresponding manipulated force to the active suspension system.

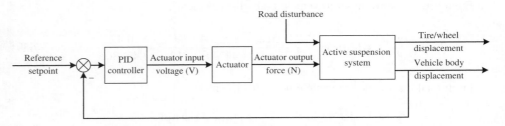

Figure 7.26 PID-based active suspension control system

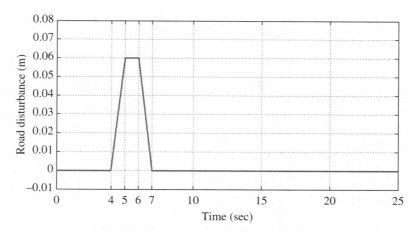

Figure 7.27   Exerted road disturbance

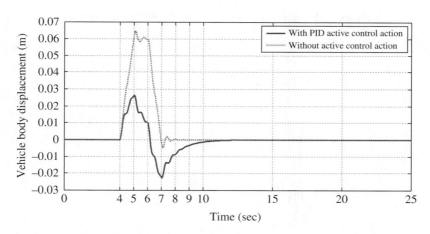

Figure 7.28   Vehicle body responses with/without PID-based active suspension control

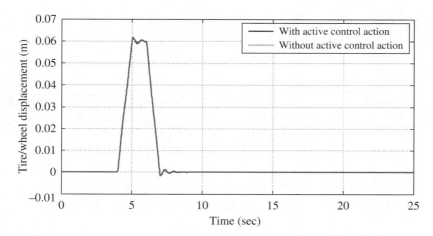

Figure 7.29   Tire/wheel responses with/without PID-based active suspension control

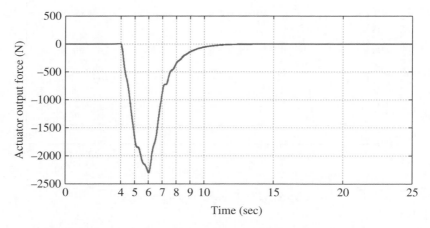

Figure 7.30   Actuator output based on the PID control strategy

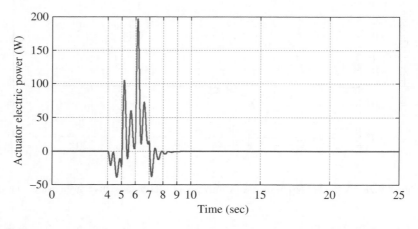

Figure 7.31   Actuator output electrical power based on the PID control strategy (+ power, − regen)

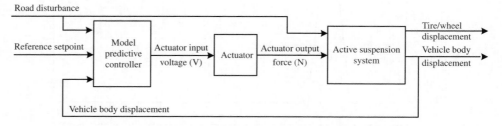

Figure 7.32   Model predictive-based active suspension control system

The set control objective is to minimize the displacement of the vehicle body, and detailed parameters for the model predictive controller are as follows:

- Sampling time: 100 µs;
- Prediction horizon: 10
- Control horizon: 2
- Weight factor of manipulated variable rate: 0.1
- Weight factor of manipulated variable: 1
- Weight factor of output variable: 1000
- Actuator input constraints: $-5(\text{V}) \leq u_{\text{input\_actuator}} \leq 5(\text{V})$
- Actuator output constraints: $-2800(\text{N}) \leq u_{\text{output\_actuator}} \leq 2800(\text{N})$

The achieved simulation results can be described as follows. When exerting the same road disturbance signal as for the PID controller, as shown in Fig. 7.27, the dynamics of the vehicle body are significantly improved by active control action, as shown in Fig. 7.33; the tire/wheel responses of the suspension system with/without active control action are shown in Fig. 7.34; the output forcer and consumed/regenerative electrical power of the actuator from the MPC control strategy are shown in Fig. 7.35 and Fig. 7.36.

The mechanism of an active suspension system means that it is very challenging to control the dynamic behavior in order to meet driving comfort and operating safety requirements. From the presented active suspension control strategies, it can be found that the control strategy plays a very important role in an active suspension system, and an advanced control strategy can effectively improve performance and efficiency of an active suspension system. Since the 1990s, advanced active suspension control systems have been studied extensively,

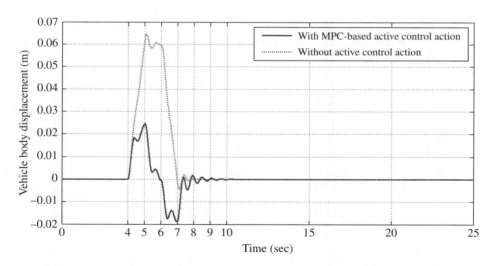

Figure 7.33 Vehicle body responses with/without MPC-based active suspension control

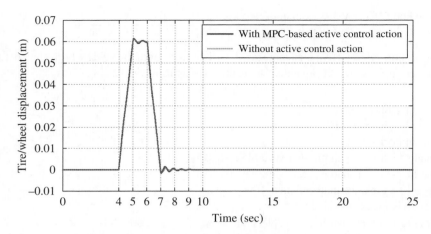

Figure 7.34 Tire/wheel responses with/without MPC-based active suspension control

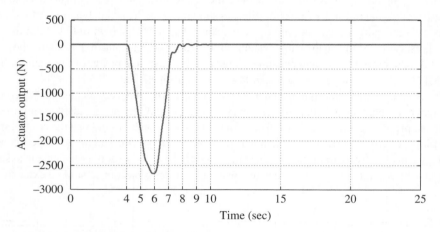

Figure 7.35 Actuator output based on the MPC control strategy

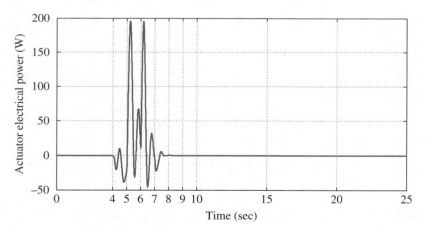

Figure 7.36 Actuator output electrical power based on the model predictive control strategy (+ power, − regen)

and most recent work has been focused on control strategy development and application, such as the optimal feedback control and adaptive control strategies, to achieve the desired system performance and overcome the degradation of component performance with age (Fischer and Isermann, 2004).

7.7 Adaptive Charge-sustaining Setpoint and Adaptive Recharge SOC Determination for PHEVs

Plug-in hybrid electric vehicles (PHEVs) and electric range-extended vehicles (EREVs) are important hybrid electric vehicles for reducing fossil fuel consumption and emissions, since they can run on electricity for a certain distance after each recharge, depending on the battery's energy capacity. A PHEV/EREV generally has two operational modes: charge-depleting (CD) mode and charge-sustaining (CS) mode. The CD mode enables the PHEV/EREV to drive a certain number of miles using the electric motor(s) only, and when the battery is depleted to a certain level, the operational mode is switched to CS, in which the vehicle operates as a traditional HEV with the ICE and electric motor(s). A PHEV/EREV is typically expected to have an electric drive range between 20 and 80 miles, thus commuters can potentially drive their daily commute exclusively with the electric motor.

The typical characteristic electrical energy usage curves are shown in Fig. 7.37 and Fig. 7.38. In the first portion of Fig. 7.38, the battery starts to deplete from the last recharge termination point. Once the battery depletes to a certain level, the ICE starts and the operational mode becomes the charge-sustaining mode in the second portion of Fig. 7.38, where the battery's SOC goes up and down within the designed CS window around the CS setpoint depending on the actual driving scenarios.

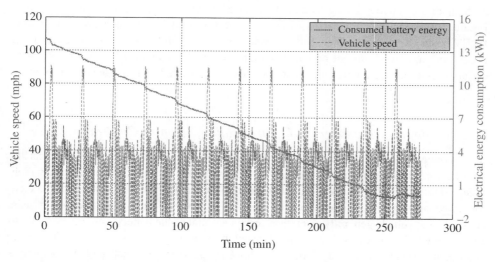

Figure 7.37   Vehicle speed and electrical energy consumption over repeated US06 cycles

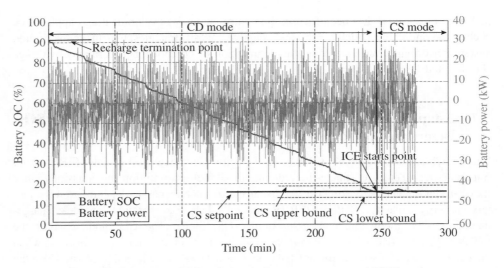

Figure 7.38 Battery SOC and electrical power over repeated US06 cycles

7.7.1 Scenarios of Battery Capacity Decay and Discharge Power Capability Degradation

As discussed in Chapter 5, a battery's capacity decays and its power capabilities fade when it ages. Detailed degradation for a given vehicle varies depending on how the vehicle has been operated and where and how the vehicle has been stored. Figure 7.39 shows the energy capacity decays at room temperature of a PHEV/EREV battery system for vehicles with high, medium, and low mileages operated in different climate regions over the service time. Figure 7.40 shows the cold discharge power capability degradation scenarios of the PHEV/EREV battery system over time. Since a battery's capacity and power capabilities naturally decay with age, working out how to maintain the performance of a PHEV/EREV over a certain period of time and for different operational environments is a crucial control issue.

7.7.2 Adaptive Recharge SOC Termination Setpoint Control Strategy

Since conventional vehicle performance degrades very slowly, customers expect that the performance degradation of an electric vehicle will also be slow, although it is widely accepted that the battery will degrade. In order to have a PHEV/BEV with a consistent electric range, and especially to avoid the obvious EV mileage loss in the first couple of years after sale, the vehicle generally reserves a certain amount of the battery's energy capacity to expend gradually to deal with the natural decay in the battery's energy capacity. The exact amount of the reserved capacity depends on design targets for the vehicle and the company's sale strategy.

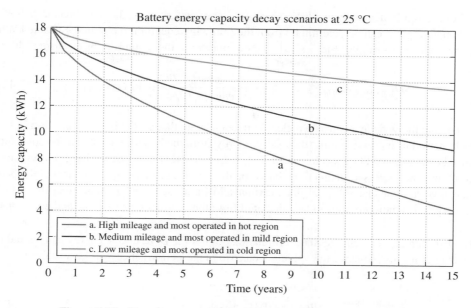

Figure 7.39 Battery energy capacity decay scenarios over service years

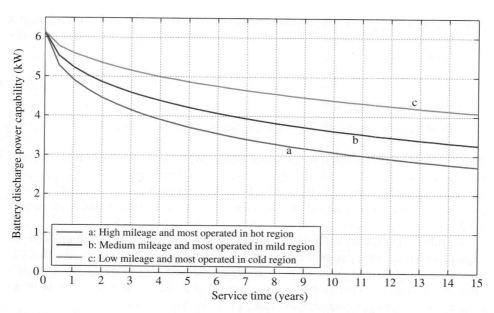

Figure 7.40 Battery ten-second discharge power capability degradation scenarios at 10% SOC at −30 °C

For a given PHEV with an x kWh energy capacity battery pack, the initial recharge SOC termination point can be determined from the following equation to obtain the usable kWh energy needed to support the desired electric mileage at the beginning of the battery life:

$$SOC_{\text{recharge\_termination\_BOL}} \cong \frac{Usable\_energy\_BOL}{Total\_energy\_capacity} + SOC_{\text{CS\_setpoint}} \qquad (7.42)$$

To give the vehicle a consistent electric mileage even though the battery ages during the early period of its service time, the usable energy from the battery pack has to be constant, thus the operational SOC window needs to be expanded; that is, the SOC charge termination point must be increased in real time, based on the actual capacity decay scenario of the battery. If we assume that SOC_{max} is the maximal allowable used SOC point, the real-time SOC recharge termination point can be calculated from the following procedures and equations.

1. Determining the recharge SOC termination point if the BOL usable energy can be maintained; in this scenario, the vehicle's electric range performance will be sustained.

$$SOC(t)_{\text{recharge\_termination}} = \frac{Usable\_energy\_BOL}{Ahr\,capacity(t) \cdot V_{\text{nominal}}}$$
$$+ SOC_{\text{CS\_setpoint}}; \text{ for } SOC(t)_{\text{recharge\_termination}} \leq SOC_{\text{max}} \qquad (7.43)$$

2. Setting the recharge SOC termination point as SOC_{max} if the BOL usable energy cannot be maintained; in this scenario, the vehicle's electric range performance will gradually degrade.

$$SOC(t)_{\text{recharge\_termination}} = SOC_{\text{max}}$$
$$Usable\_energy(t) \cong (SOC_{\text{max}} - SOC_{\text{CS\_setpoint}}) \cdot Ahr\_capacity(t) \cdot V_{\text{nominal}} \qquad (7.44)$$

In the above, $Ahr\_capacity(t)$ is the estimated battery system Ahr capacity in real time, $usable\_energy\_BOL$ is the electrical energy (kWh) delivered by the battery to support the electric range performance of the vehicle at the beginning of life of the battery, SOC_{max} is the predetermined maximal allowable SOC based on the cell chemistry and battery service life prediction, and t is the time in years.

The online Ahr capacity estimator is the key algorithm in this control strategy. It can be implemented based on the principles presented in Chapter 5. Since the battery of a PHEV is periodically charged through an AC electrical outlet, the Ahr capacity can be estimated right after plug-in charging is terminated or at the moment of the next key-up based on the information gained during plug-in charging using the following procedures:

1. Establish and store the relationship data between V_{oc} (open-circuit voltage) and capacity. As discussed in Chapter 5, the V_{oc} vs. SOC characteristics of a given battery are

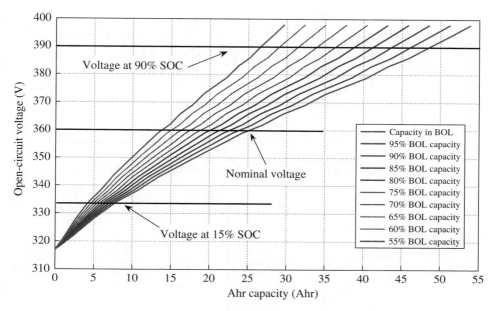

Figure 7.41 V_{oc} vs. Ahr capacity characteristics of a Li-ion battery

determined by the properties of the battery materials, and materials' properties do not significantly change over time. Therefore, the characteristics of V_{oc} vs. Ahr capacities for a given battery, shown in Fig. 7.41, can be established and stored in the vehicle controller at the BOL. The data can be stored in a form such as $\left\{ Cap\_Ah^{(j)}(k) \ \ V_{oc}^{(j)}(k) \right\}$, where $j = 1, \cdots, N$ is the jth set of the Ahr capacity characteristic data, $k = 1, \cdots, n$ is the kth data pair of Ahr capacity and open-circuit voltage.

2. Integrate the current during plug-in charging to obtain the charges Q in Ahr between V_{oc_Begin} and V_{oc_End} as:

$$Q = \frac{1}{3600} \int_{V_{oc\_Begin}}^{V_{oc\_End}} I(t)dt \cong \frac{1}{3600} \sum_{k=1}^{n} I(k)\Delta T, \quad k = 1, \cdots, n; \ \ V_t(1) = V_{oc\_Begin} \text{ and } V_t(n) = V_{oc\_End}$$

(7.45)

where V_t is the battery terminal voltage, I is the battery current, and ΔT is the sampling time in seconds.

3. Estimate V_{oc} online using a recursive least squares estimation based on the following second-order difference equation derived from the electrical equivalent circuit model shown in Fig. 7.42 at each time point. For the detailed model and recursive least squares estimation method, interested readers should refer to Chapter 5 and to the book by Ljung (1987).

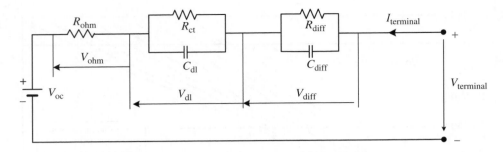

Figure 7.42 Two *RC* pair electrical circuit equivalent model of a battery

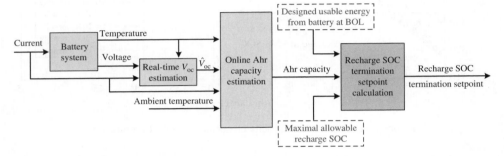

Figure 7.43 Adaptive recharge SOC termination setpoint control strategy

$$V_{\text{terminal}}(k) + a_1 V_{\text{terminal}}(k-1) + a_2 V_{\text{terminal}}(k-1) = b_0 I_{\text{terminal}}(k) + b_1 I_{\text{terminal}}(k-1)$$
$$+ b_2 I_{\text{terminal}}(k-1) + c_0 \qquad (7.46)$$
$$\hat{V}_{\text{oc}} = \frac{\hat{c}_0}{1 - \hat{a}_1 - \hat{a}_2}$$

where $\hat{a}_1$, $\hat{a}_2$, and $\hat{c}_0$ are the estimated parameter values in Eq. 7.46 from the recursive least squares estimation algorithm presented in Appendix A.

4. Find the best matched capacity curve which makes the following objective function minimal by a real-time optimization method:

$$J = \min \left(\sum_{j=1}^{N} \sum_{k=1}^{n} \left(Cap\_Ah^{(j)}(k) - Q(k) \right)^2 \Big|_{\hat{V}_{\text{oc}}(k) = V_{\text{oc}}(k)} \right), \quad j = = 1, \cdots, N, \quad k = 1, \cdots, n \quad (7.47)$$

where $\{ Q(k) \ \hat{V}_{\text{oc}}(k) \}$ are the time series data pairs of the real-time calculated plug-in charges and the estimated open-circuit voltage; $\left\{ Cap\_Ah^{(j)}(k) \ V_{\text{oc}}^{(j)}(k) \right\}$ are the stored pre-established V_{oc} and the capacity relationship data sets.

5. Use the estimated battery capacity to determine the plug-in charge termination point.

The diagram of the presented adaptive recharge SOC termination setpoint control strategy is shown in Fig. 7.43. In the application, V_{oc} is estimated in real time every sampling time, the capacity estimation algorithm is executed once every key cycle, and the control output is used to set the recharge termination point of the next plug-in charging event.

Example 7.1

A PHEV is designed to have a 55-mile electric range, and an 18.5 kWh Li-ion battery system is used to support the performance. To have 55 electric drive miles, the vehicle requires the battery system to be able to deliver 14 kWh of usable energy. The maximal operational SOC of the battery can be set to 98.5% with minor impact on the life. Based on the characteristics of the battery system, the vehicle performance requirements, and the ICE cold-cranking power requirements, the charge-sustaining point is set at 16% SOC and the battery can deliver 14 kWh of energy if the depleting window is set between 16 and 90% SOC at the beginning of life. In order to make the electric range consistent for a certain period of time, the manufacturer decided to reserve battery energy capacity of between 90 and 98.5% SOC at the assembly line and to open the reserved capacity gradually to cover the capacity decay of the battery through the algorithm presented above. This example illustrates the vehicle EV range degradation scenarios and how the reserved battery capacity is gradually opened over time based on the degradation scenario (b) shown in Fig. 7.39. The details are as follows.

1. **SOC window-expanding scenarios**

   In order to meet the EV range performance requirement, the battery needs to deliver 14 kWh of energy constantly regardless of how the battery capacity decays. To compensate for battery capacity decay over time, the SOC-depleting window is gradually expanded to 16–98.5% SOC from the initial 16–90% SOC. It should be noted that achieving EV range consistency is a trade-off whereby the potential initial EV range performance is reduced – the 16–98.5% SOC-depleting window of this battery system could support 62 electric drive miles at the beginning of life of the battery. Figure 7.44 shows the process of expanding the SOC-depleting window over repeated US06 cycles based on the presented adaptive strategy. For this usage case, the figure shows that the SOC-depleting window reaches the maximum after two years, and after that, the vehicle's EV range will degrade over time.

2. **Battery-depleting scenarios over time**

   After the SOC-depleting window reaches its maximum, the depletion rate will be faster due to the battery capacity decay over time. Figure 7.45 shows the SOC-depleting scenarios over years for repeated US06 drive cycles.

3. **Battery usable energy decay scenarios**

   Once the battery recharge termination point has adapted to the maximal point, the battery is no longer able to deliver the desired amount of usable energy due to capacity decay, which results in a reduction in the PHEV's electric range. Figure 7.46 shows the process of usable energy decline after the recharge SOC reaches the maximum of 98.5%. The figure shows that the usable energy reduces to 10 kWh from the original 14 kWh in ten years on repeated US06 drive cycles.

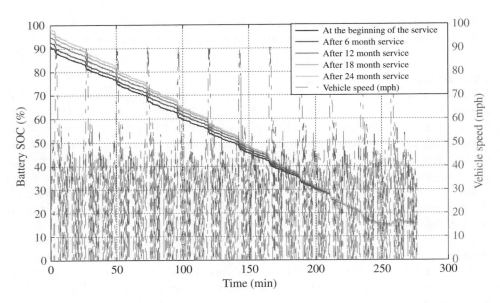

Figure 7.44 Example of battery SOC-depleting window expansion over time

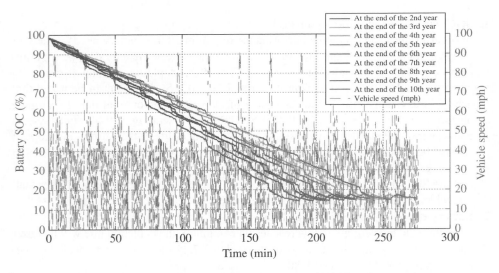

Figure 7.45 SOC-depleting scenarios on repeated US06 cycles over time

4. Vehicle electric range degradation over time

For average vehicle usage cases such as scenario (b) in Fig. 7.39, this example shows that the battery-delivered energy dwindles by about 4 kWh in ten years; correspondingly, this will result in the vehicle's electric range reducing to 39 miles from the original 55 miles. The decay curve of the usable energy from the battery over years is shown in Fig. 7.47, and the associated vehicle electric range loss curve is shown in Fig. 7.48.

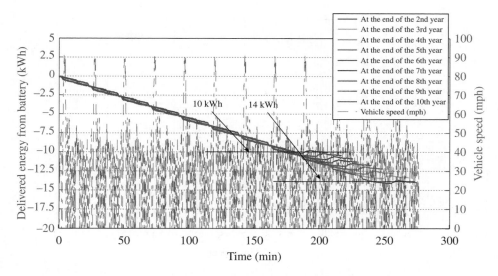

Figure 7.46   Battery usable energy declining scenarios on repeated US06 cycles over time

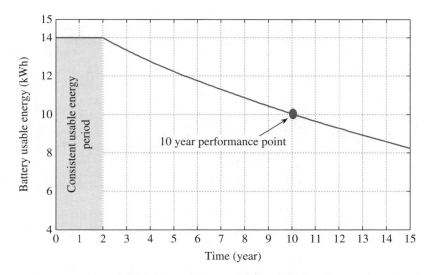

Figure 7.47   Battery usable energy decay over time

7.8   Online Tuning Strategy of the SOC Lower Bound in CS Operational Mode

7.8.1   PHEV Charge-sustaining Operational Characteristics

As previously mentioned, the design objective for PHEVs/EREVs is that the vehicle should be capable of operating exclusively with the electric motor(s) for a certain number of miles

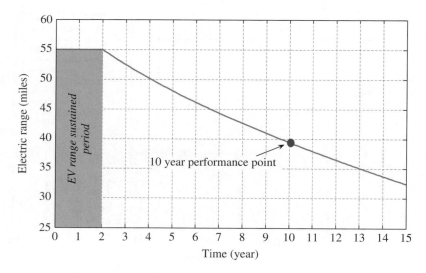

Figure 7.48   The vehicle's electric mileage loss over time

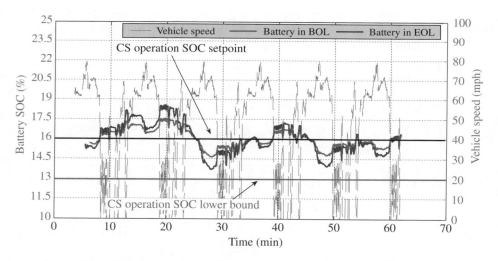

Figure 7.49   Battery charge-sustaining example on repeated US06 cycles over time

after each recharge so that no gasoline is consumed by most daily commuters, and the ICE should start once the distance traveled goes beyond the electric range, so removing the driver's anxiety on the electric range of BEVs. The normal operational sequence of a PHEV/EREV is CD mode and then CS mode, as shown in Fig. 7.37. In the CS mode, the vehicle operates as a traditional hybrid electric vehicle, and the operational characteristics are shown in Fig. 7.49.

From Fig. 7.49, it can be seen that the SOC setpoint and the lower bound in CS mode are two key design parameters for PHEVs/EREVs. The general design principles for these two parameters are:

1. The lower bound of CS-operation SOC is determined ensuring that the battery at the end of life has enough power to crank the ICE in cold weather conditions such as where the vehicle is soaked at $-30\,°C$ for a while.
2. The CS-operation setpoint is determined in such a way that the battery, between the setpoint and the lower bound in the middle of life, has enough energy to supplement the ICE to meet the vehicle's demands in most acceleration or uphill scenarios based on the optimal hybrid control strategy such as the one introduced in Chapter 6.

However, the above design principles generally result in a conservative outcome at the beginning of the vehicle's life, since the performance engineer has to balance vehicle performance at the beginning of life and at the end of life as well as operation in plain regions and in mountainous regions. For example, on the same vehicle driving cycle, Fig. 7.49 shows that the SOC swing window will increase by about 40% when the example battery reaches the tenth year of service based on nominal usage. In addition, at the beginning of its life, the battery is capable of cranking the ICE at $-30\,°C$ even at 5% SOC, but the battery SOC needs to be 13% to provide the same cold-cranking power at the end of its life.

7.8.2 PHEV Battery CS-operation SOC Lower Bound Online Tuning

The lower bound of the CS-operation SOC window is normally determined to meet the ICE cold-cranking power requirement over the vehicle's service life. Since battery cold-cranking power capability degrades with aging, as shown in Fig. 7.50, the lower bound needs to be set based on the end-of-life performance of the battery, so generally resulting in a big margin at the beginning of the system's life. Therefore, there is a present need to develop an online tuning strategy to adjust the lower bound based on the health state of the battery so as to use the battery most effectively.

From Fig. 7.50, it can be seen that battery cold-cranking capacity degrades with aging, thus identifying the cold-cranking capability at the required temperature is paramount when setting the CS-operation SOC lower bound. This section introduces a CS-operation SOC lower bound self-tuning algorithm for PHEVs/EREVs based on the battery resistance online estimation and cold-cranking power capability real-time calculation with ambient temperature feedforward compensation. The algorithm is executed once per key cycle and enabled only if the vehicle operates in CS mode. The main functional blocks of the algorithm are as follows.

7.8.2.1 Battery Resistance Online Estimation While the Vehicle is in the CS Mode of Operation

Selecting a proper battery model is essential in order to estimate battery resistance in real time. Taking both complexity and accuracy into account, the following first-order difference

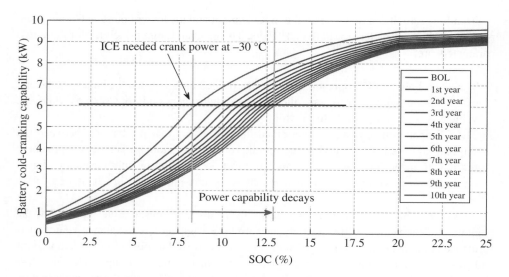

Figure 7.50   An example of battery cold-cranking capabilities over time at −20 °C

model is used to describe battery dynamic characteristics at the SOC setpoint and the operational temperature in the presented algorithm. The model is derived from a one *RC* pair electrical equivalent model and the mathematical model parameters contain information on the battery open-circuit voltage, the ohmic resistance, the dynamic resistance, and the capacitance. For details of the model, readers should refer to Section 5.2.2 in Chapter 5).

$$V_{terminal}(k) = a_1 V_{terminal}(k-1) + b_0 I_{terminal}(k) + b_1 I_{terminal}(k-1) + c_1 \qquad (7.48)$$

where a_1, b_0, b_1, and c_1 are model parameters which can be estimated in real time based on the measured battery terminal voltage and current through recursive least squares.

Once these parameters have been estimated, the electrical circuit model parameters at the operating temperature and the CS-operation SOC setpoint can be obtained by solving the following equations:

$$
\begin{cases}
R_{ohm} = \hat{b}_0 \\[2mm]
V_{oc} = \dfrac{\hat{c}_1}{1-\hat{a}_1} \\[2mm]
R_{dyn} = \dfrac{\hat{b}_1 + \hat{a}_1 \cdot \hat{b}_0}{1-\hat{a}_1} \\[2mm]
C_{dyn} = \dfrac{\Delta T}{\hat{b}_1 + \hat{a}_1 \cdot \hat{b}_0}
\end{cases}
\qquad (7.49)
$$

where ΔT is the system sampling time.

7.8.2.2 Electrical Circuit Model Parameter Adjustment

Based on the properties of Li-ion battery materials, it can be assumed that each model parameter follows its own decay mechanism over life, as shown in Fig. 7.39, regardless of temperature. Therefore, the model parameter decay rate can be obtained by comparing each estimated parameter with the BOL parameter at the same temperature, and the SOC can be found by looking at the pre-established BOL parameter table, an example of which is shown in Table 7.1. Furthermore, the model parameters needed to predict the cold-cranking power capability can be achieved by multiplying the decay rate by the BOL parameters at the required cold-cranking temperature.

7.8.2.3 Calculate Battery Cold-cranking Power Capability

Once the model parameters have been obtained, the battery cold-cranking power capability can be calculated using the following equation at the given SOC and temperature, and the output used to compare it with the needed ICE cold-cranking power. Details of the electrical equivalent-based power capability calculation are given in Section 5.3.3.

If we assume that the maximal allowable battery cold-cranking current is $I_{\text{max\_cold\_cranking}}$, the minimal allowable battery cold-cranking voltage is $V_{\text{min\_cold\_cranking}}$, and the cranking power duration needed is $t_{\text{cold\_cranking}}$ seconds, the maximal current-based battery power $P_{\text{I-limited\_cold\_cranking}}$ and the minimal voltage-based battery power $P_{\text{V-limited\_cold\_cranking}}$ are as follows:

$$P_{\text{I\_limited\_cold\_cranking}}\left(t_{\text{cold\_cranking}}\right) = V_{\text{oc}}I_{\text{max\_cold\_cranking}} - R_{\text{ohm}}I^2_{\text{max\_cold\_cranking}}$$

$$-R_{\text{dyn}}I^2_{\text{max\_cold\_cranking}}\left(1 - e^{\left(-\frac{t_{\text{cold\_cranking}}}{R_{\text{dyn}}C_{\text{dyn}}}\right)}\right) \qquad (7.50)$$

Subject to:

$$V_{\text{terminal}}(t) = V_{\text{oc}} - R_{\text{dyn}}I_{\text{max\_cold\_cranking}}\left(1 - e^{\left(-\frac{t_{\text{cold\_cranking}}}{R_{\text{dyn}}C_{\text{dyn}}}\right)}\right)$$

$$-R_{\text{ohm}}I_{\text{max\_cold\_cranking}} \geq V_{\text{min\_cold\_cranking}} \qquad (7.51)$$

$$P_{\text{V\_limited\_cold\_cranking}}\left(t_{\text{cold\_cranking}}\right) = \frac{V_{\text{min\_cold\_cranking}} - V_{\text{oc}}}{R_{\text{dyn}} + R_{\text{ohm}}}V_{\text{min\_cold\_cranking}} +$$

$$+ \frac{\left(V_{\text{min\_cold\_cranking}} - V_{\text{oc}}\right)R_{\text{dyn}}}{R_{\text{ohm}}\left(R_{\text{dyn}} + R_{\text{ohm}}\right)}V_{\text{min\_cold\_cranking}}e^{\left(-\frac{R_{\text{dyn}} + R_{\text{ohm}}}{R_{\text{dyn}}C_{\text{dyn}}R_{\text{ohm}}}t_{\text{cold\_cranking}}\right)} \qquad (7.52)$$

Table 7.1 BOL Li-ion battery pack electrical equivalent circuit model parameters

Temp (°C)	Model param.	SOC										
		0	5	10	15	20	25	30	35	40	45	50
−30	R_{ohm} (Ω)	0.666	0.660	0.655	0.650	0.646	0.642	0.639	0.636	0.634	0.632	0.631
	R_{dyn} (Ω)	1.248	1.070	0.915	0.780	0.665	0.568	0.487	0.422	0.370	0.330	0.302
	C_{dyn} (F)	0.78	0.90	1.01	1.23	1.49	2.19	2.91	3.65	4.39	5.11	5.82
	V_{oc} (V)	256	319	324	329	333	338	342	346	350	354	358
−20	R_{ohm} (Ω)	0.455	0.438	0.424	0.412	0.402	0.395	0.388	0.384	0.380	0.378	0.377
	R_{dyn} (Ω)	0.818	0.643	0.501	0.382	0.295	0.227	0.176	0.139	0.116	0.104	0.100
	C_{dyn} (F)	4.10	4.68	3.95	4.21	5.41	7.40	10.02	13.11	16.52	20.08	23.66
	V_{oc} (V)	256	321	333	336	340	344	347	350	353	356	360
−10	R_{ohm} (Ω)	0.325	0.320	0.316	0.311	0.307	0.304	0.301	0.298	0.295	0.293	0.291
	R_{dyn} (Ω)	0.331	0.273	0.226	0.188	0.158	0.136	0.121	0.111	0.107	0.106	0.108
	C_{dyn} (F)	5.23	6.18	7.50	8.54	28.77	45.64	59.40	70.34	78.70	84.76	88.78
	V_{oc} (V)	256	330	333	336	339	343	346	350	353	357	360
0	R_{ohm} (Ω)	0.219	0.213	0.208	0.202	0.197	0.193	0.189	0.185	0.182	0.179	0.176
	R_{dyn} (Ω)	0.170	0.155	0.142	0.130	0.120	0.111	0.103	0.097	0.093	0.089	0.086
	C_{dyn} (F)	39.08	53.91	68.03	80.52	91.79	102.6	112.1	120.3	127.3	132.9	137.3
	V_{oc} (V)	256	327	331	335	339	343	347	351	354	358	361

	Parameter											
	R_{dyn} (Ω)	0.248	0.186	0.140	0.108	0.085	0.069	0.060	0.055	0.053	0.055	0.058
	C_{dyn} (F)	39	55	63	96	124	146	163	174	181	184	183
	V_{oc} (V)	256	320	327	333	338	343	347	351	355	359	362
20	R_{ohm} (Ω)	0.119	0.111	0.104	0.098	0.093	0.089	0.086	0.084	0.082	0.080	0.080
	R_{dyn} (Ω)	0.062	0.062	0.061	0.061	0.061	0.060	0.060	0.060	0.060	0.059	0.059
	C_{dyn} (F)	129	144	158	170	181	190	198	205	210	214	216
	V_{oc} (V)	256	319	326	332	338	343	348	352	356	359	363
25	R_{ohm} (Ω)	0.104	0.097	0.091	0.086	0.082	0.078	0.075	0.073	0.071	0.069	0.068
	R_{dyn} (Ω)	0.058	0.056	0.055	0.054	0.054	0.053	0.052	0.052	0.051	0.051	0.050
	C_{dyn} (F)	160	172	182	194	202	211	222	230	234	235	234
	V_{oc} (V)	256	323	328	334	338	343	347	351	355	359	362
35	R_{ohm} (Ω)	0.075	0.071	0.067	0.063	0.061	0.059	0.057	0.056	0.055	0.054	0.054
	R_{dyn} (Ω)	0.061	0.056	0.051	0.047	0.044	0.043	0.041	0.041	0.041	0.041	0.042
	C_{dyn} (F)	93	126	153	188	215	234	248	255	258	257	253
	V_{oc} (V)	256	322	328	333	339	343	348	352	356	359	363

Subject to:

$$\left| I_{\text{terminal}}\left(t_{\text{cold\_cranking}}\right) \right| = \left| \frac{V_{\text{min\_cold\_cranking}} - V_{\text{oc}}}{R_{\text{dyn}} + R_{\text{ohm}}} \right.$$

$$\left. + \frac{\left(V_{\text{min\_cold\_cranking}} - V_{\text{oc}}\right)R_{\text{dyn}}}{R_{\text{ohm}}\left(R_{\text{dyn}} + R_{\text{ohm}}\right)} e^{\left(-\frac{R_{\text{dyn}} + R_{\text{ohm}}}{R_{\text{dyn}}C_{\text{dyn}}R_{\text{ohm}}} t_{\text{cold\_cranking}}\right)} \right| \qquad (7.53)$$

$$\leq I_{\text{max\_cold\_cranking}}$$

The maximal battery cold-cranking power capability in the period $t_{\text{cold\_cranking}}$ is the minimum of $P_{\text{I-limited\_cold\_cranking}}$ and $P_{\text{V-limited\_cold\_cranking}}$; that is:

$$P_{\text{max\_cold\_cranking}}\left(t_{\text{cold\_cranking}}\right) = \min\left\{ P_{\text{I-limited\_cold\_cranking}}, \ P_{\text{V-limited\_cold\_cranking}} \right\} \qquad (7.54)$$

7.8.2.4 Tuning the SOC Point to Meet the ICE Cold-cranking Power Requirement

Having obtained the estimated aged battery cold-cranking power capability, a PI control can be used to find what battery SOC is able to crank the ICE at the desired temperature. The PI governing diagram is shown in Fig. 7.51. The regulating process can be considered complete when the SOC point is bounded within 1%.

It should be noted that the above CS-operation SOC lower bound is tuned at a desired cold-cranking temperature of −30 °C or −20 °C, and the present ambient temperature and operational date may show that it is unlikely for the vehicle to be keyed up at such a cold temperature. Therefore, a feasible lowest battery temperature in the near future can be projected based on the present ambient temperature and date. In order to make more efficient use of the PHEV battery, the predicted lowest battery temperature may be considered the battery temperature for cold cranking.

Example 7.2

The ICE in a PHEV requires 8.5 kW of cranking power for 1.5 seconds to be fired up sufficiently at −30 °C, so the lower SOC bound of the battery is set at 8% SOC in the CS operational mode at the beginning of life. The battery model parameters have the values listed in

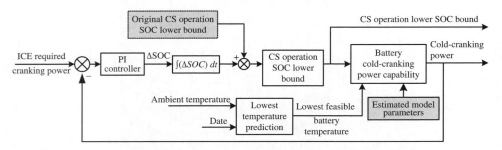

Figure 7.51 PI governing diagram of the algorithm for setting a PHEV battery lower SOC bound

Table 7.1 at the BOL. After three years of service, the model parameters are estimated as $R_{ohm} = 0.1174\,\Omega$, $R_{dyn} = 0.0684\,\Omega$, $C_{dyn} = 155\,F$, and $V_{oc} = 333\,V$ at 25 °C and 15% SOC. Assume that the minimal cold-cranking voltage is 192 V, the maximal cold-cranking current is 30 A at −30 °C, and the characteristics of ICE cold cranking do not change. Illustrate the algorithm execution process determining the SOC lower bound for CS operation with and without taking the ambient temperature and the operational date into account if the vehicle operates in July with a 35 °C ambient temperature.

Solution:

1. Obtain the model parameters at −30 °C

   The model parameters under the described operational conditions are estimated from an online parameter identification algorithm:

   $$R_{ohm} = 0.1118\,\Omega, R_{dyn} = 0.0702\,\Omega, C_{dyn} = 155\,F, \text{ and } V_{oc} = 333\,V$$

   Compared with the BOL model parameters listed in Table 7.1, the aging factors of these parameters can be calculated in real time as:

   $$F_{Rohm} = R_{ohm\_age}/R_{ohm\_BOL} = 0.1174/0.086 = 1.365; F_{Rdyn} = R_{dyn\_age}/R_{dyn\_BOL}$$
   $$= 0.0684/0.054 = 1.267; F_{Cdyn} = C_{dyn\_age}/C_{dyn\_BOL}$$
   $$= 155/194 = 0.799; F_{Voc}$$
   $$= V_{oc\_age}/V_{Voc\_BOL} = 333/334 = 1$$

   Based on an understanding of the cold-cranking power needed from the battery for the given vehicle, we can attempt to set a minimal SOC; in this example, a 10% SOC is attempted, and the associated model parameters at −30 °C and 10% SOC are obtained online as follows:

   $$R_{ohm\_aged\_-30°C} = 0.655*1.365 = 0.894\,\Omega; R_{dyn\_aged\_-30°C} = 0.915*1.267$$
   $$= 1.159\,\Omega; C_{dyn\_aged\_-30°C} = 1.01*0.799 = 0.807\,F; \text{ and } V_{oc\_aged\_-30°C} = 324\,V$$

2. The battery's cold-cranking power capability at the desired temperature and output SOC from PI control is computed online from Eqs 7.50 to 7.54 as follows:

   i. *The maximal current-limited cranking power calculation:*

   $$P_{I\_limited\_cold\_cranking}(1.5) = V_{oc}I_{max\_cold\_cranking} - R_{ohm}I^2_{max\_cold\_cranking}$$

   $$- R_{dyn}I^2_{max\_cold\_cranking}\left(1 - e^{\left(-\frac{t_{cold\_cranking}}{R_{dyn}C_{dyn}}\right)}\right)$$

   $$= 324\cdot30 - 0.894\cdot30^2 - 1.159\cdot30^2\left(1 - e^{\left(-\frac{1.5}{1.159\cdot0.807}\right)}\right)$$

   $$= 8.082\ (kW)$$

and the battery terminal voltage at a 30 A discharge current is:

$$V_{\text{terminal}}(t) = V_{\text{oc}} - R_{\text{dyn}}I_{\text{max\_cold\_cranking}}\left(1 - e^{\left(-\frac{t_{\text{cold\_cranking}}}{R_{\text{dyn}}C_{\text{dyn}}}\right)}\right) - R_{\text{ohm}}I_{\text{max\_cold\_cranking}}$$

$$= 324 - 1.159 \cdot 30\left(1 - e^{\left(-\frac{1.5}{1.159 \cdot 0.807}\right)}\right) - 0.894 \cdot 30 = 269 > 192(\text{V})$$

ii. *The maximal voltage-limited cranking power calculation:*

$$P_{\text{V\_limited\_cold\_cranking}}(1.5) = \frac{V_{\text{min\_cold\_cranking}} - V_{\text{oc}}}{R_{\text{dyn}} + R_{\text{ohm}}}V_{\text{min\_cold\_cranking}}$$

$$+ \frac{(V_{\text{min\_cold\_cranking}} - V_{\text{oc}})R_{\text{dyn}}}{R_{\text{ohm}}(R_{\text{dyn}} + R_{\text{ohm}})}V_{\text{min\_cold\_cranking}}e^{\left(-\frac{R_{\text{dyn}} + R_{\text{ohm}}}{R_{\text{dyn}}C_{\text{dyn}}R_{\text{ohm}}}t_{\text{cold\_cranking}}\right)}$$

$$= \frac{192 - 324}{1.159 + 0.894}192 + \frac{(192 - 324) \cdot 1.159}{0.894(1.159 + 0.894)}192e^{\left(-\frac{1.159 + 0.894}{1.159 \cdot 0.807 \cdot 0.894}1.5\right)}$$

$$= 12.747(\text{kW})$$

And the battery current if pushing the battery terminal voltage to the minimal voltage is:

$$|I_{\text{terminal}}(1.5)| = \left|\frac{V_{\text{min\_cold\_cranking}} - V_{\text{oc}}}{R_{\text{dyn}} + R_{\text{ohm}}} + \frac{(V_{\text{min\_cold\_cranking}} - V_{\text{oc}})R_{\text{dyn}}}{R_{\text{ohm}}(R_{\text{dyn}} + R_{\text{ohm}})}e^{\left(-\frac{R_{\text{dyn}} + R_{\text{ohm}}}{R_{\text{dyn}}C_{\text{dyn}}R_{\text{ohm}}}t_{\text{cold\_cranking}}\right)}\right|$$

$$= \left|\frac{192 - 324}{1.159 + 0.894} + \frac{(192 - 324)1.159}{0.894(1.159 + 0.894)}e^{\left(-\frac{1.159 + 0.894}{1.159 \cdot 0.894 \cdot 0.807}1.5\right)}\right|$$

$$= 66.39 > 30(\text{A}) = I_{\text{max\_cold\_cranking}}$$

Thus, at −30 °C and under a 10% SOC condition, the battery's cold-cranking power capability is 8.082 kW, which is less than the ICE's required 8.5 kW cranking power.

3. Since, at −30 °C and under a 10% SOC condition, the battery cannot supply 8.5 kW for 1.5 seconds, the PI controller finally sets the lower bound SOC at 11% to meet the ICE's cranking power requirement through the feedback regulation, the governing process of which is shown in Fig. 7.52.

4. In order to meet the ICE's cold-cranking power requirement, the battery's SOC lower bound is automatically set at 11% for CS operation without considering the ambient temperature. Since the vehicle is presently operating in July and at a 35 °C ambient temperature, it has a sufficient margin to set 0 °C as the lowest battery temperature over the next two months. At 0 °C, the battery is able to provide

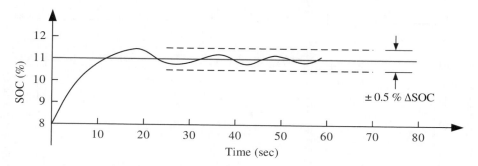

Figure 7.52 SOC regulation process for the PI feedback control strategy

8.5 kW for 1.5 seconds to crank the ICE at an 8% SOC; therefore, the CS-mode lower bound can be set at 8% SOC for the present operation of the vehicle.

7.9 PHEV Battery CS-operation Nominal SOC Setpoint Online Tuning

7.9.1 PHEV CS-operation Nominal SOC Setpoint Determination at BOL

In CS mode, a PHEV operates as an HEV; the battery provides the supplemental power to the ICE following a hybrid control command. As described in Chapter 6, the objective of hybrid control is to ensure that the vehicle operates most efficiently. Since a PHEV battery generally has enough power even at a low SOC, the CS-operation nominal SOC setpoint is mainly determined to ensure that the battery has enough energy to achieve the control objective in the most aggressive driving scenarios, such as 0 to 30 mph acceleration for stop–start, 30 to 70 mph quarter-mile acceleration ramping on a highway, and 50 to 80 mph acceleration for highway passing. When a PHEV operates in CS mode, even in these wide-open-throttle (WOT) scenarios, it is expected that the ICE will still run at its most efficient steady state and the battery will supplement the power if needed. The following example illustrates how the CS-operation nominal SOC setpoint is normally determined for PHEVs.

Example 7.3
Given a PHEV with a 120 kW electrical power capability when the battery is at 10% SOC, room temperature and BOL, if the vehicle technical specification requires that the electric powertrain is capable of powering the vehicle from initial movement to 80 mph in CS operation mode, determine the CS-operation nominal SOC setpoint to achieve a silent launching performance. Based on the battery's cold-cranking capability, the lower SOC bound of CS operation is set at 11%, the acceleration and battery power curves from initial vehicle movement (IVM) to 80 mph are shown in Fig. 7.53, and the characteristic curve of kWh energy versus SOC for the battery's BOL is shown in Fig. 7.54.

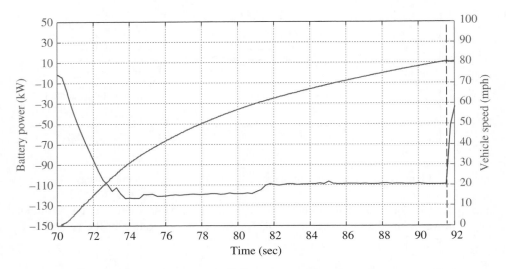

Figure 7.53   Battery power and vehicle speed curves from IVM to 80 mph acceleration

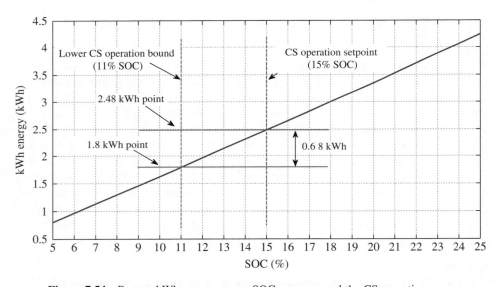

Figure 7.54   Battery kWh energy versus SOC curve around the CS-operation zone

From Fig. 7.53, the energy needed from the battery can be obtained as 0.62 kWh by inte-grating the battery power curve. Since the lower SOC bound of CS operation is set at 11%, from Fig. 7.54, it can be seen that the battery can deliver the needed 0.62 kWh energy to accelerate the vehicle electrically from IVM to 80 mph if the CS-operation nominal SOC setpoint is set at 15%.

7.9.2 Online Tuning Strategy of PHEV CS-operation Nominal SOC Setpoint

As discussed above, the CS-operation nominal SOC setpoint is set to maximally make use of the battery for boosting the vehicle in most WOT scenarios, such that the vehicle has superior performance and fuel economy. However, since the battery's energy capacity decays with age, as shown in Fig. 7.55, the CS-operation nominal setpoint needs to be set higher than is actually needed at BOL so that the battery still has enough energy to assist the ICE in achieving superior vehicle performance and fuel economy to compensate for the capacity decay, which is obviously traded off by decreasing the electric range, especially in the BOL period of the battery. Therefore, there is a need for a PHEV to adjust the CS-operation nominal SOC setpoint online based on the actual capacity of the battery and the actual vehicle need, in order to avoid rapid degradation of vehicle performance and fuel economy.

From Fig. 7.55, it is readily understood that the CS-operation nominal SOC setpoint should be set at above 18% for the vehicle to have a similar CS performance when the vehicle reaches its tenth year of service, which costs about three electric miles at the BOL. To use battery energy sufficiently and balance vehicle performance between the BOL and the EOL, an online CS-operation nominal setpoint adaption strategy is presented in Fig. 7.56. The main functions are detailed below.

7.9.2.1 Online Estimation of Battery Ahr Capacity, Open-circuit Voltage and Resistances

The actual Ahr capacity of the battery can be estimated based on the online algorithm presented in Chapter 5; it can also be determined during the plug-in charging period and the

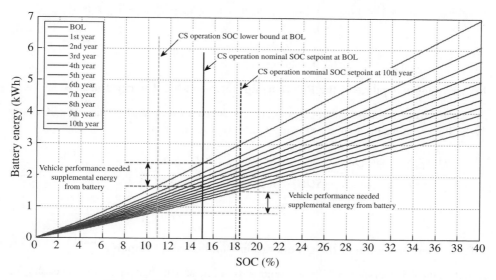

Figure 7.55 Example of a battery energy decay scenario

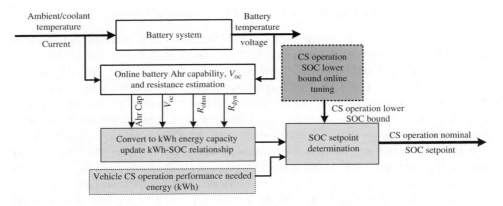

Figure 7.56 CS-operation SOC setpoint online adaption strategy

capacity information is available at the moment of the next key-up. The open-circuit voltage and battery resistances can be estimated in real time, based on the one RC pair or two RC pair electrical equivalent circuit model described in the above section.

7.9.2.2 Online Conversion of the Battery Pack Ahr Capacity to the Pack kWh Energy

The nominal kWh energy of the battery pack can be obtained from the estimated Ahr capacity and the V_{oc} and resistances at 50% SOC as:

$$kWh_{\text{battery\_pack}} = Ah\,capacity \cdot V_{\text{nominal}} = Ah\,capacity \cdot \left(V_{oc} - \left(R_{ohm} + R_{dyn}\right) \cdot I_{1c}\right)\big|_{50\%\,SOC}. \tag{7.55}$$

7.9.2.3 Online Update of the Relationship Between kWh Energy and SOC

Based on the estimated pack Ahr capacity, the model parameters, and the battery's SOC, the kWh–SOC relationship is updated using the following equation:

$$\begin{aligned} kWh_{\text{battery\_pack}}(SOC) &= Ah\,capacity \cdot V_{\text{terminal}}(SOC) \cdot SOC \\ &= Ah\,capacity \cdot \left(\hat{V}_{oc}(SOC) - \left(\hat{R}_{ohm}(SOC) + \hat{R}_{dyn}(SOC)\right) \cdot I_{1c}\right) \cdot SOC \end{aligned} \tag{7.56}$$

7.9.2.4 Determination of the CS-operation Nominal SOC Setpoint

Based on the CS-operation SOC lower bound set online and the battery energy necessary for vehicle performance, the CS-operation nominal SOC setpoint can be calculated as follows.

First, obtain the necessary look-up table kWh–SOC relationship of the battery pack, and then find out the battery pack's available kWh at the given SOC lower bound of CS operation as:

$$kWh_{\text{pack\_available}}\left(SOC_{\text{CS\_lower\_bound}}\right) = lookup(kWh - SOC)\big|_{SOC_{\text{CS\_lower\_bound}}} \tag{7.57}$$

Next, calculate the required energy of the battery pack from the CS-operation nominal SOC setpoint to the lower bound of SOC as:

$$kWh_{\text{pack\_required}}\left(SOC_{\text{CS\_setpoint}}\right) = kWh_{\text{pack\_CS\_SOC\_lower\_bound}} + kWh_{\text{pack\_CS\_operation\_needed}} \tag{7.58}$$

Finally, find the vehicle's CS-operation nominal SOC setpoint as:

$$SOC_{\text{CS\_setpoint}} = lookup(SOC - kWh)\big|_{kWh_{\text{pack\_required}}\left(SOC_{\text{CS\_setpoint}}\right)} \tag{7.59}$$

References

Abe, T., Minami, H., Atago, T. *et al.* 'Nissan's New High-drivability Vibration Control System,' SAE technical paper: 891157, Warrendale, Pennsylvania, USA, 1989.

Åstrom, K. and Hägglund, T. *PID Controllers: Theory, Design, and Tuning*, 2nd edition, Instrument Society of America, 1995.

Bellman, R. *Dynamic Programming*, Princeton University Press, 1962.

Fischer, D. and Isermann, R. 'Mechatronic Semi-Active and Active Vehicle Suspensions,' *Control Engineering Practice*, 12, 1353–1367, 2004.

Goodwin, G. and Payne, R. Dynamic System Identification: Experiment Design and Data Analysis, *Mathematics in Science and Engineering, volume 136*, Academic Press, New York, 1977.

Guzzella, L. and Onder, H. C. *Introduction to Modeling and Control of Internal Combustion Engine Systems*, Springer-Verlag, Berlin/Heidelberg, 2010.

Kim, S., Park, J., Hong, J. *et al.* 'Transient Control Strategy of Hybrid Electric Vehicle during Mode Change,' SAE 2009-01-0228, SAE International, Warrendale, Pennsylvania, USA, 2009.

Lewis, F. and Syrmos, V. *Optimal Control*, 2nd edition, John Wiley & Sons, Inc., New York, 1995.

Ljung, L. *System Identification Theory for the User*, Prentice-Hall, Inc., Englewood Cliffs, New Jersey, 1987.

Nakajima, Y., Uchida, M., Ogane, H., and Kitajima, Y. 'A Study on the Reduction of Crankshaft Rotational Vibration Velocity by Using a Motor-generator,' *JSAE Review*, 21, 335–341, 2000.

Pontryagin, L., Boltyanskii, V., Gamkrelidze, R., and Mishchenko, E. *The Mathematical Theory of Optimal Processes*, Interscience, New York, 1962.

Tamimi, M., Littlejohn, D., Kopka, M., and Seeley, G. 'Hill Hold Moding,' SAE technical Paper, 2005-01-0786, Warrendale, Pennsylvania, USA, 2005.

Utkin, V. I. *Sliding Modes in Control and Optimization*,' Springer-Verlag, Berlin/Heidelberg, 1992.

Utkin, V. I., Guldner, J., and Shi, J. *Sliding Mode Control in Electromechanical Systems*,' Taylor & Francis, London, 1999.

8

Plug-in Charging Characteristics, Algorithm, and Impact on the Power Distribution System

8.1 Introduction

During the past decade, plug-in hybrid electric vehicles (PHEVs) and pure battery electric vehicles (BEVs) have been developing and taking their place on the market. In addition to sharing the characteristics of traditional HEVs with two power sources, a PHEV is capable of driving a certain number of miles in purely electric mode, and its battery system can be recharged by a plug-in charger connecting to the electric grid. Some PHEVs with a large battery system can even power the vehicle for more than 40 miles, so that commuters can drive exclusively with the electric motor for their daily commute. What distinguishes PHEVs from HEVs are the former's all-electric driving capability and their ability to be charged by plugging in to the electric grid.

With PHEVs and BEVs going into the mainstream market, existing electric grids and power distribution systems may be impacted. In order to maximize efficiency, the plug-in charging strategies must be studied through a systematic approach, which considers not only the capabilities of the battery system and plug-in charger, but also the capability of the electric grid. There is an urgent need for such advanced charging strategies taking both vehicle battery systems and electric grid capabilities into account. The most recent results and current trends of control system and power electronic technology show that developing an optimal or smart plug-in charging control strategy is a realistic goal. This chapter briefly introduces plug-in charging characteristics, control, and the impacts on the electric grid and distribution system.

Hybrid Electric Vehicle System Modeling and Control, Second Edition. Wei Liu.
© 2017 John Wiley & Sons Ltd. Published 2017 by John Wiley & Sons Ltd.

8.2   Plug-in Hybrid Vehicle Battery System and Charging Characteristics

As mentioned above, a PHEV has a specific number of miles for which it can be driven purely on battery energy. In order to electrically drive a mid-size passenger vehicle for 20 to 60 miles, the usable energy capacity of the battery system is between 8 kWh and 25 kWh. A PHEV normally uses the battery system in two different modes: charge-depleting (CD) mode and charge-sustaining (CS) mode. When the battery is operated in CD mode, the vehicle is powered by the battery only; once the battery is drained to a specific level, the battery enters CS mode where the vehicle is powered by the engine and battery-motor system, and in this mode the power distribution from the engine and battery is determined by the energy management algorithm based on the states of the vehicle and the battery system. Furthermore, in CS mode, the battery is principally utilized to provide short boosts of power to the powertrain and capture electric braking (regenerative) energy. The changing point from CD to CS mode is set based on the design targets, such as the service life of the battery system and fuel economy, and the CS mode normally occurs between 20 and 30% SOC in a Li-ion battery system.

With regard to the plug-in charging aspect, the amount of power flowing to the battery is limited by the size of the electrical wall circuit and the capacity of the vehicle's battery system, which consequently results in different plug-in charging times. Based on the AC voltage level and electrical circuit capacity, plug-in chargers are classified as 120 V AC plug-in chargers or 240 V AC plug-in chargers (used in garages) and fast, high-voltage DC chargers for public service stations.

8.2.1   AC-120 Plug-in Charging Characteristics

An AC-120 plug-in charger is designed to charge a PHEV battery system through a 120 V AC power outlet available in a garage or parking lot. Since the electrical circuit in the garage is generally limited to 15 amps for a 120 V AC power outlet, the maximal charging power is normally designed up to 1.5 kW.

Due to the AC charging power limit, the charging current and time vary with the battery temperature and specification of the actual battery system used in a PHEV. For example, the charging characteristics of a PHEV battery system with 45 Ahr capacity and supporting 40 electric miles can be obtained at room temperature as follows:

1. The maximal DC charging power limit

$$P_{chg\_DC} = \eta \cdot P_{chg\_AC} \tag{8.1}$$

For example, if the maximal AC charging power, $P_{chg\_AC} = 1.5\text{kW}$ and the charging efficiency $\eta = 0.92$, then the maximal achievable DC charging power $P_{chg\_DC} = \eta \cdot P_{chg\_AC} = 0.92 \cdot 1.5 = 1.38\,\text{kW}$

2. Charging current profile of the battery

   To charge the battery system as quickly as possible, the AC 120 V plug-in charger normally charges the battery system in constant-power charging mode first and then in constant-voltage charging mode. The charging current can be calculated based on the following formula:

$$
I_{chg}(t) = \begin{cases} \dfrac{P_{chg\_DC}}{V_{terminal}(SOC, I_{chg})} & \text{if } V_{terminal} < V_{threshold} \text{ (Constant power mode)} \\ I(t)_{V_{terminal} = V_{threshold}} & \text{if } V_{terminal} = V_{threshold} \text{ (Constant voltage mode)} \\ 0 & \text{if } I \leq I_{threshold} \end{cases}
$$

$$(8.2)$$

   In the constant-power charging mode, the actual charging current is determined based on the charging power limit and the actual battery terminal voltage; in the constant-voltage charging mode, the charging current is determined to keep the battery terminal voltage equal to the threshold voltage which reflects the SOC that the battery will be charged to, and the charging will be stopped once the charging current is equal to or less than the designed threshold.

3. The required charging time varies with battery capacity and the initial value of the battery SOC. For example, it takes six to ten hours to charge a 16 kWh battery system from 30% SOC to 80% SOC using an AC 120 V charger.

   Hybrid vehicle system design and calibration engineers need to know the charging current curve, the terminal voltage curve, the charging power curve, and the SOC varying curve while the battery system is being charged by a plug-in AC charger from the given initial SOC to the final SOC. For example, when a Li-ion PHEV battery system with 45 Ahr capacity and 360 nominal voltage is charged to 80% SOC from 30% SOC by an AC 120 V charger, the corresponding charging curves are shown in Figs 8.1, 8.2, 8.3, and 8.4. It takes 7.2 hours for an AC-120 charger to charge the battery to the target SOC at room temperature.

8.2.2 AC-240 Plug-in Charging Characteristics

An AC-240 plug-in charger is designed to charge the battery system of a PHEV/BEV through a 240 V AC power outlet, where available. As the electrical circuit is generally limited to 40 amps for an AC 240 V power outlet, the charging power can be up to 9.6 kW; thus, the battery system can be charged much faster by an AC-240 plug-in charger.

For the PHEV Li-ion battery system presented in the previous section, when an AC-240 charger is plugged in to charge the battery to 80% SOC from 30% SOC at room temperature, it only takes about two hours to complete the charge, and the corresponding charging curves are shown in Figs 8.5, 8.6, 8.7, and 8.8.

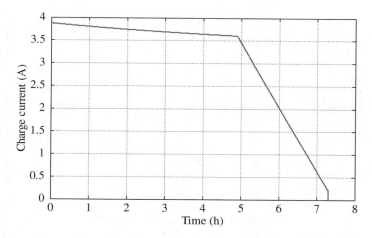

Figure 8.1 Current profile of a PHEV battery system when charged by an AC-120 charger

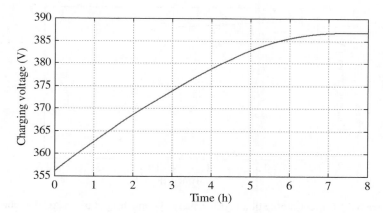

Figure 8.2 Terminal voltage profile of a PHEV battery system when charged by an AC-120 charger

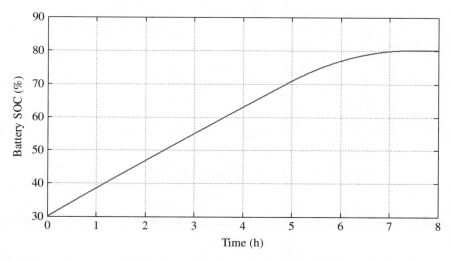

Figure 8.3 SOC charging process of a PHEV battery system when charged by an AC-120 charger

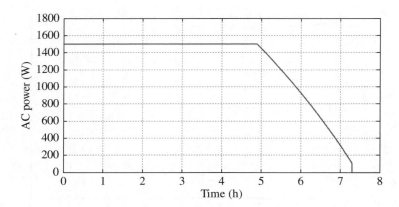

Figure 8.4   AC power-charging profile

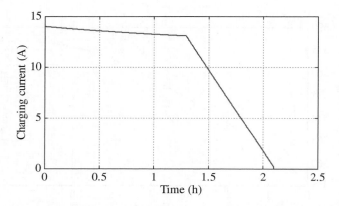

Figure 8.5   Current profile of a PHEV battery system charged by an AC-240 charger

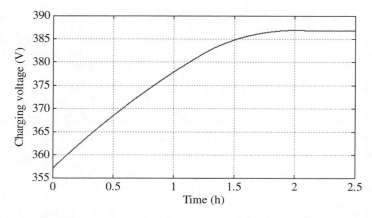

Figure 8.6   Terminal voltage profile of a PHEV battery system when charged by an AC-240 charger

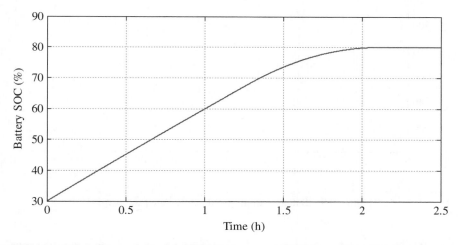

Figure 8.7 SOC charging process of a PHEV battery system when charged by an AC-240 charger

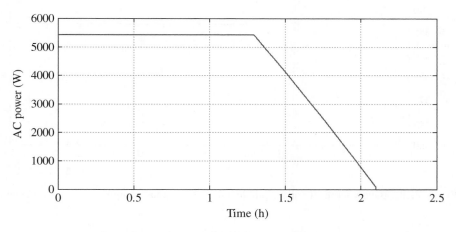

Figure 8.8 AC power taken from an AC-240 charger

8.2.3 DC Fast-charging Characteristics

If the AC power capacity is high enough, a PHEV/BEV battery system can be charged at a higher C rate. For example, at a $3C$ rate or even a $5C$ rate, the battery system may be charged to a certain level of SOC in several minutes at room temperature. Figures 8.9, 8.10, and 8.11 show the $3C$ (135 A) rate charging characteristics of the Li-ion battery system presented in the previous section. The figures also show that it only takes 11 minutes to charge the battery to 80% SOC from 25% SOC, which means that drivers can use electrical energy extensively if plug-in charging stations are set up at regular gas stations.

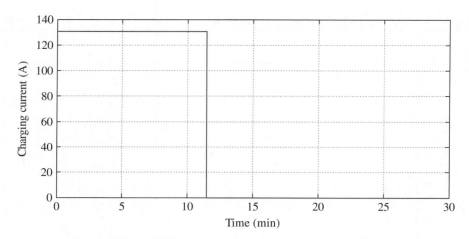

Figure 8.9 Battery current profile at 3*C*-rate fast charging

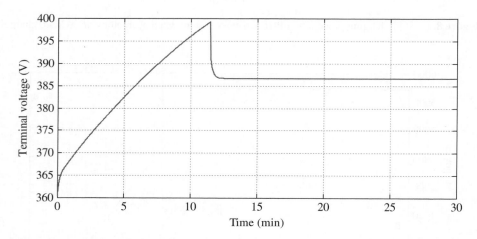

Figure 8.10 Battery terminal voltage profile during 3*C*-rate fast charging at room temperature

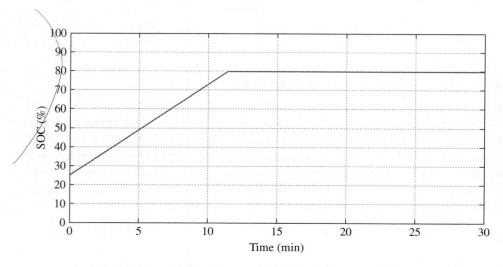

Figure 8.11 Battery SOC changing curve at 3*C*-rate fast charging

8.3 Battery Life and Safety Impacts of Plug-in Charging Current and Temperature

Based on the chemistry of a Li-ion battery, the plug-in charge current and temperature play crucial roles in battery life and operational safety, and excessive charging current could result in the following two major problems:

1. Lithium plating
   During the charging process, the induced electromotive force makes Li^+ ions move to the graphite negative electrode (anode) from the cathode. If the charging current is very high, the transport rate of Li^+ ions to the anode might exceed the rate at which Li^+ ions can be inserted. In this case, Li^+ ions are deposited as metallic lithium on the surface of the anode, and this is known as *lithium plating*. Lithium plating results in an irreversible capacity loss due to a reduction in free Li^+ ions. Excessive lithium plating can ultimately lead to a short circuit, and, in addition, metallic lithium is highly flammable. Based on the Arrhenius Law, the Li^+ ion insertion rate into the graphite is significantly low at cold temperatures. Therefore, lithium plating can often occur if the plug-in charge current is not properly controlled. For example, a PHEV battery can be plug-in charged at $1C$ rate at $0\,°C$ without any problem, but lithium plating is observed using $0.2C$-rate charging at $-30\,°C$.

2. Overheating
   Temperature impacts not only on battery performance but also on battery life. Heat is a major battery killer, either an excess of it or a lack of it. Excessive plug-in charge current causes an increased temperature in the cell. In hot weather conditions, plug-in charging at high temperatures brings on a different set of problems which can result in the destruction of the cell. At high temperatures, although the Arrhenius effect helps the battery to output higher power by increasing the reaction rate, the higher charging current generates higher I^2R heat. If the plug-in charge current is not properly controlled and the heat cannot be removed as fast as it is generated, a thermal runaway could ensue, which is a serious safety concern and can permanently damage battery cells.

8.4 Plug-in Charging Control

8.4.1 AC Plug-in Charge Control

As previously discussed, plug-in charging strength is limited by the AC power outlet capability and the maximal allowable transport rate of Li^+ ions at different temperatures and cell voltages. Since the AC power outlet capability is determined by the size of the AC electrical circuit, the maximal input power from an AC outlet can be set offline and plug-in charging needs to be controlled based on the battery temperature, SOC, and voltage. The general plug-in charging control diagram is shown in Fig. 8.12.

In plug-in charging, the constant-voltage (CV) mode setpoint is set based on the desired charging SOC termination point, such as a 90% SOC, and the battery's chemical

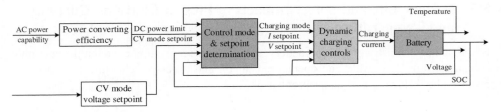

Figure 8.12   AC plug-in charging control

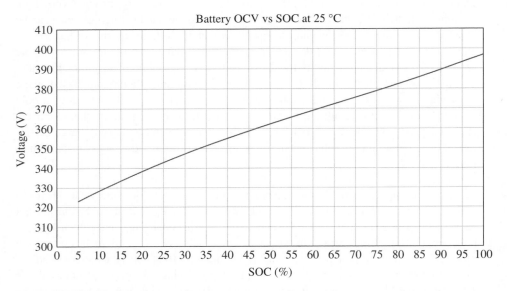

Figure 8.13   OCV–SOC curve of an NMC-based Li-ion battery system at room temperature

characteristics. For example, an NMC-based Li-ion battery pack has the OCV–SOC characteristic shown in Fig. 8.13 at room temperature. If the plug-in charge SOC termination is specified based on the vehicle's technical requirements, such as the battery life and electric range requirements, the CV-mode voltage setpoint can be determined from the OCV–SOC curve; for example, if the desired plug-in charge SOC is 85% SOC, the CV-mode voltage setpoint needs to be set at 386 V.

For control, the charging current setpoint in constant-current (CC) mode and the charging voltage setpoint in CV mode should be determined based on the present battery temperature and voltage as well as the charger's AC power limit. In order to avoid lithium metal plating, the plug-in charging current must be properly set for different battery temperatures and SOC conditions. Table 8.1 shows the maximal allowable charging current at different temperatures and SOCs of a PHEV battery pack as an example.

In principle, the plug-in charge mode of the presented control algorithm is determined based on the actual battery voltage, whereby plug-in charging starts from CC mode and then

Table 8.1　Maximal plug-in charge current of a PHEV battery pack

SOC (%)	Maximal plug-in charge current (A)									
	−30 °C	−20 °C	−10 °C	0 °C	10 °C	20 °C	25 °C	45 °C	50 °C	55 °C
100	0	0	0	0	0	0	0	0	0	0
95	0.5	1	2.5	4	7	10	10	8	2	0
90	2	4	10	16	28	40	40	32	8	0
70	5	10	25	40	70	100	100	80	20	0
0	5	10	25	40	70	100	100	80	20	0

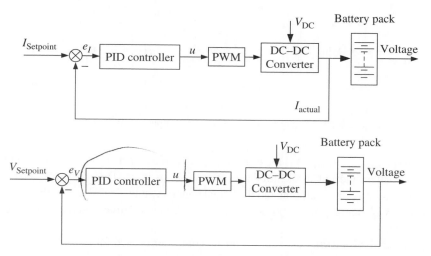

Figure 8.14　PID plug-in charging algorithm

switches to CV mode once the battery voltage reaches the charge termination voltage setpoint. In reality, the charge control algorithm must ensure that no large overshoot occurs when the battery voltage is approaching the CV-mode setpoint. Either in CC or CV mode, the PID control algorithm shown in Fig. 8.14 can be used to govern the charging current to have either a charging current with the current setpoint in CC mode or a charging voltage with the voltage setpoint in CV mode.

In the PID control diagram, the output of the PID controller is calculated from the following equation:

$$u(t) = K\left(e(t) + \frac{1}{T_i}\int_0^t e(\tau)d\tau + T_d\frac{de(t)}{dt} \right) \tag{8.3}$$

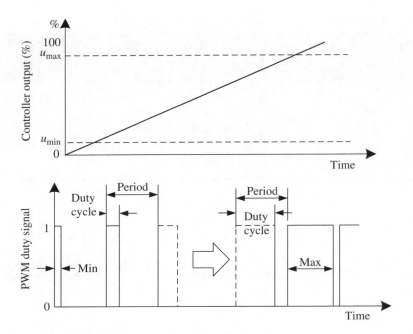

Figure 8.15   Conversion from controller output to PWM duty signal

where $e_I(t) = I_{\text{setpoint}} - I_{\text{actual}}(t)$ is the error between the plug-in charging current setpoint and the actual battery current in CC mode, $e_V(t) = V_{\text{setpoint}} - V_{\text{actual}}(t)$ is the error between the plug-in charging voltage setpoint and the actual battery voltage in CV mode, the calibration parameters K, T_i, and T_d are the proportional gain, the integral time, and the derivative time, respectively.

In practice, the following discrete-form PID controller is used for hybrid electric vehicle application:

$$u(k) = k_p e(k) + k_i \cdot T_s \sum_{i=0}^{k} e(i) + \frac{k_d}{T_s}(e(k) - e(k-1)) \tag{8.4}$$

where T_s is the sampling period of time, k is the discrete step which equals t in the continuous time domain, and $k_p = K$, $k_i = k/T_i$, and $k_d = K \cdot T_d$.

As described in Chapter 4, the PID controller outputs a control signal within the 0–100 range, and the pulse width modulation (PWM) block converts it to an actual PWM duty signal to control power electronic switches, as illustrated in Fig. 8.15.

8.4.2   DC Fast-charging Control

With the increasing range of battery-powered electric vehicles, DC fast-charging control becomes more and more important. In order to charge a BEV/PHEV battery rapidly while avoiding lithium plating and overheating of the battery, DC fast charging must be controlled well for every scenario in the real world. Fundamentally, the state-space-based control strategy is necessary to regulate the DC fast-charging action, and the following factors need to be considered during the fast-charging process:

1. Maximal and minimal cell temperatures
   Since the cell units in the battery system of a BEV/EREV/PHEV are connected in series, the cell temperatures will be different from each other. Therefore, in cold weather conditions, the control action will be restricted by the coldest cell, while the control will be restricted by the hottest cell in hot weather conditions.
2. Maximal cell SOC
   Since the cell capacities in a battery system are not identical and each cell is connected with a different parasitic load, the cell SOCs are also different from each other. The fast-charge control action should also be regulated based on the cell with the maximal SOC rather than the pack average SOC to avoid lithium plating.
3. Maximal cell voltage
   In order to avoid lithium plating and overcharging of the battery, the fast-charging control action needs to be regulated based on the maximal cell voltage rather than the pack voltage.
4. Rate of change of cell voltage
   The cell voltage is smoothly increased during the normal charging process, and the charging current should be reduced once a rapid increase in cell voltage is observed.

A state-space-based DC fast-charging control diagram is shown in Fig. 8.16, where the fast-charging current setpoint is optimally determined based on the max/min cell temperature, the maximal cell voltage, the pack voltage and pack SOC, and the maximal cell SOC. The actual charging current is regulated via state feedback control, where the states of the battery are estimated by a state observer. For the design principles of state feedback control

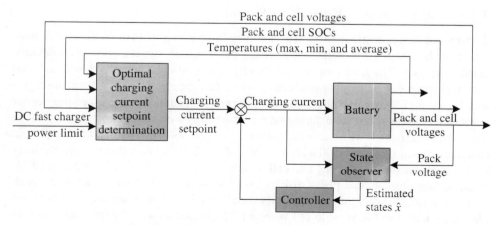

Figure 8.16   State-space-based DC fast-charging control

and state observers, interested readers should refer to Appendix B or to Åstrom and Wittern-mark (1984). Depending on the detailed Li-ion chemistry, the states of the battery can be defined as follows:

$$x_1 : \text{resistance polarization voltage}$$

$$x_2 : \text{activation polarization voltage}$$

$$x_3 : \text{concentration polarization voltage}$$

The above states generally reflect the insertion capability of Li ions at the present charging status; therefore, lithium plating will be avoided if the charging current can be effectively adjusted based on the battery's internal states. Furthermore, since these battery states are not measurable online, a state observer is needed to estimate the states in real time, and it is very challenging to design an accurate and robust state observer due to the complexity and non-linearity of electrochemical reactions of Li-ion batteries.

8.5   Impacts of Plug-in Charging on the Electricity Network

If a PHEV/BEV is driven 40 miles a day for seven days a week, approximately 3000 kWh of electrical energy is consumed by plug-in charging outlets per year. Based on a report by the DOE (Department of Energy, USA), average monthly residential electricity consumption was 936 kWh and the average price was 9.65 cents per kWh in 2007 (DOE, 2010). This amount would increase by at least 20% per household if a PHEV/BEV was plugged in for charging once a day, which would have a big impact on existing power distribution systems and the electric grid. Such an emerging demand spike for electric grids from individual households has not occurred since air-conditioners became widely used in residences in the 1950s (Hadley and Tsvetkova, 2008; Meliopoulos *et al.*, 2009).

8.5.1   Impact on the Distribution System

With regard to the distribution system, the plug-in charging of PHEVs/BEVs will have effects on the size of the feeder, branch circuits, panel-boards, switches, protective devices, transformers, etc. The following analysis briefly shows how the household load and load pattern changes caused by plug-in charging will affect the electric power distribution system. Figure 8.17 shows five household daily winter and summer power-usage profiles for residential consumers based on the regional average daily winter and summer power usage measured in the field. If we assume that there are two PHEVs in five households and plug-in charging starts at 17:45 and 0:00 with AC-120 and AC-240 chargers, respectively, the daily load demand curves for charging two PHEVs are shown in Figs 8.18, 8.19, 8.20, and 8.21. These figures illustrate that PHEVs/BEVs have a significant effect on the power demand; however, setting the charging time properly could effectively reduce the impact. For example, the peak load will increase to 11 kW from 8.15 kW in summer if the household starts to

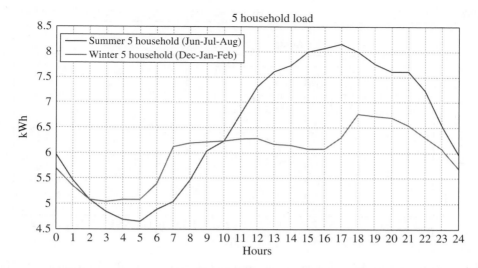

Figure 8.17 Averaged five-household daily load profiles in summer and winter

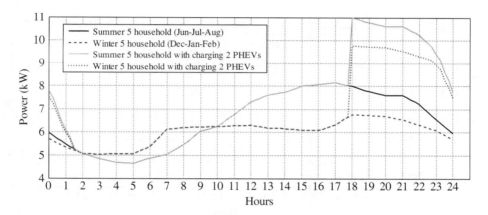

Figure 8.18 Averaged five-household daily load plus charging of two PHEVs using an AC-120 charger starting at 17:45 in summer and winter

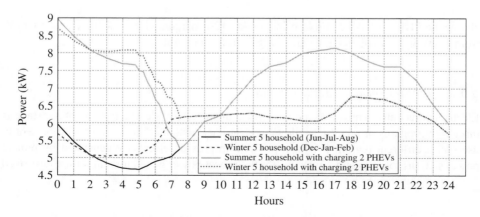

Figure 8.19 Averaged five-household daily load plus charging of two PHEVs using an AC-120 charger starting at 00:00 in summer and winter

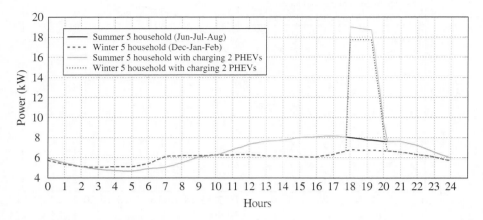

Figure 8.20 Averaged five-household daily load plus charging of two PHEVs using an AC-240 charger starting at 17:45 in summer and winter

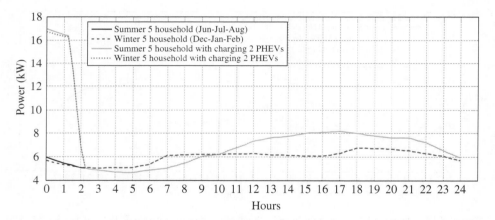

Figure 8.21 Averaged five-household daily load plus charging of two PHEVs using an AC-240 charger starting at 00:00 in summer and winter

charge the PHEVs/BEVs at 17:45 using an AC-120 charger, while it will increase to 9 kW from 6 kW if the PHEVs/BEVs are charged from midnight.

8.5.2 Impact on the Electric Grid

Electric grids, also known as power grids, interconnect electricity-generating stations throughout regions or countries, allowing electricity generated in one region to be sent to users in a different region. It is also possible to provide electricity generated by remote power stations to cities and towns in case their local power generators fail or are destroyed by an accident or sabotage. In the interconnection, every generator or synchronous machine connected to the grid operates in synchronicity with every other generator; that is, each generator operates at the same synchronous speed or frequency while a delicate balance between the input mechanical power and the output electrical power is maintained. The electrical speed or frequency chosen for this operation is 60 cycles per second, commonly referred to as 60 hertz in North America. Whenever generation is less than the actual customer load, the system frequency falls; on the other hand, when generation exceeds load, the system frequency rises.

For a more detailed explanation, an electric generator is a device that converts mechanical energy into electrical energy based on Faraday's electromagnetic principle, called Faraday's law. The device providing the mechanical energy for each generator is called a turbine, which may be powered by steam generated by burning fossil fuel, by heat derived from nuclear fission, or by the kinetic energy of water or wind, etc. The generators are connected with each other and with loads through high-voltage transmission lines also called power transmission lines. The electricity on the transmission line is stepped down from high voltage to 120/240 volts for home or office use.

An important feature of the electrical power system is that electricity has to be generated when it is needed because it cannot be efficiently stored. Hence, it is necessary for the generator to be scheduled for every hour in a day to match with loads through a sophisticated load-forecasting procedure. In addition, some generators are also placed on active standby to provide electricity in case of emergency. Since power grids interconnect such a large number of generators and loads, they routinely undergo a variety of disturbances, and even the act of switching on or off the range in a kitchen can be regarded as a small disturbance. Each power system has a stability margin so that it can tolerate a certain level of disturbance. If high or unbalanced demands suddenly present on the power grid, stability may not be kept. In some cases, the disturbance can cause power outage in a grid section, or can even ripple through the whole grid resulting in generators shutting down one after another. The stability of the power grid will be in jeopardy once additional high demands are added in periods of already high demand, for example, in the evenings in summer, as grids across the whole region will be placed under increasing stress and there is very limited spare power.

With a number of PHEVs/BEVs added to grids, the overall power demand and demand pattern will change. This could significantly impact the electric grids, depending on the times at which battery systems are charged through wall outlets. The common method of handling changes in power demand and load pattern is to add new electrical capacity and reschedule generators' operational times; otherwise the reserve margins could be reduced if the electrical capacity of the grid does not keep up with the added demand or changes in the demand pattern. Depending on the power level, time, and duration of PHEV/BEV connections to the grid, they could have various impacts on power grids. Therefore, it is important to explore and understand the consequences of a number of PHEVs/BEVs connecting to the grid, and the impact analysis should focus on the changes in power demand and pattern resulting from PHEVs/BEVs. It is also important to analyze and compute real and reactive power flow on each branch and the impact on each component.

8.6   Optimal Plug-in Charging Strategy

As discussed above, not only do plug-in chargers need to be designed to be efficient, cost-effective, and to use as much off-peak grid time as possible, but the charging strategy should also take the grid characteristics into account. The following charging strategies may be applied to PHEVs/BEVs:

- **Instant charging strategy:** Also called an uncontrolled charging strategy, by which the battery system is immediately charged when the charger cable is plugged into a power outlet. The charging normally starts in the evening after the vehicle has been driven back home. It takes between two and ten continuous hours to charge the battery system to the target level depending on the initial SOC and the capacity of the battery system and available charging power.
- **Predetermined start time charging strategy:** Also called a delayed charging strategy, by which the battery is not immediately charged when the charger is plugged into a power outlet, and the charging action will not start until the set time, for example, 11:00 pm.
- **Automatic off-peak charging strategy:** Unlike the predetermined start time strategy, the charging action is controlled by a utility company and occurs once the charger receives the signal for off-peak time from the utility company. In order to support this strategy, the plug-in charger has to be able to communicate with the utility company.
- **The optimal charging strategy:** Also called a smart charging strategy, by which the charging power is adapted in real time based on the states of the battery system and the electric grid and the set charging objective subject to the target SOC and the required duration to complete the charge. This section addresses two optimal plug-in charging strategies: optimally setting the charging target, and a charging strategy with minimal charging cost.

8.6.1   The Optimal Plug-in Charge Back Point Determination

As discussed in Chapter 5, each type of battery has an optimal SOC setpoint which enables the battery to have its longest lifetime. For example, the best setpoint for NiMH batteries is around 60% SOC, while it is about 50% SOC for Li-ion batteries. A PHEV/BEV generally has a 20–50 mile electric range; however, no matter how many miles the vehicle actually drives, the battery system is normally charged to full or the designed SOC level when the vehicle is back in the garage.

The optimal PHEV plug-in charge back point determination method illustrated in this section is based on a PHEV that has a 40-mile electric range with a 45 Ahr Li-ion battery system, and the curve for battery SOC usage versus electric mile range is shown in Fig. 8.22. The corresponding end SOC after each drive cycle and the designed charge back point are shown in Fig. 8.23. According to the mechanism of battery life degradation, the best operational setpoint of SOC for the battery life is 50% SOC; hence, the optimal plug-in charge back point should be set as in Fig. 8.24 for the given driven miles of the vehicle. A diagram

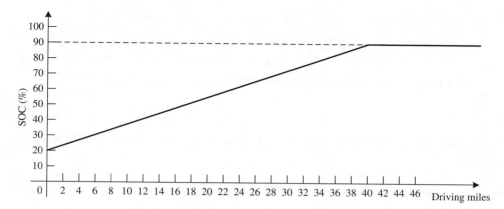

Figure 8.22 SOC usage vs. electric mileage for a PHEV

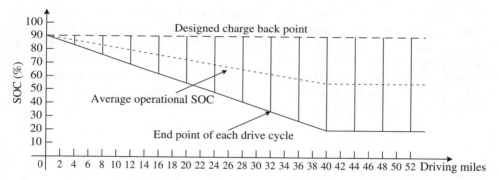

Figure 8.23 The actual operational SOC setpoint, end SOC point, and charge back SOC point

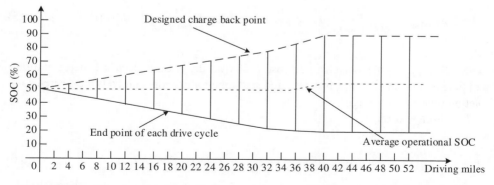

Figure 8.24 Optimal operational SOC setpoint, end SOC point, and charge back SOC point

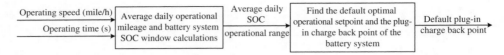

Figure 8.25 The optimal plug-in charge end point determination algorithm for PHEVs

of the optimal plug-in charge back point determination algorithm for a PHEV is shown in Fig. 8.25.

8.6.2 Cost-based Optimal Plug-in Charging Strategy

The primary goal in developing a plug-in charging algorithm is to obtain the lowest cost for PHEV/BEV owners to charge the battery system back to the desired SOC level, while the utility company will set the electricity price based on the power demand and the electrical power availability. Hence, the objective of the plug-in charging strategy is to determine the best charging power and time, subject to the constraints of the capabilities of the charger and the battery system, the required energy (target SOC), and the charging timeframe. This section presents a cost-function-based optimal plug-in charging algorithm, the flow chart for which is shown in Fig. 8.26.

8.6.2.1   Cost Function

The cost function for optimizing charging cost can be defined as:

$$J = Cost_{energy} = \frac{1}{3600 \cdot 1000} \int_{t_i}^{t_f} Price(t) \cdot Power(t) dt = \frac{1}{3600 \cdot 1000} \int_{t_i}^{t_f} Price(t) v(t) i(t) dt \quad (8.5)$$

where $Cost_{energy}$ is the total charging cost, $Price(t)$ is the price of electrical energy at different time periods in \$/kWh, t_i is the desired start time for charging, and t_f is the desired completion time for charging.

The discrete form is:

$$J = Cost_{energy} = \frac{1}{3600 \cdot 1000} \sum_{i=1}^{N} Price(i) \cdot Power(i) \cdot \Delta T \quad (8.6)$$

where ΔT is the control time interval and N is the number of intervals, $N = \dfrac{t_f - t_i}{\Delta T}$.

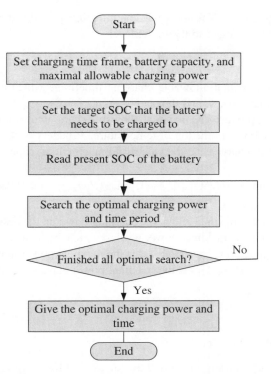

Figure 8.26 Optimal plug-in charging algorithm flow chart

8.6.2.2 Operational Constraint

In order to charge the battery optimally, it is necessary to consider the timing requirement and physical limitations of the charger, the battery system, and the power outlet. These considerations can be defined as constraints when performing an optimization.

- **Meet energy requirements:** The total charged energy should be equal to the required energy; that is:

$$
\begin{aligned}
Energy_{chg} &= \frac{1}{3600 \cdot 1000} \int_{t_i}^{t_f} Power(t) \cdot \eta_{bat}(t) \cdot \eta_{charger}(t) dt \\
&= Energy_{req'd} \quad \Rightarrow \quad SOC_{init} \rightarrow SOC_{target}
\end{aligned}
\tag{8.7}
$$

where $Energy_{chg}$ is the total charged energy in kWh, $Energy_{req'd}$ is the energy required to charge the battery from the initial SOC_{init} to the desired target SOC_{target}, η_{bat} and $\eta_{charger}$ are the efficiencies of the battery system and the charger, respectively, η_{bat} is a function of temperature and battery SOC, that is, $\eta_{bat} = f(SOC, T)$, and $Power(t) = V(t) \cdot I(t)$ is the charging power between the completion time, t_f, and the start time, t_i.

- **Allowable charging time:** The allowable charging time period is the timeframe in which the battery has to be charged to the target level:

$$T_{max} = t_f - t_i \tag{8.8}$$

where T_{max} is the maximal allowable charging timeframe which equals the difference between the set final time, t_f, and the start time, t_i.

- **Maximal charging power:** The maximal allowable charging power from the charger and battery system, which is a function of temperature, T, battery SOC, and the power limitation of the plug-in charger and the electrical branch circuit capacity:

$$Power_{max} = \min\left\{\ \max\left(Pwr_{bat(SOC,T)}\right),\ \max\left(Pwr_{chg}\right),\ \max(Pwr_{branch})\ \right\} \tag{8.9}$$

where $Power_{max}$ is the maximal allowable charging power in the desired charging period, $\max(Pwr_{bat(SOC,T)})$ is the maximal power level to which the battery can be charged at the given SOC and temperature, $\max(Pwr_{chg})$ is the maximal power that the plug-in charger can deliver, and $\max(Pwr_{branch})$ is the maximal power capacity of the household power outlet.

8.6.2.3   Optimization Problem

The objective of the optimization is to find the lowest cost for charging the battery, and the problem can be mathematically stated as follows:

$$\text{Minimize}(J) = \text{Minimize}\left(Cost_{energy}\right) = \frac{1}{3600 \cdot 1000}\text{Minimize} \cdot \left(\sum_{i=1}^{N} Price(i) \cdot Power(i) \cdot \Delta T\right) \tag{8.10}$$

where $Cost_{energy}$ is the total charging cost, $Price(i)$ is the price of electrical energy at different time periods in \$/kWh, ΔT is the control time interval, and N is the number of intervals, $N = \dfrac{t_f - t_i}{\Delta T}$, t_i is the desired start time for charging, and t_f is the desired completion time for charging.

Subject to:

$$SOC_f = SOC_{target} \tag{8.11}$$

That is:

$$Energy_{chg} = \frac{1}{3600 \cdot 1000}\int_{t_i}^{t_f} Power(t) \cdot \eta_{bat}(t) \cdot \eta_{chg}(t)dt \doteq \frac{1}{3600 \cdot 1000}\sum_{i=1}^{N} Power(i) \cdot \Delta T = Energy_{req'd} \tag{8.12}$$

Satisfying the constraints:

$$Power(i) \leq \min\left\{ \max\left(Pwr_{bat(SOC,T)}(i)\right), \; \max\left(Pwr_{chg}(i)\right), \; \max\left(Pwr_{branch}(i)\right) \right\} \quad (8.13)$$

where $\max(Pwr_{bat(SOC,T)}(i))$ is the maximal power level to which the battery can be charged during the ith charging period, $\max(Pwr_{chg}(i))$ is the maximal power that the plug-in charger can deliver during the ith charging period, and $\max(Pwr_{branch}(i))$ is the maximal power capacity of the household power outlet during the ith charging period.

8.6.2.4 The Procedures Determining the Optimal Charging Schedule

Once $Price(i)$ has been obtained for each period of time, the optimal charging schedule is determined by the dynamic programming method. As described in Appendix B, the dynamic programming method was developed by Richard Bellman in the early 1950s, and is a mathematical method well suited to the optimization of multi-stage decision problems (Bellman, 1957). The application diagram for the dynamic programming method to solve this problem is shown in Fig. 8.27. At each time point, the costs to charge the battery and the temperature and SOC of the battery are computed for different charging power levels. For example, the SOC and temperature values of the battery system shown in Table 8.2 are calculated during the ith charging period. These values can be used to determine the charger and battery efficiency and the optimal charging power.

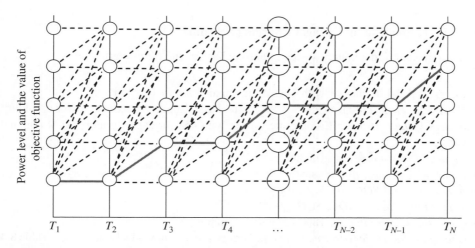

Figure 8.27 Process of optimizing operational points using the dynamic programming method

Table 8.2   Example of the calculated battery SOC and temperature corresponding to different charging power during the ith charging period

Charging power (kW)	0	0.2	0.4	0.6	0.8	1.0	1.1	1.2	1.3	1.4	1.5
SOC (%)	57	57	62	65	56	59	61	62	63	59	58
Temperature (°C)	23	24	25	26	26	27	26	28	27	28	29

Table 8.3   Operational efficiency of the given plug-in charger

Charging power (kW)	0.2	0.5	0.7	0.9	1.0	1.1	1.2	1.3	1.4	1.5
Efficiency	0.85	0.88	0.9	0.91	0.92	0.95	0.93	0.92	0.90	0.89

Table 8.4   Charging efficiency of the given battery system

Temperature (°C)	−40	−30	−20	−10	0	10	25	35	45	50
10% SOC	0.75	0.80	0.85	0.87	0.88	0.90	0.91	0.92	0.91	0.90
20% SOC	0.76	0.81	0.86	0.88	0.89	0.91	0.92	0.93	0.92	0.91
30% SOC	0.79	0.82	0.87	0.89	0.91	0.92	0.93	0.94	0.93	0.92
40% SOC	0.80	0.83	0.88	0.89	0.91	0.93	0.94	0.94	0.94	0.92
50% SOC	0.81	0.84	0.88	0.89	0.91	0.93	0.94	0.94	0.94	0.9
60% SOC	0.82	0.84	0.87	0.89	0.91	0.93	0.94	0.94	0.94	0.92
70% SOC	0.80	0.83	0.86	0.89	0.91	0.93	0.94	0.93	0.93	0.92
80% SOC	0.79	0.82	0.85	0.88	0.90	0.92	0.94	0.92	0.92	0.90
90% SOC	0.77	0.81	0.85	0.88	0.89	0.90	0.92	0.91	0.90	0.89

8.6.2.5   Example of an Optimal Charging Schedule

The parameters of the battery system, plug-in charger, and electric grid are:

- *Charger*: AC-120 charger with 1.5 kW maximal charging capability
- *Household electrical branch*: 15 A AC current and 1.65 kW maximal power capability
- *Battery system*: 16 kWh Li-ion battery pack with 45 Ahr capacity and 360 V nominal voltage
- *Charging start time, t_i*: 18:00 pm
- *Charging end time, t_f*: 6:00 am
- *Battery's initial SOC, SOC_{init}*: 30%
- *Target (end) battery SOC, SOC_{end}*: 80%
- *Calculation time interval*: 1800 seconds (30 minutes)
- *Charging efficiencies of the charger and battery system*: shown in Table 8.3 and Table 8.4
- *Electric usage price*: shown in Table 8.5

Table 8.5 Given price of electrical energy

Time	18:00	18:30	19:00	19:30	20:00	20:30	21:00	21:30	22:00	22:30	23:00	23:30	0:00	0:30
Price ($/kWh)	0.16	0.16	0.16	0.16	0.14	0.14	0.14	0.14	0.12	0.12	0.10	0.10	0.08	0.08
Time	1:00	1:30	2:00	2:30	3:00	3:30	4:00	4:30	5:00	5:30	6:00	6:30	7:00	7:30
Price ($/kWh)	0.08	0.08	0.08	0.08	0.08	0.08	0.08	0.08	0.12	0.12	0.14	0.14	0.16	0.16

Table 8.6 Calculated charging schedule

Time	18:00	18:30	19:00	19:30	20:00	20:30	21:00	21:30	22:00	22:30	23:00	23:30	0:00	0:30
Power (kW)	0	0	0	0	0	0	0	0	0	0	1.0	1.0	1.1	1.1
Time	1:00	1:30	2:00	2:30	3:00	3:30	4:00	4:30	5:00	5:30	6:00	6:30	7:00	7:30
Power (kW)	1.1	1.1	1.1	1.1	1.1	1.1	1.1	1.1	0.5	0.2	0	0	0	0

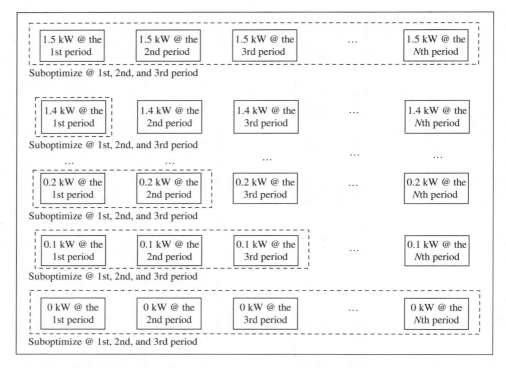

Figure 8.28 Principle of the optimal decision made using the dynamic programming method

The calculation results show that the optimized charge cost is $0.56, while the uncontrolled charge cost is $1.25 if charging starts at 18:00 pm; both use a smart grid price. Compared with an uncontrolled plug-in charging schedule, the owner saves more than 50% financially if the optimized charging algorithm is employed. The optimized charging schedule is shown in Table 8.6 and the decision-making process is shown in Fig. 8.28.

References

Åstrom, K. J. and Wittenmark, B. *Computer Controlled Systems – Theory and Design*, Prentice-Hall Inc., Englewood Cliffs, N.J., 1984.

Bellman, R. E. *Dynamic Programming*, Princeton University Press, Princeton, NJ, 1957.

DOE 'Electric Power Monthly January 2010 With Data for October 2009,' Department of Energy report, U.S. Energy Information Administration, Office of Coal, Nuclear, Electric and Alternate Fuels, U.S. Department of Energy, DOE/EIA-0226, 2010.

Hadley, W. S. and Tsvetkova, A. 'Potential Impacts of Plug-in Hybrid Electric Vehicles on Regional Power Generation,' report of Oak Ridge National Laboratory, Oak Ridge, Tennessee 37831, 2008.

Meliopoulos, S., Meisel, J., Cokkinides, G., and Overbye, T. 'Power System Level Impacts of Plug-In Hybrid Vehicles,' final project report, Power Systems Engineering Research Center, Georgia Institute of Technology and University of Illinois at Urbana-Champaign, 2009.

9

Hybrid Electric Vehicle Vibration, Noise, and Control

Vehicle noise and vibration is a fairly complex and broad subject, and concern associated with it has been increasing as hybrid electric vehicles have taken their place on the market. Each new architecture and new technology applied in hybrid vehicle systems generally brings unique noise and vibration challenges. To articulate the challenges surrounding the noise and vibration in hybrid electric vehicle system integration, this chapter briefly introduces the basics of vibration and noise, and addresses the unique vibration and noise characteristics and issues associated with electric powertrains, drivelines, and vehicle vibrations. The details include electrified component-specific vibration and noise, such as accessory whine, gear rattle, and motor/generator electromagnetic vibration and noise.

9.1 Basics of Noise and Vibration

9.1.1 Sound Spectra and Velocity

Vibration is the oscillation of particles, while sound is a periodic mechanical vibration in an elastic medium by means of which a certain amount of energy is carried through the medium. The term *sound* usually refers to the form of energy which produces hearing in humans, while the term *noise* is defined as any unwanted sound. Sound propagation generally involves the transfer of energy but it does not cause the transport of matter.

1. **Sound spectra**

   In the real world, a sound wave is normally composed of a combination of single frequencies and hardly ever has only one frequency. A combination of a finite number of

Hybrid Electric Vehicle System Modeling and Control, Second Edition. Wei Liu.
© 2017 John Wiley & Sons Ltd. Published 2017 by John Wiley & Sons Ltd.

Table 9.1 Sound velocities and wavelengths in different materials commonly used in road vehicles

Material/medium	Velocity of sound (m/s)	Wavelength at 1000 Hz (m)
Air, standard temperature and pressure (20 °C, 1014 kPa)	343	0.343
Water (fresh, 10 °C)	1440	1.44
Rubber (according to hardness)	60–1500	0.06–1.5
Aluminum (rod)	5100	5.1
Steel (rod)	5000	5.0
Steel (sheet)	6100	6.1

tones is said to have a *line spectrum*, while a combination of an infinite number of tones has a *continuous spectrum*. The characteristic of *random noise* comes from a continuous spectrum for which the amplitude versus time occurs with a normal (Gaussian) distribution.

2. **Velocity of sound**

   The sound velocity is the velocity of propagation of a sound wave, which depends on the properties of the medium. In the air, the sound velocity is a function of the elasticity and density of the air and can be calculated as:

   $$c = \sqrt{\frac{1.4P_s}{\rho}} \, (\text{m/s}) \tag{9.1}$$

   where P_s = actual atmospheric pressure (Pa) and ρ = density of air (kg/m^3).

   The sound wavelength (λ) is a term associated with sound velocity, and is defined as the distance traveled by the sound wave during a period (T), expressed as:

   $$\lambda = cT = \frac{c}{f} \, (\text{m}) \tag{9.2}$$

   where c is the velocity of the sound (m/s) and f is the frequency of the sound (Hz). Some sound velocities and wavelengths in materials commonly used in road vehicles are listed in Table 9.1.

9.1.2 Basic Quantities Related to Sound

1. **Sound pressure and sound pressure level (SPL)**

   The sound pressure is the local pressure difference between the ambient atmospheric pressure and the pressure produced by a sound wave. The sound pressure fluctuates with time and can be positive or negative with respect to the ambient atmospheric pressure. In air, the sound pressure can be measured using a microphone in the units of pascals (Pa).

Since both the magnitude and frequency of sound pressure vary with time, the root-mean-square of sound pressure is usually used in engineering analysis and design. If the sound pressure at any instant, t, is $p(t)$, the root-mean-square (rms) pressure, P_{rms}, over the time interval T is defined as follows:

$$P_{rms} = \sqrt{\frac{1}{T} \int_0^T p^2(t)\, dt} \qquad (9.3)$$

For a pure tone sound, for which the pressure is given by $p(t) = P\sin(\omega t)$, the root-mean-square sound pressure is:

$$P_{rms} = \frac{P}{\sqrt{2}} \qquad (9.4)$$

The sound pressure is an absolute value which does not describe the strength of the sound at a given location. In road vehicle noise analysis and refinement, the term most often used in measuring the strength of the sound is a relative quantity called the *sound pressure level* (SPL) or *sound level* (L_p), which is the ratio between the actual sound pressure and a fixed reference pressure on a logarithmic scale and measured in decibels (dB), as follows:

$$L_p = 10 \log_{10}\left(\frac{P_{rms}^2}{P_{ref}^2}\right) = 20 \log_{10}\left(\frac{P_{rms}}{P_{ref}}\right) \quad (dB) \qquad (9.5)$$

Since the human ear detects a large dynamic range of sound pressure fluctuations between 20×10^{-6} Pa and 60 Pa, P_{ref} is equal to 20 µPa. Figure 9.1 shows some sound pressure levels of typical sounds.

Example 9.1
If the root-mean-square of sound pressure (P_{rms}) doubles, what is the change in sound pressure level?

Solution:
Since P_{rms} doubles, $P_{2,rms} = 2P_{1,rms}$. Therefore, the change in sound pressure level is:

$$\Delta L_p = L_{P2} - L_{P1} = 10 \log_{10}\left(\frac{P_{2,rms}^2}{P_{ref}^2}\right) - 10 \log_{10}\left(\frac{P_{1,rms}^2}{P_{ref}^2}\right)$$

$$= 20 \log_{10}\left(\frac{P_{2,rms}}{P_{1,rms}}\right) = 20 \log(2) \approx 6 (dB)$$

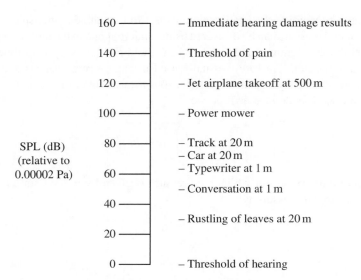

	160 —	– Immediate hearing damage results
	140 —	– Threshold of pain
	120 —	– Jet airplane takeoff at 500 m
	100 —	– Power mower
SPL (dB)	80 —	– Track at 20 m
(relative to		– Car at 20 m
0.00002 Pa)	60 —	– Typewriter at 1 m
		– Conversation at 1 m
	40 —	
		– Rustling of leaves at 20 m
	20 —	
	0 —	– Threshold of hearing

Figure 9.1 Some typical sound pressure levels, SPLs (dB). Reproduced from *Handbook of Noise and Vibration Control*, edited by Malcolm J. Crocker, p. 11, 2007 with permission from John Wiley & Sons

Example 9.2
If there are two equal sound sources with a sound level of 70 (dB), what is the total sound pressure level?

Solution:

$$L_{P2} = L_{P1} = 70(\text{dB}) = 10 \log_{10}\left(\frac{P_{\text{rms}}^2}{P_{\text{ref}}^2}\right)$$

Therefore:

$$P_{2,\text{rms}}^2 = P_{1,\text{rms}}^2 = P_{\text{ref}}^2 \cdot 10^7 = \left(20 \times 10^{-6}\right)^2 \cdot 10^7 = 0.004 \ \left(\text{Pa}^2\right)$$

The total square of the sound pressure is:

$$P_{\text{total,rms}}^2 = P_{2,\text{rms}}^2 + P_{1,\text{rms}}^2 = 0.008 \ \left(\text{Pa}^2\right),$$

and the total pressure level is:

$$L_{\text{total,p}} = 10 \log_{10}\left(\frac{P_{\text{total,rms}}^2}{P_{\text{ref}}^2}\right) = 10 \log_{10}\left(\frac{0.008}{4 \times 10^{-10}}\right) = 10 \log\left(2 \cdot 10^7\right) \approx 73(\text{dB})$$

2. Sound power and sound power level

While the sound level describes what can be heard, which is dependent on distance and the acoustic environment, sound power gives the total sound energy emitted by a source, which is independent of distance and the acoustic environment. In the same way that sound level is used to measure the strength of a sound, the sound power of a sound source is described by the sound power level, L_w, which is also expressed as a ratio of the sound power amplitude to the reference sound power of 10^{-12} W:

$$L_w = 10 \log_{10}\left(\frac{W}{W_{\text{ref}}}\right) \quad \text{dB} \tag{9.6}$$

where W is the sound power of a source and $W_{\text{ref}} = 10^{-12}$ (W) = 1 (pW) is the reference sound power.

Sound power uniquely describes the strength of a sound source, but it cannot be measured directly. It must be calculated based on quantities related to the sound field which surrounds the source. Some typical sound power levels are given in Fig. 9.2.

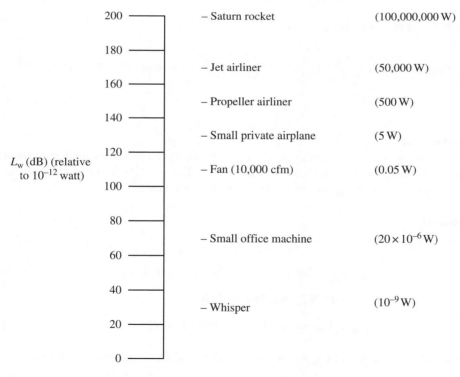

Figure 9.2   Some typical sound power levels, L_w (dB). Reproduced from *Handbook of Noise and Vibration Control*, edited by Malcolm J. Crocker, p. 12, 2007 with permission from John Wiley & Sons

Example 9.3

If a hybrid powertrain has components with the following sound power levels: engine – 82 (dB), transmission – 78 (dB), and motor/generators – 70 (dB), what is the total sound power level of the powertrain?

Solution:
Since:

$$W_{\text{engine}} = W_{\text{ref}} \cdot 10^{\frac{82}{10}} = 10^{-12} \cdot 10^{8.2} = 10^{-3.8} = 1.6 \times 10^{-4} (\text{W})$$

$$W_{\text{trans}} = W_{\text{ref}} \cdot 10^{\frac{78}{10}} = 10^{-12} \cdot 10^{7.8} = 10^{-4.2} = 6.3 \times 10^{-5} (\text{W})$$

$$W_{\text{MG}} = W_{\text{ref}} \cdot 10^{\frac{70}{10}} = 10^{-12} \cdot 10^{7} = 10^{-5} = 1 \times 10^{-5} (\text{W})$$

the total sound power is:

$$W_{\text{total}} = W_{\text{engine}} + W_{\text{trans}} + W_{\text{MG}} = 2.33 \times 10^{-4} \ (\text{W})$$

and the total power level is:

$$L_{\text{total, w}} = 10 \log_{10} \left(\frac{W_{\text{total}}}{W_{\text{ref}}} \right) = 10 \log_{10} \left(\frac{2.33 \times 10^{-4}}{1 \times 10^{-12}} \right) = 10 \log \left(2.33 \cdot 10^{8} \right) \approx 83.7 (\text{dB})$$

Example 9.4

If the sound power level target for a hybrid powertrain is 65 dB and the main components take the following percentages: engine – 50%, transmission – 25%, and motor/generators – 25%, what is the sound power level target for each component?

Solution:
Since:

$$W_{\text{total}} = W_{\text{ref}} \cdot 10^{\frac{65}{10}} = 10^{-12} \cdot 10^{6.5} = 10^{-5.5} = 3.16 \times 10^{-6} (\text{W})$$

$$W_{\text{engine}} = 0.5 \cdot W_{\text{total}} = 1.58 \times 10^{-6} (\text{W})$$

$$W_{\text{trans}} = W_{\text{MG}} = 0.25 \cdot W_{\text{total}} = 0.79 \times 10^{-6} (\text{W})$$

the sound power levels of the engine, transmission, and motor/generator need to be:

$$L_{\text{engine, w}} = 10 \log_{10} \left(\frac{W_{\text{engine}}}{W_{\text{ref}}} \right) = 10 \log_{10} \left(\frac{1.58 \times 10^{-6}}{1 \times 10^{-12}} \right) = 10 \log \left(1.58 \cdot 10^{6} \right) \approx 62 (\text{dB})$$

$$L_{\text{trans,w}} = 10 \log_{10}\left(\frac{W_{\text{trans}}}{W_{\text{ref}}}\right) = 10 \log_{10}\left(\frac{0.79 \times 10^{-6}}{1 \times 10^{-12}}\right) = 10 \log\left(0.79 \cdot 10^{6}\right) \approx 59(\text{dB})$$

$$L_{\text{MG,w}} = 10 \log_{10}\left(\frac{W_{\text{MG}}}{W_{\text{ref}}}\right) = 10 \log_{10}\left(\frac{0.79 \times 10^{-6}}{1 \times 10^{-12}}\right) = 10 \log\left(0.79 \cdot 10^{6}\right) \approx 59(\text{dB})$$

3. Sound intensity and sound intensity level

In engineering acoustics, another commonly measured quantity is the sound intensity, which is defined as the continuous flow of power carried by a sound wave through an incrementally small area at a point in space and the unit is watts per square meter (W/m^2).

Corresponding to the sound pressure level and the power level, the sound intensity is described by the sound intensity level, L_{I}. The sound intensity level, L_{I}, is expressed as a ratio of the sound intensity amplitude to the reference sound intensity standardized as 10^{-12} (W/m^2):

$$L_{\text{I}} = 10 \log_{10}\left(\frac{I}{I_{\text{ref}}}\right) \quad (\text{dB}) \tag{9.7}$$

where I is the sound intensity of a sound source and $I_{\text{ref}} = 10^{-12}$ (W/m^2) is the reference for sound intensity.

Since sound intensity is related to the total power radiated into the air by a sound source, it bears a fixed relation to the sound pressure at a point in free space. It should be noted that since the sound pressure level, the power level, and intensity levels are all expressed in decibels and all of them are measured acoustic quantities, it is easy to get confused. The sound power level is a measure of the total acoustic power radiated by a source in watts. The sound intensity level and sound pressure level specify the acoustic quantities produced at a point removed from the source and their levels depend on the distance from the source, losses in the intervening air path, and environmental effects.

4. Sound energy density

In road vehicle situations, where the sound waves are propagated in a closed space, the desired quantity is sound energy density rather than sound intensity. The sound energy density is the sound energy per unit volume of space. The energy transported by a sound wave is composed of the kinetic energy of the moving particles and the potential energy of the air. The instantaneous sound energy density, D', is:

$$D' = \frac{p^2}{\rho c^2} \tag{9.8}$$

where $\rho =$ density of air (kg/m^3), $p =$ sound pressure (Pa), and $c =$ sound velocity (m/s).

Corresponding to the instantaneous sound energy density, the following space-averaged sound energy density, D, is commonly used in engineering acoustics:

$$D = \frac{p_{\text{rms}}^2}{\rho c^2} = \frac{p_{\text{rms}}^2}{1.4 P_{\text{s}}} \quad (N/m^2) \tag{9.9}$$

where p_{rms}^2 = mean-square sound pressure (Pa^2), and P_s = actual atmospheric pressure (Pa).

5. **Sound transmission loss and transmission coefficient**

In engineering noise and vibration control, understanding how sound waves are transmitted through panels is also essential, and an associated measured quantity is the sound transmission loss (*TL*) through a panel, which is defined as:

$$TL = 10 \log_{10} \left(\frac{1}{\tau} \right) \tag{9.10}$$

where τ is the ratio of the transmitted to the incident sound intensities on a surface. The ratio τ is also referred to as the *transmission coefficient*. In principle, when a sound wave is incident upon a surface, the sound energy is split into three parts: the first part is reflected, the second part is transmitted through the surface, and the third part is dissipated within the surface. A material with a small transmission coefficient means that only a small portion of the sound energy is transmitted through the material, and most energy is either reflected or dissipated. Conversely, when a material has a large transmission coefficient, it implies that most of the incident sound is transmitted through the material, and only a small portion of the energy is either reflected or dissipated within the material.

9.1.3   Frequency Analysis Bandwidths

The normal audible frequency range is from 15 Hz to 16 kHz, and the ultrasonic region starts at about 18 kHz; the human ear is not sensitive to ultrasonic frequencies. In road vehicle system engineering, the vibration signal frequency of interest can extend down to 0.1 Hz, and noise and vibration signals are always analyzed in frequency bands; customary bandwidths are one-third octave and one octave, as shown in Fig. 9.3.

1. **One-octave band**

Octave bands, implying halving and doubling a frequency, are the widest bands that are used for frequency analysis. Since 1000 Hz is the internationally accepted reference frequency for an octave band, it is normally defined as the central frequency of the one-octave band, while its cut-off frequencies are obtained by multiplying and dividing the central frequency by a factor of $\sqrt{2}$, so that the upper cut-off frequency of an octave band is equal to twice the lower cut-off frequency.

If we denote the central frequency as f_C, the upper cut-off frequency as f_U, the lower cut-off frequency as f_L, and the bandwidth as Δf, we have the following relations:

$$f_L = \frac{f_C}{\sqrt{2}} \quad \text{and} \quad f_U = \sqrt{2} f_C \tag{9.11}$$

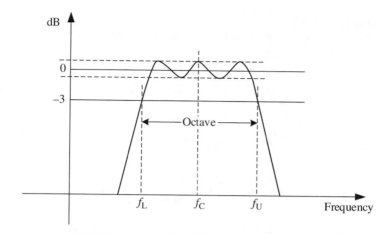

Figure 9.3 Frequency response of a one-octave-band filter

Table 9.2 Central and approximate cut-off frequencies (Hz) of one-octave bands

Band number	Lower cut-off frequency	Central frequency	Upper cut-off frequency
15	22	31.5	44
18	44	63	88
21	88	125	177
24	177	250	355
27	355	500	710
30	710	1000	1420
33	1420	2000	2840
36	2840	4000	5680
39	5680	8000	11,360
42	11,360	16,000	22,720

$$f_C = \sqrt{f_L f_U} \quad \text{and} \quad \frac{f_U}{f_L} = 2 \qquad (9.12)$$

$$\Delta f = f_U - f_L = f_C \left(\sqrt{2} - \frac{1}{\sqrt{2}} \right) = \frac{f_C}{\sqrt{2}} \cong 71\%(f_C) \qquad (9.13)$$

Applying the above principle, we can obtain the central frequencies of other octave bands, as shown in Table 9.2. It can be seen that the human hearing range is covered by ten one-octave bands and their central frequencies are 31.5, 63, 125, 250, 500, 1000, 2000, 4000, 8000, and 16,000 Hz.

2. One-third-octave band

For one-third-octave bands, the cut-off frequencies, f_U and f_L, are defined as follows:

$$f_L = \frac{f_C}{\sqrt[6]{2}} \quad \text{and} \quad f_U = \sqrt[6]{2} f_C \tag{9.14}$$

$$f_C = \sqrt{f_L f_U} \quad \text{and} \quad \frac{f_U}{f_L} = \sqrt[3]{2} \tag{9.15}$$

$$\Delta f = f_U - f_L = f_C \left(\sqrt[6]{2} - \frac{1}{\sqrt[6]{2}} \right) \cong 23\% (f_C) \tag{9.16}$$

So, from the above we can obtain the central, lower cut-off, and upper cut-off frequencies of one-third-octave bands, as shown in Table 9.3. It can be seen that the central frequencies of one-third-octave bands are related by $\sqrt[3]{2}$, and ten cover a decade, while 30 one-third-octave bands are needed to cover the human hearing range: 20, 25, 31.5, 40, 50, 63, 80, 100, 125, 160, 2000, …, 16,000 Hz.

9.1.4   Basics of Vibration

Compared with noise, which relates to oscillatory motion in fluids, vibration deals with oscillatory motion in solids. The inertia and elasticity properties of a body are the necessary conditions causing oscillatory motion. The inertia allows an element within the body to transfer momentum to adjacent elements, while the elasticity tends to force the displaced element back to the equilibrium position. Since all real physical systems are bounded in space, any real physical system contains natural frequencies and modes of vibration. In this section, the concepts of natural frequencies, modes of vibration, forced vibration, and resonance will be briefly described based on the simple free mass–spring systems shown in Fig. 9.4(a) and Fig. 9.4(b).

1. Simple free mass–spring vibration system without damping

For the free, undamped vibration system shown in Fig. 9.4(a), suppose that a mass of M is placed on a spring of stiffness K. If the mass is displaced upwards from its equilibrium position and is then released, the spring force will accelerate the mass downwards based on Newton's second law of motion and Hooke's law.

At any instant of time after release, when the mass has a distance y from it its equilibrium position, from Hooke's Law, the force acting on the mass is as follows:

$$F = -Ky \tag{9.17}$$

Table 9.3 Central and approximate cut-off frequencies (Hz) of one-third-octave bands

Band number	Lower cut-off frequency	Central frequency	Upper cut-off frequency
13	17.8	20	22.4
14	22.4	25	28.2
15	28.2	31.5	35.5
16	35.5	40	44.7
17	44.7	50	56.2
18	56.2	63	70.8
19	70.8	80	89.1
20	89.1	100	112
21	112	125	141
22	141	160	178
23	178	200	224
24	224	250	282
25	282	315	355
26	355	400	447
27	447	500	562
28	562	630	708
29	708	800	891
30	891	1000	1122
31	1122	1250	1413
32	1413	1600	1778
33	1778	2000	2239
34	2239	2500	2818
35	2818	3150	3548
36	3548	4000	4467
37	4467	5000	5623
38	5623	6300	7079
39	7079	8000	8913
40	8913	10,000	11,220
41	11,220	12,500	14,130
42	14,130	16,000	17,780
43	17,780	20,000	22,390

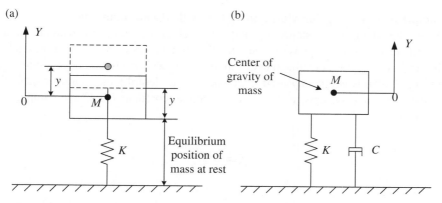

Figure 9.4 Simple free mass–spring vibration system. (a) Simple free mass-spring system without damping, (b) Simple free mass-spring system with damping

Newton's second law of motion states that the force F is equal to the mass (M) times acceleration (a); that is:

$$-Ky = M\frac{d^2y}{dt^2}$$ (9.18)

or

$$M\ddot{y} + Ky = 0 \quad \Rightarrow \quad \ddot{y} + \frac{K}{M}y = 0$$ (9.19)

If we assume that Eq. 9.19 has the following solution form:

$$y = A\sin(\omega t + \varphi)$$ (9.20)

then, substituting Eq. 9.20 into Eq. 9.19, yields:

$$-A\omega^2\sin(\omega t + \varphi) + \frac{K}{M}A\sin(\omega t + \varphi) = 0$$ (9.21)

Solving Eq. 9.21, we obtain:

$$\omega = \sqrt{\frac{K}{M}} \quad (\text{rad/s})$$ (9.22)

The above equations show that the system in Fig. 9.4(a) freely vibrates at an angular frequency ω (rad/s). The frequency ω depends only on the body properties of mass (M) and stiffness (K), so it is called the natural angular frequency and is normally symbolized by ω_n. The corresponding natural frequency is f_n and is equal to:

$$f_n = \frac{1}{2\pi}\sqrt{\frac{K}{M}} \quad (\text{Hz})$$ (9.23)

2. Simple free mass–spring vibration system with viscous damping

Since all real physical systems exhibit damping characteristics accompanying vibration, a certain amount of energy is lost, and the vibration decays with time after the excitation is removed. The free body diagram of a mass–spring with viscous damper system, shown in Fig. 9.4(b), is commonly used in engineering to describe the characteristics of real systems. In such systems, if the viscous damping coefficient is C, the damping force is equal to:

$$f_C = -C\frac{dy}{dt}$$ (9.24)

Based on Newton's second law of motion, we have the following homogenous second-order differential equation:

$$-C\frac{dy}{dt} - Ky = M\frac{d^2y}{dt^2} \quad \Rightarrow \quad M\ddot{y} + C\dot{y} + Ky = 0 \tag{9.25}$$

From differential equation theory, Eq. 9.25 has the following solution form:

$$y(t) = Ae^{s_1 t} + Be^{s_2 t} \tag{9.26}$$

where A and B are arbitrary constants which are evaluated from the initial conditions, and s_1 and s_2 are the roots of the characteristic equation of the system:

$$s_{1,2} = \frac{1}{2M}\left(-C \pm \sqrt{C^2 - 4MK}\right) \tag{9.27}$$

If we define the following terms:

$$\text{(i)} \ \omega_n^2 = \frac{K}{M}; \quad \text{(ii)} \ 2\xi\omega_n = \frac{C}{M}; \quad \text{(iii)} \ \xi = \frac{C}{2\sqrt{MK}} \tag{9.28}$$

then, Eq. 9.25 can be rewritten as:

$$\ddot{y} + 2\xi\omega_n\dot{y} + \omega_n^2 y = 0 \tag{9.29}$$

and the solutions in Eq. 9.27 can be expressed as:

$$s_{1,2} = -\xi\omega_n \pm \omega_n\sqrt{\xi^2 - 1} \tag{9.30}$$

where ω_n is the natural frequency and ζ is the ratio of the viscous damping coefficient to a critical viscous damping coefficient.

When the two roots of the above equation are real and identical, the system is said to be critically damped. From Eq. 9.30, we see that critical damping occurs when $\xi = 1$. Furthermore, three cases of interest arise, which are: (i) $\xi > 1$, (ii) $\xi < 1$, and (iii) $\xi = 1$.

If $\zeta > 1$, the system has two real negative roots and the system is described as being over-damped. As shown in Fig. 9.5(a), in this case, the motion is not oscillatory regardless of the initial conditions. Because the roots are negative, the motion diminishes with increasing time and is aperiodic.

If $\zeta < 1$, the roots are complex conjugate and the system is described as being under-damped. As shown in Fig. 9.5(b), because the real part of two roots are negative, in this case, the motion is oscillatory with diminishing amplitude, and the radian frequency of the damped oscillation is:

$$\omega_d = \omega_n\sqrt{1 - \xi^2} \tag{9.31}$$

(a)   (b)

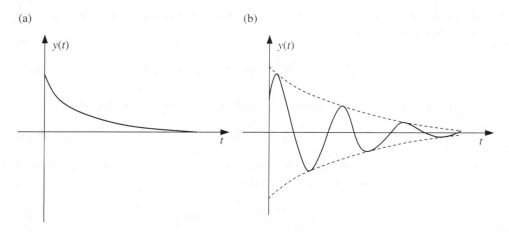

Figure 9.5 Viscous damped mass–spring vibration system. (a) An over-damped viscous-damped motion, (b) A under-damped viscous-damped motion

If $\zeta = 1$, both roots are identical and are equal to $-\omega_n$. In this case, the system is in transition between oscillatory and aperiodic damped motions.

Beyond the above-described motion of viscous damped mass–spring systems, in practical engineering, there also exist many different types of damping, and two of the most common forms are Coulomb (dry-friction) damping and hysteretic (structural) damping. The Coulomb-damping force is constant in magnitude and is in phase with the velocity. In bolted or riveted structures, Coulomb damping is the major damping form. When structural materials such as steel and rubber are cyclically stressed, a hysteretic loop is formed, which results in a damping action. Actually, hysteretic damping is most relevant to engineering noise and vibration control. Engineering experiments show that hysteretic force is proportional to the displacement but it is in phase with the velocity. The analysis of these damping types is very complex; the interested reader should refer to Norton and Karczub (2003).

3. **Simple forced mass–spring vibration system with damping**

If a viscous damped mass–spring system is excited by a harmonic force, $F = F_0 \sin(\omega t)$, the following second-order linear differential equation of motion can be readily obtained by applying Newton's second law to the body, as shown in Fig. 9.6(a):

$$M\ddot{y} + C\dot{y} + Ky = F_0 \sin(\omega t) \qquad (9.32)$$

Using the notation of Eq. 9.28, the above equation can be rewritten as:

$$\ddot{y} + 2\xi\omega_n\dot{y} + \omega_n^2 y = \frac{F_0}{M}\sin(\omega t) \qquad (9.33)$$

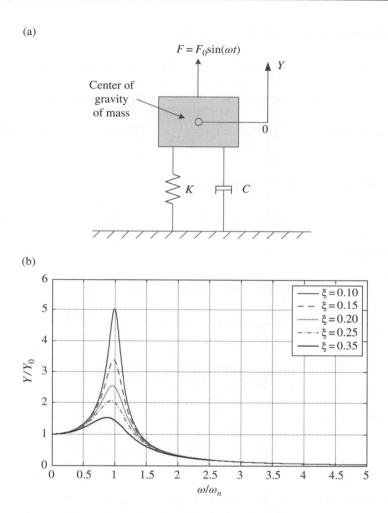

Figure 9.6 Simple viscous damped mass–spring system with harmonic excitation. (a) Simple viscous-damped mass-spring system with harmonic excitation, (b) Displacement amplitude of simple viscous-damped mass-spring system with harmonic excitation

Differential equation theory states that the general solution of Eq. 9.33 is the sum of its complementary function, the solution of the homogeneous equation $\ddot{y} + 2\xi\omega_n\dot{y} + \omega_n^2 y = 0$, and the particular integral. Since the complementary function actually describes the free vibration of a viscous damped mass–spring system, it decays with time and eventually is equal to zero; therefore, the general solution only leaves the particular integral when the system reaches a steady state in which the system oscillates at the exciting force frequency, ω.

Thus, we can assume that the particular integral has the following form:

$$y(t) = Y_S \sin(\omega t - \phi) \tag{9.34}$$

and, substituting Eq. 9.34 into Eq. 9.33 yields:

$$-MY_S\omega^2\sin(\omega t-\phi)+CY_S\omega\cos(\omega t-\phi)+KY_S\sin(\omega t-\phi)=F_0\sin(\omega t) \tag{9.35}$$

Now, let's introduce complex algebra and define the following items:

$$F_0\sin(\omega t)=\mathrm{Im}\left(\mathbf{F_0}e^{j\omega t}\right);\ Y_S\sin(\omega t-\phi)=\mathrm{Im}\left(Y_Se^{j\omega t-\phi}\right)=\mathrm{Im}\left(Y_Se^{-j\phi}e^{j\omega t}\right)=\mathrm{Im}\left(\mathbf{Y}e^{j\omega t}\right) \tag{9.36}$$

If we let $\mathbf{Y}=Y_se^{-j\phi}$, Eq. 9.32 can be written as:

$$-M\omega^2\mathbf{Y}e^{j\omega t}+iC\omega\mathbf{Y}e^{j\omega t}+K\mathbf{Y}e^{j\omega t}=\mathbf{F_0}e^{j\omega t} \tag{9.37}$$

Therefore, we have:

$$K\mathbf{Y}-M\omega^2\mathbf{Y}+iC\omega\mathbf{Y}=\mathbf{F_0}\ \Rightarrow\ \mathbf{Y}=\frac{\mathbf{F_0}}{K-M\omega^2+iC\omega} \tag{9.38}$$

Thus, the amplitude of the output displacement is:

$$|\mathbf{Y}|=Y=\frac{|\mathbf{F_0}|}{\sqrt{(K-M\omega^2)^2+(C\omega)^2}}=\frac{F_0}{\sqrt{(K-M\omega^2)^2+(C\omega)^2}} \tag{9.39}$$

and the phase angle is:

$$\angle\mathbf{Y}=-\phi=\tan^{-1}\left(-\frac{C\omega}{K-M\omega^2}\right)\ \Rightarrow\ \phi=\tan^{-1}\left(\frac{C\omega}{K-M\omega^2}\right) \tag{9.40}$$

Using the notation of Eq. 9.28 and defining $Y_0={F_0}/{K}$, we obtain:

$$\frac{Y}{Y_0}=\frac{1}{\sqrt{\left(1-\left({\omega}/{\omega_\mathrm{n}}\right)^2\right)^2+\left(2\xi\omega/{\omega_\mathrm{n}}\right)^2}}\ \text{and}\ \phi=\tan^{-1}\left(\frac{2\xi\omega/{\omega_\mathrm{n}}}{1-\left({\omega}/{\omega_\mathrm{n}}\right)^2}\right) \tag{9.41}$$

Furthermore, the derivative of Eq. 9.34 can be obtained as:

$$\frac{d\left({Y}/{Y_0}\right)}{d\left({\omega}/{\omega_\mathrm{n}}\right)}=\frac{2\left(1-2\xi^2-\left({\omega}/{\omega_\mathrm{n}}\right)^2\right)\left({\omega}/{\omega_\mathrm{n}}\right)}{\left\{\left(1-\left({\omega}/{\omega_\mathrm{n}}\right)^2\right)^2+\left(2\xi\omega/{\omega_\mathrm{n}}\right)^2\right\}^{3/2}} \tag{9.42}$$

Therefore, we can note the following based on Eqs 9.41 and 9.42:

1. When $\omega/\omega_n = \sqrt{1-2\xi^2}$, the output of displacement Y is maximal and $Y_{max} = \frac{Y_0}{2\xi\sqrt{1-\xi^2}}$, with the corresponding phase angle $\phi|_{Y_{max}} = \tan^{-1}\left(\frac{\sqrt{1-2\xi^2}}{\xi}\right)$. This condition is called *amplitude resonance*.

2. When $\omega/\omega_n = 1$, the phase angle $\phi = 90°$, and the corresponding amplitude $Y|_{\phi=90°} = \frac{F_0}{2K\xi}$, and this condition is referred to as *phase resonance*.

3. Both the amplitude and phase angle of the output displacement are functions of damping ratio ξ and $\omega-\omega_n$, the phase angle ϕ can vary from $0°$ to $180°$, but the displacement amplitude could be very large depending on the damping ratio, ξ. Figure 9.6(b) shows how the displacement amplitude changes with ξ and ω/ω_n based on Eq. 9.39.

9.1.5 Basics of Noise and Vibration Control

Noise always originates from a source, i.e. some energy input of some sort, then travels through a path, and finally reaches a receiver; therefore, noise and vibration problems are generally described by using the simple source–path–receiver model shown in Fig. 9.7. Furthermore, a noise and vibration control problem can also be spilt into these three components. However, since the receiver is the human ear in road vehicle applications, the noise and vibration levels are mainly controlled from source and path control approaches.

1. Vibration source control approaches

As is well known, noises in road vehicle systems mainly come from two sources: structure-borne noise associated with vibrating structural components, or aerodynamic noise. Modifying a noise source is the best practical way to control the vibration level (noise source) but sometimes this is very costly or difficult to accomplish, while vibration control is not only important in terms of minimizing structural vibrations and any associated fatigue, but it is also important in terms of reducing noise.

In vehicle engineering, the techniques and methods for controlling noise through vibration source control include improving bearing alignment, increasing damping for some noise sources, reducing system dynamics, changing the natural frequencies of main subsystems, properly balancing moving components, minimizing aerodynamics by smoothing the vehicle body and reducing the frontal area of the vehicle, and choosing quieter movers. For example, in order to damp noise effectively, Helmholtz resonators,

Figure 9.7 Simple source–path–receiver model

acting as reflection sound absorbers, are commonly used for vehicle exhaust and engine intake manifold systems.

2. Noise path control approaches

Common noise-reduction techniques using noise path control include installing vibration isolators, noise barriers, and reactive/dissipative mufflers. These techniques are also referred to as passive noise control approaches.

Vibration isolators are generally used in situations either where the source is producing vibration that is desired to prevent vibrational energy flowing to a supporting structure or where a delicate piece of equipment must be protected from vibration in the structure. The principle of vibration isolation is to insert an isolator that can attenuate the exciting force from the vibrational source. Springs, elastomeric and pneumatic isolators are basic types of isolator. In practice, the isolation applied is normally a combination of two or three of them, and a noise reduction of about 30 dB can be expected by applying isolation techniques.

One common and effective method of reducing noise using a noise path control approach is to mount an acoustic enclosure around the noise source to encapsulate the source. Acoustic enclosures can be classified into close-fitting and not-close-fitting types. The performance analysis for not-close-fitting enclosures is relatively simple; however, due to space limitations, most enclosures installed in hybrid electric vehicles are close-fitting enclosures and are normally installed within 0.5 m of the surface of any major vibrational source. Close-fitting enclosures produce complicated physical effects such as cavities which could result in enclosure resonance, so the design and analysis must be carefully carried out. The actual effectiveness of inserting an enclosure is restricted by the extent to which it can isolate the vibrational source, the presence of air gaps and leaks, and transmission loss of the panels which are used to construct it, and generally a noise reduction of about 40 dB can be expected from installing an enclosure. It should be noted that enclosures do not eliminate or reduce the source of the noise, they just contain it, and therefore it is good engineering practice to consider enclosures only as a last resort (Norton and Karczub, 2003).

Another useful low-cost noise-reduction technique is to place acoustic barriers between a noise source and a receiver. The principle of this approach is that the barrier has some sound wave reflected so that the sound pressure level is reduced at the receiver position. In practice, this approach is very effective for high-frequency noise reduction, and less effective for low frequencies, since the length of low-frequency sound waves may exceed the barrier height. Generally, the barriers are made from materials with high absorption coefficients as they absorb the sound waves and significantly reduce the reflected energy.

As is well known, when a sound wave is incident upon a surface, part of the sound energy is reflected, part of it is transmitted through the surface, and part of it is dissipated within the surface. If there is no energy reflected, the sound is said to be absorbed, which means that the sound is either transmitted through the material or dissipated in the material. The property of a material absorbing sound is described by the sound absorption coefficient, which is defined as the ratio of the absorbed to incident sound energies.

The sound absorption coefficient has a value somewhere between 0 and unity. When a material has a small sound absorption coefficient, most of the incident sound is reflected back and only a small portion is either transmitted or dissipated; by contrast, when a material has a large sound absorption coefficient, most of the incident sound is either transmitted through the material or dissipated as heat by the material. Thus, in the natural acoustic environment, people receive sound more or less from a direct path or by reflection depending on the distance from the noise source. In order to control the noise level, sound-absorbing materials are often installed on noise transmission paths, close to the source of the noise or the receiver, and from a practical perspective, sometimes this is a very effective way to control the noise level.

9.2   General Description of Noise, Vibration, and Control in Hybrid Electric Vehicles

Noise, vibration, and harshness (NVH) refinement is an important traditional task in powertrain development and the vehicle integration process. The emergence of an electrified powertrain has led to significant changes in the vehicle design and integration process. Each hybrid electric vehicle configuration brings unique NVH challenges resulting from a variety of sources. This section briefly discusses some unique vehicle NVH challenges inherent in the operation of hybrid electric vehicles.

Figure 9.8 shows a typical driveline architecture for hybrid electric vehicles. In hybrid transmissions, the torque converter in traditional automatic transmission is normally replaced by a motor. Most hybrid vehicles can be driven by the motor/generator B alone in the electric drive mode, by the engine alone in mechanical drive mode, or by all the motor/generators A and B and the engine together in hybrid mode. These drive modes are achieved through controlling the corresponding clutches in the transmission to meet the high-level requirements of drivability and fuel economy.

As is well known, one main function of the torque converter in traditional automatic transmission is to attenuate torque ripple generated by the engine; however, in most hybrid vehicle systems, the engine and a motor/generator are directly connected through a planetary gear set as a result of removing the torque converter. While this improves system responsiveness and efficiency, it also easily causes twisting of the driveshaft and produces torsional vibration, and this becomes even worse during engine start and stop. Compared to a

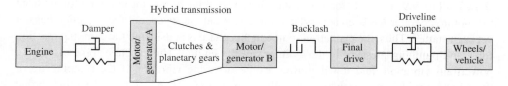

Figure 9.8   Driveline dynamics of hybrid electric vehicles

traditional vehicle, strong vibration and gear rattle sometimes occur on the output shaft of the transmission, and new noise paths and patterns are generated due to additional inertia and component backlash introduced in a hybrid system.

Other NVH issues in hybrid vehicle systems include electromechanical vibration and noise generated by electric motors, electromagnetic vibration and noise generated by power converters and inverters, and more complex vibration and noise generated by the energy storage system. These unique NVH characteristics and associated control techniques are addressed in the following sections.

9.2.1 Engine Start/Stop Vibration, Noise, and Control

One distinguishing characteristic of hybrid vehicles is that the engine frequently starts and stops during city driving. In a conventional ICE-powered vehicle, the engine start-up occurs only at the beginning of the vehicle's operation and the resulting vibration, due to cranking the engine, is expected by the driver; in addition, the engine start/stop is usually performed with the transmission in the parking position. For hybrid electric vehicles, start-up of the engine is linked to factors such as the state of charge of the battery and the driver's torque demand, which can result in 'unexpected' vehicle vibration. Depending on the layout of the driveline, the start-up vibrations for hybrid vehicles can be further complicated, since the start-up of the engine does not occur when the transmission is in neutral, which may cause excitation of driveline torsional modes.

Another distinguishing characteristic of hybrid vehicles is that most hybrid vehicles run in an electric mode during initial operation of the vehicle before a certain speed is reached, for example, 15 km/h, so that all interior noise resulting from road noise, wind noise, and electric motor noise is shared. Once the engine starts up, the interior noise is dominated by noise from the engine, causing a sudden change in the sound quality in terms of frequency of content and absolute noise level. Since the ICE in HEV/PHEV/BEV applications is typically downsized, the torsional fluctuations and noise and vibration levels can be significantly increased in comparison to conventional vehicles, so that it is essential to reduce the vibration and noise in hybrid electric vehicles when the engine starts up.

9.2.1.1 Engine Start Vibration and Measuring Method

The engine start vibration normally occurs in the cranking and initial combustion phases shown in Fig. 9.9 at −25 °C for a cold start. In the cranking phase, the cranking torque from the motor/generator has been applied to ramp up the engine speed, but fuel injection has not yet been started. Thus, the excitation source in this phase is considered the cranking reaction force and pumping pressure in the cylinders of the engine. This reaction force can excite powertrain vibration, and it will further excite driveline vibration if the ramping up of the engine speed is not properly controlled. On the other hand, the rapid engine torque increase after ignition can also excite powertrain vibration.

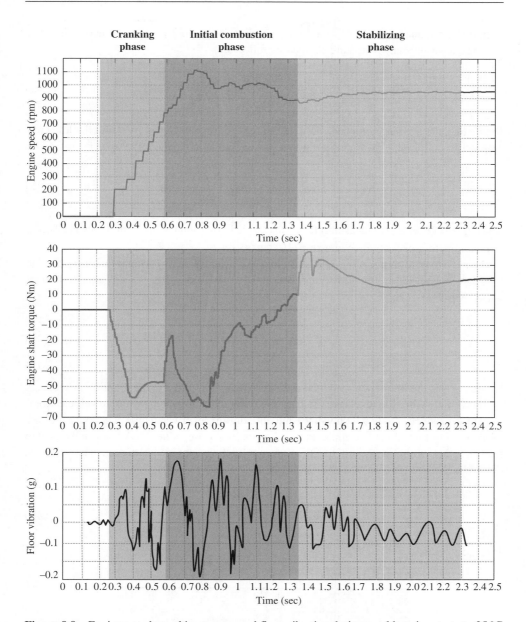

Figure 9.9 Engine speed, cranking torque, and floor vibration during a cold engine start at −25 °C

In order to quantify vehicle vibrations, acceleration, velocity, and displacement levels (dB) are usually used, which are defined as follows:

$$\text{Acceleration level}: L_a = 20 \log_{10}\left(\frac{a}{a_{\text{ref}}}\right) \quad \text{(dB)} \tag{9.43}$$

In practice, the acceleration is normally measured from rms levels in m/s^2 and the level reference is usually taken as one micrometer per second square rms (10^{-6} m/s^2).

$$\text{Velocity level}: L_v = 20 \log_{10}\left(\frac{v}{v_{\text{ref}}}\right) \quad \text{(dB)} \tag{9.44}$$

The velocity level is also measured from rms levels in m/s and the level reference is taken as one nanometer per second square rms (10^{-9} m/s).

We need to note here that it is necessary to declare the frequency weight factors applied to the data when these quantities are used. The weighted rms (m/s^2) acceleration is expressed as:

$$a_w = \left[\frac{1}{T}\int_0^T a_w^2(t)dt\right]^{1/2} \tag{9.45}$$

where $a_w(t)$ is the frequency weighted acceleration in m/s^2 and T is the duration in seconds.

The vehicle vibration levels are often assessed based on human exposure to vehicle vibrations; for instance, Table 9.4 shows the human comfort level versus the acceleration of the vibration. For hybrid vehicle applications, the vibration dose value (VDV) is usually used to measure the vibration during transient events such as engine start/stop. The VDV is defined as:

$$\text{VDV} = \sqrt[4]{\int_0^T (a_w(t))^4 dt} \tag{9.46}$$

Table 9.4 Comfort levels for various acceleration levels

Acceleration level	Comfort level
<0.35 m/s^2	Not uncomfortable
0.3–0.65 m/s^2	A little uncomfortable
0.6–1.0 m/s^2	Fairly uncomfortable
0.85–1.65 m/s^2	Uncomfortable
1.2–2.6 m/s^2	Very uncomfortable
>2.5 m/s^2	Extremely uncomfortable

where $a_w(t)$ is the weighted acceleration in m/s^2, T is the duration in seconds, and the unit of VDV is m/s$^{1.75}$.

It should be appreciated that the measurement of VDV is complex as it involves the gathering and weighting of acceleration time histories and their subsequent integration. The VDV results in a single number metric, which is normally used for vehicle benchmarking, target setting, and to track engine start/stop performance in an HEV/PHEV/BEV during its development phase.

9.2.1.2 Engine Start-up and Shut-down Vibration Mechanism and Reduction Control

Since ICE start-up and shut-down vibration is one of the significant contributors to overall vehicle vibration, the following controls are often taken in hybrid electric vehicles to reduce the vibration during ICE start-up and shut-down events.

Basis of ICE Ignition and Valve Timing Control
In a reciprocating engine, the position at which a piston is farthest from the crankshaft is called the *top dead center* (TDC), while the position at which a piston is nearest to the crankshaft is called the *bottom dead center* (BDC). For a four-stroke ICE, a piston completes one single thermodynamic cycle per two revolutions of the engine's crankshaft. These four strokes are called the intake stroke, the compression stoke, the power or combustion stroke, and the exhaust stroke. The gas exchange process of a four-stroke ICE is shown in Fig. 9.10.

Since only every second shaft rotation of a four-stroke process is used to generate work, an ICE is normally configured with multiple cylinders, and the pistons may reach TDC simultaneously or at different times depending on the detailed configuration. For example, in the straight-four configuration shown in Fig. 9.11, the two end pistons (pistons 1 and 4) reach TDC simultaneously, as do the two center pistons (pistons 2 and 3), but these two pairs reach TDC with an angular displacement of 180°. Similar patterns can be found in almost all straight ICEs with even numbers of cylinders, with the two end pistons and two middle pistons moving together and the intermediate pistons moving in pairs in a mirror image around the ICE center. In practice, ICE manufacturers often mark the TDC for cylinder one on the crankshaft pulley, the flywheel, or the dynamic balancer, or both, with adjacent timing marks showing the recommended ignition timing settings determined during engine development. These timing marks can be used to set the ignition timing.

Unlike a diesel ICE, a gas ICE cannot completely burn out fuel at the moment of the spark. It takes a period of time for the combustion gases to expand, and the detailed burning and expansion timeframe varies with the angular or rotational speed of the engine. Therefore, it is necessary to advance the ignition time for gas ICEs, and the angle before top dead center (BTDC) is usually used to describe how much ignition time is advanced for the engine. Since the purpose of the power stroke in the engine is to force the combustion chamber to expand, it needs to be emphasized that advancing the spark BTDC means that the gas is ignited prior to the point where the combustion chamber reaches its minimal size. For

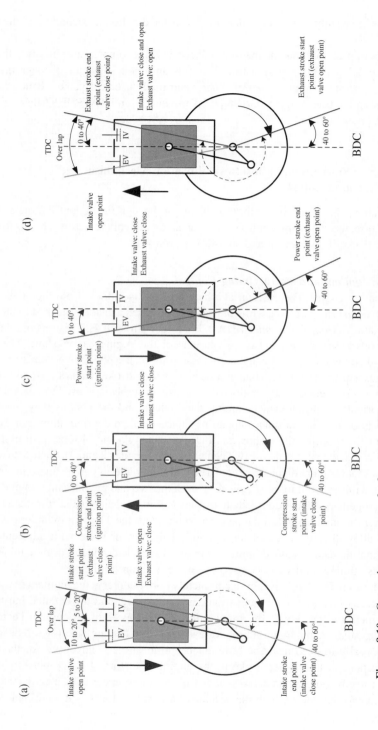

Figure 9.10 Gas exchange process of a four-stroke ICE. (a) Intake stroke; (b) compression stroke; (c) power stroke; (d) exhaust stroke

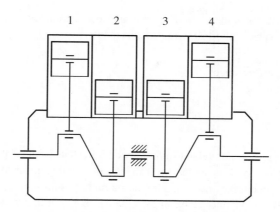

Figure 9.11 Straight-four engine configuration

a gas ICE, sparks occurring after top dead center (ATDC) usually result in some negative effects such as backfiring and knocking, so this should be avoided unless it is necessary to continue the spark prior to the exhaust stroke.

The ignition timing of modern vehicles is adjustable to advance or retard the ignition timing either based on the optimized ignition-advance curves or on actual vehicle load circumstances. For example, when the acceleration pedal is fully depressed and the throttle is wide open (WOT), along with increasing engine speed, ignition takes place earlier in order to maintain the combustion pressure at the levels required for optimal engine performance; by contrast, during part-throttle situations, leaner air and fuel mixtures are encountered, which results in more difficult ignition. Since light-load circumstances require more time for ignition, it must be triggered earlier, with the timing further advanced.

Another important controllable variable affecting the performance of an engine is the valve opening time. Having large overlaps between the inlet valve and outlet valve, resulting from early opening of the inlet valve, increases the internal exhaust gas recirculation which helps to reduce NOx emissions. However, the valve opening timing must be properly set as the recirculated exhaust gas displaces the fresh air-fuel mixture, which leads to a reduction in maximal torque generation; furthermore, excessively high exhaust gas recirculation can lead to combustion misfire at idle and, in turn, can cause an increase in HC emissions.

Therefore, setting the correct ignition and valve timing is crucial in the performance of an engine, and affects many variables including engine longevity, fuel economy, emissions, and engine power. Sparks occurring too soon or too late in the engine cycle often result in excessive vibration and even engine damage, while opening a valve too soon or too late may cause less power output or excessive emissions. There are many factors that influence proper ignition and valve opening times for a given engine, and the ignition timing and valve timing can be optimally controlled corresponding to the changes in the engine's operational states. Actually, the ignition and valve timing of modern engines is set through a closed-loop control, which mainly takes the engine's speed and load into account in real time, and the control objective is to meet the following requirements:

- Maximal engine performance;
- Low fuel consumption;
- No engine knock;
- Clean exhaust gas.

In addition, either ignition or valve timing may drift from its initial setting over the course of time if proper maintenance is neglected. If the ignition timing shifts towards a later firing point, a gradual drop in engine power and increased fuel consumption and emissions will be noticed, while excessive advance may, in extreme cases, result in serious damage to spark plugs or to the engine if the engine knocks. Therefore, it is critical to realize real-time ignition and valve timing control, as detailed in Chapter 7.

Hybrid Vehicle ICE Ignition and Valve Timing Control to Reduce Engine Start/Stop Vibration

As mentioned in the previous section, several milliseconds elapse between the initial ignition and the completion of combustion, so the ignition spark must be triggered early enough to ensure that main combustion (the combustion-pressure peak) occurs shortly after the piston reaches TDC, and the ignition angle should move farther in the advance direction along with increasing engine speed. We also know that the correct ignition and valve timing significantly affect the performance of an engine, fuel economy, and emissions. In fact, they also impact on the pressure of the combustion chamber, which therefore results in vibration of the powertrain. Figure 9.12 shows the curves for combustion chamber pressure in a four-stroke engine with correct and incorrect ignition timing.

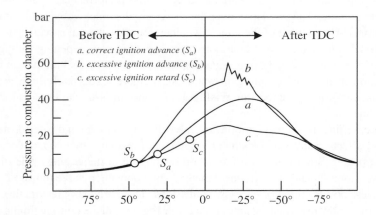

Figure 9.12 Example of combustion pressure vs. ignition firing process

   The engine start-up request for a well-developed hybrid electric vehicle is associated with many factors of other subsystems, such as the state of charge of the battery and the driver's torque demand. Depending on the layout of the driveline, the start-up vibrations of a hybrid vehicle can be further complicated by the fact that the start-up of the engine does not normally occur when the transmission is in neutral or park, which is different from a conventional ICE-powered vehicle in which the engine start-up occurs only at the beginning of the vehicle's operation. The engine in hybrid vehicles is designed to frequently start/stop during traffic jams or when waiting at traffic lights to minimize fuel consumption and emissions, which can cause excitation of driveline torsional modes unless special measures are taken.

   Commonly applied techniques to reduce engine start/stop vibration in a hybrid vehicle include controlling the ignition timing, the intake valve opening/closing timing, the piston stop position, and the volume of fuel injected at each phase of the engine's operation. For example, in order to reduce engine start vibration in the crank phase, the timing of the intake valve closing is normally determined between 100° and 120°, and the cranking torque increase rate is properly controlled once the electric motor is able to provide supplemental torque to meet the vehicle's demand; in addition, in the engine shut-off phase, the piston is controlled to stop at a position after the intake valve close point and before the TDC, as shown in Fig. 9.13. To reduce engine start vibration in the combustion phase, the following three measures are normally taken: richly inject the fuel right after the cranking action to help stabilize combustion, delay the ignition timing to reduce any rapid increase in engine torque, and smoothly relieve the intake air volume to avoid pressure disturbance. Toyota has made great efforts to reduce engine start/stop vibrations in its hybrid electric vehicles. The interested reader should refer to Yoshioka and Sugita (2001) and Kawabata *et al.* (2007).

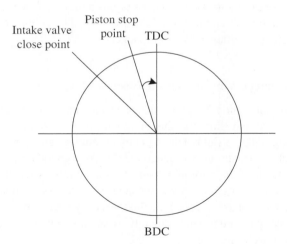

Figure 9.13   Piston stop position for engine start-up vibration reduction

9.2.2   Electric Motor Noise, Vibration, and Control

Electric motors/generators are electrical–mechanical machines to convert electrical energy into mechanical energy and vice versa, and they can be subdivided into direct current (DC) and asynchronous (AC) machines. Since an electric motor/generator is the main power source in hybrid and all-electric vehicles, the noise and vibration characteristics of electric motors/generators play an important role in hybrid and all-electric vehicle system integration. The main sources of noise and vibration in motors/generators are as follows:

- Electromagnetic forces;
- Imbalance of the rotor;
- Bearings;
- Aerodynamic forces.

Among these noise and vibration sources, electromagnetic forces are active in the air gap between the stator and the rotor and are characterized by rotating or pulsating power waves. The vibration caused by electromagnetic force is also called electromagnetic vibration, and it results from slight periodic deformations in the frame of the machine caused by certain magnetic attraction between the stator and the rotor. Depending on the angular speed of a machine, the electromagnetic vibration frequency is typically in the range of 100 to 4000 Hz and the noise from a motor manifests in the form of a whining noise or tonal noise, the frequency range of which is typically between 400 and 2000 Hz. Mechanical imbalance of rotors may also excite an appreciable amount of vibration, particularly when the machine operates at high speed. The intensity level of noise generated by aerodynamic forces depends on the construction of the fan and ventilation channels of the machine, and relatively large airflow and irregular air gaps in the machine may produce aerodynamic noise spectra with pure tone components. The intensity of the noise and vibration in bearings is determined by the quality of manufacturing, the accuracy of machining of the bearing seats, and the vibroacoustic properties of the end brackets.

9.2.2.1   Electromagnetic Vibration of the Electric Motor

The windings of the stator and rotor create their own electromagnetic fields in the machine, rotating with one speed and shifted in the circle with respect to one another for a certain displacement angle, θ. These fields tend to compensate one another to create a working moment of the machine. By design, the generated electromagnetic forces act not on the conductors but on the teeth of the rotor and stator. Further studies show that electromagnetic vibrations in electric motors are mainly associated with electrodynamic, electromagnetic, and magnetostrictive oscillatory forces, while in the steady state of operation, the dominant vibration and audible noise are produced mainly by stator ovalization and breathing modes caused by radial Maxwell pressure.

The cross-section of an electric motor is shown in Fig. 9.14. The electromagnetic forces acting in the motor have a radial direction and angular symmetry, and the sum of the

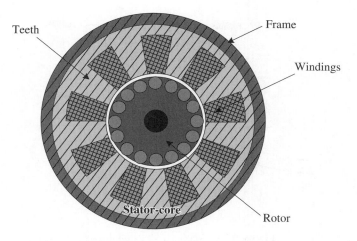

Figure 9.14   Cross-section of the motor stator and rotor

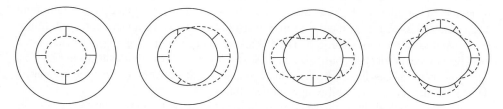

Figure 9.15   Maxwell pressure distribution decomposition in sinusoidal force

electromagnetic forces acting on the rotor is equal to zero in such a defect-free machine. However, the stator is affected by the electromagnetic forces tending to deform it. The spatial shape of these forces is a circular wave characterized by a number of waves, called the order of stator oscillations, appearing on the circumference of the stator. Since mechanical rigidity of the stator grows quickly with growth in the oscillation order, normally only low-order deformations are taken into consideration during analysis of vibrations.

Figure 9.15 illustrates stator ovalization and breathing modes under radial Maxwell pressure. Such stator breathing generally excites vibration in the electric drivetrain and audible noise may result. For a defect-free machine with four magnetic poles, it is safe to expect that only the fourth-order ovalization and breathing mode will be excited. However, since the electric motor of a hybrid electric vehicle cannot have perfect cylindrical symmetry due to manufacturing tolerance and the connection with the reducer, any order of ovalization may appear to be excited; for example, even with four magnetic poles, second-order ovalization may appear to be excited.

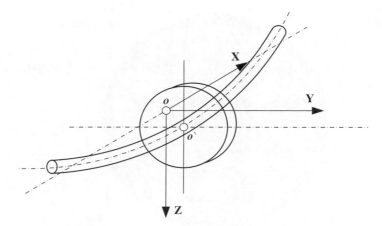

Figure 9.16   Maxwell stress on the rotor

On the other hand, electromagnetic forces also have an impact on the radial vibration behaviors of the rotor due to the eccentric position of the rotor with respect to the stator bore. The radial forces acting on the rotor, often referred to as *unbalanced magnetic pull* (ump), are generated in the motor air gap. The unbalanced magnetic pull not only excites radial vibration in the motor shaft, but also gives rise to a bending of the shaft, shown in Fig. 9.16, resulting in the necessity to increase shaft diameter and bearing size as well as sharply decreasing the motor's reliability and service life.

The tonal nature of the whining noise from the electric motor may annoy the customer, and tonal noise issues can even play a larger role in hybrid and all-electric vehicles due to the lack of sufficient noise masking, as implemented for conventional ICE powertrains. For a hybrid vehicle, if we take all the electric powertrain components into account, the vibration behaviors are more complex. The spectral content of electric powertrain noise and vibration can be up to 8000 Hz, versus 1500 Hz for a conventional ICE powertrain. In addition, the motor's fast response to driveshaft load and associated torque ripple may also excite electric powertrain vibrations in hybrid electric vehicles.

9.2.2.2   Vibration Caused by Rotor Imbalance

Another vibrational source in an electric motor is an imbalance of the rotor, and this can even excite vibration in the entire electric powertrain. It is a fundamental factor to be considered in hybrid and/or electric motor design. Balancing of rotors prevents excessive loading of bearings and reduces fatigue failures; thus, having a smoothly rotating, balanced rotor will enhance the reliability and increase the service life of the motor.

If the inertia axis (the center of mass) of the rotor is out of alignment with the geometric axis (the center of rotation), a rotating imbalance of an electric motor occurs, which results in vibration when the rotor turns; that is, imbalance in a rotor is a result of an uneven distribution of mass around an axis of rotation. In theory, the vibration is produced by the interaction of an imbalanced mass component with the radial acceleration due to rotation, which together generate a centrifugal force. Since the mass component rotates, the force also rotates and tries to move the rotor along the line of action of the force. Once the vibration occurs, it will be transmitted to the rotor's bearings, and any point on the bearing will experience this force once per revolution. Therefore, some other mechanical problems of electric motors, such as noise and the short life of bearings, may also result from rotational imbalance. Rotational imbalance of an electric motor can be further classified as static, couple, or dynamic imbalance, as follows.

Static rotor imbalance is characterized by eccentricity in the center of gravity of the rotor mass, which is caused by a point mass at a certain radius from the rotor's geometric spin axis, as shown in Fig. 9.17. From Fig. 9.17, it is easy to understand that static balancing involves resolving the primary forces into one plane and adding a correction mass in that plane only. This type of rotational imbalance is also commonly found on engine flywheels and car wheels.

Couple imbalance occurs when two equal masses are placed symmetrically about the center of gravity, but positioned at 180° from each other, as shown in Fig. 9.18, which also shows that the end masses balance each other when the rotor is stationary. However, when it

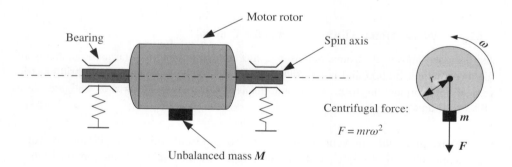

Figure 9.17   Static imbalance of the rotor

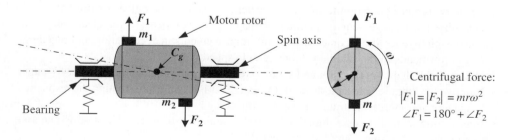

Figure 9.18   Couple imbalance of the rotor

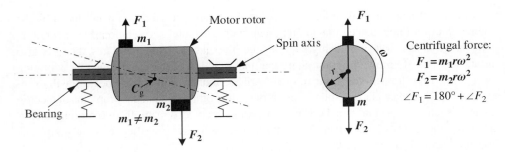

Figure 9.19 Dynamic imbalance of the rotor

rotates, the two masses cause a shift in the inertia axis and a strong imbalance is experienced, which leads to strong vibrations in the bearings. In a hybrid electric powertrain, couple imbalance is found frequently where the motor shaft is connected to other, imbalanced powertrain components.

Dynamic imbalance, shown in Fig. 9.19, is a combination of static and couple imbalance, and it is the most common type of imbalance found in rotors in hybrid electric powertrains. To correct dynamic imbalance, it is necessary to make vibration measurements while the machine is running and to add balancing masses in two planes.

9.2.2.3 Electric Motor Vibration and Noise Control

As discussed above, electromagnetic force, mechanical imbalance, and aerodynamics are the three main vibration and noise sources for hybrid vehicle systems. In hybrid electric vehicle applications, the following countermeasures are generally used to mitigate these vibrational and noise phenomena:

- Develop a motor vibration control strategy to reduce electric drivetrain vibration by compensating the engine torque pulsation during engine starts and moderate torsional vibration of the driveshaft by utilizing the fast-response characteristics of the electric motor to mechanical torque ripples.
- Optimize the motor's operational characteristics of speed vs. torque and optimally coordinate two-motor operation to improve acceleration and regenerative performance.
- Optimize the arrangement of permanent magnets to compensate for electromagnetic noise, which generally becomes worse with improvements in acceleration and regenerative performance in the middle to high vehicle speed range.
- Optimize the shapes of the rotor and stator to reduce torque ripple in the high speed range.
- Improve the vibration characteristics of the electric drivetrain by dissipating the resonance frequency and increasing transaxle case stiffness.
- Improve the acoustic transfer function by adding sound-absorbing and insulating material around the center tunnel and the dash panel.
- Improve radiated noise characteristics by installing a dynamic damper on the transmission.

9.2.3   Power Electronics Noise and Control

A hybrid electric vehicle is normally equipped with several switching power converters and inverters such as a three-phase DC–AC inverter, a single-phase DC–AC inverter, and a DC–DC converter. To achieve high operational efficiency, a power converter/inverter employs switching devices, energy storage elements and transformers, and relies on appropriate modulation of the switches to convert electrical power from one form to another, as required by the load. The switching frequency of most power converters/inverters applied in a hybrid electric powertrain is usually designed up to 25 kHz, and the switching frequency is lowered under light-load and no-load circumstances to reduce switching loss in the power electronics, so that the actual switching frequency of most power converters/inverters could fall in the audible range mentioned in previous sections.

Since the switching frequency of power electronic devices could be within the audible frequency range, vibrations may be excited and a disturbing tone may be emitted. In fact, the audible noise in switched-mode power converters/inverters comes from two main sources: the cooling system (liquid-cooled or air-cooled) and magnetic components such as transformers, input filter inductors, and power-factor-correction (PFC) chokes. For an air-cooled system, cooling noise is mainly caused by air turbulence generated by the fins and is dominant under heavy and medium loads; the noise level can be controlled by optimizing air-flow channels and the speed of the cooling fan. The noise from a liquid-cooled system comes mainly from the cooling pump, and the noise level can be controlled by the methods described in previous sections.

On the other hand, the noise from magnetic components consists mainly of two parts depending on the excitation mechanisms. The first and most dominant part of the noise is caused by magnetization of the core. As is well known, one property of ferromagnetic materials is called *magnetostriction*, which causes the magnetic core to change shape or dimension during the process of magnetization. Therefore, magnetostriction can cause a mechanical interaction between the core and the windings that leads to vibration. The second part of the noise from magnetic components is caused by electromagnetic forces created by the magnetic field of the currents in the components' windings.

In theory, the length of a magnetic core changes in the direction of the applied field, whereas the volume stays approximately constant. This relative change in length can be described by the following equation:

$$\lambda = \frac{\Delta l \, \mu \mathrm{m}}{l \quad \mathrm{m}} = \frac{\Delta l}{l} \lambda_s \tag{9.47}$$

where Δl is the change in the length, l. A measure of the magnetostriction is the so-called 'saturation magnetostriction', λ_s which describes the relative longitudinal change in length of a magnetically saturated material. The saturation magnetostriction differs for different materials and is in the range of a few λ_s.

The switched magnetic field in a power converter/inverter leads to a periodic deformation of the core with twice the switching frequency. The reason for doubling the frequency is that the change in length of the core is the same for both magnetization directions. The deformation of the magnetic core is shown in Fig. 9.20. It is obvious that the periodic deformation of the magnetic core will excite mechanical vibrations which are closely related to the magnetic flux swing. In order to minimize noise, some air gaps are added in practice to damp the vibrations.

The mechanical vibration of a magnetic component generates audible noise in power electronic devices. Such noise is undesirable in hybrid electric vehicle applications because these power devices are normally placed close to the user and the noise characterization is obviously different from a traditional vehicle. Therefore, it is necessary to study such switching-mode power electronic devices in order that appropriate measures may be taken to reduce the audible noise when designing hybrid vehicles.

As shown in Fig. 9.21, the switching-mode power electronic devices utilized in hybrid electric vehicles are designed with variable frequency operation to increase the efficiency; that is, the switching frequency decreases as the load increases, so the switching frequency can drop below the upper threshold of the unmasked audible range.

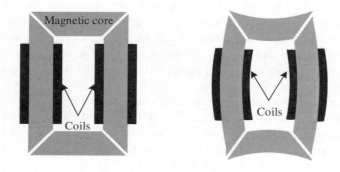

Figure 9.20   Periodic deformation of a magnetic core

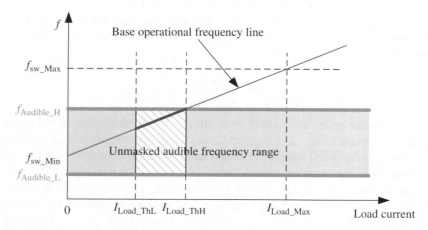

Figure 9.21   Base operational frequency of switching mode power electronic devices

In practice, engineers usually use both mechanical and electrical approaches to reduce the magnetic-component-related audible noise in switching-mode power electronic devices. The mechanical approach is based on techniques that prevent or damp vibrations by mechanical means, such as varnishing, gluing, and potting. Although these techniques are successful in some applications, they generally involve extra cost and manufacturing steps if trying to eliminate the noise completely. As a matter of fact, using only mechanical approaches cannot fully mask noise in a switching-mode power electronic device. Supplemental to using mechanical techniques, electrical techniques are also employed to eliminate the unmasked noise, and in some cases, electrical methods are preferred since they are more successful and cost-effective (Huber and Jovanović, 2011).

An advanced technique for reducing power electronic device noise is to prevent the switching frequency from dropping below the upper threshold of the unmasked audible range, which can be achieved by a closed-loop control strategy shown in Fig. 9.22. By feeding back and forward the output current of the power electronic device, the switching frequency can be maintained within the masked audible frequency range, and therefore the audible noise can be reduced. The corresponding diagram of switching-mode frequency operation is shown in Fig. 9.23.

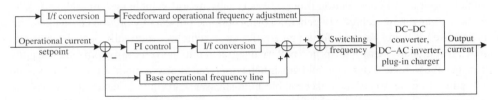

Figure 9.22 PI feedback with feedforward control strategy to adjust the switching frequency of power electronic devices

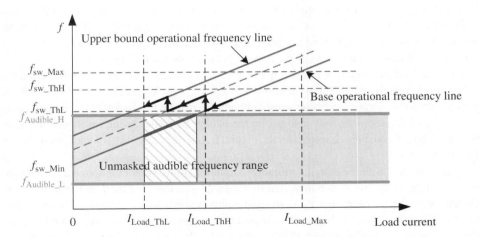

Figure 9.23 Switching frequency adjusted by a closed-loop control strategy

It should be noted that under very light loads and no loads, an increased switching frequency will result in a significant increase in switching losses. Consequently, in these situations, the switching frequency for noise reduction should just decrease to the value at which the power and efficiency requirements are still met. Thus, mechanical methods should be applied to eliminate the audible noise under very light loads as well as under no-load operation.

9.2.4 Battery System Noise, Vibration, and Control

Unlike other subsystems in an electric vehicle, the battery system interacts with, for example, chemical, electrical/electronic, mechanical, and thermal elements. This unique characteristic of a battery provides several technical challenges when trying to integrate a reliable battery system in order to meet increased stringent performance requirements of long-range electric vehicles.

Without an internal combustion engine and complicated mechanical transmission, battery-powered electric vehicles are quieter and more comfortable to drive, so their noise and vibration problems may not be taken as seriously during vehicle construction. In actuality, noise and vibration are important aspects of hybrid powertrain development and the vehicle design and construction process. Every new hybrid vehicle system architecture as well as applied new technology generally brings unique noise and vibration challenges; this is because the battery system for long-electric-range vehicles is normally packed with rigid metal trays. Such a heavy, unsprung battery integrated into the vehicle not only significantly changes the vehicle's dynamic and crash behaviors, but is also a main source of noise and vibration. In order to reduce unacceptably high stresses, carrying out detailed analysis and calculations is an essential task of vehicle engineering, and the existing chassis frame is often modified and sometimes a completely new design is required.

From the overview of vehicle vibrations, all low-frequency vibrations from the chassis/body, mid-frequency vibrations from the driveline, and high-frequency vibrations from the high-voltage bus could simultaneously act on the battery system, both from outside and within the battery pack. Furthermore, the situation will be worse if the external vibrations propagate to the inside of the battery pack and work together with internal electromagnetic vibration to excite resonance with the battery pack's own natural frequency.

Figure 9.24 shows that low-frequency flexural vibrations from the chassis/body act on the battery pack, which may result in deflection of the battery pack, and may excite resonance if the chassis/body vibrational frequencies approach the natural frequency of the battery pack.

The driveline vibrations may also act on the battery pack and excite resonance if one of the vibrational frequencies approaches the natural frequency of the battery pack, as shown in Fig. 9.25.

In addition to the above external vibration excitation, unlike other components or subsystems, the battery system is also impacted by internal vibrations. Figure 9.26 shows irregular current on the high-voltage bus of a battery-powered vehicle with a 100-mile EV range during a US06 drive schedule. Figure 9.27 shows two test results of high-voltage ripple during the wide-open-throttle event of a PHEV with a 40-mile EV range. The associated current

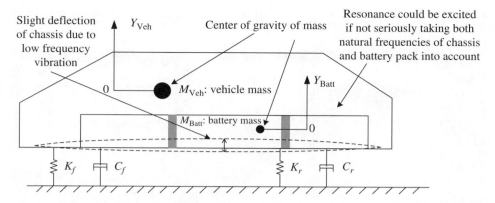

Figure 9.24 Low-frequency vibrations act on a vehicle chassis/body and battery pack

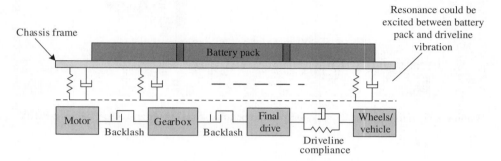

Figure 9.25 Driveline dynamics act on the battery pack

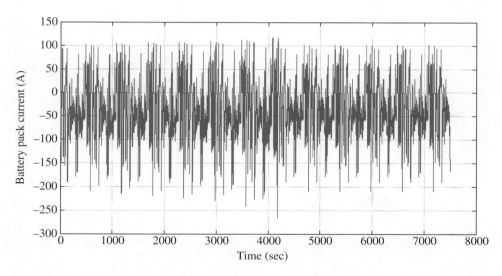

Figure 9.26 High-voltage bus current of a battery-powered electric vehicle during a repeated US06 drive cycle (positive: charge, negative: discharge)

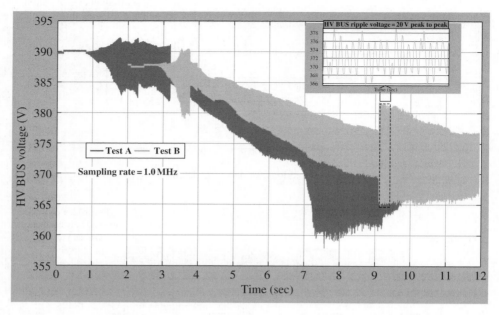

Figure 9.27   High-frequency voltage ripples superimposed on the high-voltage bus of a PHEV

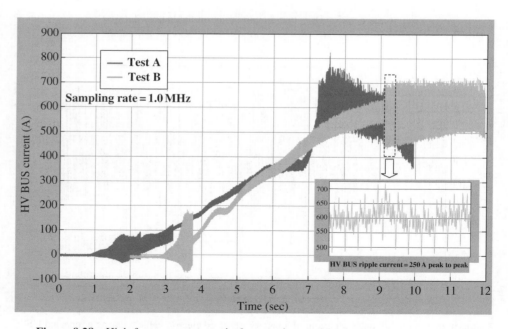

Figure 9.28   High-frequency current ripples superimposed on the DC current of a PHEV

ripple data on the high-voltage bus are shown in Fig. 9.28. The test data show that 250 A peak-to-peak current ripples are superimposed on the 600 A DC current; these are generated by the traction power inverter module (TPIM) due to power electronic switching.

Based on Ampere's force law, two parallel wires carrying electrical current act on each other, and the acting force is proportional to the magnitude of the current. Therefore, the electromagnetic force generated by the charge/discharge current and associated ripples will act on the electrical components such as the battery disconnect unit (BDU), the metal shell, and the cell-to-cell connection nodes; furthermore, it may excite resonance inside the battery pack and result in material fatigue of system components and damage to electrical connections. Since the traction power inverter module (TPIM) is the root cause of the high-voltage bus current-ripple-related noise and vibration, TPIM operational performance needs to be improved if the generated noise and vibration are not acceptable. Generally, increasing the size of the TPIM filter capacitor or increasing the power electronic switching frequency will reduce the magnitude of current/voltage ripples, but these measures often result in a cost increase.

Battery system engineering practice also shows that vibration is a root cause of most coolant leakage, high-voltage isolation loss, electrical parts failure, and connection damage. It is a priority agenda item to enhance battery system vibration analysis and testing in order to improve battery system reliability and durability, lower warranty costs, and extend system service life. In addition, since carrying out vibration analysis is complicated and very challenging in terms of generating accurate analysis results, one urgent task is to build up a test facility in cooperation with chassis experts, where all vibrational modes can be sufficiently excited under different vehicle operational environments. It is imperative to perform comprehensive tests emulating real scenarios of battery operation in the design and prototype stages of the battery pack.

References

Crocker, M. J. *Handbook of Noise and Vibration Control*, John Wiley & Sons, Inc., New Jersey, 2007.

Huber, L. and Jovanović, M. M. 'Methods of Reducing Audible Noise Caused by Magnetic Components in Variable-Frequency-Controlled Switch-Mode Converters,' *IEEE Transactions on Power Electronics*, **26**(6), 1673–1681, 2011.

Kawabata, N., Komada, M., and Yoshioka, T. 'Noise and Vibration Reduction Technology in the Development of Hybrid Luxury Sedan with Series/Parallel Hybrid System,' SAE technical paper, 2007-01-2232, SAE International, Warrendale, PA, 2007.

Norton, M. P. and Karczub, D. *Fundamentals of Noise and Vibration Analysis for Engineers*, 2nd edition, Cambridge University Press, Melbourne, Australia, 2003.

Yoshioka, T. and Sugita, H. 'Noise and Vibration Reduction Technology in Hybrid Vehicle Development,' SAE technical paper, 2001-01-1415, SAE International, Warrendale, PA, 2001.

10

Hybrid Electric Vehicle Design and Performance Analysis

Modeling and simulation have widely been used in support of hybrid electric vehicle design and performance analysis. Such simulation systems can be used to undertake various studies and assessments under given driving conditions. This chapter briefly introduces a hybrid vehicle simulation system, typical driving cycles, and the principles of sizing the system's main components based on the individual models presented in Chapter 3.

10.1   Hybrid Electric Vehicle Simulation System

A hybrid vehicle simulation system can be simple or complicated depending on the requirements and objectives, but the basic simulation system should consist of an engine, a generator/motor set, a power split device, a transmission, a DC–DC converter, a gearbox, a final drive, wheels/tires, a vehicle, and a driver. An example of a hybrid vehicle simulation system structure is shown in Fig. 10.1.

In this simulation system, the driver model outputs the vehicle's power demand to the energy management system based on the difference between the desired vehicle speed and the actual vehicle speed. The energy management system determines the required power from the engine and the motor/generator. The power split device implements the power distribution for the engine and the motor/generator and sets the corresponding speeds. The transmission model calculates the torque and speed at the output shaft. The vehicle body model, the wheel/tire model, the final drive model, and the gear reduction model are used to calculate the required torque and speed of each component to follow the given driving schedule. The energy storage system provides the battery's SOC and terminal voltage to the DC–DC converter, the motor/generator, and the energy management system. The

Hybrid Electric Vehicle System Modeling and Control, Second Edition. Wei Liu.
© 2017 John Wiley & Sons Ltd. Published 2017 by John Wiley & Sons Ltd.

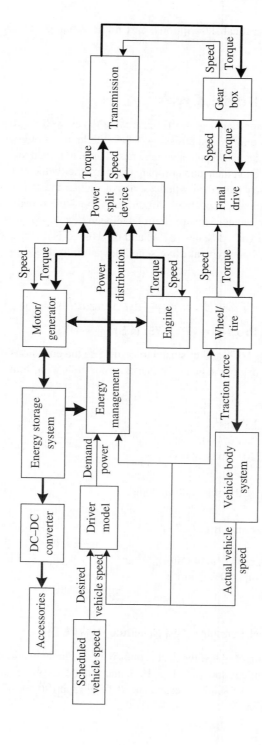

Figure 10.1 Basic hybrid electric vehicle simulation system

detailed function, mathematical descriptions, and implementation of each model can be found in Chapter 3.

10.2   Typical Test Driving Cycles

Since environmental factors such as traffic and weather as well as road conditions can strongly affect the fuel economy and emissions of a vehicle, it is not good practice to evaluate the fuel economy and emissions of a vehicle based on actual road tests. In order to judge the performances of a given vehicle and compare them to other vehicles fairly, governments and the automobile industry have developed a series of standard tests whereby a vehicle can be measured under completely repeatable conditions. The drive profiles of these tests are called drive schedules or cycles. Most of them have been developed by the Environmental Protection Agency of the United States (EPA) and the United Nations Economic Commission (UNECE or ECE). Some typical drive schedules are introduced below.

10.2.1   Typical EPA Fuel Economy Test Schedules

10.2.1.1   Federal Test Procedure (FTP-75) (US)

The FTP is shown in Fig. 10.2, and it is used to certificate the emissions of light-duty vehicles. The entire FTP-75 driving schedule consists of the following four segments:

- Cold start phase (0–505 s);
- Transient phase (506–1372 s);
- Hot soak (min 540 s, max 660 s);
- Hot start phase (min 540 s, max 660 s).

The basic parameters of the schedule include:

- Distance traveled: 11.04 miles (17.77 km);
- Duration: 1874 s;
- Average speed: 21.21 mph (34.14 km/h);
- Maximal speed: 56.7 mph (91.25 km/h).

10.2.1.2   Highway Fuel Economy Test Schedule (HWFET)

The HWFET was developed to test the fuel economy of a vehicle in a highway driving scenario. The driving schedule is shown in Fig. 10.3, and it represents a 10.26-mile (16.51-km) route with 48.27 mph (77.68 km/h) average speed, 59.9 mph (96.4 km/h) maximal speed, and a 765-second duration.

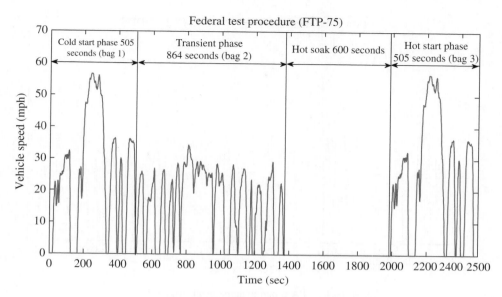

Figure 10.2   FTP-75 test schedule

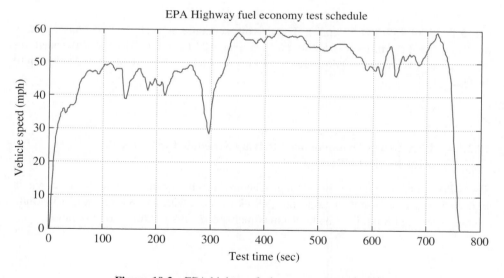

Figure 10.3   EPA highway fuel economy test schedule

10.2.1.3   EPA Urban Dynamometer Driving Schedule (UDDS)

The EPA urban dynamometer driving schedule (UDDS) is also called the FTP-72 cycle and is shown in Fig. 10.4. The driving schedule simulates an urban route of 7.45 miles (12 km) with frequent stops. The maximal speed is 56.7 mph (91.25 km/h) and the average

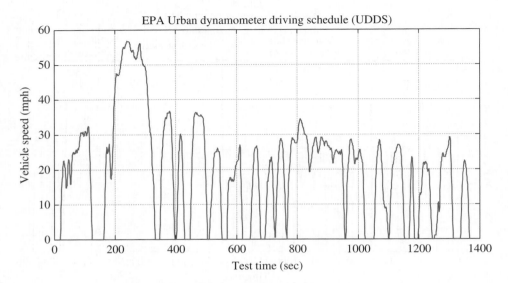

Figure 10.4 EPA urban dynamometer drive schedule

speed is 19.59 mph (31.53 km/h). The driving schedule consists of two phases: phase one runs for 3.59 miles (5.78 km) at an average 25.65 mph (41.28 km/h) speed for 505 s; phase two runs for 3.86 miles (6.21 km) for 864 s with a 16.06 mph (25.85 km/h) average speed.

10.2.1.4 EPA Urban Dynamometer Driving Schedule for Heavy-duty Vehicles (HD-UDDS)

The EPA urban dynamometer driving schedule for heavy-duty vehicles is shown in Fig. 10.5, and it represents a 5.55-mile (8.94-km) route with an 18.86 mph (30.35 km/h) average speed, a 58 mph (93.34 km/h) maximal speed, and a 1060-second duration.

10.2.1.5 New York City Driving Cycle (NYCC)

The NYCC was developed to simulate the driving scenario in a metro area. The entire driving schedule is shown in Fig. 10.6, and it has a 1.18-mile (1.9-km) route with a 7.1 mph (11.43 km/h) average speed, a 27.7 mph (44.58 km/h) maximal speed, and a 598-second duration. A summary of the parameters of FTP, HWFET, UDDS, HD-UDDS, and NYCC driving schedules is given in Table 10.1.

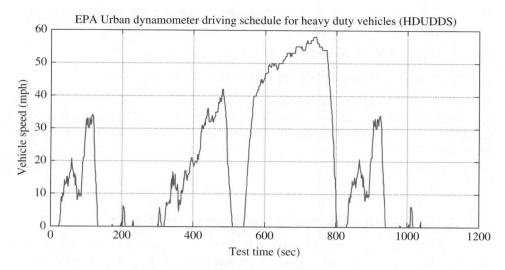

Figure 10.5 EPA HD-UDDS drive schedule for heavy-duty vehicles

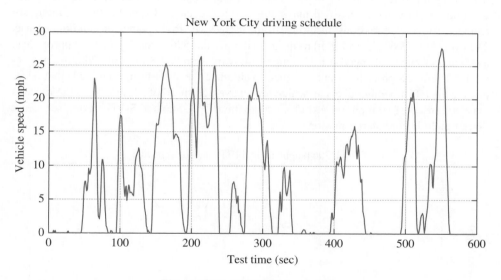

Figure 10.6 NYCC drive schedule

Table 10.1 Summary of the UDDS, HD-UDDS, FTP-75, HWFET, and NYCC driving schedules

Characteristics	Unit	FTP-75	HWFET	UDDS	HD-UDDS	NYCC
Distance	miles (km)	11.04 (17.77)	10.26 (16.51)	7.45 (12)	5.55 (8.94)	1.18 (1.9)
Duration	seconds	1874	765	1369	1060	598
Average speed	mph (km/h)	56.7 (91.25)	48.27 (77.68)	19.59 (31.53)	58 (93.34)	7.1 (11.43)
Maximal speed	mph (km/h)	21.21 (34.14)	59.9 (96.4)	56.7 (91.25)	18.86 (30.35)	27.7 (44.58)

10.2.2   Typical Supplemental Fuel Economy Test Schedules

10.2.2.1   US06 Supplemental Federal Test Procedure (SFTP-US06)

The SFTP-US06 driving schedule was developed to represent aggressive, high-speed, and high-acceleration driving behaviors and is supplemental to the FTP test schedule. The schedule is shown in Fig. 10.7, and it represents an 8.01-mile (12.89-km) route with a 48.05 mph (77.33 km/h) average speed, an 80.3 mph (129.2 km/h) maximal speed, and a 600-second duration.

10.2.2.2   SC03 Supplemental Federal Test Procedure (SFTP-SC03)

The SFTP-SC03 driving cycle, shown in Fig. 10.8, is a chassis dynamometer test performed with the vehicle's A/C unit turned on, at a lab temperature of 95 °F (35 °C). The SFTP-SC03 was developed to quantify and control vehicle emissions not accounted for by the FTP, so it is designed to represent the engine load and emissions associated with the use of air-conditioning in vehicles certified under the FTP test cycle and specifically to capture so-called 'off-cycle' emissions resulting from aggressive driving and air-conditioner use. The cycle has a 3.6-mile (5.8-km) route and a duration of 596 seconds, and has approximate average and maximal speeds of 21.6 mph (34.8 km/h) and 54.8 mph (88.2 km/h), respectively. These test procedures are designed to simulate the short-term high loads that occur outside the FTP operational cycle, which has an average speed of 21 mph (33.8 km/h) and a maximal speed of 59 mph (94.9 km/h). Unlike the FTP, US06 and SC03 driving schedules do not include cold-start emissions.

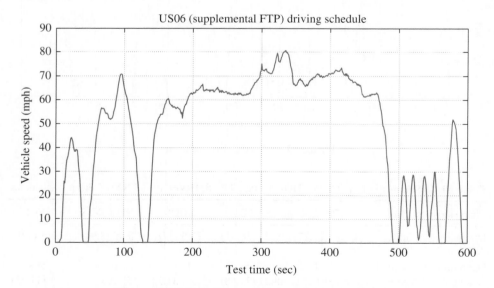

Figure 10.7   SFTP-US06 supplemental drive schedule

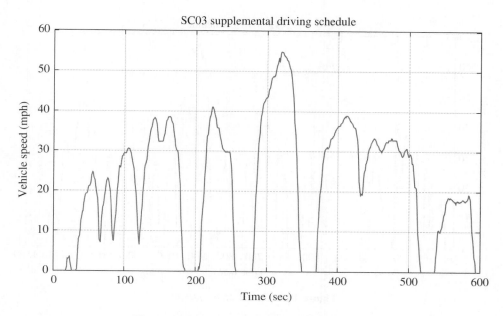

Figure 10.8   SC03 supplemental drive schedule

The SFTP-SC03 cycle currently applies to light-duty vehicles through 8500 pounds (lb) gross vehicle weight rating (GVWR) and manufacturers are required to certify applicable vehicles to 4000-mile SFTP-SC03 exhaust emission standards. New SC03 emission standards began to be implemented at the start of the 2015 model year for passenger cars, light-duty trucks, and medium-duty passenger vehicles, with a phase-in through to the 2025 model year. Emission standards for medium-duty vehicles from 8501 through 14,000 lb GVWR will be phased in from 2016 through to the 2025 model year.

10.2.2.3   California Unified Cycle Driving Schedule (UC, LA92)

The California unified cycle is a dynamometer driving schedule for light-duty vehicles developed by the California Air Resources Board. The cycle has also been referred to as the LA92 cycle and is often called 'Unified LA92'. Compared with the FTP, the LA92 schedule, shown in Fig. 10.9, is a more aggressive driving cycle with a higher speed, higher acceleration, fewer stops per mile, and less idle time. The cycle consists of four segments: cold start (bag 1), stabilized or running portion (bag 2), hot soak, and hot start (bag 3). The UC test is performed through the following procedures:

1. Run bags 1 and 2 consecutively.
2. Have a ten-minute hot soak.
3. Run bag 3 by duplicating bag 1.

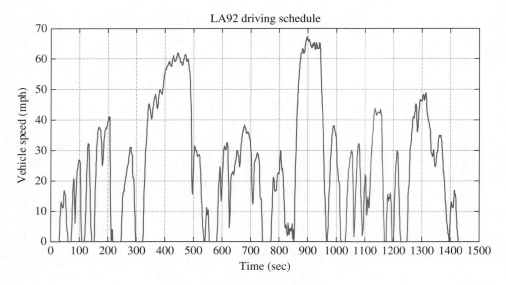

Figure 10.9 LA92 drive schedule

Table 10.2 Summary of US06, SC03, and LA92 driving schedules

Characteristics	Unit	US06	SC03	LA92
Distance	miles (km)	8.0 (12.9)	3.6 (5.8)	9.8 (15.7)
Duration	seconds	600	596	1435
Average speed	mph (km/h)	48.1 (77.3)	21.6 (34.8)	24.6 (39.6)
Maximal speed	mph (km/h)	80.3 (129.2)	54.8 (88.2)	67.2 (108.1)

The emissions over the cycle are calculated by using the same weighted formula as the FTP calculation and by taking actual mileage from the UC into account. Some UC characteristic parameters and emission test segments are as follows. A summary of the parameters of US06, SC03, and LA92 driving schedules is given in Table 10.2.

- Unified cycle:
  Duration: 1435 seconds
  Total distance: 9.8 miles (15.7 km)
  Average speed: 24.6 mph (39.6 km/h)
  Maximal speed: 67.2 mph (108.1 km/h)
- Bag 1:
  Duration: 300 seconds
  Total distance: 1.2 miles (1.9 km)
- Bag 2:
  Duration: 1135 seconds
  Total distance: 8.6 miles (13.8 km)

- Hot soak:
  Duration: 600 seconds
- Bag 3:
  Duplicate of Bag 1

10.2.3 Other Typical Test Schedules

10.2.3.1 The United Nations Economic Commission for Europe Elementary Urban Driving Schedule (UN/ECE) and Extra Urban Driving Schedule (EUDC)

Four repeats of the ECE + EUDC driving schedule is called the New European Driving Cycle (NEDC), which is performed on a chassis dynamometer for emissions certification of light-duty vehicles in Europe. The ECE segment is shown in Fig. 10.10 and the EUDC segment is shown in Fig. 10.11. The EUDC segment is a very aggressive, high-speed driving schedule. The EUDC involves a constant speed of 120 km/h with 10-second duration. The vehicle is allowed to soak for more than six hours at the test temperature before performing the test. It is also allowed to idle for 40 s before the test. This test procedure is often called a type I test. The warmed-up idle tailpipe CO test, called a type II test, is conducted right after the type I test. The crankcase emission determination test is performed by a two-mode (idle and 50 km/h) chassis dynamometer test, and this is called a type III test. For some vehicles that do not have enough power to follow a type I test, an alternative EUDC with a 90 km/h maximal speed, shown in Fig. 10.12, can be used to perform the same tests. The entire NEDC test schedule includes four repeated ECE segments without interruption plus

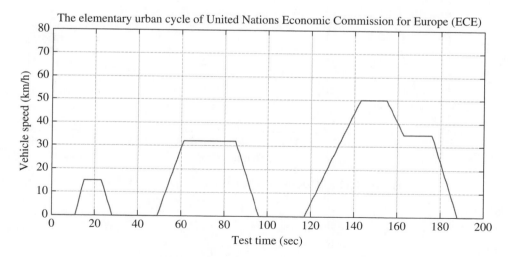

Figure 10.10 ECE test drive schedule

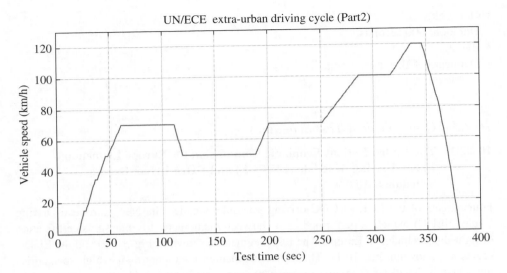

Figure 10.11   EUDC drive schedule

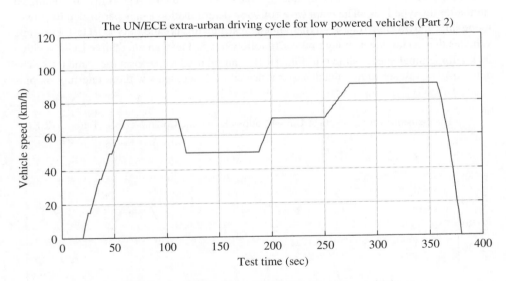

Figure 10.12   EUDC drive schedule for low-power vehicles

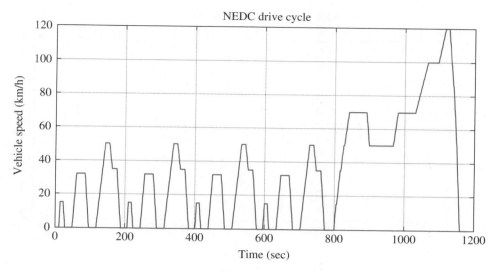

Figure 10.13 NEDC drive schedule

Table 10.3 Summary of the ECE, EUDC, and EUDC-LP driving schedules

Characteristics	Unit	ECE	EUDC	NEDC	EUDC-LP
Distance	miles (km)	0.618 (0.994)	4.32 (6.955)	6.79 (10.93)	4.11 (6.61)
Duration	seconds	195	400	1180	400
Average speed (include stops)	mph (km/h)	11.40 (18.35)	38.89 (62.59)	20.73 (33.35)	36.96 (59.48)
Maximal speed	mph (km/h)	31.07 (50)	74.56 (120)	74.56 (120)	55.92 (90)

an EUDC segment, which is shown in Fig. 10.13. The parameters of the ECE and EUDC driving schedules are given in Table 10.3.

10.2.3.2 Worldwide Harmonized Light Vehicles Test Procedure

The worldwide harmonized light vehicles test procedure (WLTP) has been developed by a UN ECE working group. As for the NEDC, the WLTP has also been designed as a chassis dynamometer test schedule for the determination of emissions and fuel consumption for light-duty vehicles and is expected to replace the NEDC procedure introduced in the previous section. The WLTP procedure is further categorized into three classes corresponding to different power (kW) to curb mass (ton) ratios (PMRs), and each WLTP test includes three test schedules for the following different classes of vehicles:

- Class 1: For low-power vehicles with PMR $\leq$ 22;
- Class 2: For vehicles with 22 < PMR $\leq$ 34;
- Class 3: For high-power vehicles with PMR > 34.

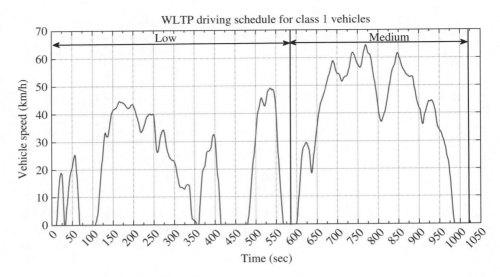

Figure 10.14   WLTP drive schedule for Class 1 vehicles

The curb weight of a vehicle is the total weight of the vehicle with standard equipment, all necessary operating fuel and fluids such as engine oil, transmission oil, coolant, refrigerants, and a full tank of fuel, but not loaded with passengers or cargo. In general, most passenger cars nowadays have around 40–70 kW/ton, so they belong to Class 3, while most vans and buses belong to Class 2.

In order to represent real-world vehicle operation on urban and extra-urban roads and highways, the WLTP requires several driving tests to be performed. The WLTP driving schedule for a Class 1 vehicle, shown in Fig. 10.14, has low- and medium-speed parts and is performed in the sequence low–medium–low. If the maximal speed of the vehicle is less than 70 km/h, the WLTP allows the medium-speed part of the test to be replaced by the low-speed part. The WLTP driving schedule for a Class 2 vehicle, shown in Fig. 10.15, is made up of low-, medium-, and high-speed parts and is performed in the sequence low–medium–high. If the maximal speed of the vehicle is less than 90 km/h, the high-speed part is replaced by the low-speed part. The WLTP driving schedule for a Class 3 vehicle, shown in Fig. 10.16, consists of low-, medium-, high-, and extra-high-speed parts and is performed in the sequence low–medium–high–extra high. If the maximal speed of the vehicle is less than 135 km/h, the extra-high-speed part is replaced by the low-speed part. A summary of WLTP driving schedules is given in Table 10.4.

10.2.3.3   Single Cycle Range and Energy Consumption Test for Battery-powered Vehicles (SCT)

In order to determine either the total city or the total highway range and the associated energy consumption for a battery-powered vehicle (BEV), the SAE proposed a single cycle

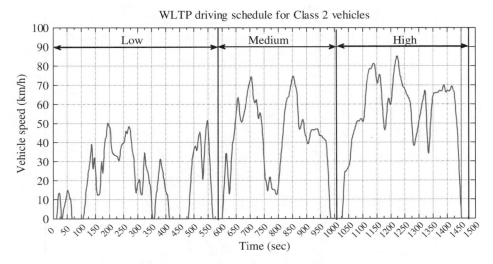

Figure 10.15 WLTP drive schedule for Class 2 vehicles

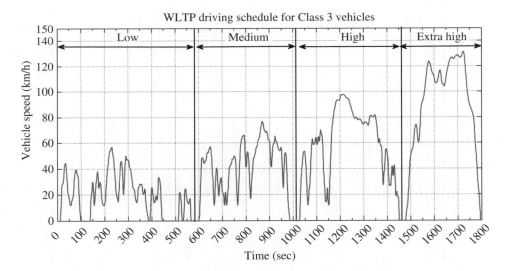

Figure 10.16 WLTP drive schedule for Class 3 vehicles

range and energy consumption test cycle (SCT) operated on a chassis dynamometer over repeated UDDS or HWFET driving cycles, respectively. Therefore, the SCT is a full depletion test where the vehicle is driven until the useable energy content of the vehicle's battery is completely exhausted. The SCT is suitable for battery-powered vehicles with a range capability of less than 60 miles (97 km). The SCT actually requires that the vehicle is

Table 10.4 Summary of WLTP driving schedules

Characteristics	Unit	Class 1 vehicle		Class 2 vehicle			Class 3 vehicle			
		low	medium	low	medium	high	low	medium	high	extra high
Distance	miles	2.066	2.963	1.946	2.929	4.239	1.924	2.956	4.449	5.13
	(km)	(3.324)	(4.767)	(3.132)	(4.712)	(6.82)	(3.095)	(4.756)	(7.158)	(8.254)
Total distance	miles	5.029 (8.091)		9.114 (14.664)			14.457 (23.262)			
	(km)									
Duration	seconds	589	433	589	433	455	589	433	455	323
Total duration	seconds	1022		1477			1800			
Stop duration	seconds	155	48	155	48	30	156	48	31	7
Total stop duration	seconds	203		233			242			
Stop percentage	%	26.3	11.1	26.3	11.1	6.6	26.5	11.1	6.8	2.2
Total stop percentage	seconds	19.9		15.8			13.4			
Average speed (exclude stops)	mph	17.154	27.719	16.159	27.408	35.923	15.973	27.657	37.787	58.421
	(km/h)	(27.6)	(44.6)	(26.0)	(44.1)	(57.8)	(25.7)	(44.5)	(60.8)	(94.0)
Total average speed (exclude stops)	mph	22.126 (35.6)		26.352 (42.4)			33.437 (53.8)			
	(km/h)									
Average speed (include stops)	mph	12.617	24.612	11.871	24.363	33.561	11.746	24.549	35.177	57.178
	(km/h)	(20.3)	(39.6)	(19.1)	(39.2)	(54.0)	(18.9)	(39.5)	(56.6)	(92.0)
Total average speed (include stops)	mph	17.713 (28.5)		22.188 (35.7)			28.9 (46.5)			
	(km/h)									
Maximal speed	mph	30.515	40.025	31.945	46.426	52.952	35.115	47.607	60.534	81.603
	(km/h)	(49.1)	(64.4)	(51.4)	(74.7)	(85.2)	(56.5)	(76.6)	(97.4)	(131.3)
Maximal acceleration	m/s²	0.8	0.6	0.9	1.0	0.8	1.5	1.6	1.6	1.0
Maximal deceleration	m/s²	1.0	0.6	1.1	1.0	1.1	1.5	1.5	1.5	1.2

(a)

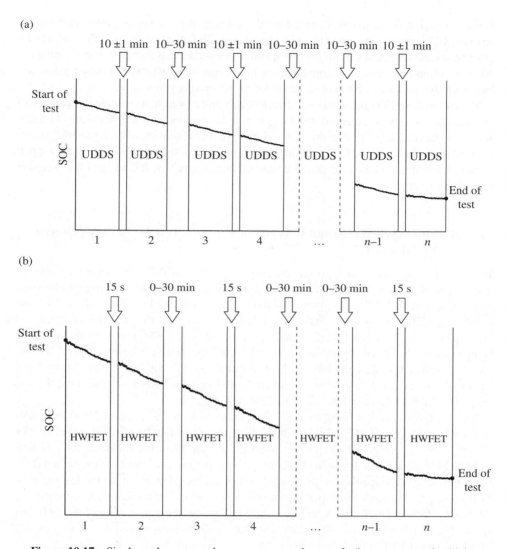

Figure 10.17 Single cycle range and energy consumption test for battery-powered vehicles

repeatedly operated over one of the following pairs of drive cycles until the end-of-test criterion has been satisfied. The entire test procedure is shown in Fig. 10.17(a) and (b):

- City test: 2 × UDDS;
- Highway test: 2 × HWFET

Based on the test procedure, the vehicle needs to be soaked for 10 to 30 minutes for the city test and for 0 to 30 minutes for the highway test between each cycle pair. Furthermore, during all soak periods, the SCT procedure also requires that the key or power switch must be in the OFF position, the hood must be closed, test cell fans must be OFF, the brake pedal must be released, the test site ambient temperature should be maintained within the range of 20 to 30 °C, and the total discharged ampere-hour capacity (Ahr) must be measured during the entire dynamometer test procedure including all driving phases and soaks.

The end-of-test (EOT) criterion is defined as the point at which the vehicle is incapable of maintaining the scheduled speed due to its power limitations with a certain level of tolerance. For the detailed EOT criterion defined by SAE, the reader should refer to SAE standard J1634 (2012). Once the EOT criterion has been satisfied, the vehicle needs to be stopped within 15 seconds, and the total distance traveled once the vehicle has reached zero speed is the vehicle's range.

10.2.3.4 Multi-cycle Range and Energy Consumption Test for Battery-powered Vehicles (MCT)

With advances in battery technology, the range capability of BEVs has increased dramatically. It would be very expensive and time-consuming to determine the range and energy consumption of such BEVs using SCT methodology. For example, a BEV with a 150-mile unadjusted UDDS range would take about 18.5 hours of total dynamometer test time to accomplish an SCT test. As described in the previous section, the SCT requires that a vehicle is repeatedly driven over the same drive cycles until the vehicle's battery energy is completely exhausted. In order to effectively determine both the city and highway range of a BEV with a range capability greater than 60 miles (97 km), the SAE developed a multi-cycle range and energy consumption test procedure (MCT).

The MCT consists of a fixed number of dynamic drive cycles combined with constant-speed driving cycles (CSCs). The detailed MCT, shown in Fig 10.18, is made up of four UDDS cycles, two HWFET cycles, and two CSC cycles. The dynamic drive cycle UDDS and HWFET are used to determine the energy consumption associated with specified driving patterns. The constant-speed driving schedules are intended to reduce the test duration by depleting the battery more rapidly, to minimize the impact of driving style variations on the energy determination, and to enhance the consistency of the triggering of the EOT criterion. The entire test consists of the following four distinct segments and the test sequence is shown in Fig. 10.19.

- Initial UDDS–HWFET–UDDS sequence (S_1);
- Mid-test CSC depletion phase (CSC_M);
- Second UDDS–HWFET–UDDS sequence (S_2);
- End-of-test CSC (CSC_E).

In the MCT, the UDDS and HWFET cycles make up 50.52 miles (81.31 km) of the driving distance, and the remaining range capability of the vehicle is used up in the CSC phases, which deplete battery energy more rapidly than the UDDS and HWFET cycles. The CSC_M phase is

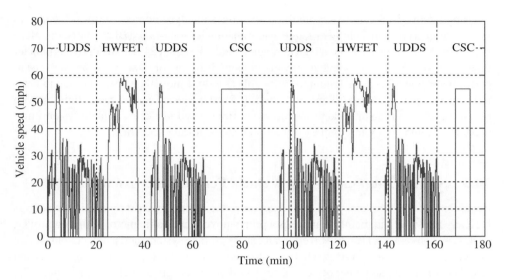

Figure 10.18   Multi-cycle range test schedule

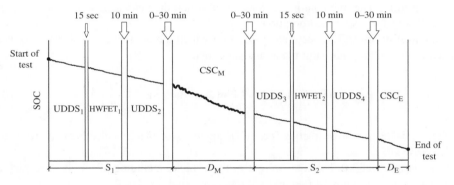

Figure 10.19   Multi-cycle range test sequence

placed between S_1 and S_2, and the CSC_M distance (D_M) is required to ensure that S_2 is conducted at a 'substantially' lower SOC condition than S_1. Therefore, depending on the range capability of the test vehicle, good engineering judgment should be applied in order to select an appropriate CSC_M distance; and it should be noted that the CSC_M for vehicles with a relatively short range capability may be omitted, as the required UDDS and HWFET cycles and recommended maximal CSC_E distance can cover the range capability of such vehicles. In general, a D_M should be selected such that it will result in a CSC_E distance (D_E) that is not more than 20% of the total driven distance of the MCT, and the SAE recommends a method for determining the length of D_M both prior to and during a test (SAE standard J1634, 2012).

Using the MCT procedure can significantly reduce the dynamometer test time needed to produce both a city and a highway range determination for long-range BEVs. For example, a BEV with a 200-mile unadjusted UDDS range would take about 24½ hours of dynamometer time to perform SCT tests, while it only takes about 5½ hours of total dynamometer time to perform a single MCT. In addition to being able to reduce more than 75% of dynamometer test time compared to the SCT method, the MCT test sequence can be easily built up with new test cycles and can provide further opportunities for hybrid electric vehicle system engineers to accomplish more complex or comprehensive testing scenarios.

10.3 Sizing Components and Vehicle Performance Analysis

Regardless of the type of hybrid vehicle system, the hybrid electric propulsion system is basically the same and mainly comprises a prime mover, an electric motor with a controller, a DC–DC converter, a DC–AC inverter, an energy storage system, and a power split device as well as a transmission system. This section addresses how to size system components and calculate drivability and other performances based on the system model and simulation. The used terms and expressions are:

- **Gradeability:** The maximal percentage grade which the vehicle can go over at the specified speed for a specified duration.
- **Gradeability limit:** The percentage grade on which the vehicle can just move forward, determined by the following mathematical equation:

$$\text{Percentage Gradeability Limit} = 100 \tan\left(\sin^{-1}\left(\frac{F}{m_v g}\right)\right) \tag{10.1}$$

where: F is the measured traction force (N), and m_v is the manufacturer's rated gross vehicle mass (kg).

- **Base vehicle weight:** The total weight of the vehicle, including all fluids necessary for normal operation, without fuel and without a payload.
- **Base vehicle mass:** The base vehicle weight divided by the gravitational constant.
- **Curb weight:** The base vehicle weight plus a full tank of fuel.
- **Payload:** The weight of the driver, passengers, and cargo under the given load condition.
- **Load condition:** A description of the loaded state of the vehicle being analyzed or tested. It includes descriptions of fuel load and the payload, and the location(s) of the payload.
- **Vehicle operating weight:** The total weight of the vehicle under a given load condition, including the base vehicle weight, fuel load weight, and the payload weight.
- **Cut-off terminal voltage:** The manufacturer-recommended minimal voltage under the load condition below which battery system damage could occur.
- **Rolling resistance:** The lumped contact force between tire and road, which is tangential to the test surface and parallel to the tire plane.
- **Rolling resistance coefficient:** The ratio of the rolling resistance to the load on the tire.

- **Loaded radius:** The perpendicular distance from the axis of rotation of the loaded tire to the surface on which it is rolling.
- **Driveshaft ratio:** The ratio between a vehicle's driveshaft, also called the propeller shaft, and its wheel axle, for example, a 4:1 driveshaft ratio means that the driveshaft turns four times for every one time the tires spin.

10.3.1 Drivability Calculation

The drivability performances of a hybrid electric vehicle generally include maximal speed, acceleration, and gradeability. Given the vehicle system components, it is necessary for a vehicle system engineer to be able to calculate the vehicle's maximal speed, acceleration, and gradeability under different driving conditions.

10.3.1.1 Maximal Speed

Based on the given system components, the theoretical maximal speed of the vehicle can be determined from the following equation:

$$\text{Veh\_spd\_max} = \max\{\text{spd\_eng\_max} \cdot g_e, \ \text{spd\_mot\_max} \cdot g_m\} \cdot g_r \cdot f_d \cdot r_{wh} \cdot 3.6 \quad (10.2)$$

where Veh_spd_max is the maximal speed of the vehicle in km/h, spd_eng_max is the maximal speed of the engine in rad/s, g_e is the gear reduction ratio from the engine shaft to the hybrid shaft, spd_mot_max is the maximal speed of the motor in rad/s, g_m is the gear reduction ratio from the motor shaft to the hybrid shaft, g_r is the gear reduction ratio from the hybrid shaft to the final drive, f_d is the final drive ratio, and r_{wh} is the radius of the wheel (m).

10.3.1.2 Acceleration

Acceleration is one of the most important performance merits. The SAE provides a standardized means of repeatedly measuring launch response and maximal acceleration performance of passenger cars and light-duty trucks (SAE standard J1491, 1985, 2006). Since road and environmental factors such as the road condition, wind conditions, and the altitude at which the vehicle operates affect the vehicle's acceleration performance, it is necessary to take such factors into account when performing the maximal acceleration calculation for a given hybrid vehicle, and the SAE standardizes the road and environmental factors (SAE standard J2188, 1996, 2012).

For a given vehicle, the acceleration time, t_{acc} (s), of the vehicle from an initial speed, v_i, to the final speed, v_f, is calculated based on the following equation:

$$v_f = v_i + \int_0^{t_{acc}} a(t)dt = v_i + \frac{3.6}{m_v} \int_0^{t_{acc}} \left(F_{traction} - \left(F_{rolling} + F_{aero} + F_{grade}\right)\right)dt = \text{required speed}$$

$$(10.3)$$

where v_i is the initial speed, v_f is the target speed (km/h) for the acceleration calculation, a is the vehicle's acceleration (m/s^2), $F_{traction}$ is the traction force on the wheel from the drivetrain and is calculated by:

$$F_{traction} = r_{wh} \cdot \left(\eta_{pt\_eng} \cdot \tau_{eng}(eng\_spd) + \eta_{pt\_mot} \cdot \tau_{mot}(mot\_spd, SOC) \right) \tag{10.4}$$

where r_{wh} is the effective wheel rolling radius (m), $\tau_{eng}(eng\_spd)$ is the traction torque (Nm) on the wheel from the engine, which is a function of engine speed, $\tau_{mot}(mot\_spd, SOC)$ is the traction torque (Nm) on the wheel from the electric motor, which is a function of motor speed and the state of charge of the battery, η_{pt_eng} is the engine power drivetrain efficiency, and η_{pt_mot} is the electric motor power drivetrain efficiency. The details of Eq. 10.4 are given in Chapter 3.

In Eq. 10.3, m_v is the vehicle mass (kg), $F_{rolling}$, F_{aero}, F_{grade} are the rolling resistance, aero drag resistance, and grade weight forces, which are calculated using the following equations:

$$F_{rolling} \approx k_{rrc} k_{sc} m_v g \tag{10.5}$$

$$F_{aero} = k_{aero} v^2$$
$$k_{aero} = \frac{1}{2} C_a \cdot A_d \cdot C_d \cdot F_a \tag{10.6}$$

$$F_{grade} = m_v g \sin(\alpha) \tag{10.7}$$

where k_{rrc} is the rolling resistance coefficient, k_{sc} is the road surface coefficient, k_{aero} is the aero drag factor, C_a is the air density correction coefficient for the altitude, A_d is the air mass density (kg/m^3), C_d is the vehicle's aerodynamic drag coefficient (N$\cdot$s^2/kg$\cdot$m), F_a is the vehicle's frontal area (m^2), and α is the road incline angle (rad).

In addition, if a vehicle has a stepped gearbox, the gear ratio should be taken into account and the relationship between the engine speed (rpm) and the maximal vehicle speed for each gear ratio can be illustrated as shown in Fig. 10.20.

10.3.1.3   Gradeability Calculation

The percentage gradeability versus vehicle speed is defined as:

$$Pct = 100 \cdot \tan \left(\sin^{-1} \left(\frac{P_{fd} \cdot \eta_{fd} - (P_R + P_A)}{m_v g \cdot V_{spd}} \right) \right) \tag{10.8}$$

where Pct is the calculated gradient (%) (vertical climb/travel distance horizontal projection), P_R is the power required to overcome the rolling resistance (W), P_A is the power required to overcome air drag (W), P_{fd} is the power on the final driveshaft at the given speed (W), η_{fd} is the final driveshaft efficiency, m_v is the gross mass (kg) of the vehicle, and V_{spd} is the operational speed (m/s) of the vehicle. P_R and P_A are calculated by the following equations:

$$P_R = F_{rolling} \cdot V_{spd} \approx k_{rrc} k_{sc} m_v g V_{spd} \tag{10.9}$$

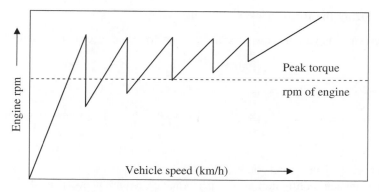

Figure 10.20   Engine speed vs. vehicle speed

$$P_A = F_{aero} \cdot V_{spd} = \left(k_{aero} V_{spd}^2 \right) \cdot V_{spd} = k_{aero} V_{spd}^3 \tag{10.10}$$

where k_{rrc} is the rolling resistance coefficient, k_{sc} is the road surface coefficient, and k_{aero} is the aero drag factor for the given vehicle.

If the vehicle is a series hybrid, P_{fd} will be $\eta_{drive\_train} P_{mot}$; that is, $P_{fd} = \eta_{drive\_train} P_{mot}$ where η_{drive_train} is the lumped efficiency from the motor to the final drive through the gearbox and P_{mot} is the mechanical power on the motor shaft.

The theoretical maximal gradient on which the vehicle can operate at the given speed can be calculated by the following equation:

$$Pct_{max} = 100 \cdot \tan \left(\sin^{-1} \left(\frac{P_{fd}^* \cdot \eta_{fd} - (P_R + P_A)}{m_v g \cdot V_{spd}} \right) \right) \tag{10.11}$$

where P_{fd}^* is the maximal achievable power on the driveshaft at the given vehicle speed. If the vehicle is a series hybrid electric vehicle, $P_{fd}^* = \eta_{drive\_train} \cdot P_{mot}^*$, P_{mot}^* is the maximal mechanical power on the motor shaft at the given speed.

10.3.2   *Preliminary Sizing of the Main Components of a Hybrid Electric Vehicle*

Since most hybrid vehicles are modified versions of well-designed conventional vehicles, this section addresses the procedure of preliminary sizing of the main hybrid components such as the prime mover, the electric motor/generator unit, the gear ratio, and the energy storage system. To perform such calculations, it is necessary to have vehicle data including the vehicle curb weight, the frontal area, the air drag coefficient, the tire size, the driveshaft ratio, the rolling resistance coefficient, etc. In addition, the vehicle performance parameters, such as the desired acceleration time, the maximal percentage gradeability, the electric mileage, and the tow capability, are needed.

10.3.2.1   Sizing the Prime Mover

To size the prime mover, the power required on the driveshaft needs to be calculated first, as follows:

$$P_{fd} = P_R + P_A + P_{acc} + P_{grade} = F_{rolling} \cdot V_{spd} + F_{aero} \cdot V_{spd} + F_{acc} \cdot V_{spd} + F_{grade} \cdot V_{spd}$$

$$= \left(F_{rolling} + F_{aero} + F_{acc} + F_{grade}\right) \cdot V_{spd} = \left(k_{rrc}k_{sc}m_v g + k_{aero}V_{spd}^2 + m_v a + m_v g \sin(\alpha)\right) \cdot V_{spd}$$

$$(10.12)$$

where a is the vehicle's acceleration (m/s^2).

Once the vehicle configuration, the performance requirements, and other component parameters have been determined, the required prime mover can be sized.

Series Hybrid Vehicle System

The prime mover in a series hybrid electric vehicle is the electric motor. The required motor power is equal to the power required at the driveshaft, as follows:

$$P_{fd} = P_{veh}/\eta_{fd}$$
$$P_{mot} = P_{fd}/\eta_{drive\_train}$$

$$(10.13)$$

where P_{mot} is the required motor power, P_{veh} is the vehicle's required power, P_{fd} is the calculated required power at the driveshaft, η_{drive_train} is the drivetrain efficiency, and η_{fd} is the efficiency of the final drive.

If the wheel, tire, and drivetrain ratios are given, the motor speed versus vehicle speed can be calculated from the following equation:

$$\omega_{mot} = \frac{V_{spd}}{R_{wh} \cdot r_{drive\_train}} \cdot \frac{60}{2\pi}$$

$$(10.14)$$

where ω_{mot} is the motor speed (rpm), V_{spd} is the vehicle speed (m/s), R_{wh} is the radius of the wheel (m), and r_{drive_train} is the drivetrain ratio from the motor shaft to the propulsion wheel.

Once the motor speed has been backward calculated from the vehicle speed, its torque can be calculated as:

$$\tau_{mot} = {}^{P_{mot}}\!\big/\!{}_{\omega_{mot}}$$

$$(10.15)$$

From Eqs 10.13, 10.14, and 10.15, the curves of torque versus speed and/or power versus speed can be obtained. The required characteristics of the motor are then determined based on these curves.

In series hybrid vehicle systems, the engine/generator set (genset) has to be able to generate enough electrical power so that it can power the motor to drive the vehicle and charge the energy storage system simultaneously; therefore, the power needed from the genset can be calculated as:

$$P_{\text{genset}} = \frac{P_{\text{mot}} \cdot \left(1 + k_{\text{chg}}(P_{\text{veh}})\right)}{\eta_{\text{gen}} \cdot \eta_{\text{mot}} \cdot \eta_{\text{genset}}} \tag{10.16}$$

where P_{genset} is the power (W) from the engine at the given vehicle speed, η_{mot} is the efficiency of the motor, η_{gen} is the efficiency of the generator, η_{genset} is the efficiency of the engine/generator set, and $k_{\text{chg}}(P_{\text{veh}})$ is the charging power margin factor under the given operational conditions.

The operational speed of the genset can be determined based on the characteristics of power versus speed and taking the efficiencies of the engine, generator, and gearbox into account. The required characteristics of these components can be further determined optimally in series hybrid vehicle system design.

Parallel Hybrid Vehicle System

Since most parallel hybrid electric vehicle systems only have one motor/generator set, the power required by such a vehicle needs to be split between the motor and the engine; therefore, the combined power from motor and engine has to meet the power requirement of the vehicle; that is:

$$\begin{aligned} P_{\text{fd}} &= P_{\text{veh}}/\eta_{\text{fd}} \\ P_{\text{fd}} &= \left(P_{\text{eng}} + P_{\text{mot}}\right) \cdot \eta_{\text{drive\_train}} \end{aligned} \tag{10.17}$$

$$\begin{aligned} P_{\text{eng\_shaft}} &= k_{\text{split}}(P_{\text{fd}}) \cdot \frac{P_{\text{fd}}}{\eta_{\text{eng\_lump}}} \\ P_{\text{mot\_shaft}} &= \left(1 - k_{\text{split}}(P_{\text{fd}})\right) \cdot \frac{P_{\text{fd}}}{\eta_{\text{mot\_lump}}} \end{aligned} \tag{10.18}$$

where P_{veh} is the vehicle's required power, P_{fd} is the calculated required power at the driveshaft, P_{eng} is the power required from the engine at the joint driveshaft, P_{mot} is the power required from the motor at the joint driveshaft, $P_{\text{eng\_shaft}}$ is the power required at the engine output shaft, $P_{\text{mot\_shaft}}$ is the power required at the motor output shaft, k_{split} is the split coefficient, $\eta_{\text{eng\_lump}}$ is the lumped efficiency from the engine shaft to the joint driveshaft, η_{fd} is the efficiency of the final drive, and $\eta_{\text{mot\_lump}}$ is the lumped efficiency from the motor shaft to the joint driveshaft.

The curves of the required engine power versus vehicle speed and the curves of the required motor power versus vehicle speed can be achieved from Eqs 10.17 and 10.18. If the transmission and power split device are determined, the speeds of the engine and motor can be backward calculated from the scheduled vehicle speed, so the curve of engine power versus speed and the curve of motor power versus speed can be obtained. The engine and motor can be further sized based on these characteristics.

10.3.2.2 Sizing the Transmission – Number of Speeds and Gear Ratios

Once the movers have been selected, the number of speeds and gear ratios of the transmission can be determined and scaled to the standard ratio to enable the characteristics of the mover to match the power required by the vehicle. For step gear transmissions, the factors outlined in the following subsections need to be taken into account.

Determination of the First Gear

When determining the first gear, the following factors need to be taken into account:

- Launching performance requirements, including maximal acceleration (peak *g*s), time to peak *g*s, and 0→100 km/h acceleration time;
- Satisfaction in the 1–2 shift;
- Wheel slip;
- Launching energy and inertial effects;
- Driving quality.

The typical vehicle performance requirements for sizing first and reverse gears are listed in Table 10.5.

Determination of the Top Gear

The top gear is determined based on the following aspects:

- Providing the desired gradeability at top speed, such as at 90, 105, 120 km/h;
- Maximizing fuel economy;
- Top speed requirement;
- Highway passing performance requirement.

The typical top gear gradeability requirements are listed in Table 10.6.

Table 10.5 Typical automatic transmission first and reverse gear launch requirements for drivability

Direction	Region/country	Grade (%) requirement at vehicle speed >16 km/h	Conditions			
			Altitude (m)	Load	Ambient temp (°C)	AC on/off
Forward	Europe	22	1000	GVW	30	ON
	USA, Asia, Middle East	30	670	GVW	30	ON
	Mexico, Central America	25	2800	GVW	30	ON
Reverse	Europe	17	1000	GVW	30	ON
	USA, Asia, Middle East	25	670	GVW	30	ON
	Mexico, Central America	20	2800	GVW	30	ON

Table 10.6   Typical automatic transmission top gear gradeability requirements

Vehicle speed (km/h)	Vehicle catalog	Grade (%) requirement at the vehicle speed	Conditions			
			Altitude (m)	Load	Ambient temp (°C)	AC on/off
120	A & B segment	2.0	0	TWC	25	ON
	Large truck/SUV/Van	2.3	0	TWC	25	ON
	Eco (fuel economy)	2.4	0	TWC	25	ON
	Normal passenger cars	2.8	0	TWC	25	ON
105	A & B segment	2.6	0	TWC	25	ON
	Large truck/SUV/Van	3.0	0	TWC	25	ON
	Eco (fuel economy)	3.0	0	TWC	25	ON
	Normal passenger cars	3.4	0	TWC	25	ON
90	A & B segment	2.7	0	TWC	25	ON
	Large truck/SUV/Van	3.3	0	TWC	25	ON
	Eco (fuel economy)	3.1	0	TWC	25	ON
	Normal passenger cars	3.5	0	TWC	25	ON

Determination of the Number of Speeds

One of the main transmission objectives is to ensure that the engine power is delivered maximally to the wheels so that the vehicle has the best performance. The number of transmission speeds has a significant impact on the vehicle's tractive capability and performance. For example, consider a vehicle with a 6070 N traction limit and the road load curve on a level road shown in Fig. 10.21; the engine's torque–speed characteristics are shown in Fig. 10.22. If a one-speed transmission with 200 (km/h) to engine speed (rpm) ratio (N/V) is used in this vehicle, the maximal speed of the vehicle is only 80 km/h, as shown in Fig. 10.23. Figure 10.24 shows that the vehicle is able to reach a theoretical maximal speed of 270 km/h if a two-speed transmission is used, but the engine's propulsion power cannot be used maximally. Figure 10.25 shows that the traction performance is improved if a three-speed transmission is used in the vehicle. Figure 10.26 shows that a four-speed transmission can basically give the vehicle the maximal propulsion power.

Determination of Intermediate Gear Ratios

Once the number of speeds of a transmission, the first gear N/V, and the top gear N/V have been determined, the overall ratio spread (OAR) can be calculated as:

$$\text{Overall ratio spread} = \text{First gear } N/V \div \text{Top gear } N/V \qquad (10.19)$$

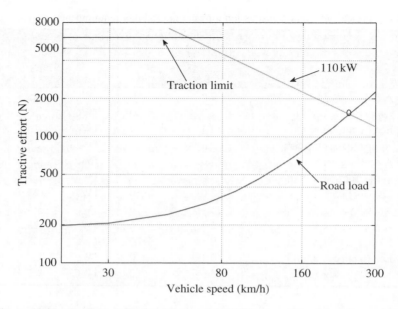

Figure 10.21   Road load, traction limit, and 110 kW power line of the given vehicle

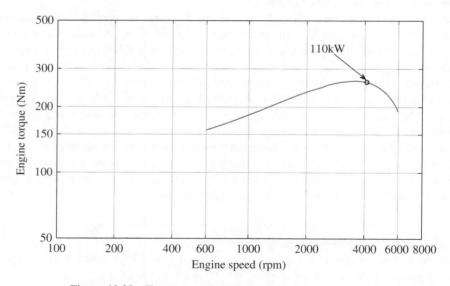

Figure 10.22   Torque–speed characteristics of the given engine

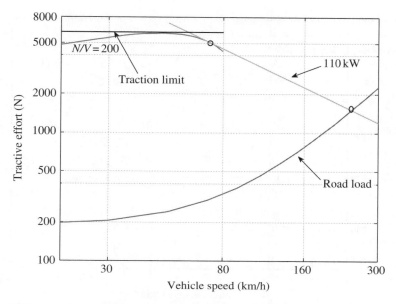

Figure 10.23 Traction capability of the given vehicle equipped with a one-speed transmission

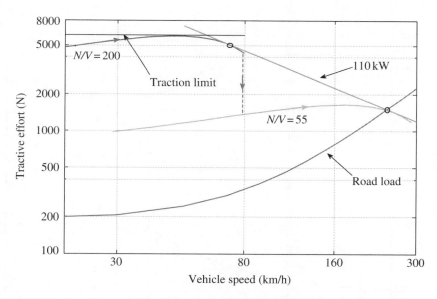

Figure 10.24 Traction capability of the given vehicle equipped with a two-speed transmission

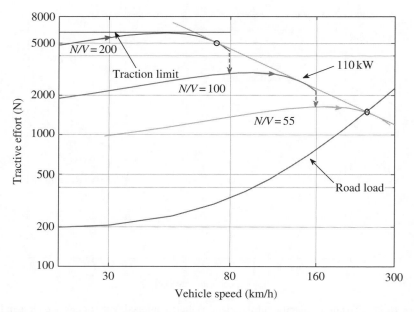

Figure 10.25   Traction capability of the given vehicle equipped with a three-speed transmission

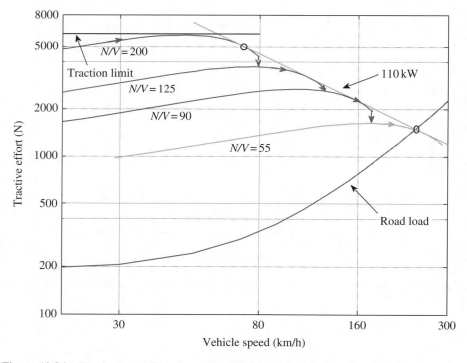

Figure 10.26   Traction capability of the given vehicle equipped with a four-speed transmission

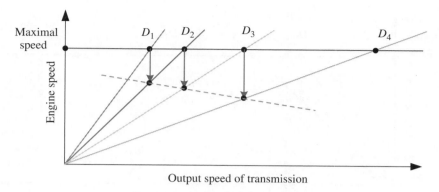

Figure 10.27 Diagram of arithmetic progression

After the desired overall ratio spread has been obtained, the intermediate gear ratios can be further selected based on the following three principles:

- **Arithmetic gear ratios**

  An arithmetic gear ratio is shown in Fig. 10.27. In this progression, the gear ratio change is constant; that is:

$$R_{\text{gear}} - R_{\text{gear}-1} = \text{Constant} \tag{10.20}$$

If we assume that R_1 = first gear ratio, R_n = final gear ratio, and n = number of ratios (number of speeds), then the ith arithmetic gear ratio is:

$$R_i = R_{i-1} - (i-1)a, \quad i = 2, \cdots, n$$
$$a = \frac{R_1 - R_n}{n} \tag{10.21}$$

- **Geometric gear ratios**

  A diagram of geometric gear ratios is presented in Fig. 10.28. In this progression, the engine speed change is constant; that is, $\Delta N_{\text{eng}} = \text{Constant}$. If we assume that the first gear ratio is R_1 and the final gear ratio is R_n, then the ith geometric gear ratio is:

$$R_i = R_{i-1}a = R_1 a^{i-1}, \quad (i = 2, \cdots, n)$$
$$a = 1 - \frac{\Delta N_{\text{eng}}}{N_{\text{eng\_max}}} \tag{10.22}$$

- **Harmonic gear ratios**

  A diagram of harmonic gear ratios is given in Fig. 10.29. In this progression, the change in output speed of the transmission is constant; that is, $\Delta N_{\text{output}} = \text{Constant}$. If we assume that the first gear ratio is R_1 and the final gear ratio is R_n, then the ith harmonic gear ratio is:

$$\frac{1}{R_i} = (i-1)c + \frac{1}{R_1} \implies R_i = \frac{R_1}{1 + (1-i)cR_1}, \quad (i = 2, \cdots, n)$$
$$c = \text{constant, i.e., } c = \frac{1}{R_i} - \frac{1}{R_{i-1}} \tag{10.23}$$

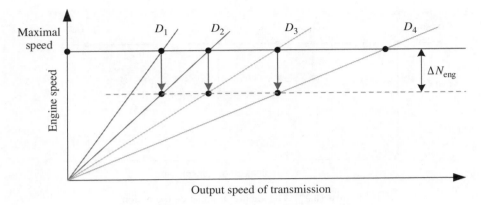

Figure 10.28   Diagram of geometric progression

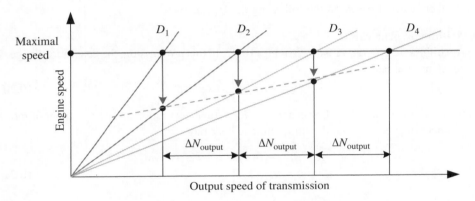

Figure 10.29   Diagram of harmonic progression

The engine speed change at the shift point is:

$$\Delta N_{\text{eng}\_i-1} = N_{\text{eng}\_\text{max}} \frac{R_i - R_{i-1}}{R_i}, \quad (i = 2, \cdots, n) \tag{10.24}$$

Comparing these gear ratios, we can see that arithmetic progression has an increasing step size (= $R_{\text{gear}}/R_{\text{gear-1}}$), geometric progression has a constant step size, and harmonic progression has a decreasing step size. In practice, the actual gear ratios often result from a fine adjustment process taking into account overall considerations of manufacturability, vehicle performance, fuel economy, and driving satisfaction in shift spacing.

Example 10.1
Determine the target gear ratios for a final design based on the vehicle described in Table 10.7 and a V8 engine with the wide-open-throttle curves shown in Fig. 10.30.

Table 10.7   Vehicle parameters for Example 10.1

Vehicle parameters		Unit	Value
Curb weight		kg	1950
Wheel base		m	2.95
Rolling radius (r)		m	0.311
Final drive ratio (fd)			3.08
Final drive efficiency (η_{fd})		%	95
Gear efficiency (η_{gr})		%	93
Drive type			RWD
Torque converter stall K factor			105
Torque converter stall torque ratio			1.67
Vehicle acceleration requirement		g	0.9
Coast down coefficient	Rolling resistance (a_0)	N	251
	Viscous coefficient (a_1)	N/(m/s)	3.22
	Aero coefficient (a_2)	N/(m/s)^2	0.427

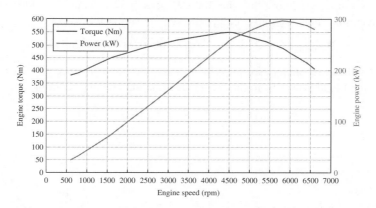

Figure 10.30   Wide-open-throttle torque and power for the engine of Example 10.1

Solution:

1. Estimate first gear

Since the vehicle needs to have 0.9 g acceleration performance, the torque at the wheel is:

$$\tau_{\text{wheel}} = F \cdot r = m \cdot a \cdot r = 1950 \cdot 0.9 \cdot g \cdot 0.311 = 5354 \, \text{Nm} \tag{10.25}$$

In order to achieve this, it is necessary to supply a torque at the driveshaft of:

$$\tau_{\text{trans\_out}} = \frac{\tau_{\text{wheel}}}{fd \cdot \eta_{\text{fd}}} = \frac{5354}{3.08 \cdot 0.95} = 1830 \, \text{Nm} \tag{10.26}$$

upstream of the gear set, to get the necessary torque at the input shaft of the gearbox as:

$$\tau_{\text{trans\_input}} = \frac{\tau_{\text{trans\_out}}}{R_{\text{gear}} \cdot \eta_{\text{gr}}} = \frac{1830}{R_{\text{gear}} \cdot 0.93} = \frac{1968}{R_{\text{gear}}} \text{Nm} \tag{10.27}$$

Now, working from the other end, from Fig. 10.30 we know that the engine is capable of 490 Nm at 2400 rpm. Based on the definition of the K factor, the given 105 K converter would stall at 2324 rpm with the given engine at 490 Nm; correspondingly, the torque into the transmission gearbox would be 1.67 times the engine output torque, thus this input shaft torque is 818 Nm.

Returning to Eq. 10.27, we can solve for the necessary first gear ratio as $R_{\text{gear}} = \dfrac{1968}{818} = 2.41$. Considering engine and drivetrain inertia, we can scale up about 25% to set the first gear ratio to 3.0.

2. **Estimate top gear**

Since the top speed occurs in top gear, the top gear ratio can be determined based on the top speed performance requirement. From the given engine torque/power capability curve, we know that the 300 kW peak power occurs at 5900 rpm. Taking drivetrain efficiency into account, the peak power at the wheel is:

$$P_{\text{wheel\_peak}} = P_{\text{eng\_peak}} \cdot \eta_{\text{gr}} \cdot \eta_{\text{fd}} = 300 \cdot 0.93 \cdot 0.95 = 265 \text{ kW} \tag{10.28}$$

On the other side, based on vehicle coast down test parameters, we can calculate the road load power from the following equation and as shown in Fig. 10.31:

$$P_{\text{Roadload}} = V_{\text{spd}} \times \left(a_0 + a_1 \cdot V_{\text{spd}} + a_2 \cdot V_{\text{spd}}^2 \right) \tag{10.29}$$

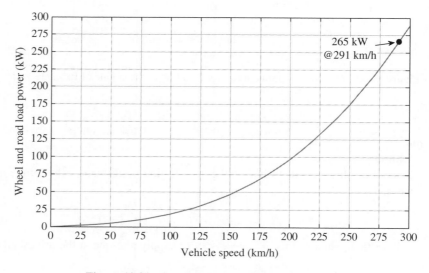

Figure 10.31   Road load power of the given vehicle

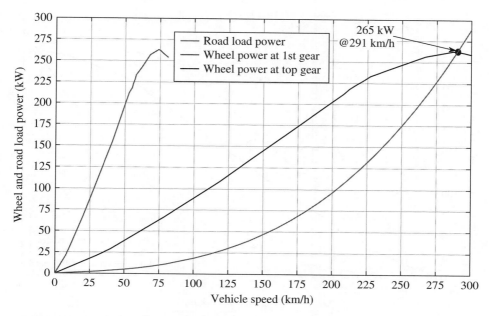

Figure 10.32 Road load power and wheel power in first and top gears of the given vehicle

In order to achieve the maximal speed, we want the engine to operate at 5900 rpm at 291 km/h. Thus, we can calculate the necessary top gear ratio from the following equation:

$$V_{spd}(m/s) = \frac{Eng_{spd}(rpm)\frac{\pi}{30}}{R_{gear}fd} \cdot r_{wheel} \Rightarrow R_{gear} = \frac{5900\frac{\pi}{30}}{\frac{291}{3.6} \cdot 3.08} \cdot 0.311 = 0.77 \qquad (10.30)$$

The wheel power curves in first gear and top gear are shown in Fig. 10.32.

Determining the Intermediate Gear Ratios

From the first gear ratio and the top gear ratio, we can obtain the overall ratio spread of 3.9, which is typical for a four-speed transmission. Now we need to determine the second and third gears. In this example, the gear ratios are set in harmonic progression enabling the change in output shaft speed to be constant, as follows:

$$\frac{1}{R_i} = (i-1)c + \frac{1}{R_1} \Rightarrow c = \frac{1}{i-1}\left(\frac{1}{R_i} - \frac{1}{R_1}\right) = \frac{1}{3}\left(\frac{1}{0.77} - \frac{1}{3}\right) \cong 0.32 \qquad (10.31)$$

Thus, we get the gear ratios as [3.0 1.54 1.01 0.77]. Finally, after a slight adjustment, the target gear ratios of [3.0 1.65 1.0 0.77] are obtained for the next phase in the design process. The overall wheel and road load power curves are shown in Fig. 10.33.

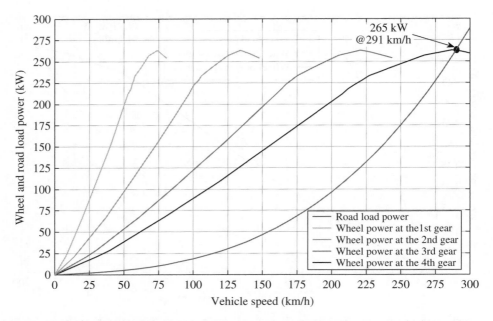

Figure 10.33 Overall road load power and wheel power at the preliminary gear ratios for the given vehicle

10.3.2.3 Sizing the Energy Storage System

The required specification for the energy storage system varies depending on the configuration and performance requirements of a hybrid vehicle system. Once a vehicle has been specified, the energy storage system is normally sized based on the drivability, fuel economy, and electric mileage requirements of the vehicle. In addition to the usage life, efficiency, weight, and cost, the operating voltage, energy storage capacity (Ahr), and adequate peak power (kW) also need to be sized. For example, the energy storage system of a series hybrid vehicle system can be sized based on a given driving schedule and the desired operating temperature.

From a given driving schedule, the power required, P_{ESS}, from the energy storage system can be calculated as:

$$P_{ESS}(t) = \begin{cases} -\dfrac{|P_{fd}|}{\eta_{mot} \cdot \eta_{bat\_dischg}} & \text{if } P_{fd} \geq 0 \\[4mm] \eta_{mot} \cdot \eta_{bat\_chg} \cdot |P_{fd}| & \text{if } P_{fd} < 0 \end{cases} \tag{10.32}$$

where P_{fd} is the power required (W) from the vehicle driveshaft (positive is propulsive power and negative is regenerative braking power), η_{mot} is the motor's efficiency, η_{bat_dischg} is the battery discharge efficiency, and η_{bat_chg} is the battery charge efficiency.

Sizing the Capacity of the ESS Based on the Electric Mileage Requirement

For a given driving schedule undertaken at the designed temperature, T, if we assume that the vehicle can run x miles during period of time $[0 \quad t_f]$, then the required operational energy capacity of the energy storage system is:

$$Cap_{energy}(T) = \frac{1}{3600} \int_0^{t_f} P_{ESS}(t)dt \quad (\text{Whr}) \qquad (10.33)$$

If the designed SOC operational window $\Delta SOC(T) = SOC_{initial}(T) - SOC_{final}(T)$ during the period $[0 \quad t_f]$, then the required Ahr capacity of the energy storage system can be calculated from the following equation:

$$Cap_{Ahr}(T) = \frac{Cap_{energy}(T)}{\Delta SOC(T) \cdot V_{normal}(T)} \qquad (10.34)$$

Sizing the Maximal Power Capability Based on the Drivability Requirement for a Given Driving Schedule

The power capability of the energy storage system can be sized as:

$$P_{max\_chg} = \min\left\{ V_{max} \cdot I_{chg} \quad V_{chg} \cdot I_{chg\_max} \right\} \geq \max\left\{ P_{ESS\_chg}(t) \right\} \quad t \in (0 \quad t_f)$$
$$P_{max\_dischg} = \min\left\{ V_{min} \cdot |I_{dischg}| \quad V_{dischg} \cdot |I_{dischg\_max}| \right\} \qquad (10.35)$$
$$\geq \max\left\{ |P_{ESS\_dischg}(t)| \right\} \quad t \in (0 \quad t_f)$$

where P_{max_chg} is the maximal charge power capability, P_{max_dischg} is the maximal discharge power capability, V_{max} is the maximal charge voltage, V_{min} is the minimal discharge voltage, I_{chg_max} is the maximal charge current, I_{dischg_max} is the maximal discharge current of the energy storage system, $P_{ESS\_chg}(t)$ and $P_{ESS\_dischg}(t)$ are the required charge power capability and discharge power capability of the energy storage system to meet the vehicle's drivability requirements.

Determining the Nominal Operational Voltage and Configuring the Energy Storage System

The nominal operational voltage is the average operational voltage of the electric powertrain of a hybrid vehicle system, which is systematically determined by considering the overall efficiency of the electric powertrain and the cost and availability of motor/generators and the energy storage system. Once the nominal operational voltage has been determined, the energy storage system can be configured to meet the requirements of electric mileage and drivability. For example, based on the preliminary design, the energy storage system of a hybrid vehicle system needs to provide 310 V nominal operating voltage, 70 kW peak

Table 10.8 Specification of the selected battery cell

Cell shape	Cylindrical
Nominal voltage (V)	3.65
Nominal capacity (Wh)	16.5
Nominal power (W)	500
Peak discharge current (A)	200
Peak charge current (A)	150
Maximal operating voltage (V)	4.1
Minimal operating voltage (V)	2.7
Diameter (mm)	40
Length (mm)	92
Cell weight (g)	260

discharge power and 50 kW peak charge power with the electric powertrain, and if the battery cell is selected with the specifications shown in Table 10.8, the energy storage system needs be configured as follows:

1. Required number of cells in a string:

$$N_{cell} = \frac{V_{system\_nominal}}{V_{cell\_nominal}} = \frac{310}{3.65} \cong 85 \tag{10.36}$$

2. Required number of strings (number of cells in parallel):

$$N_{string} = \max\left\{ \frac{P_{system\_chg\_max}}{P_{cell\_chg\_max} \cdot N_{cell}}, \ \frac{P_{system\_dischg\_max}}{P_{cell\_dischg\_max} N_{cell}} \right\}$$

$$= \max\left\{ \frac{70000}{200 \cdot 2.7 \cdot 85}, \ \frac{50000}{150 \cdot 4.1 \cdot 85} \right\} \cong 2 \tag{10.37}$$

Therefore, the required energy storage system should consist of a string with 85-cell double-parallel strings. In addition, since battery performance degrades with age, the aging factor needs to be taken into account when sizing the energy storage system.

10.3.2.4 Design Examples

Example 10.2

A hybrid electric vehicle system design example is given in this section. The design is based on the following vehicle parameters and performance requirements:

1. **Vehicle and operating environment parameters:**
   Vehicle operational weight: 2400 (kg)
   Vehicle frontal area: 2.79 (m^2)

Hybrid vehicle system configuration: series hybrid

Vehicle rolling resistance coefficient: 0.012

Vehicle wheel radius: 0.335 (m)

Vehicle final drive ratio: 4.1:1

Road type: good asphalt

Road grade: level

Vehicle aerodynamic drag coefficient: 0.46 $(N \cdot s^2/kg \cdot m)$

Air mass density: 1.202 (kg/m^3)

2. **Vehicle performance requirements:**

   Ability to follow FTP city and FTP highway driving schedules

   Acceleration time from 0 to 60 km/h: $\leq$ 13s

   Maximal grade at 50 (km/h) speed: $\geq 15\%$

   Maximal speed on a good, level road: 130 km/h

   Electric operating range: up to 2.5 (km)

3. **Size of hybrid system components:**

   - The calculated speed, torque, and power at the driveshaft during FTP city and FTP highway driving schedules are shown in Fig. 10.34(a), (b), and (c), and Fig. 10.35 (a), (b), and (c).

     From the above calculations, we know that the peak driveshaft speed is 3110 rpm, the peak propulsion torque is 400 Nm, the peak brake torque is −300 Nm, the peak propulsion power is 45 kW, and the peak brake power is −40 kW. Based on the above information, the curves for power versus speed and torque versus speed can be obtained, as shown in Fig. 10.36(a) and (b) and Fig. 10.37(a) and (b).

   - Size of the electric motor:

     From the above calculations, the electric motor/inverter set can be selected initially based on the characteristics of the motor set, as shown in Fig. 10.38.

   - Size of the gear ratio:

     After the motor/generator set has been initially selected, the gear ratio can be determined to match the characteristics of the motor's torque with the torque appearing at the driveshaft. In this example, the gear ratio is set as 1.9:1.

   - Drivability verification:

     Based on the established sizes of the motor and gear ratio, the drivability of the designed vehicle needs to be verified using Eqs 10.2 to 10.11. For this example, the achieved results are: (i) the vehicle can be accelerated to 60 km/h from still in 10 seconds on a level road; (ii) the top speed of the vehicle is 135 km/h; (iii) the vehicle is able to navigate a 15% grade with 50 km/h speed. Thus, the above design meets the drivability requirements, as shown in Fig. 10.39.

   - Size of the energy storage system:

     i. Configure the battery pack

        Based on the availability of motor and generator sets, the nominal operational voltage of the electric powertrain is set as 320 V. At this voltage level, if the battery cell shown in Table 10.8 is selected, a 40 kW power capability is needed, based on Figs 10.34 to 10.38, which means that a maximal 150 A discharge current is required

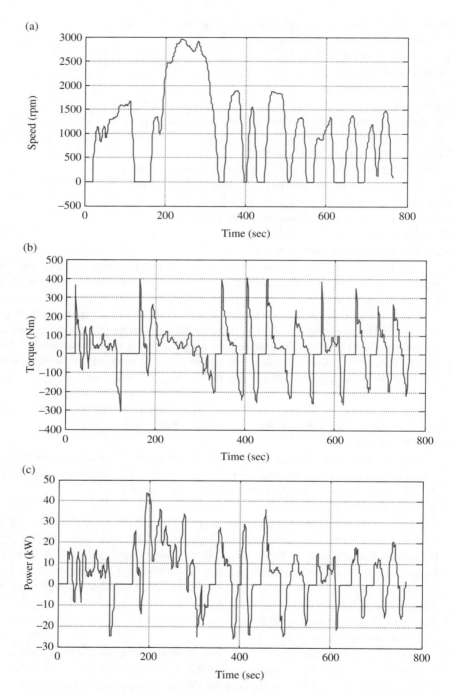

Figure 10.34 Driveshaft speed, torque, and power during FTP city drive schedule

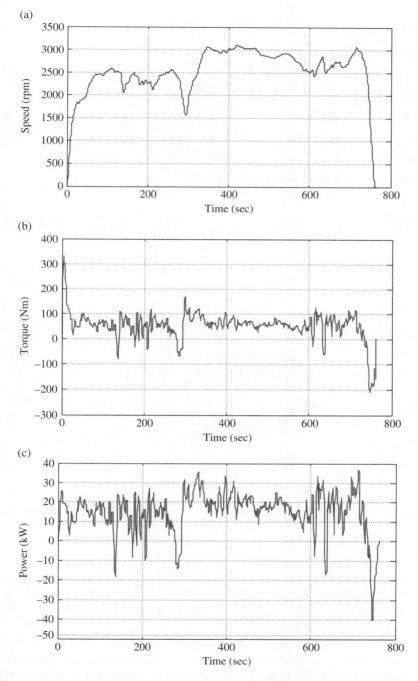

Figure 10.35   Driveshaft speed, torque, and power during FTP highway drive schedule

(a)

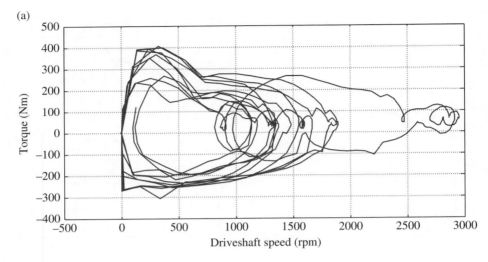

(b)

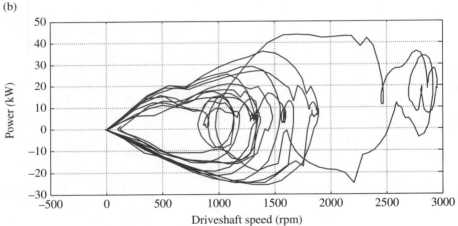

Figure 10.36   Driveshaft torque and power vs. speed during FTP city drive schedule

at a 50% SOC from Eq. 10.35. Considering the aging factor, one string is good enough to deliver 40 kW, and the number of cells in the string should be 88 from Eq. 10.36 to meet the nominal operational voltage requirement.

ii. Determine the cell capacity

If we assume that the hybrid vehicle system can drive electrically for 2.5 km in the FTP driving schedule, 1.2 kWh energy is required; therefore, the required cell capacity is 13.64 Wh calculated from Eq. 10.33. Since the selected cell capacity is 16.5 Wh, the configured energy storage system can meet the electric mileage requirement even if we take the aging factor into account.

(a)

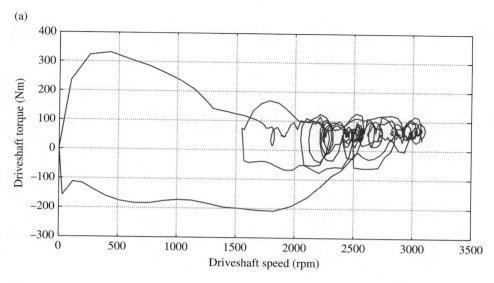

(b)

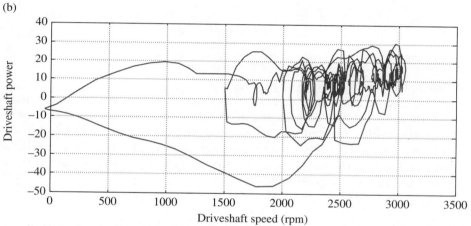

Figure 10.37 Driveshaft torque and power vs. speed during FTP highway drive schedule

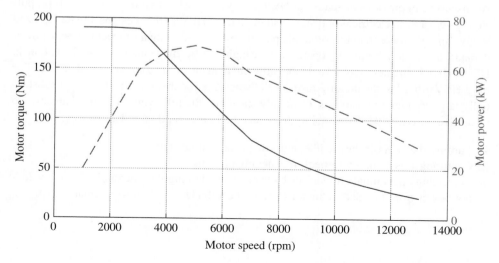

Figure 10.38 Characteristics of the initially selected electric motor set

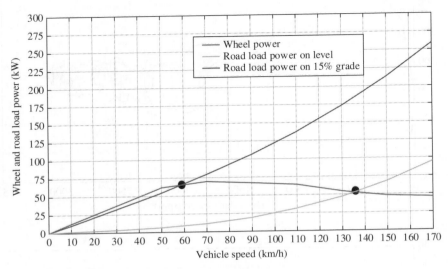

Figure 10.39   Wheel and road power of the vehicle described in Example 10.2

- Size of the genset:

    Selecting the genset for a series hybrid vehicle is relatively flexible. The genset is sized to make the total electrical power (ESS output + genset output) equal to or greater than the vehicle's required peak power. In this design example, a 35 kW genset is selected to meet the 70 kW power requirement to accelerate the vehicle to 60 km/h from still in 13 s. The scaled genset is a 40 kW, 1.2 L turbo-charged gasoline engine coupled with a 35 kW generator.

10.4   Fuel Economy, Emissions, and Electric Mileage Calculation

10.4.1   Basics of Fuel Economy and Emissions Calculation

As discussed in previous chapters, an HEV/PHEV consists of two movers: an ICE propulsion system and an electric propulsion system. In most HEV/PHEV designs, the ICE is sized to give slightly more than the vehicle's power demand rather than to meet the peak power demand, and it more frequently operates at the highest efficiency range. Hence, the ICE in an HEV/PHEV is smaller, lighter, and more efficient than the one in a conventional vehicle. For most HEVs/PHEVs, the motor is also specially designed so that it is able to deliver substantially large torque at low speeds. Generally speaking, the following three mechanisms are what give an HEV/PHEV the ability to reduce fuel consumption and emissions:

1. Turning the ICE off during idle and low-power-output states;
2. Capturing waste energy through regenerative braking;
3. Enhancing operational efficiencies by using a smaller, highly efficient ICE, and by adding power from the electric motor to meet the vehicle's peak power demand.

Once an HEV/PHEV design has been completed, it is necessary to perform fuel economy and emission tests to understand its performance (Bosch, 2005). The official fuel economy and emission test takes place on a dynamometer, and it is quite complex and costly to perform such physical tests. Simulation provides an effective way to estimate the fuel economy and emissions of an HEV/PHEV at concept and pre-design stages. The fuel economy and emissions of an HEV/PHEV can be estimated through simulations based on a given driving schedule such as the FTP drive schedule, and the calculation procedures are:

1. Set simulation driving schedules, for example, the FTP driving schedule.
2. Ensure that the energy management strategy properly splits the vehicle's required power between the ICE and the electric motor and has the start SOC and the end SOC of the energy storage system equal for the given driving schedule; that is:

$$SOC(t_i) = SOC(t_f) \qquad (10.38)$$

3. Calculate the split power P_{eng} and P_{mot} for the ICE and the electric motor to propel the vehicle from the energy management strategy; that is:

$$P_{fd} = P_{veh}/\eta_{fd}$$
$$P_{eng} + P_{mot} = P_{fd}/\eta_{drive\_train} = \left(k_{split} \cdot P_{fd} + \left(1 - k_{split}\right) \cdot P_{fd}\right)/\eta_{drive\_train} \qquad (10.39)$$

$$P_{eng\_shaft} = k_{split} \cdot \frac{P_{fd}}{\eta_{eng\_lump}\eta_{drive\_train}}$$

$$P_{mot\_shaft} = \left(1 - k_{split}\right) \cdot \frac{P_{fd}}{\eta_{mot\_lump}\eta_{drive\_train}} \qquad (10.40)$$

where P_{veh} is the vehicle's power demand, P_{fd} is the required power at the driveshaft, P_{eng} is the power required from the engine at the joint driveshaft, P_{mot} is the power required from the motor at the joint driveshaft, P_{eng_shaft} is the power required at the engine output shaft, P_{mot_shaft} is the power required at the motor output shaft, k_{split} is the split coefficient calculated from energy management, η_{eng_lump} is the lumped efficiency from the engine shaft to the joint driveshaft, η_{fd} is the efficiency of the final drive, η_{mot_lump} is the lumped efficiency from the motor shaft to the joint driveshaft, and η_{drive_train} is the lumped efficiency of the drivetrain.

4. Calculate the fuel economy and emissions based on the split engine output shaft power P_{eng_shaft}, the fuel consumption map, and emissions maps for the given driving schedule:
   i. Fuel consumption:

$$FC = \frac{1}{3600}\int_0^{t_f} fc\_fuel\_map\_gpkWh\left(P_{eng\_shaft}, \omega_{eng\_shaft}\right)dt \qquad (10.41)$$

where $fc\_fuel\_map\_gpkWh$ is the fuel consumption map indexed by the engine power (kW) and the speed (rad/s) and FC is the fuel consumption in grams (g) for the whole cycle.

Further, this can be converted into liters of fuel per 100 km drive:

$$\text{Liters per } 100\,\text{km} = \frac{FC \cdot 10^5}{F_{sg} \cdot \displaystyle\int_0^{t_f} Veh_{spd}(t)\,dt} \tag{10.42}$$

where FC is the calculated fuel consumption (g) from Eq. 10.42, F_{sg} is fuel-specific gravity (kg/m^3) (F_{sg} = 720–775 for unleaded gasoline, F_{sg} = 815–885 for diesel), Veh_{spd} is the scheduled vehicle speed (m/s).

ii. Fuel economy:

The fuel economy of a vehicle is normally rated in miles per gallon, and it can be converted from Eq. 10.42 as:

$$\text{mile/gallon(US)} = \frac{235.21}{\left(^L/_{100}\text{km}\right)}; \quad \text{mile/gallon(UK)} = \frac{282.48}{\left(^L/_{100}\text{km}\right)} \tag{10.43}$$

iii. Emissions absolute values (g):

$$\text{HC} = \frac{1}{3600} \int_0^{t_f} HC\_map\_gpkWh\left(P_{eng\_shaft}, \omega_{eng\_shaft}\right) dt \tag{10.44}$$

$$\text{CO} = \frac{1}{3600} \int_0^{t_f} CO\_map\_gpkWh\left(P_{eng\_shaft}, \omega_{eng\_shaft}\right) dt \tag{10.45}$$

$$\text{CO}_2 = \frac{1}{3600} \int_0^{t_f} CO_2\_map\_gpkWh\left(P_{eng\_shaft}, \omega_{eng\_shaft}\right) dt \tag{10.46}$$

$$\text{NOx} = \frac{1}{3600} \int_0^{t_f} NOx\_map\_gpkWh\left(P_{eng\_shaft}, \omega_{eng\_shaft}\right) dt \tag{10.47}$$

iv. Emissions relative values (g/km):

$$\text{HC(g/km)} = \frac{\displaystyle\int_0^{t_f} HC\_map\_gpkWh\left(P_{eng\_shaft}, \omega_{eng\_shaft}\right) dt}{3.6 \cdot \displaystyle\int_0^{t_f} Veh_{spd}(t)\,dt} \tag{10.48}$$

$$\text{CO(g/km)} = \frac{\displaystyle\int_0^{t_f} CO\_map\_gpkWh\left(P_{eng\_shaft}, \omega_{eng\_shaft}\right) dt}{3.6 \cdot \displaystyle\int_0^{t_f} Veh_{spd}(t)\,dt} \tag{10.49}$$

$$\text{CO}_2\text{(g/km)} = \frac{\displaystyle\int_0^{t_f} CO_2\_map\_gpkWh\left(P_{eng\_shaft}, \omega_{eng\_shaft}\right) dt}{3.6 \cdot \displaystyle\int_0^{t_f} Veh_{spd}(t)\,dt} \tag{10.50}$$

$$\text{NOx(g/km)} = \frac{\int_0^{t_f} NOx\_map\_gpkWh\left(P_{\text{eng\_shaft}}, \omega_{\text{eng\_shaft}}\right)dt}{3.6 \cdot \int_0^{t_f} Veh_{\text{spd}}(t)dt} \tag{10.51}$$

where HC is hydrocarbons, CO is carbon monoxide, CO_2 is carbon dioxide, and NOx is nitrous oxides.

10.4.2 EPA Fuel Economy Label Test and Calculation

In order to reflect more realistic values for a vehicle's fuel economy that customers may experience under real-world operation in terms of driving patterns (speeds/accelerations), air-conditioning and heater usage, and hot and cold temperatures, the Environmental Protection Agency of the US (EPA) has been refining testing methods and fuel economy label calculations. The test methods generally consist of driving a vehicle on a chassis dynamometer through prescribed speed vs. time drive cycles and determining a fuel economy rating in miles per gallon (mpg) to predict how much fuel the vehicle can be expected to use under real-world driving conditions.

With continuing advances in vehicle technology, improved vehicle architecture and performance result in driving style changes over time; therefore, revisions of the test methods are needed in response to changes in vehicle characteristics and driver behavior. The most recent revision of the testing methods for fuel economy label calculations was implemented in 2011, and this weights test results from five FTP cycles (UDDS at 75 F, US06 at 75 F, HWFET at 75 F, SC03 at 95 F, and UDDS at 20 F); the previous method is sometimes referred to as the two-cycle method, as it weighted FTP highway and urban cycles. For hybrid electric vehicles, the revised method causes additional complexities as the balance of the state of charge of the applied battery system has to be taken into account.

To help readers understand the EPA fuel economy test procedure and method, this section briefly introduces the method that vehicle manufacturers use to provide fuel economy labels, and details the test procedures and adaption of the five-cycle test and calculation for hybrid vehicles through a practical example.

10.4.2.1 Corporate Average Fuel Economy (CAFE)

The corporate average fuel economy (CAFE) regulations were enacted by the US Congress in 1975 to reduce energy consumption by increasing the fuel economy of cars and light trucks produced for sale in the US. The CAFE is the production-weighted harmonic mean fuel economy expressed in miles per gallon (mpg). The CAFE regulations require vehicle manufacturers to comply with the CAFE standards, increased every year, for passenger cars and light-duty trucks with a gross vehicle weight rate (GVWR) of 8500 pounds or less, but they also cover medium-duty passenger vehicles with a GVWR up to 10,000 pounds. In a given model year, if the average fuel economy of a manufacturer's annual fleet of vehicles cannot meet the CAFE defined standard, the manufacturer must either apply sufficient CAFE credits to cover the shortfall or pay a penalty.

Unlike the EPA measuring vehicle fuel efficiency, the National Highway Traffic Safety Administration (NHTSA) regulates the CAFE standards, which are maximally set by taking the following factors into account:

- Technological feasibility;
- Economic practicality;
- The effect of other standards on fuel economy;
- The need for the nation to conserve energy.

Since small and lightweight vehicles naturally have higher fuel efficiency, the CAFE system inherently encourages manufacturers to change the types of vehicles they produce; however, the NHTSA has expressed concerns that more fuel-efficient small vehicles on the road may lower traffic safety. In 2011, the NHTSA updated the standards, specifying different requirements based upon the vehicle footprint determined by multiplying the vehicle's wheelbase by its average track width. Therefore, the fleet CAFE requirement for a particular vehicle manufacturer currently is the sum of its sales-weighted vehicle footprint requirements.

Compared to the EPA-label mpg test and calculation methods using the arithmetic mean, CAFE mpg is based on FTP urban and highway testing, often referred to as two-cycle testing, and is calculated using the following weighted harmonic mean, or the reciprocal of the average of the reciprocal values:

$$CAFE\ fuel\ economy = \frac{n_A + n_B + \cdots + n_X}{\dfrac{n_A}{FE_A} + \dfrac{n_B}{FE_B} + \cdots + \dfrac{n_X}{FE_X}}\ (mpg) \tag{10.52}$$

where n_A is the total produced number for vehicle A, n_B is the total produced number for vehicle B, and n_X is the total produced number for vehicle X; FE_A is the fuel economy in mpg for vehicle A, FE_B is the fuel economy in mpg for vehicle B, and FE_X is the fuel economy in mpg for vehicle X.

Example 10.3

For a fleet composed of 2000 small-sized cars with 70 mpg fuel economy, 1000 medium-sized cars with 55 mpg fuel economy, 1500 full-sized SUVs with 45 mpg fuel economy, and 2500 light-duty trucks with 22 mpg fuel economy, the CAFE fuel economy of the fleet can be calculated as:

$$CAFE\ fuel\ economy = \frac{2000 + 1000 + 1500 + 2500}{\dfrac{2000}{70} + \dfrac{1000}{55} + \dfrac{1500}{45} + \dfrac{2500}{22}} = \frac{7000}{193.72} = 36\ (mpg)$$

It should be noted that the CAFE mpg ratings are based upon the unadjusted dynamometer test results from FTP urban (UDDS) and HWFET and are weighted by 55% and 45%, respectively. Therefore, the CAFE fuel economy values will be much higher than the EPA fuel economy label values.

10.4.2.2   Vehicle-specific Five-cycle Method

The objective of labeling the fuel economy for a vehicle is to give vehicle buyers an idea of the fuel economy they can expect during normal driving on the highway and in the city. Before model year 2013, the fuel economy label values were based on the two-cycle method, which measures fuel economy over UDDS and HWFET driving cycles and then adjusts these measured values downward by 10% and 22%, respectively, to compensate for the differences between the ways in which vehicles are driven on the road and over the test cycles. However, it is commonly recognized that the EPA fuel economy label ratings are still above the real-world averages even though the raw test results are adjusted.

In order to improve the accuracy of fuel economy labeling, the EPA proposed a five-cycle test method and started to apply it to 2013 model year vehicles. The purpose of the new method is to better account for factors including real road driving patterns, air-conditioning, and colder weather environments, and the details include: (i) the US06 drive cycle has been introduced so that the maximal test speed is up to 80 mph and acceleration rates have risen to 8.4 mph per second from a 60 mph maximal speed and 3.3 mph per second maximal acceleration in the two-cycle test; (ii) given that drivers often use air-conditioning in warm, humid conditions, the SC03 cycle has been included in the new method and is performed at 95 °F with the A/C on during the tests; (iii) since vehicles are also often driven in cold weather, the new method adds the cold UDDS test portion to the fuel economy label values, reflecting the additional fuel needed to start up an engine in a cold winter. The key characteristics of the five-cycle emission and fuel economy tests are listed in Table 10.9.

In 2012, the EPA released the technical support document presenting the complete five-cycle fuel economy label calculation equations developed by combining the results of the sections on start fuel use, running fuel use, air-conditioning, cold temperatures, and non-dynamometer effects (EPA, 2012). This section details the five-cycle fuel economy label calculation formulae for a vehicle tested over a three-bag FTP.

Table 10.9   Key characteristics of the five-cycle emission and fuel economy tests

Test	Driving characteristics	Ambient temperature	Engine condition at start	Accessory state
UDDS	Low speed	75 °F (24 °C)	Cold and Hot	None
HWFET	Medium speed	75 °F (24 °C)	Hot	None
US06	Aggressive, low and high speed	75 °F (24 °C)	Hot	None
SC03	Low speed	95 °F (35 °C)	Hot	A/C on
Cold UDDS	Low speed	20 °F (−7 °C)	Cold and Hot	None

Five-cycle City Fuel Economy Calculation

$$City\ fuel\ economy = 0.905 \times \frac{1}{Start\ FC + Running\ FC}\ (\text{mpg}) \tag{10.53}$$

where 0.905 is the overall adjustment for non-dynamometer effects and *Start FC* and *Running FC* are the start and running fuel consumption, respectively:

$$Start\ FC = 0.33 \times \left(\frac{(0.76 \times StartFuel_{75}) + (0.24 \times StartFuel_{20})}{4.1} \right)\ (\text{gallons per mile}) \tag{10.54}$$

where:

$$StartFuel_x = 3.6 \times \left(\frac{1}{Bag1\ FE_x} - \frac{1}{Bag3\ FE_x} \right)\ (\text{gallons per mile}) \tag{10.55}$$

where *BagN* FE_x is the fuel economy in miles per gallon during the specified bag of the FTP test conducted at an ambient temperature of 75 °F or 20 °F.

$$Running\ FC = 0.82 \times \left(\frac{0.48}{Bag2_{75}\ FE} + \frac{0.41}{Bag3_{75}\ FE} + \frac{0.11}{US06\ City\ FE} \right)$$
$$+ 0.18 \times \left(\frac{0.5}{Bag2_{20}\ FE} + \frac{0.5}{Bag3_{20}\ FE} \right) \tag{10.56}$$
$$+ 0.133 \times 1.083 \times (A/C\ FE)\ (\text{gallons per mile})$$

where:

$$A/C\ FE = \frac{1}{SC03\ FE} - \left(\frac{0.61}{Bag3_{75}\ FE} + \frac{0.39}{Bag2_{75}\ FE} \right) \tag{10.57}$$

where:

US06 *City FE* = fuel economy in miles per gallon over the city portion of the US06 test;
$BagN_x$ *FE* = fuel economy in miles per gallon during the specified *Bag N* of the FTP test
    conducted at an ambient temperature of 75 °F or 20 °F;
SC03 *FE* = fuel economy in miles per gallon over the SC03 test.

Five-cycle Highway Fuel Economy Calculation

$$Highway\ fuel\ economy = 0.905 \times \frac{1}{Start\ FC + Running\ FC}\ (\text{mpg}) \tag{10.58}$$

where the dynamometer effective adjustment is 0.905 as for the city economy calculation, and the start and running fuel consumption are defined as follows:

$$Start\ FC = 0.33 \times \left(\frac{(0.76 \times StartFuel_{75}) + (0.24 \times StartFuel_{20})}{60} \right) \text{ (gallons per mile)}$$

(10.59)

$$Running\ FC = 1.007 \times \left(\frac{0.79}{US06\ Highway\ FE} + \frac{0.21}{HWFET\ FE} \right) + 0.133 \times 0.377 \times (A/C\ FE)$$

(10.60)

In the above, US06 *Highway FE* and *HWFET FE* are the fuel economy in miles per gallon over the highway portion of the US06 test and the HWFET test, respectively. All other symbols have the same definitions as for the city fuel economy calculation described in the previous section.

It should be noted that the five-cycle fuel economy calculations introduced above are based on a three-bag test, and the mentioned 'bags' are literally used to collect the produced emissions during each portion of the test. Bags 1 and 3 are used for the exact same drive pattern, but Bag 2 is for a different pattern. From three-bag five-cycle test procedures, Bag 1 begins with an engine start after an overnight soak, which means the vehicle has been sitting with the engine off for at least 12 hours, so it is referred to as a 'cold' start. Bag 3 begins with an engine start after the vehicle has been sitting with the engine off for only ten minutes, which represents a short stop to refuel or a short shopping trip, thus this is referred to as a 'hot' start, since the engine is still essentially at its fully warmed-up temperature.

It also needs to be mentioned that the five-cycle fuel economy test separately calculates the fuel necessary to start and warm up the engine as part of the term *Start FC*. For a given vehicle, the fuel needed to warm up the engine depends primarily on the ambient temperature and the length of time that the vehicle has been sitting. It is obvious that the longer the trip, the less significant is the start fuel use.

10.4.2.3 Net Energy Change Tolerances of Fuel Economy Labeling Tests

It is common sense that the electrical energy consumed by the electric motor will affect the actual liquid fuel consumption of hybrid electric vehicles; and a net discharge of the battery will result in less use of fuel, while a net recharge of the battery will require extra fuel for a drive cycle. Therefore, it is readily understood that a valid condition of the five-cycle fuel economy test and calculation method for HEVs, PHEVs, and EREVs is to have the net energy change (NEC) of the energy storage system equal to zero for a drive cycle. The NEC condition when perfectly fulfilled means that the state of charge of the battery at the end of a test has returned to the same level as it was at the beginning of the test.

Since certain hybrid electric vehicles such as PHEVs and EREVs use charged electrical energy, the EPA requires additional miles per gallon equivalent (mpge) and all-electric range (AER) labels in charge-depleting (CD) mode for such hybrid vehicles, while the EPA five-cycle fuel economy test/calculation method presented in the previous section is only used to calculate the fuel economy in charge-sustaining (CS) mode with the zero NEC condition.

In order to ensure that the fuel consumption test result on an HEV represents a valid CS-operation result, the SAE has developed a standard to set a boundary criterion for the NEC condition of the battery in an HEV (SAE-J1711, 2010). For valid CS operation, the standard requires that the NEC of the battery must be less than 1% of the total used energy. It should specifically note that the net change in electrical energy of the battery is the quantity obtained by multiplying the cumulative current flow at the battery terminals by a representative system voltage, rather than by integrating the power flow at the battery terminals, and the representative system voltage is normally the open-circuit voltage of the battery system.

Since NEC tolerance is a critical criterion for an HEV fuel economy test to be valid, the tolerance must be defined small enough to ensure that the vehicle is operated as near as possible to CS operation over the test cycle to achieve repeatable test results. However, due to the limitations of current rechargeable energy storage systems and measurement technologies, it is impossible to produce repeatable fuel economy test results under a zero NEC tolerance requirement, thus some tolerance must be allowed in practice. In order to ensure the measured fuel consumption is within ±3% of the vehicle's true value, the SAE specifies that the change in RESS stored electrical energy over the fuel economy test cycle should be limited to ±1% of the total fuel energy consumed over the same cycle based on analysis and test experience.

$$NEC\ tolerance: \left| \frac{Net\ battery\ energy\ change}{Total\ consumed\ fuel\ energy} \right| \leq 1\% \tag{10.61}$$

where:

$$Net\ battery\ energy\ change = \left((A \cdot h)_{final} - (A \cdot h)_{initial} \right) \cdot V_{system}, \quad (W \cdot h) \tag{10.62}$$

$$Total\ consumed\ fuel\ energy = NHV_{fuel} \cdot m_{fuel} \tag{10.63}$$

where:

$(A \bullet h)_{initial}$ = Battery ampere-hours stored at the beginning of the test phase
$(A \bullet h)_{final}$ = Battery ampere-hours stored at the end of the test phase
V_{system} = Battery's DC nominal system voltage (normally, the battery system's open-circuit voltage is used) at the CS SOC level
NHV_{fuel} = Net heating value (per consumable fuel analysis), in joules per kilogram
m_{fuel} = Total mass of fuel consumed over the test phase, in kilograms

From Eqs 10.61, 10.62, and 10.63, the maximal and minimal allowed final battery net energy changes over the test cycle can be solved as:

$$(A \cdot h_{\text{final}})_{\text{max}} = (A \cdot h_{\text{initial}}) + \frac{NHV_{\text{fuel}} \cdot m_{\text{fuel}}}{V_{\text{system}} \cdot K_{\text{cf}}} \cdot 0.01 \tag{10.64}$$

$$(A \cdot h_{\text{final}})_{\text{min}} = (A \cdot h_{\text{initial}}) - \frac{NHV_{\text{fuel}} \cdot m_{\text{fuel}}}{V_{\text{system}} \cdot K_{\text{cf}}} \cdot 0.01 \tag{10.65}$$

where:

$(A \bullet h_{\text{final}})_{\text{max}}$ = Maximal allowed stored battery ampere-hours at the end of the test phase
$(A \bullet h_{\text{final}})_{\text{min}}$ = Minimal allowed stored battery ampere-hours at the end of the test phase
K_{cf} = Conversion factor = 3600 (s/h)

10.4.3 Electrical Energy Consumption and Miles per Gallon Gasoline Equivalent Calculation

10.4.3.1 Energy Consumption Test Cycle, Transition Cycle, and End-of-test Criterion for Charge-depleting Modes

Energy Consumption Test Cycle for CD Modes
The energy consumption of PHEVs in charge-depleting (CD) mode is usually tested using the UDDS, HWFET, US06, and SC03 schedules. As described earlier, in the five-cycle fuel economy test, the UDDS is also used under cold weather conditions (−7 °C). Some fuel economy test layouts are briefly described below, and readers are referred to SAE J1711 for detailed test procedures.

Since the duration of each FTP schedule is relatively short, the vehicle needs to be driven over multiple continuous FTP schedules on the dynamometer to deplete the battery's energy, followed by one or more corresponding charge-sustaining cycles until the end-of-test criterion is met. Once the vehicle propulsion system has started, the initial and final SOC readings and DC energy in watt-hours (W·h) for each cycle needs to be recorded.

Figure 10.40 shows an energy consumption test layout for a UDDS schedule. Using the SAE J1711 test procedure, it is necessary to add intra-test pauses between the UDDS cycles; consistent 10 ± 1 minute pauses are preferred, but 10–30 minute intra-test pauses may be allowed depending on the conditions at the test facility. During these pauses, the vehicle should be keyed down with the hood closed and no brake pedal depressed, test cell fan(s) should be turned off, no recharging RESS from an external electrical energy source is allowed, but more important is that the SOC instrumentation should not be turned off or reset to zero. If the ampere-hour meter measurement is used, the integration should remain active throughout the entire energy consumption test until it is concluded.

As a part of the test procedure, the vehicle has to be preconditioned at its CS SOC level, then the battery should be charged to the desired level and the total AC energy (W·h) needs to be counted as energy consumed while the vehicle is soaking. After 12 to 36 hours of soaking and when the battery is charged, the vehicle should be pushed or towed into position on the dynamometer and it should not be rolled more than 1.6 km (1 mile) after the end of the

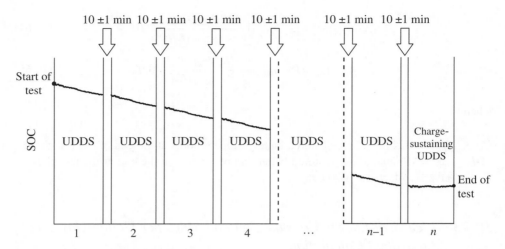

Figure 10.40   Examples of UDDS energy consumption test layouts for CD modes of PHEVs

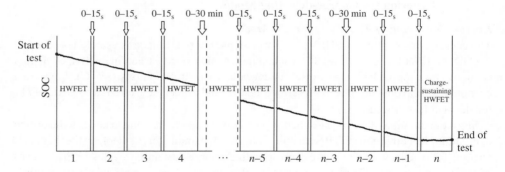

Figure 10.41   Examples of HWFET energy consumption test layouts for CD modes of PHEVs

soak period. At the start point of the test, shown in Fig. 10.40, the ambient temperature should be no less than 20 °C (68 °F) and no more than 30 °C (86 °F), and the vehicle drivetrain should be in a 'cold' condition; during dynamometer driving, all accessories of the vehicle should be shut down, and a cooling fan with a fixed speed should direct cooling air towards the vehicle.

An example of an energy consumption test layout is shown in Fig. 10.41 for an HWFET schedule. By the SAE J1711 test procedure, it is also necessary to precondition the vehicle before starting the energy economy test. While the vehicle is soaking, its battery should be charged to the desired SOC level. After about a 12-hour soak, the vehicle should be pushed or towed into position on the dynamometer and should not be rolled more than 1.6 km (1 mile) after the end of the soak period. Before starting the test, the test site temperature should be set to no less than 20 °C (68 °F) and no more than 30 °C (86 °F) as for the UDDS cycle. Since the HWFET only lasts 765 seconds, the vehicle needs to be driven over many

HWFET cycles in CD mode, preferably separated by a 15-second key-on idle rest, which means there is no pause between the HWFET cycles. However, for some PHEVs with large batteries, test pauses may be required to halt testing and reinitialize the test system. If this is the case, the test can be paused after several repeats of the HWFET cycle, and the duration of these pauses should be less than 30 minutes.

About the Transition Cycle and Transitional Range of the CD-mode Energy Consumption Test for PHEVs

In the electrical energy consumption test for PHEVs, an FTP standard cycle will be repeated until the monitored battery energy indicates that a transition is occurring from the CD mode to a CS mode. This FTP cycle is defined as the transition cycle which will end in CS mode but will be net depleting, then the following cycle will be a full CS cycle. The definition of the transition cycle for the PHEV energy consumption test is illustrated in Fig. 10.42.

Furthermore, the range of the charge-depleting cycle (Rcdc) for a given energy consumption test is defined as the total number of cycles driven in CD mode and transition multiplied by the cycle distance. The counted range is the total distance, measured between the start point of the energy consumption test, through any subsequent CD test cycles, and the CS SOC point in the transitional cycle. This range is defined as the actual charge-depleting range (Rcda). Correspondingly, the transitional range for a PHEV in the energy consumption test is defined as the distance traveled between CD and CS operational modes. An important note is that the travel distance in the CS mode of operation should not be included in the Rcdc. These defined ranges are also illustrated in Fig. 10.42.

End-of-test Criterion (EOT)

During an electrical energy consumption test for a given PHEV, a test cycle is repeated until the battery energy is depleted to a desired level. While the battery SOC is close to or even slightly below the designed charge-sustaining point, it is very challenging for the tester to decide whether the test result should be determined or whether the test should continue for a

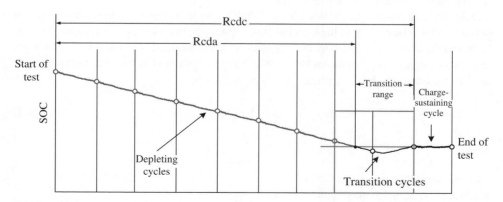

Figure 10.42 Transition cycle, Rcdc, and Rcda definitions in energy consumption tests for PHEVs

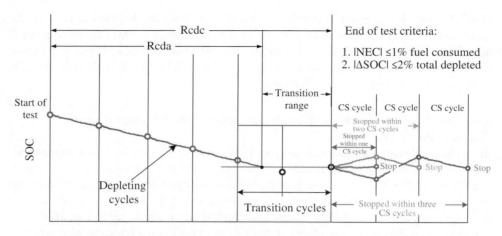

Figure 10.43   End-of-test scenarios in energy consumption tests for PHEVs

short period, as the vehicle may only temporarily operate in charge-sustaining mode. In order to achieve repeatable test results, SAE J1711 sets the following end-of-test criterion and suggests that the vehicle manufacturer inform the tester if the vehicle only temporarily operates in charge-sustaining mode to avoid stopping the test too early.

SAE J1711 states two EOT criteria to terminate an energy consumption test. One is based on the 1% fuel consumed NEC tolerance and the other one requires the absolute delta SOC to be less than 2% within one or more CS cycles following CD-mode operation. Due to control dynamics, in practice, multiple CS-mode operational cycles may be needed to satisfy the EOT criterion and terminate the test. SAE J1711 gives several example scenarios, shown in Fig. 10.43.

10.4.3.2   AC Recharge Energy, DC Discharge Energy, and the Derived Five-cycle Electrical Energy Consumption Adjustment

AC Recharge and DC Discharge Energy
The AC recharge energy (E_{ac}) for BEVs and PHEVs is the energy taken from the AC power outlet to return the battery to the targeted state of charge after a test. E_{ac} is measured in AC W·h, and the energy consumed by the charging device is included. The DC discharge energy (E_{dc}) is the output energy of the battery in W·h, but it is calculated by integrating the terminal power of the battery during vehicle operation, as follows:

$$E_{dc} = \frac{\Delta T}{3600} \times \sum_{i=1}^{n} \left(V_t(i) \cdot I_t(i) \right) \quad (\text{W} \cdot \text{h}) \qquad (10.66)$$

where:

ΔT = the sampling time (s)
V_t = battery terminal voltage (V)
I_t = battery terminal current (A)

Full AC Recharge Energy, Full Depletion DC Discharge Energy, and Recharge Allocation Factor

The full recharge energy is the AC recharge energy needed to return the battery to the state of charge that provides the DC energy needed to complete a full depletion test, and is defined as:

$$E_{FR} = E_{ac}\big|_{\text{full depletion}} \quad (W \cdot h) \tag{10.67}$$

The full depletion DC discharge energy (E_{FD}) of the battery is the total deliverable energy supporting the vehicle to complete an entire charge-depleting test cycle and is defined as:

$$E_{FD} = E_{dc}\big|_{\text{full depletion}} \quad (W \cdot h) \tag{10.68}$$

The ratio of the full AC recharge energy to the full depletion DC energy is defined as the recharge allocation factor, R_{AF}. The recharge allocation factor is normally used to allocate the AC recharge energy needed for the consumed DC discharge energy in the phase test; thus, this factor enables the use of a single recharging event to determine the AC energy needed for multiple drive cycles. The R_{AF} is expressed as:

$$R_{AF} = \frac{E_{FR}}{E_{FD}} \tag{10.69}$$

Electrical Energy Consumption Calculations

Electrical energy consumption can be classified into AC and DC electrical energy consumption, which may be defined, respectively, as the used AC and DC electrical energy in W·h over the distance traveled in miles or km by the vehicle; that is:

$$EC_{AC} = \frac{E_{ac}}{\text{Traveled miles (or km)}} \tag{10.70}$$

$$EC_{DC} = \frac{E_{dc}}{\text{Traveled miles (or km)}} \tag{10.71}$$

Since the BEV or PHEV energy consumption test normally consists of multiple cycles, it is necessary to determine the electrical energy consumption of each individual cycle, and the ith cycle energy consumption is determined as:

$$EC_{DC\_cycle}(i) = \frac{E_{dc\_cycle}(i)}{\textit{Traveled miles\_cycle}(i)} \tag{10.72}$$

The full depletion DC discharge energy consumption of the test cycle is equal to the summation of the DC electrical energy consumption of individual cycles; that is:

$$EC_{DC\_FD} = \sum_{i=1}^{n} EC_{DC\_cycle}(i) \tag{10.73}$$

The corresponding full AC recharge energy is:

$$EC_{AC\_FD} = R_{AF} \cdot EC_{DC\_FD} \qquad (10.74)$$

Similarly, city cycle energy consumption and highway cycle energy consumption are determined as follows:

$$EC_{DC\_city} = \sum_{i=1}^{n} EC_{DC\_UDDS}(i) \qquad (10.75)$$

$$EC_{AC\_city} = R_{AF} \cdot EC_{DC\_city} \qquad (10.76)$$

$$EC_{DC\_highway} = \sum_{i=1}^{n} EC_{DC\_HWFET}(i) \qquad (10.77)$$

$$EC_{AC\_highway} = R_{AF} \cdot EC_{DC\_highway} \qquad (10.78)$$

Derived Five-cycle Adjustment for Electrical Energy Consumption in Electric Vehicles

As described in previous sections, the EPA regulations now require the use of the five-cycle test method to generate fuel economy labels. Since the five-cycle test is quite complex and expensive, the EPA provides a derived five-cycle adjustment method to calculate the energy consumption of electric vehicles for the window sticker, and often only the UDDS and HWFET tests are run, with correction factors applied to estimate the effect of the other three cycles. For electric vehicles, the following formulae are used to adjust the electrical energy consumption and all-electric range based on FTP, UDDS, and HWFET cycles, capped at the maximal adjustments of 30% (that is, the unadjusted values are multiplied by 0.7):

$$\text{Derived five-cycle city fuel economy} = \frac{1}{\left(\{City\ intercept\} + \dfrac{\{City\ slope\}}{MT\ FTP\ FE}\right)} \qquad (10.79)$$

$$\text{Derived five-cycle highway fuel economy} = \frac{1}{\left(\{Highway\ intercept\} + \dfrac{\{Highway\ slope\}}{MT\ HWFET\ FE}\right)} \qquad (10.80)$$

where:

City intercept = 0.003259
City slope = 1.1805
Highway intercept = 0.001376
Highway slope = 1.3466

For electric vehicles, the EPA regulations require the use of weighting factors of 55% for the UDDS, and 45% for the HWFET to determine the combined electrical energy consumption value for urban and highway conditions as:

$$E_{DC\_combined} = 0.55 \cdot E_{DC\_city} + 0.45 \cdot E_{DC\_highway} \qquad (10.81)$$

$$E_{AC\_combined} = R_{AF} \cdot E_{DC\_combined} \qquad (10.82)$$

10.4.3.3 Miles per Gallon Gasoline Equivalent Labeling Calculation

To compare the energy consumption of alternative fuel vehicles such as BEVs and PHEVs with the fuel economy of conventional ICE vehicles expressed in miles per US gallon (mpg), the EPA introduced the metric of miles per gallon gasoline equivalent (mpge) and first put it on the window sticker of the Nissan Leaf electric car and the Chevrolet Volt plug-in hybrid in November 2010. Now, all new BEVs and PHEVs sold in the US are required to have this label showing the EPA's estimate of fuel economy; Figures 10.44 and 10.45 show examples for an electric vehicle and a plug-in hybrid electric vehicle, respectively (EPA, 2011).

The US Department of Energy (DOE) provides the following gasoline-equivalent energy content of electricity factor (GPO Federal Digital Systems, 2000):

$$E_g = (T_g \cdot T_t \cdot C)/T_p = (0.328 \cdot 0.924 \cdot 33705)/0.830 = 12,307 \ (\text{W} \cdot \text{h/gal}) \qquad (10.83)$$

where:

E_g = gasoline-equivalent energy content of electricity (W·h/gal)
T_g = US average fossil-fuel electricity generation efficiency = 0.328
T_t = US average electricity transmission efficiency = 0.924
T_p = Petroleum refining and distribution efficiency = 0.830
C = Watt-hours of energy per gallon of gasoline conversion factor = 33,705 (W·h/gal)

However, at the time of writing, the EPA's mpge ratings only consider the wall-to-wheel energy consumption, and petroleum refining and distribution, electricity generation, and power transmission required energy are not included; thus, the EPA mpge calculation uses 33,705 W·h/gal as the conversion factor. For a given PHEV or BEV, the mpge number is calculated based on the following formula, converting the vehicle consumption per unit distance from the AC electrical energy (W·h) into a gasoline equivalent:

$$\text{mpge} = \frac{E_G}{E_M \cdot E_E} = \frac{11,500}{E_M \cdot 3.412} = \frac{33,705}{E_M} \ (\text{gal}) \qquad (10.84)$$

where:

E_G = energy content per gallon of gasoline = 11,500 BTUs/gallon, set by the DOE and reported by the Alternative Fuel Data Center (GPO Federal Digital Systems, 2000)
E_E = energy content per watt-hour of electricity = 3.412 BTUs/W·h, set by the DOE and reported by the Alternative Fuel Data Center (GPO Federal Digital Systems, 2000)
E_M = Wall-to-wheel electrical energy consumed per mile (W·h/mi), measured through the EPA's fuel economy tests.

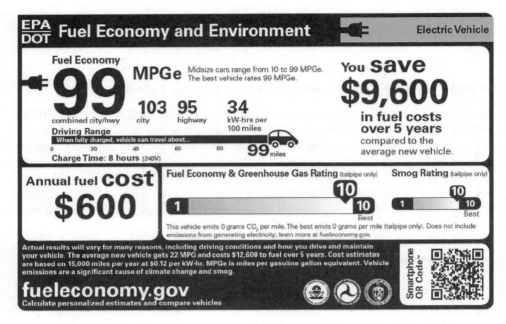

Figure 10.44 Example of a window sticker for an electric vehicle (EPA, 2011)

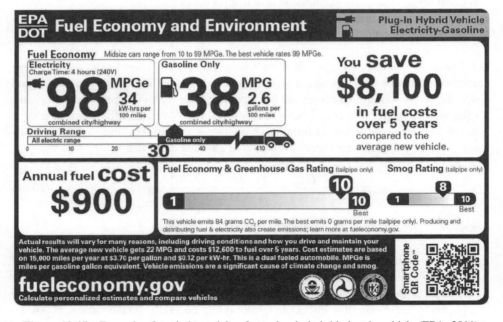

Figure 10.45 Example of a window sticker for a plug-in hybrid electric vehicle (EPA, 2011)

For BEVs, the miles per gallon equivalent values in the city and on the highway need to be calculated separately, as follows:

$$\text{mpge}\big|_{\text{City}} = \frac{E_G}{E_{\text{M\_city}} \cdot E_E} = \frac{11,500}{E_{\text{M\_city}} \cdot 3.412} = \frac{33,705}{E_{\text{M\_city}}} \quad \text{(gal)} \tag{10.85}$$

$$\text{mpge}\big|_{\text{Highway}} = \frac{E_G}{E_{\text{M\_highway}} \cdot E_E} = \frac{11,500}{E_{\text{M\_highway}} \cdot 3.412} = \frac{33,705}{E_{\text{M\_highway}}} \quad \text{(gal)} \tag{10.86}$$

For PHEVs, the miles per gallon equivalent value is calculated by using the combined electrical energy consumed per mile, as follows:

$$\text{mpge} = \frac{E_G}{E_{\text{M\_combined}} \cdot E_E} = \frac{11,500}{E_{\text{M\_combined}} \cdot 3.412} = \frac{33,705}{E_{\text{M\_combined}}} \quad \text{(gal)} \tag{10.87}$$

where:

$$E_{\text{M\_city}} = \frac{E_{\text{AC\_city}}}{R_{\text{City\_adjusted}}} \tag{10.88}$$

$$E_{\text{M\_Highway}} = \frac{E_{\text{AC\_Highway}}}{R_{\text{Highway\_adjusted}}} \tag{10.89}$$

$$E_{\text{M\_combined}} = \frac{E_{\text{AC\_combined}}}{R_{\text{combined\_adjusted}}} \tag{10.90}$$

where:

$E_{\text{AC\_city}}$, $E_{\text{AC\_highway}}$ and $E_{\text{AC\_combined}}$ are the total AC electrical energy consumption in W·h
$R_{\text{City\_adjusted}}$, $R_{\text{Highway\_adjusted}}$ and $R_{\text{combined\_adjusted}}$ are the adjusted full depletion electric mileage tested under FTP city, FTP highway, and combined city and highway cycles for the BEV and PHEV.

Example of an EPA FE Window Labeling Test and Calculation

The following example is based on a PHEV which has a 52 Ahr/18.5 kWh Li-ion battery pack and the designed full depletion window is about 91–16% SOC, supporting about a 50-mile all-electric range. To complete the EPA FE labeling calculation, the following tests and measurements are performed:

- The city electrical energy consumption test is completed over multiple days and consists of 12 phases of UDDS including ten full charge-depletion phases, one transition phase, and one charge-sustaining phase;
- 16.70 kWh of AC energy is measured during recharge following the above urban charge-depleting test;

- The highway electrical energy consumption test is also completed over multiple days and consists of eight phases of HWFET including six full charge-depletion phases, one transition phase, and one charge-sustaining phase;
- 16.38 kWh of AC energy is measured during recharge following the above highway charge-depleting test.

The first phase and the last three phases of the UDDS test results are shown in Figs 10.46 to 10.52; the first phase and the last three phases of the HWFET test results are shown in Figs 10.53 to 10.59.

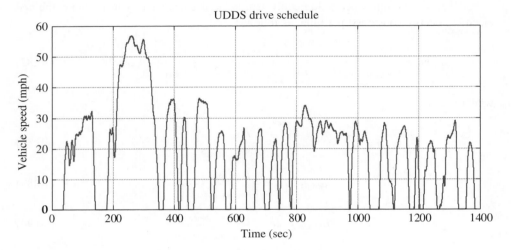

Figure 10.46   FTP UDDS drive schedule

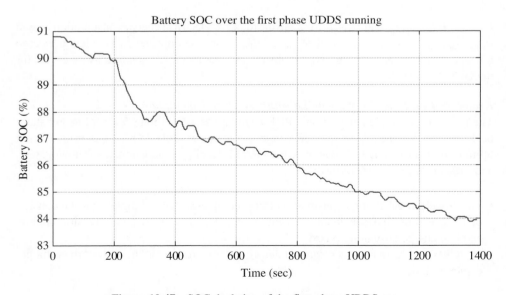

Figure 10.47   SOC depletion of the first phase UDDS run

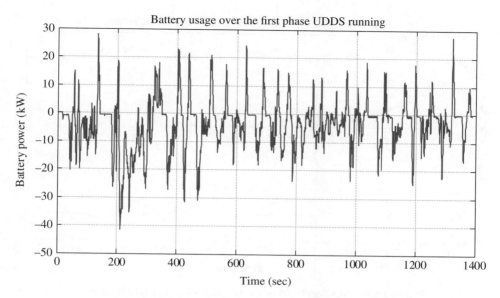

Figure 10.48 Battery usage of the first phase UDDS run

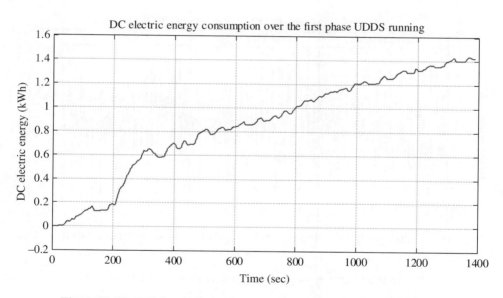

Figure 10.49 DC electrical energy consumption of the first phase UDDS run

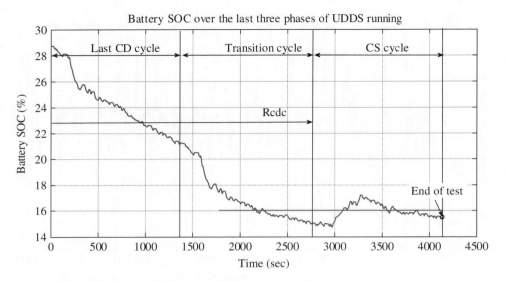

Figure 10.50   SOC depletion of the last three phases of UDDS running

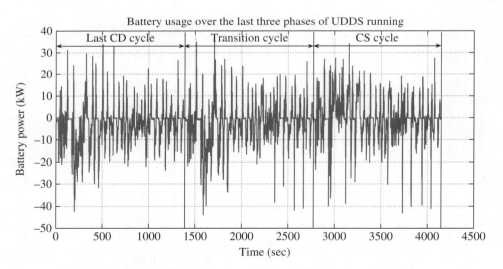

Figure 10.51   Battery usage of the last three phases of UDDS running

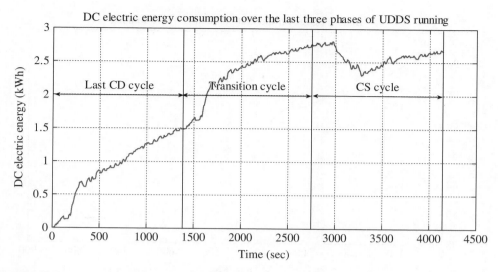

Figure 10.52 DC electrical energy consumption of the last three phases of UDDS running

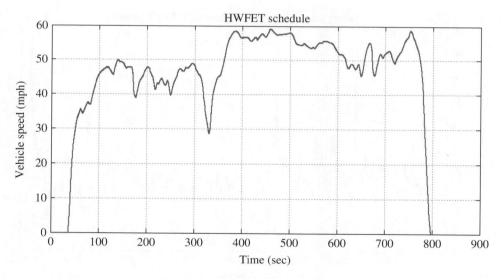

Figure 10.53 HWFET drive schedule

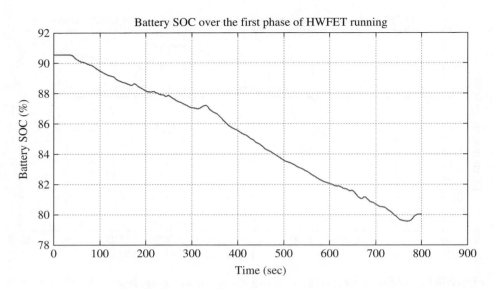

Figure 10.54   SOC depletion of the first phase HWFET run

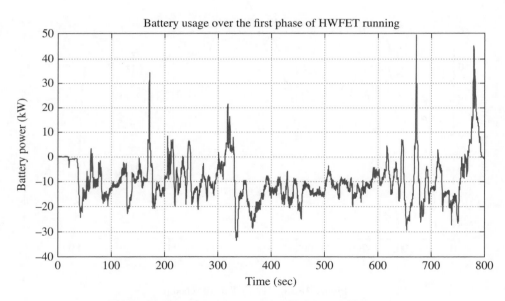

Figure 10.55   Battery usage of the first phase HWFET run

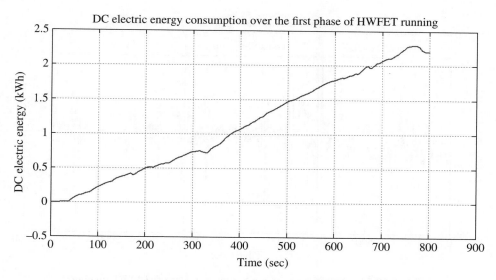

Figure 10.56   DC electrical energy consumption of the first phase HWFET run

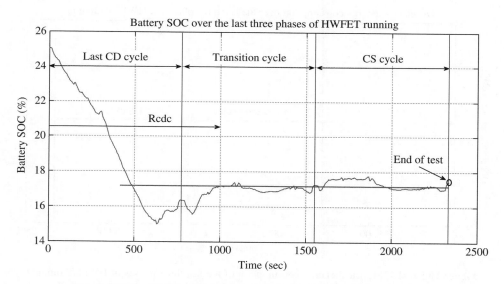

Figure 10.57   SOC depletion of the last three phases of HWFET running

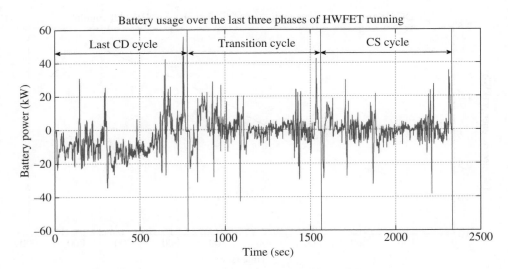

Figure 10.58 Battery usage of the last three phases of HWFET running

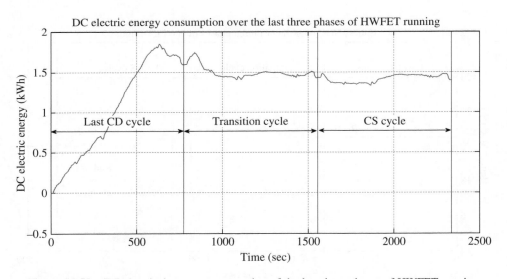

Figure 10.59 DC electrical energy consumption of the last three phases of HWFET running

References

Bosch, R. *Bosch Automotive Handbook*, 7th edition, 2005.

EPA 'New Fuel Economy and Environment Labels for a New Generation of Vehicles,' U.S. Environmental Protection Agency, 2011 (http://www.epa.gov).

EPA 'Fuel Economy Labeling of Motor Vehicles: Revisions to Improve Calculation of Fuel Economy Estimates,' U.S. Environmental Protection Agency, 2012 (http://www.epa.gov).

GPO Federal Digital Systems '65 FR 36986 – Electric and Hybrid Vehicle Research, Development, and Demonstration Program: Petroleum-Equivalent Fuel Economy Calculation, Electric and Hybrid Vehicle Research, Development, and Demonstration Program,' Federal Register, RIN number: 1904–AA40, U.S. Government Publishing Office, 2000 (http://www.gpo.gov/fdsys/pkg/FR-2000-06-12/pdf/00-14446.pdf).

SAE Standard J1491 'Vehicle Acceleration Measurement,' issued in June 1985 and revised in July 2006, Society of Automotive Engineering (SAE) (http://www.sae.org).

SAE Standard J1634 'Battery Electric Vehicle Energy Consumption and Range Test Procedure,' issued in May 1993 and revised in October 2012, Society of Automotive Engineering (SAE) (http://www.sae.org).

SAE Standard J2188 'Commercial Truck and Bus SAE Recommended Procedure for Vehicle Performance Prediction and Charting,' issued in March 1996 and revised in July 2012, Society of Automotive Engineering (SAE) (http://www.sae.org).

SAE Standard J1711 'Recommended Practice for Measuring the Exhaust Emissions and Fuel Economy of Hybrid-Electric Vehicles, Including Plug-in Hybrid Vehicles,' issued in March 1999 and revised in June 2010, Society of Automotive Engineering (SAE) (http://www.sae.org).

Appendix A

System Identification, State and Parameter Estimation Techniques

Building mathematical models of subsystems and components is one of the most important tasks in the analysis and design of hybrid vehicle systems. There are two approaches to building a mathematical model: one is based on the principles and mechanism as well as the relative physical and chemical laws describing the characteristics of a given system, the other is based on the observed system behaviors. In engineering practice, the architecture of the mathematical model is usually determined from the first approach, and detailed model parameters are determined from the second approach. In this appendix, we introduce basic theories and methodologies used to build a mathematical model and estimate the parameters of the model.

A.1 Dynamic System and Mathematical Models

A.1.1 Types of Mathematical Model

The models mentioned in this book are one or a set of mathematical equations that describe the relationship between inputs and outputs of a physical system. These mathematical equations may have various forms, such as algebraic equations, differential equations, partial differential equations, or state-space equations. Mathematical models can be classified as:

- **Static vs. dynamic models:** A static model does not include time; that is, the behavior of the described system does not vary with time, while a dynamic model does. Dynamic

Hybrid Electric Vehicle System Modeling and Control, Second Edition. Wei Liu.
© 2017 John Wiley & Sons Ltd. Published 2017 by John Wiley & Sons Ltd.

models are typically described by a differential equation or a difference equation. Laplace and Fourier transforms can be applied to time-invariant dynamic models.

- **Time-varying vs. time-invariant models:** For a time-varying model, the input–output characteristics vary with time; that is, the model parameters differ with time, while a time-invariant model does not. Laplace and Fourier transforms cannot be applied to time-variant systems.
- **Deterministic vs. stochastic models:** A deterministic model equation is uniquely determined by parameters and previous states of the variables, but different initial conditions may result in different solutions in such models. In contrast, the variables of a stochastic model can only be described by a stochastic process or probability distribution.
- **Continuous vs. discrete models:** A continuous model is one in which the variables are all functions of the continuous time variable, t. A discrete model differs from the continuous model in that one or more variables of the model are in the form of either a pulse train or a digital code. In general, a discrete system receives data or information only intermittently at specific instants of time. For HEV/EV analysis and design, discrete models are mostly used.
- **Linear vs. nonlinear models:** If a mathematical model is linear, it means that all operators in the model present linearity; otherwise the model is nonlinear. A linear model satisfies the principles of superposition, which takes homogeneity and additivity rules together.
- **Lumped-parameter vs. distributed-parameter models:** A lumped-parameter model is a model that can be represented by an ordinary differential or difference equation with time as the independent variable. A distributed-parameter system model is a model that is described by a partial differential equation with space variable and time variable dependence. Heat flow, diffusion processes, and long transmission lines are typical distributed-parameter systems. Given certain assumptions, a distributed-parameter system can be converted into a lumped-parameter system through a finite analysis method such as a finite difference method.

A.1.2   Linear Time-continuous Systems

A.1.2.1   Input–Output Model of a Linear Time-invariant and Time-continuous System

For a linear time-invariant and time-continuous system, like the one shown in Fig. A.1, the input–output relationship is normally described by a linear differential equation:

$$\frac{d^n y}{dt^n} + a_{n-1}\frac{d^{n-1} y}{dt^{n-1}} + \cdots a_1\frac{dy}{dt} + a_0 y = b_m\frac{d^m u}{dt^m} + b_{m-1}\frac{d^{m-1} u}{dt^{m-1}} + \cdots + b_1\frac{du}{dt} + b_0 u \quad (n \geq m)$$

Initial conditions at $t = t_0$: $\left.\frac{d^i y}{dt^i}\right|_{t=t_0}$ $i = 0,\ 1,\ 2, \cdots, n-1$

$$(A.1)$$

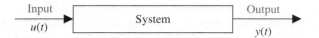

Figure A.1   System with input and output

where u is the input variable and y is the output variable, coefficients a_i, b_i are real constant, and independent of u and y.

For a dynamic system, once the input for $t \geq t_0$ and the initial conditions at $t = t_0$ are specified, the output response at $t \geq t_0$ is determined by solving Eq. A.1.

The transfer function is another way of describing a dynamic system. To obtain the transfer function of the linear system represented by Eq. A.1, we simply take the Laplace transform on both sides of the equation and assume zero initial conditions. The result is:

$$\left(s^n + a_{n-1}s^{n-1} + \cdots + a_1 s + a_0\right)Y(s) = \left(b_m s^m + b_{m-1}s^{m-1} + \cdots + b_1 s + b_0\right)U(s) \quad (n \geq m)$$

$$(A.2)$$

The transfer function between $u(t)$ and $y(t)$ is defined as the ratio of $Y(s)$ and $U(s)$; therefore, the transfer function of the system shown in Fig. A.1 is:

$$G(s) = \frac{Y(s)}{U(s)} = \frac{b_m s^m + b_{m-1}s^{m-1} + \cdots + b_1 s + b_0}{s^n + a_{n-1}s^{n-1} + \cdots + a_1 s + a_0} \tag{A.3}$$

From Eq. A.3, it can be seen that the transfer function is an algebraic equation, by which it will be much easier to analyze the system performance. The transfer function (Eq. A.3) has the following properties:

- The transfer function (Eq. A.3) is defined only for a linear time-invariant system;
- All initial conditions of the system are assumed to be zero;
- The transfer function is independent of the input and output.

Another way of modeling a linear time-invariant system is the impulse response (or weighting function). The impulse response of a linear system is defined as the output response $g(\tau)$ of the system when the input is a unit impulse function $\delta(t)$.

The output of the system shown in Fig. A.1 can be described by its impulse response (or weighting function) $g(\tau)$ as follows:

$$y(t) = \int_{\tau=0}^{t} g(\tau)u(t-\tau)d\tau = g(t)*u(t) = g(t) \text{ convolves into } u(t) \tag{A.4}$$

Knowing $\{g(\tau)\}|_{\tau=0}^{\infty}$ and $u(t)$ for $\tau \leq t$, we can consequently compute the corresponding output $y(t)$, $\tau \leq t$ for any input. Thus, the impulse response is a complete characterization of the system.

The impulse response also directly leads to the definition of transfer functions of linear time-invariant systems. Taking the Laplace transform on both sides of Eq. A.4, we have the following equation from the real convolution theorem of Laplace transformation:

$$G(s) = L(y(t)) = Y(s) = L(g(t)*u(t)) = L(g(t)) \cdot L(u(t)) = G(s)U(s) \qquad (A.5)$$

Equation A.5 shows that the transfer function is the Laplace transform of the impulse response $g(t)$, and the Laplace transform of the output $Y(s)$ is equal to the product of the transform function $G(s)$ and the Laplace transform of the input $U(s)$.

A.1.2.2   State-space Model of a Linear Time-invariant and Time-continuous System

In the state-space form, the relationship between the input and output is written as a first-order differential equation system using a state vector, $\mathbf{x}(t)$. This description of a linear dynamic system became a primary approach after Kalman's work on modern prediction control (Goodwin and Payne, 1977). It is especially useful for hybrid vehicle system design in that insight into physical mechanisms of the system can more easily be incorporated into a state-space model than into an input–output model.

A linear time-invariant and time-continuous system can be described by the following state-space equation:

$$\left.\begin{array}{ll} \text{State equations}: & \dot{\mathbf{x}} = \dfrac{d\mathbf{x}}{dt} = Ax + \mathbf{B}\mathbf{u}(t) \\[2mm] \text{Output equations}: & \mathbf{y}(t) = \mathbf{C}x + \mathbf{D}\mathbf{u}(t) \\[2mm] \text{Initial conditions}: & \mathbf{x}(\mathbf{0}) = \mathbf{x_0} \end{array}\right\} \qquad (A.6)$$

where $\mathbf{x}$ is the $n \times 1$ state vector; $\mathbf{u}$ is the $p \times 1$ input vector; $\mathbf{y}$ is the $q \times 1$ output vector; $\mathbf{A}$ is an $n \times n$ coefficient matrix with constant elements:

$$\mathbf{A} = \begin{bmatrix} a_{11} & a_{12} & \cdots & a_{1n} \\ a_{21} & a_{22} & \cdots & a_{2n} \\ \vdots & \vdots & & \vdots \\ a_{n1} & a_{n2} & \cdots & a_{nn} \end{bmatrix} \qquad (A.7)$$

$\mathbf{B}$ is an $n \times p$ coefficient matrix with constant elements:

$$\mathbf{B} = \begin{bmatrix} b_{11} & b_{12} & \cdots & b_{1p} \\ b_{21} & b_{22} & \cdots & b_{2p} \\ \vdots & \vdots & & \vdots \\ b_{n1} & b_{n2} & \cdots & b_{np} \end{bmatrix} \qquad (A.8)$$

C is a $q \times n$ coefficient matrix with constant elements:

$$
C = \begin{bmatrix} c_{11} & c_{12} & \cdots & c_{1n} \\ c_{21} & c_{22} & \cdots & c_{2n} \\ \vdots & \vdots & & \vdots \\ c_{q1} & c_{q2} & \cdots & c_{qn} \end{bmatrix}
\tag{A.9}
$$

D is a $q \times p$ coefficient matrix with constant elements:

$$
D = \begin{bmatrix} d_{11} & d_{12} & \cdots & d_{1p} \\ d_{21} & d_{22} & \cdots & d_{2p} \\ \vdots & \vdots & & \vdots \\ d_{q1} & d_{q2} & \cdots & d_{qp} \end{bmatrix}
\tag{A.10}
$$

1. Relationship between the state-space equation and differential equation of a dynamic system

Let us consider a single-input, single-output, linear time-invariant system described by an nth-order differential equation (Eq. A.1). The problem is to represent the system (Eq. A.1) by a first-order differential equation system. Since the state variables are internal variables of a given system, they may not be unique and may be dependent on how they are defined. Let us seek a convenient way of assigning the state variables and ensure they have the form of Eq. A.6, as follows.

For Eq. A.6, a way of defining the state variables for Eq. A.1 is:

$$
\dot{x}_1 = \frac{dx_1}{dt} = x_2
$$
$$
\dot{x}_2 = \frac{dx_2}{dt} = x_3
$$
$$
\cdots \cdots \cdots \cdots \cdots
\tag{A.11}
$$
$$
\dot{x}_{n-1} = \frac{dx_{n-1}}{dt} = x_n
$$
$$
\dot{x}_n = \frac{dx_n}{dt} = -a_0 x_1 - a_1 x_2 - a_2 x_3 - \cdots - a_{n-2} x_{n-1} - a_{n-1} x_n + u
$$

where the last state equation is obtained by the highest-order derivative term to the rest of Eq. A.6. The output equation is the linear combination of state variables and input:

$$
y = (b_0 - a_0 b_m) x_1 + (b_1 - a_1 b_m) x_2 + \cdots + (b_{m-1} - a_0 b_m) x_m + b_m u
\tag{A.12}
$$

In vector-matrix form, Eqs A.11 and A.12 are written as:

$$\left.\begin{array}{l} \text{State equations}: \quad \dot{\mathbf{x}} = A x + \mathbf{B} u(t) \\ \text{Output equations}: \quad y(t) = \mathbf{C} \mathbf{x} + \mathbf{D} u(t) \end{array}\right\} \qquad (A.13)$$

where $\mathbf{x}$ is the $n \times 1$ state vector, u is the scalar input, and y is the scalar output. The coefficient matrices are:

$$\mathbf{A} = \begin{bmatrix} 0 & 1 & 0 & \cdots & 0 \\ 0 & 0 & 1 & \cdots & 0 \\ & \cdots & \cdots & \cdots & \\ 0 & 0 & 0 & \cdots & 1 \\ -a_0 & -a_1 & -a_2 & \cdots & -a_{n-1} \end{bmatrix}_{n \times n} \qquad \mathbf{B} = \begin{bmatrix} 0 \\ 0 \\ \vdots \\ 0 \\ 1 \end{bmatrix}_{n \times 1} \qquad (A.14)$$

$$\mathbf{C} = \begin{bmatrix} b_0 - a_0 b_m & b_1 - a_1 b_m & \cdots & b_{m-1} - a_{m-1} b_m & 0 & \cdots & 0 \end{bmatrix}_{1 \times n} \quad \mathbf{D} = [b_m]_{1 \times 1}$$

Example A.1

Consider the input–output differential equation of a dynamic system:

$$\dddot{y} + 6\ddot{y} + 41\dot{y} + 7y = 6u \qquad (A.15)$$

Define a state-space equation and output equation of the system.

Solution:

Define the state variables as:

$$x_1 = \frac{y}{6}, \quad x_2 = \frac{\dot{y}}{6}, \quad x_3 = \frac{\ddot{y}}{6},$$

then, we have the following equations:

$$\dot{x}_1 = \frac{\dot{y}}{6} = x_2, \quad \dot{x}_2 = \frac{\ddot{y}}{6} = x_3$$

$$\dot{x}_3 = \frac{\dddot{y}}{6} = -\ddot{y} - \frac{41}{6}\dot{y} - \frac{7}{6}y + u = -7x_1 - 41x_2 - 6x_3 = u$$

$$y = 6x_1$$

That is, the defined state-space and output equations of the system are:

$$
\begin{bmatrix} \dot{x}_1 \\ \dot{x}_2 \\ \dot{x}_3 \end{bmatrix} = \begin{bmatrix} 0 & 1 & 0 \\ 0 & 0 & 1 \\ -7 & -41 & -6 \end{bmatrix} \begin{bmatrix} x_1 \\ x_2 \\ x_3 \end{bmatrix} + \begin{bmatrix} 0 \\ 0 \\ 1 \end{bmatrix} u
$$

$$
y = \begin{bmatrix} 6 & 0 & 0 \end{bmatrix} \begin{bmatrix} x_1 \\ x_2 \\ x_3 \end{bmatrix}
$$

2. Relationship between the state-space equation and transfer function of a dynamic system

Consider that a linear time-invariant system is described by the state-space model:

$$
\begin{aligned}
&\text{State-space equations:} \quad \dot{\mathbf{x}} = A x + \mathbf{B} u(t) \\
&\text{Output equations:} \qquad \mathbf{y}(t) = \mathbf{C} \mathbf{x} + \mathbf{D} u(t)
\end{aligned}
\tag{A.16}
$$

where $\mathbf{x}(t)$ is the $n \times 1$ state vector, $\mathbf{u}(t)$ is the $p \times 1$ input vector, $\mathbf{y}(t)$ is the $q \times 1$ output vector, and $\mathbf{A}$, $\mathbf{B}$, $\mathbf{C}$, and $\mathbf{D}$ are the coefficient matrices with appropriate dimensions.

Taking the Laplace transform on both sides of Eq. A.16 and solving for $\mathbf{X}(s)$, we have:

$$
s\mathbf{X}(s) - \mathbf{x}(0) = \mathbf{A}\mathbf{X}(s) + \mathbf{B}U(s) \quad \Rightarrow \quad (s\mathbf{I} - \mathbf{A})\mathbf{X}(s) = \mathbf{x}(0) + \mathbf{B}U(s)
\tag{A.17}
$$

Furthermore, this can be written as:

$$
\mathbf{X}(s) = (s\mathbf{I} - \mathbf{A})^{-1}\mathbf{x}(0) + (s\mathbf{I} - \mathbf{A})^{-1}\mathbf{B}U(s)
\tag{A.18}
$$

Assume that the system has zero initial conditions, $\mathbf{x}(0) = 0$, then Eq. A.18 will be:

$$
\mathbf{X}(s) = (s\mathbf{I} - \mathbf{A})^{-1}\mathbf{B}U(s)
\tag{A.19}
$$

The Laplace transform of output equation A.16 is:

$$
Y(s) = \mathbf{C}\mathbf{X}(s) + \mathbf{D}U(s)
\tag{A.20}
$$

Substituting Eq. A.19 into Eq. A.20, we have:

$$
Y(s) = \mathbf{C}(s\mathbf{I} - \mathbf{A})^{-1}\mathbf{B}U(s) + \mathbf{D}U(s) = \left[\mathbf{C}(s\mathbf{I} - \mathbf{A})^{-1}\mathbf{B} + \mathbf{D} \right] U(s)
\tag{A.21}
$$

Thus, the transfer function is defined as:

$$G(s) = \frac{Y(s)}{U(s)} = \mathbf{C}(s\mathbf{I} - \mathbf{A})^{-1}\mathbf{B} + \mathbf{D} \tag{A.22}$$

which is a $q \times p$ matrix corresponding to the dimensions of the input and output variables of the system.

Example A.2

Consider the state-space equation of a dynamic system:

$$\begin{bmatrix} \dot{x}_1 \\ \dot{x}_2 \\ \dot{x}_3 \end{bmatrix} = \begin{bmatrix} 0 & 1 & 0 \\ -2 & -3 & 0 \\ -1 & 1 & -3 \end{bmatrix} \begin{bmatrix} x_1 \\ x_2 \\ x_3 \end{bmatrix} + \begin{bmatrix} 0 \\ 1 \\ 2 \end{bmatrix} u$$

$$y = \begin{bmatrix} 0 & 0 & 1 \end{bmatrix} \begin{bmatrix} x_1 \\ x_2 \\ x_3 \end{bmatrix} \tag{A.23}$$

Determine the transfer function of the system.

Solution:

The corresponding system matrix, input and output vectors are:

$$\mathbf{A} = \begin{bmatrix} 0 & 1 & 0 \\ -2 & -3 & 0 \\ -1 & 1 & -3 \end{bmatrix}, \mathbf{B} = \begin{bmatrix} 0 \\ 1 \\ 2 \end{bmatrix}, \mathbf{C} = \begin{bmatrix} 0 & 0 & 1 \end{bmatrix}$$

Since

$$(s\mathbf{I} - \mathbf{A})^{-1} = \begin{bmatrix} s & -1 & 0 \\ 2 & s+3 & 0 \\ 1 & -1 & s+3 \end{bmatrix}^{-1} = \frac{1}{s^3 + 6s^2 + 11s + 6} \begin{bmatrix} s^2 + 6s + 9 & s+3 & 0 \\ -2(s+3) & s(s+3) & 0 \\ s+5 & s-1 & s^2 + 3s + 2 \end{bmatrix}$$

From Eq. A.22, the transfer function of the system is:

$$G(s) = \frac{Y(s)}{U(s)} = \mathbf{C}(s\mathbf{I} - \mathbf{A})^{-1}\mathbf{B} = \frac{\begin{bmatrix} 0 & 0 & 1 \end{bmatrix}}{s^3 + 6s^2 + 11s + 6} \begin{bmatrix} s^2 + 6s + 9 & s+3 & 0 \\ -2(s+3) & s(s+3) & 0 \\ s+5 & s-1 & s^2 + 3s + 2 \end{bmatrix} \begin{bmatrix} 0 \\ 1 \\ 2 \end{bmatrix}$$

$$= \frac{2s^2 + 7s + 3}{s^3 + 6s^2 + 11s + 6}$$

3. Controllability and observability of a dynamic system

Since state variables are internal variables of a dynamic system, it is necessary to ask whether the state variables are controllable by system inputs, as well as whether they are observable from system outputs. The system controllability and observability will answer these questions, and they are defined as follows.

The states of a dynamic system are controllable if there exists a piecewise continuous control $u(t)$ which will drive the state to any arbitrary finite state $x(t_f)$ from an arbitrary initial state $x(t_0)$ in a finite time $t_f - t_0$.

Correspondingly, the states of a dynamic system are completely observable if the measurement (output) $y(t)$ contains the information which can completely identify the state variables $x(t)$ in a finite time $t_f - t_0$.

The concepts of controllability and observability are very important in theoretical and practical aspects of modern control theory. The following theorems provide the criteria for judging whether the states of a system are controllable and observable or not.

Theorem A.1:   For the system described by the system state-space equation (Eq. A.13) to be completely state controllable, it is necessary and sufficient that the following $n \times np$ matrix has a rank of n:

$$\mathbf{M} = \begin{bmatrix} \mathbf{B} & \mathbf{AB} & \mathbf{A^2B} & \cdots & \mathbf{A}^{n-1}\mathbf{B} \end{bmatrix} \qquad (A.24)$$

This theorem shows that the condition of controllability depends on the coefficient matrices $\mathbf{A}$ and $\mathbf{B}$ of the system described by Eq. A.13. The theorem also gives a way to test the controllability of a given system.

Theorem A.2:   For the system described by Eq. A.13 to be completely observable, it is necessary and sufficient that the following $qn \times n$ matrix has a rank of n:

$$\mathbf{N} = \begin{bmatrix} \mathbf{C} \\ \mathbf{CA} \\ \mathbf{CA^2} \\ \vdots \\ \mathbf{CA}^{n-1} \end{bmatrix} \qquad (A.25)$$

This theorem also gives a way to test state observability of a given system. The concepts of controllability and observability of dynamic systems were first introduced by R. E. Kalman in the 1960s. However, although the criteria of state controllability and observability given by the above theorems are quite straightforward, they are not very easy to implement for a multiple-input system (Ljung, 1987).

A.1.3   Linear Discrete Systems and Modeling

In contrast to the continuous system, the information for discrete time systems is acquired at the sampling moment. If the original signal is continuous, the sampling of the signal at discrete times is a form of signal modulation. A discrete system is usually described by a difference equation, impulse response, discrete state space, or an impulse transfer function.

For a linear time-invariant discrete system, its input–output relationship is described by the following linear difference equation:

$$y(k) + a_1 y(k-1) + \cdots + a_n y(k-n) = b_0 u(k) + b_1 u(k-1) + \cdots + b_m u(k-m) \quad (n \geq m)$$

$$\text{or} \quad y(k) + \sum_{j=1}^{n} a_j y(k-j) = \sum_{j=0}^{m} b_j u(k-j) \quad (n \geq m)$$

$$(A.26)$$

where $u(k)$ is the input variable and $y(k)$ is the output variable, coefficients a_i, b_i are real constants and are independent of $u(k)$ and $y(k)$.

If we introduce

$$A(q) = 1 + a_1 q^{-1} + \cdots + a_n q^{-n} \quad \text{and} \quad B(q) = b_0 + b_1 q^{-1} + \cdots + b_m q^{-m} \quad (A.27)$$

then Eq. A.26 can be written in the form:

$$A(q^{-1}) y(k) = B(q^{-1}) u(k) \tag{A.28}$$

Taking the z-transform on both sides of the equation and assuming zero initial conditions, we will obtain a z-transfer function:

$$G(z) = \frac{Y(z)}{U(z)} = \frac{B(z^{-1})}{A(z^{-1})} = \frac{b_0 + b_1 z^{-1} + \cdots + b_m z^{-m}}{1 + a_1 z^{-1} + \cdots + a_n z^{-n}} \tag{A.29}$$

The state-space model of a linear time-invariant discrete system is as follows:

$$\left. \begin{array}{l} \text{State equations:} \quad \mathbf{x}(k+1) = A x(k) + \mathbf{B} \mathbf{u}(k) \\ \text{Output equations:} \quad \mathbf{y}(k) = \mathbf{C} \mathbf{x}(k) + \mathbf{D} \mathbf{u}(k) \end{array} \right\} \tag{A.30}$$

where $\mathbf{x}(k)$ is an $n \times 1$ state vector; $\mathbf{u}(k)$ is a $p \times 1$ input vector, $\mathbf{y}(k)$ is a $q \times 1$ output vector, and $\mathbf{A}$ is an $n \times n$ coefficient matrix with constant elements:

$$\mathbf{A} = \begin{bmatrix} a_{11} & a_{12} & \cdots & a_{1n} \\ a_{21} & a_{22} & \cdots & a_{2n} \\ \vdots & \vdots & & \vdots \\ a_{n1} & a_{n2} & \cdots & a_{nn} \end{bmatrix} \tag{A.31}$$

B is an $n \times p$ coefficient matrix with constant elements:

$$\mathbf{B} = \begin{bmatrix} b_{11} & b_{12} & \cdots & b_{1p} \\ b_{21} & b_{22} & \cdots & b_{2p} \\ \vdots & \vdots & & \vdots \\ b_{n1} & b_{n2} & \cdots & b_{np} \end{bmatrix} \tag{A.32}$$

C is a $q \times n$ coefficient matrix with constant elements:

$$\mathbf{C} = \begin{bmatrix} c_{11} & c_{12} & \cdots & c_{1n} \\ c_{21} & c_{22} & \cdots & c_{2n} \\ \vdots & \vdots & & \vdots \\ c_{q1} & c_{q2} & \cdots & c_{qn} \end{bmatrix} \tag{A.33}$$

and **D** is a $q \times p$ coefficient matrix with constant elements:

$$\mathbf{D} = \begin{bmatrix} d_{11} & d_{12} & \cdots & d_{1p} \\ d_{21} & d_{22} & \cdots & d_{2p} \\ \vdots & \vdots & & \vdots \\ d_{q1} & d_{q2} & \cdots & d_{qp} \end{bmatrix} \tag{A.34}$$

A.1.4 Linear Time-invariant Discrete Stochastic Systems

Hybrid vehicle design and analysis engineers deal exclusively with observations of inputs and outputs in discrete form. In this section, we introduce the fundamentals of discrete systems.

A.1.4.1 Sampling and Shannon's Sampling Theorem

Because of the discrete-time nature of the hybrid vehicle controller, sampling is a fundamental problem affecting control algorithm design. Shannon's sampling theorem presents the conditions whereby the information in the original signal will not be lost during the sampling. It states that the original continuous-time signal can be perfectly reconstructed if the sampling frequency is equal to or greater than two times the maximal frequency in the original continuous-time signal spectrum; that is:

$$\omega_s = \frac{2\pi}{T} \geq 2\omega_{max} \tag{A.35}$$

The following consequences can be drawn from the theorem:

- In order to ensure a perfect reconstruction of the original signal, the lower bound of the sampling angular frequency is $2\omega_{max}$ for the original signal with the highest frequency component, ω_{max}.
- Or, if the sampling frequency, ω_s, is determined, the highest frequency component of the original signal should be less than $\omega_s/2$ for it to be reconstructed perfectly.
- The frequency $\omega_s/2$ plays an important role in signal conversions. It is also called the *Nyquist frequency*.

In the design of discrete-time systems, selecting an appropriate sampling time (T_s) is one of the most important design steps. Shannon's sampling theorem gives the conditions by which to ensure that the contained information in the original signal will not be lost during the sampling process, but does not say what happens when the conditions and procedures are not exactly met; therefore, a system design engineer who deals with sampling and reconstruction processes needs to understand the original signal thoroughly, particularly in terms of the frequency content. To determine the sampling frequency, the engineer also needs to comprehend how the signal is reconstructed through interpolation and the requirement for the reconstruction error, including aliasing and interpolation error. Generally speaking, the smaller T_s is, the closer the sampled signal is to the continuous signal. But if T_s is very small, the actual implementation may be more costly. If T_s is too large, inaccuracies may occur and much information about the true nature will be lost.

A.1.4.2 Disturbances on a System

Based on Eq. A.28, the output can be calculated exactly once the input is known, but this is unrealistic in most cases. The inputs, outputs, and parameters of a system may vary randomly with time. This randomness is called disturbance, and it may be in the nature of noise. In most cases, such random effects can be described by adding a lumped item at the output of a regular system model, see Fig. A.2 and Eq. A.36.

$$A(q^{-1})y(k) = B(q^{-1})u(k) + \gamma(k) \tag{A.36}$$

A system involving such disturbance is called a stochastic system, in which noise and uncontrollable inputs are the main sources and causes of the disturbance. The most

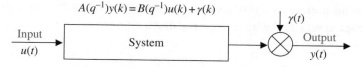

Figure A.2 System with disturbance

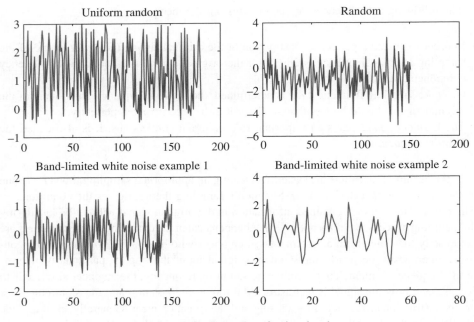

Figure A.3   Examples of noise signals

distinctive feature of a disturbance is that its value cannot be predicted exactly. However, information about past disturbance could be important for making quantified guesses about upcoming values. Hence, it is natural to employ a probability method to describe the statistical features of a disturbance. A special case is where the disturbance term follows a normal distribution; here, the statistical features are uniquely described by the mean value μ and the standard deviation σ of the disturbance. Some examples of noise signals are shown in Fig. A.3. In stochastic system control design, control algorithm design engineers must understand the characteristics of the noise signal, and it is necessary to identify whether the behavior of the system disturbance/noise is stationary or not. For a stationary stochastic process, the probability distribution is the same over time or position; therefore, some parameters obtained by sufficient tests are valid to describe this type of stochastic process.

In practice, the mean value, the standard deviation or the variance, and the peak-to-peak values are the simple features by which to characterize a stationary stochastic process, although the spectral density function, $\varphi(\omega)$, which characterizes the frequency content of a signal, is a better representation of the time behavior of a stationary signal. The value $\frac{1}{2\pi}\varphi(\omega)\Delta\omega$ is the average energy of a signal in a narrow band of width $\Delta\omega$ centered around ω. The average energy in the wide range is defined as:

$$\sigma^2 = \frac{1}{2\pi}\int_{-\infty}^{\infty} \varphi(\omega)d\omega \tag{A.37}$$

(a)   (b)

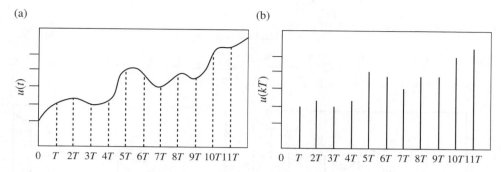

Figure A.4   Continuous and sampled data functions

A signal where $\varphi(\omega)$ is constant is called *white noise*. Such a signal has its energy equally distributed among all frequencies.

In HEV control algorithm design, engineers frequently work with signals described as stochastic processes with deterministic components. This is because the input sequence of a system or component is deterministic, or at least partly deterministic, but the disturbances on the system are conveniently described by random variables, so the system output becomes a stochastic process with deterministic components.

A.1.4.3   Zero-order Hold and First-order Hold

In hybrid vehicle systems, most original signals are continuous. These continuous signals need to be sampled and then sent to the processor at discrete times. With a uniform sampling period, the continuous signal, $u(t)$, shown in Fig. A.4(a), will be sampled at the instances of time 0, $2T$, $3T \cdots$, and the sampled values, shown in Fig. A.4(b), constitute the basis of the system information. They are expressed as a discrete-time function, $u(kT)$, or simplified as $u(k)$. The principle of the sample and hold system is shown in Fig. A.5(a); ideally, the sampler may be regarded as a switch which closes and opens in an infinitely short time, at which time $u(t)$ is measured. In practice, this assumption is justified when the switching duration is very short compared with the sampling interval, T_s, of the system. Since the sampled signal $u(kT)$ is a set of spikes, a device is needed to hold them so that the controller is able to process them. If the signal is held constant over a sampling interval, it is called a *zero-order hold*. If the signal is linearly increasing and decreasing over a sampling interval, it is called a *first-order hold*. The input–output relationship of the zero-order hold and first-order hold are illustrated by Fig. A.5(b) and Fig. A.5(c). For a zero-order hold, the output $u^*(t)$ holds a constant value during the sampling time period T_s, while a first-order hold generates a ramp signal $u_h(t)$ during the sampling time period T_s. Although higher-order holds are able to generate more complex and more accurate wave shapes between the samples, they will complicate the whole system and make it difficult to analyze and design; as a matter of fact, they are seldom used in practice.

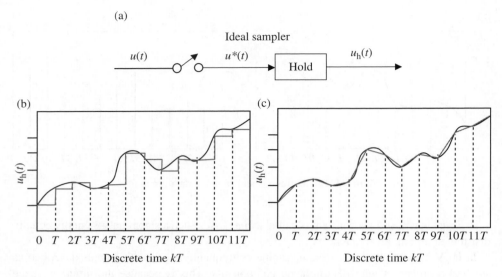

(a)

Ideal sampler

$u(t)$ $u*(t)$ $u_\mathrm{h}(t)$

Hold

(b)   (c)

$u_\mathrm{h}(t)$ $u_\mathrm{h}(t)$

0   T $2T$ $3T$ $4T$ $5T$ $6T$ $7T$ $8T$ $9T$ $10T$ $11T$ 0 T $2T$ $3T$ $4T$ $5T$ $6T$ $7T$ $8T$ $9T$ $10T$ $11T$

Discrete time kT Discrete time kT

Figure A.5   Input and output of sampler and holder

A.1.4.4   Input–Output Model of a Stochastic System

A linear time-invariant stochastic system can be described by the following input–output difference equation:

$$
\begin{aligned}
y(k) + a_1 y(k-1) + \cdots + a_{n_a} y(k-n_a) \\
= b_0 u(k) + b_1 u(k-1) + \cdots + b_{n_b} u(k-n_a) + \xi(k)
\end{aligned}
\tag{A.38}
$$

where $\{\xi(k)\}$ is a white-noise sequence which directly indicates the error in the difference equation.

If we introduce

$$
\begin{aligned}
\theta &= [a_1 \ a_2 \ \cdots \ a_{n_a} \ b_1 \ \cdots \ b_{n_b}]^\mathrm{T} \\
A(q^{-1}) &= 1 + a_1 q^{-1} + \cdots + a_{n_a} q^{-n_a} \\
B(q^{-1}) &= b_0 + b_1 q^{-1} + \cdots + b_{n_b} q^{-n_b}
\end{aligned}
\tag{A.39}
$$

a transfer-function-form model of Eq. A.38 can be obtained as:

$$
G(q^{-1}) = \frac{B(q^{-1})}{A(q^{-1})}, \ \text{and} \ H(q^{-1}) = \frac{1}{A(q^{-1})}
\tag{A.40}
$$

The model in Eq. A.38 or Eq. A.40 is called an ARX model, where AR refers to the auto-regressive part and X refers to the extra input, $B(q^{-1})u(t)$, that is called the *exogenous variable*. If a certain degree of flexibility is added to describe the white-noise error in Eq. A.38, such as a moving average of white noise, the following model is given:

$$
\begin{aligned}
y(k) + a_1 y(k-1) + \cdots + a_{n_a} y(k-n) \\
= b_0 u(k) + b_1 u(k-1) + \cdots + b_{n_b} u(k-n) + w(k) + c_1 w(k-1) + \cdots + c_{n_c} w(k-n_c)
\end{aligned} \tag{A.41}
$$

This can also be written in the form:

$$
A\!\left(q^{-1}\right) y(k) = B\!\left(q^{-1}\right) u(k) + C\!\left(q^{-1}\right) w(k) \tag{A.42}
$$

where $u(k)$ is the system input, $y(k)$ is the output, $w(k)$ is independent white noise, and $A(q^{-1})$, $B(q^{-1})$, $C(q^{-1})$ are:

$$
\begin{aligned}
A(q^{-1}) &= 1 + a_1 q^{-1} + \cdots + a_{n_a} q^{-n_a} \\
B(q^{-1}) &= b_0 + b_1 q^{-1} + \cdots + b_{n_b} q^{-n_b} \\
C(q^{-1}) &= 1 + c_1 q^{-1} + \cdots + c_{n_c} q^{-n_c} \\
\theta &= \begin{bmatrix} a_1 & a_2 & \cdots & a_{n_a} & b_0 & \cdots & b_{n_b} & c_1 & \cdots & c_{n_c} \end{bmatrix}^{\mathrm{T}}
\end{aligned} \tag{A.43}
$$

The model in Eq. A.41 is called an ARMAX model, which refers to the autoregressive moving average model with exogenous inputs. ARMAX models are usually used to estimate system parameters online based on real-time measured series data.

A.1.4.5 State-space Model of a Stochastic System

The state-space model describing a linear time-invariant stochastic system is:

$$
\begin{aligned}
x(k+1) &= \mathbf{A}x(k) + \mathbf{B}u(k) + w(k) \\
y(k) &= \mathbf{C}x(k) + \mathbf{D}u(k) + v(k)
\end{aligned} \tag{A.44}
$$

where $\mathbf{A}$, $\mathbf{B}$, $\mathbf{C}$, and $\mathbf{D}$ are the coefficient matrices with appropriate dimensions, $\{w(k)\}$ and $\{v(k)\}$ are two uncorrelated white-noise sequences with covariance Q and R, respectively.

Based on the superposition principle, the model in Eq. A.44 can be expressed in terms of two components as:

$$
y(k) = \bar{y}(k) + \eta(k) \tag{A.45}
$$

where $\bar{y}(k)$ is the output of the following deterministic model:

$$
\begin{aligned}
\bar{x}(k+1) &= \mathbf{A}\bar{x}(k) + \mathbf{B}u(k) \\
\bar{y}(k) &= \mathbf{C}\bar{x}(k) + \mathbf{D}u(k)
\end{aligned} \tag{A.46}
$$

where $\eta(k)$ is a zero-mean stochastic process with the following spectral density:

$$\Phi_\eta(z) = \mathbf{C}(z\mathbf{I}-\mathbf{A})^{-1}Q(z^{-1}\mathbf{I}-\mathbf{A}^\mathrm{T})^{-1}\mathbf{C}^\mathrm{T} + R \tag{A.47}$$

A.2 Parameter Estimation for Dynamic Systems

In this section, we turn to the problem of parameter estimation for a dynamic system. There are many different methods that can be used to determine the parameters of a model, and there are also different criteria as to which method should be selected, but we will only briefly introduce the basic principle of the least squares estimation method, which is used widely in engineering.

A.2.1 Least Squares

Least squares is a classic method for dealing with experimental data to predict the orbits of planets and comets that was first described by Gauss in 1795. The principle is that unknown parameters of a model should be chosen in such a way that the sum of the squares of the difference between actually observed and computed values is a minimum. If we assume the computed output, $\hat{y}$, is given by the model, the least squares principle can be mathematically described as follows:

$$\hat{y}(k) = \theta_1 x_1(k) + \theta_2 x_2(k) + \cdots + \theta_n x_n(k) = \phi(k)\boldsymbol{\theta} \tag{A.48}$$

where x_1, x_2, $\cdots$, x_n are known inputs, y is the output, θ_1, θ_2, $\cdots$, θ_n are unknown parameters, and $\phi(k) = \left[x_1^\mathrm{T}(k) \ x_2^\mathrm{T}(k) \ \cdots \ x_n^\mathrm{T}(k) \right]^\mathrm{T}$, $\boldsymbol{\theta} = [\theta_1 \ \theta_2 \ \cdots \ \theta_n]^\mathrm{T}$.

The pairs of observations $\{(x_i \ y_i), i = 1, 2, \cdots, N\}$ are obtained from an experiment or test. According to Gauss's principle, the estimated parameters should make the following cost function minimal:

$$J(\theta) = \sum_{i=1}^{N} \varepsilon_i^2 = \sum_{i=1}^{N} (y_i - \hat{y}_i)^2 = \sum_{i=1}^{N} (y_i - \phi(i)\boldsymbol{\theta})^2 = \min \tag{A.49}$$

Applying partial derivatives to Eq. A.49 and letting them be equal to zero, i e., $\dfrac{\partial J(\theta)}{\partial \boldsymbol{\theta}} = 0$, the solution to the least squares is:

$$\hat{\boldsymbol{\theta}} = \left(\Phi^\mathrm{T}\Phi\right)^{-1}\Phi^\mathrm{T}Y \tag{A.50}$$

where $\boldsymbol{\Phi} = [\phi(1) \ \cdots \ \phi(N)]^\mathrm{T}$ and $\mathbf{Y} = [y(1) \ \cdots \ y(N)]^\mathrm{T}$.

The above least squares method can be used to estimate the parameters in the dynamic system described by Eq. A.38 or A.41 with $C(q^{-1}) = 1$. If we assume that a sequence of input $\{u(1), u(2), \cdots, u(N)\}$ has been applied to the system and the corresponding sequence of output $\{y(1), y(2), \cdots, y(N)\}$ has been measured, the following vectors can be configured for the least squares described by Eq. A.50, and the unknown parameters are:

$$\boldsymbol{\theta} = [a_1 \cdots a_{n_a} b_0 \cdots b_{n_b}]^{\mathrm{T}} \tag{A.51}$$

$$\phi = [-y(k-1) \cdots -y(k-n_a-1) \ u(k) \cdots u(k-n_b)]^{\mathrm{T}} \tag{A.52}$$

$$\boldsymbol{\Phi} = \begin{bmatrix} \phi^{\mathrm{T}}(1) \\ \vdots \\ \phi^{\mathrm{T}}(N) \end{bmatrix} \text{ and } \mathbf{Y} = \begin{bmatrix} y(1) \\ \vdots \\ y(N) \end{bmatrix} \quad N \geq n_a + n_b + 1 \tag{A.53}$$

A.2.2 Statistical Property of Least Squares Estimator

If we assume that the data are generated from the following model:

$$\mathbf{Y} = \boldsymbol{\Phi}\boldsymbol{\theta}_0 + \boldsymbol{\varepsilon} \tag{A.54}$$

where $\boldsymbol{\theta}_0 \in R^n$ is the vector of theoretical true values of the model parameters, $\boldsymbol{\varepsilon} \in R^n$ is a vector of white noise with zero mean and variance σ^2, that is, $E\{\boldsymbol{\varepsilon}\} = 0$ and $E\{\boldsymbol{\varepsilon}\boldsymbol{\varepsilon}^{\mathrm{T}}\} = \sigma^2\mathbf{I}$, then the least squares estimate of $\boldsymbol{\theta}_0$ given by Eq. A.50 has the following properties:

1. The bias (expectations)

The bias of an estimator is defined as the difference between the true value of the estimated parameter and the expected value of the estimate. If the difference is zero, the estimator is called unbiased; otherwise it is said to be biased. The introduced LS is an unbiased estimator if the noise is independent and with zero mean.i.e.

$$E\{\hat{\boldsymbol{\theta}}\} = \boldsymbol{\theta}_0 \tag{A.55}$$

This can be proven from:

$$E\{\hat{\boldsymbol{\theta}}\} = E\{(\boldsymbol{\Phi}^{\mathrm{T}}\boldsymbol{\Phi})^{-1}\boldsymbol{\Phi}^{\mathrm{T}}\mathbf{Y}\} = E\{(\boldsymbol{\Phi}^{\mathrm{T}}\boldsymbol{\Phi})^{-1}\boldsymbol{\Phi}^{\mathrm{T}}(\boldsymbol{\Phi}\boldsymbol{\theta}_0 + \boldsymbol{\varepsilon})\}$$

$$= (\boldsymbol{\Phi}^{\mathrm{T}}\boldsymbol{\Phi})^{-1}\boldsymbol{\Phi}^{\mathrm{T}}E\{(\boldsymbol{\Phi}\boldsymbol{\theta}_0 + \boldsymbol{\varepsilon})\} = \boldsymbol{\theta}_0$$

2. Variances

The LS is the minimal variance estimator; that is:

$$\text{Var}\left(\hat{\boldsymbol{\theta}}\right) = E\left\{\left(\hat{\boldsymbol{\theta}} - \boldsymbol{\theta}_0\right)\left(\hat{\boldsymbol{\theta}} - \boldsymbol{\theta}_0\right)^{\text{T}}\right\} = \sigma^2\left(\boldsymbol{\Phi}^{\text{T}}\boldsymbol{\Phi}\right)^{-1} \tag{A.56}$$

This is derived from:

$$\begin{aligned}
E\left\{\left(\hat{\boldsymbol{\theta}} - \boldsymbol{\theta}_0\right)\left(\hat{\boldsymbol{\theta}} - \boldsymbol{\theta}_0\right)^{\text{T}}\right\} &= E\left\{\left[\left(\boldsymbol{\Phi}^{\text{T}}\boldsymbol{\Phi}\right)^{-1}\boldsymbol{\Phi}^{\text{T}}\mathbf{Y} - \boldsymbol{\theta}_0\right]\left[\left(\boldsymbol{\Phi}^{\text{T}}\boldsymbol{\Phi}\right)^{-1}\boldsymbol{\Phi}^{\text{T}}\mathbf{Y} - \boldsymbol{\theta}_0\right]^{\text{T}}\right\} \\
&= \left(\boldsymbol{\Phi}^{\text{T}}\boldsymbol{\Phi}\right)^{-1}\boldsymbol{\Phi}^{\text{T}}E\left\{(\mathbf{Y} - \boldsymbol{\Phi}\boldsymbol{\theta}_0)(\mathbf{Y} - \boldsymbol{\Phi}\boldsymbol{\theta}_0)^{\text{T}}\right\}\boldsymbol{\Phi}\left(\boldsymbol{\Phi}^{\text{T}}\boldsymbol{\Phi}\right)^{-1} \quad (\text{A.57}) \\
&= \left(\boldsymbol{\Phi}^{\text{T}}\boldsymbol{\Phi}\right)^{-1}\boldsymbol{\Phi}^{\text{T}}\sigma^2\mathbf{I}\boldsymbol{\Phi}\left(\boldsymbol{\Phi}^{\text{T}}\boldsymbol{\Phi}\right)^{-1} = \sigma^2\left(\boldsymbol{\Phi}^{\text{T}}\boldsymbol{\Phi}\right)^{-1}
\end{aligned}$$

3. Consistency

The consistency property of an estimator means that if the observation data size N is sufficiently large, the estimator is able to find the value of $\boldsymbol{\theta}_0$ with arbitrary precision. In mathematical terms, this means that as N goes to infinity, the estimate $\hat{\boldsymbol{\theta}}$ converges to $\boldsymbol{\theta}_0$.

The following proof shows that the LS estimator is a consistent estimator; that is, the LS estimate $\hat{\boldsymbol{\theta}}$ converges to $\boldsymbol{\theta}_0$ as the observation size N tends to infinity. In mathematical terms, if we define $\lim_{N \to \infty}\left\{\left(\frac{1}{N}\boldsymbol{\Phi}^{\text{T}}\boldsymbol{\Phi}\right)\right\} = \Gamma$ and Γ is nonsingular, then the estimate $\hat{\boldsymbol{\theta}}$ converges to the true value $\boldsymbol{\theta}_0$; that is, $\lim_{N \to \infty} E\left\{\left(\hat{\boldsymbol{\theta}} - \boldsymbol{\theta}_0\right)\left(\hat{\boldsymbol{\theta}} - \boldsymbol{\theta}_0\right)^{\text{T}}\right\} = 0$.

Proof:

$$\begin{aligned}
\lim_{N \to \infty} E\left\{\left(\hat{\boldsymbol{\theta}} - \boldsymbol{\theta}_0\right)\left(\hat{\boldsymbol{\theta}} - \boldsymbol{\theta}_0\right)^{\text{T}}\right\} &= \lim_{N \to \infty}\left\{\sigma^2\left(\boldsymbol{\Phi}^{\text{T}}\boldsymbol{\Phi}\right)^{-1}\right\} = \lim_{N \to \infty}\left\{\frac{\sigma^2}{N}\left(\frac{1}{N}\boldsymbol{\Phi}^{\text{T}}\boldsymbol{\Phi}\right)^{-1}\right\} \\
&= \lim_{N \to \infty}\frac{\sigma^2}{N} \cdot \lim_{N \to \infty}\left\{\left(\frac{1}{N}\boldsymbol{\Phi}^{\text{T}}\boldsymbol{\Phi}\right)^{-1}\right\} = \lim_{N \to \infty}\frac{\sigma^2}{N} \cdot \Gamma = 0
\end{aligned}$$

$$\tag{A.58}$$

Example A.3

Determine the model parameters a_1, a_2, b_1, b_2 by the least squares estimation method based on the observed input and output data $\{u(1),\ u(2),\ \cdots,\ u(N)\}$ and $\{y(1),\ y(2),\ \cdots,\ y(N)\}$.

$$y(k) + a_1 y(k-1) + a_2 y(k-2) = b_1 u(k-1) + b_2 u(k-2)$$

Solution:

Compared with the least squares formula given in Eq. A.50, we have:

$$
\mathbf{\Phi} = \begin{bmatrix} -y(2) & -y(1) & u(2) & u(1) \\ -y(3) & -y(2) & u(3) & u(2) \\ \vdots & \vdots & \vdots & \vdots \\ -y(N-2) & -y(N-3) & u(N-2) & u(N-3) \\ -y(N-1) & -y(N-2) & u(N-1) & u(N-2) \end{bmatrix}, \quad \mathbf{Y} = \begin{bmatrix} y(3) \\ y(4) \\ \vdots \\ y(N-1) \\ y(N) \end{bmatrix}, \quad \mathbf{\theta} = \begin{bmatrix} a_1 \\ a_2 \\ b_1 \\ b_2 \end{bmatrix}
$$

and

$$
\mathbf{\Phi}^{\mathrm{T}}\mathbf{\Phi} = \begin{bmatrix} \sum\limits_{i=2}^{N-1} y^2(i) & \sum\limits_{i=2}^{N-1} y(i)y(i-1) & -\sum\limits_{i=2}^{N-1} y(i)u(i) & -\sum\limits_{i=2}^{N-1} y(i)u(i-1) \\ \sum\limits_{i=2}^{N-1} y(i)y(i-1) & \sum\limits_{i=2}^{N-1} y^2(i-1) & -\sum\limits_{i=2}^{N-1} y(i-1)u(i) & -\sum\limits_{i=2}^{N-1} y(i-1)u(i-1) \\ -\sum\limits_{i=2}^{N-1} y(i)u(i) & -\sum\limits_{i=2}^{N-1} y(i-1)u(i) & \sum\limits_{i=2}^{N-1} u^2(i) & -\sum\limits_{i=2}^{N-1} u(i)u(i-1) \\ -\sum\limits_{i=2}^{N-1} y(i)u(i-1) & -\sum\limits_{i=2}^{N-1} y(i-1)u(i-1) & \sum\limits_{i=2}^{N-1} u(i)u(i-1) & \sum\limits_{i=2}^{N-1} u^2(i-1) \end{bmatrix}_{4\times 4},
$$

$$
\mathbf{\Phi}^{\mathrm{T}}\mathbf{Y} = \begin{bmatrix} -\sum\limits_{i=2}^{N-1} y(i)y(i+1) \\ -\sum\limits_{i=2}^{N-1} y(i-1)y(i+1) \\ \sum\limits_{i=2}^{N-1} u(i)y(i+1) \\ \sum\limits_{i=2}^{N-1} u(i-1)y(i+1) \end{bmatrix}
$$

If the matrix $\mathbf{\Phi}^{\mathrm{T}}\mathbf{\Phi}$ is nonsingular, the estimated parameter $\mathbf{\theta} = [a_1, a_2, b_1, b_2]^{\mathrm{T}}$ is $\hat{\mathbf{\theta}} = (\mathbf{\Phi}^{\mathrm{T}}\mathbf{\Phi})^{-1}\mathbf{\Phi}^{\mathrm{T}}\mathbf{Y}$. If persistent excitation is not imposed on the input signals, the matrix $\mathbf{\Phi}^{\mathrm{T}}\mathbf{\Phi}$ will be singular, as a result of which no unique estimate can be found by least squares.

A.2.3 Recursive Least Squares Estimator

In most practical applications, the observed data are obtained sequentially. If the least squares estimation has to be solved for N observations, the estimate not only wastes computational resources but also overoccupies limited memory. It is necessary to estimate the model parameters in such a way that the $N + 1$ estimate is performed based on the results obtained for N observations. Parameter estimation techniques that comply with this requirement are called *recursive* estimation methods. As long as the measured input–output data are processed sequentially, they become available. Recursive estimation methods are also referred to as online or real-time estimation.

Based on N observations, $y_1, y_2, \cdots, y_N$, the least squares estimate of parameter $\hat{\theta}$ is given by Eq. A.50 as:

$$\hat{\theta}_N = \left(\Phi_N^T \Phi_N\right)^{-1} \Phi_N^T Y_N \tag{A.59}$$

where

$$\Phi_N = \begin{bmatrix} \phi^T(1) \\ \vdots \\ \phi^T(N) \end{bmatrix} \quad \text{and} \quad Y_N = \begin{bmatrix} y(1) \\ \vdots \\ y(N) \end{bmatrix} \tag{A.60}$$

In order to achieve a recursive least squares algorithm, we first assume that the parameters $\hat{\theta}_N$ have been estimated based on the N known observations. The objective here is to achieve $\hat{\theta}_{N+1}$ based on $\hat{\theta}_N$ and just one extra observation, y_{N+1}.

First, let's define

$$P_N = \left(\Phi_N^T \Phi_N\right)^{-1} \tag{A.61}$$

Then, we have:

$$P_{N+1} = \left(\Phi_{N+1}^T \Phi_{N+1}\right)^{-1} = \left(\begin{bmatrix} \Phi_N \\ \phi^T(N+1) \end{bmatrix}^T \begin{bmatrix} \Phi_N \\ \phi^T(N+1) \end{bmatrix} \right)^{-1} \tag{A.62}$$

$$= \left[\Phi_N^T \Phi_N + \phi(N+1)\phi^T(N+1) \right]^{-1}$$

Based on the matrix inversion lemma introduced later, we have:

$$P_{N+1} = \left[\Phi_N^T \Phi_N + \phi(N+1)\phi^T(N+1) \right]^{-1}$$

$$= \left(\Phi_N^T \Phi_N\right)^{-1} - \left(\Phi_N^T \Phi_N\right)^{-1}\phi^T(N+1)\left[I + \phi(N+1)\left(\Phi_N^T \Phi_N\right)^{-1}\phi^T(N+1)\right]^{-1}\phi(N+1)\left(\Phi_N^T \Phi_N\right)^{-1}$$

$$= P_N - P_N\phi^T(N+1)\left[I + \phi(N+1)P_N\phi^T(N+1)\right]^{-1}\phi(N+1)P_N$$

$$\tag{A.63}$$

Let

$$\mathbf{K}_N = \mathbf{P}_N \phi^{\mathrm{T}}(N+1)\left[\mathbf{I} + \phi(N+1)\mathbf{P}_N \phi^{\mathrm{T}}(N+1)\right]^{-1} \tag{A.64}$$

Then

$$\mathbf{P}_{N+1} = \mathbf{P}_N - \mathbf{K}_N \phi(N+1)\mathbf{P}_N \tag{A.65}$$

Returning to Eq. A.59, the following results are achieved from the above equations:

$$\hat{\boldsymbol{\theta}}_{N+1} = \left(\boldsymbol{\Phi}_{N+1}^{\mathrm{T}}\boldsymbol{\Phi}_{N+1}\right)^{-1}\boldsymbol{\Phi}_{N+1}^{\mathrm{T}}\mathbf{Y}_{N+1} = \left[\mathbf{P}_N - \mathbf{K}_N \phi(N+1)\mathbf{P}_N\right]\begin{bmatrix}\boldsymbol{\Phi}_N \\ \phi^{\mathrm{T}}(N+1)\end{bmatrix}^{\mathrm{T}}\begin{bmatrix}\mathbf{Y}_N \\ y_{N+1}\end{bmatrix}$$

$$= \left[\mathbf{P}_N - \mathbf{K}_N \phi(N+1)\mathbf{P}_N\right]\left[\boldsymbol{\Phi}_N^{\mathrm{T}}\mathbf{Y}_N + \phi^{\mathrm{T}}(N+1)y_{N+1}\right]$$

$$= \mathbf{P}_N \boldsymbol{\Phi}_N^{\mathrm{T}}\mathbf{Y}_N + \mathbf{P}_N \phi^{\mathrm{T}}(N+1)y_{N+1} - \mathbf{K}_N \phi(N+1)\mathbf{P}_N \boldsymbol{\Phi}_N^{\mathrm{T}}\mathbf{Y}_N - \mathbf{K}_N \phi(N+1)\mathbf{P}_N \phi^{\mathrm{T}}(N+1)y_{N+1}$$

$$= \hat{\boldsymbol{\theta}}_N - \mathbf{K}_N \phi(N+1)\hat{\boldsymbol{\theta}}_N + \mathbf{K}_N y_{N+1}$$

$$= \hat{\boldsymbol{\theta}}_N + \mathbf{K}_N \left(y_{N+1} - \phi(N+1)\hat{\boldsymbol{\theta}}_N\right) \tag{A.66}$$

so the recursive least squares estimation method is obtained and summarized as:

$$\hat{\boldsymbol{\theta}}(k+1) = \hat{\boldsymbol{\theta}}(k) + \mathbf{K}(k)\left(y(k+1) - \phi(k+1)\hat{\boldsymbol{\theta}}(k)\right) \tag{A.67}$$

$$\mathbf{K}(k) = \mathbf{P}(k)\phi^{\mathrm{T}}(k+1)\left[\mathbf{I} + \phi(k+1)\mathbf{P}(k)\phi^{\mathrm{T}}(k+1)\right]^{-1} \tag{A.68}$$

$$\mathbf{P}(k+1) = \mathbf{P}(k) - \mathbf{K}(k)\phi(k+1)\mathbf{P}(k) \tag{A.69}$$

Remark 1: The estimate $\hat{\boldsymbol{\theta}}(k+1)$ is achieved by adding a correction to the previous estimate, $\hat{\boldsymbol{\theta}}(k)$. The correction is proportional to the difference between the measured output value of $y(k+1)$ and the prediction $\hat{y}(k+1)$ of $y(k+1)$ based on the previous estimate. The components of the gain vector, $\mathbf{K}(k)$, reflect how to correct the previous estimate, $\hat{\boldsymbol{\theta}}(k)$, based on the new observed data.

Remark 2: If $\boldsymbol{\Phi}_0$ and $\mathbf{Y}_0$ can be obtained from an initial set of data, the starting values $\mathbf{P}_0$ and $\hat{\boldsymbol{\theta}}_0$ may be obtained by evaluating $\left(\boldsymbol{\Phi}_0^{\mathrm{T}}\boldsymbol{\Phi}_0\right)^{-1}$ and $\hat{\boldsymbol{\theta}}_0 = \left(\boldsymbol{\Phi}_0^{\mathrm{T}}\boldsymbol{\Phi}_0\right)^{-1}\boldsymbol{\Phi}_0^{\mathrm{T}}\mathbf{Y}_0$, respectively. If there is no way to obtain enough initial observations, $\mathbf{P}_0$ may be set as $\mathbf{I}\rho^2$; that is, $\mathbf{P}_0 = \mathbf{I}\rho^2$ where ρ is a very large number and $\hat{\boldsymbol{\theta}}_0$ may be arbitrary. For large N, the choice of the initial values of $\mathbf{P}_0$ and $\boldsymbol{\theta}_0$ is unimportant.

Matrix Inversion Lemma

The matrix inversion lemma states that:

$$(\mathbf{A} + \mathbf{BCD})^{-1} = \mathbf{A}^{-1} - \mathbf{A}^{-1}\mathbf{B}(\mathbf{C}^{-1} + \mathbf{DA}^{-1}\mathbf{B})^{-1}\mathbf{DA}^{-1} \tag{A.70}$$

where $\mathbf{A}$, $\mathbf{C}$, and $(\mathbf{A} + \mathbf{BCD})$ are regular square matrices of appropriate size.

Proof:

Multiply the left side by the right side of Eq. A.70. If the result equals the identity matrix, the lemma is proven. Thus, we have:

$$(\mathbf{A} + \mathbf{BCD})\left[\mathbf{A}^{-1} - \mathbf{A}^{-1}\mathbf{B}(\mathbf{C}^{-1} + \mathbf{DA}^{-1}\mathbf{B})^{-1}\mathbf{DA}^{-1}\right]$$

$$= \mathbf{AA}^{-1} - \mathbf{AA}^{-1}\mathbf{B}(\mathbf{C}^{-1} + \mathbf{DA}^{-1}\mathbf{B})^{-1}\mathbf{DA}^{-1} + \mathbf{BCDA}^{-1} - \mathbf{BCDA}^{-1}\mathbf{B}(\mathbf{C}^{-1} + \mathbf{DA}^{-1}\mathbf{B})^{-1}\mathbf{DA}^{-1}$$

$$= \mathbf{I} + \mathbf{BCDA}^{-1} - \mathbf{B}(\mathbf{C}^{-1} + \mathbf{DA}^{-1}\mathbf{B})^{-1}\mathbf{DA}^{-1} - \mathbf{BCDA}^{-1}\mathbf{B}(\mathbf{C}^{-1} + \mathbf{DA}^{-1}\mathbf{B})^{-1}\mathbf{DA}^{-1}$$

$$= \mathbf{I} + \mathbf{BCDA}^{-1} - \mathbf{B}\left[(\mathbf{C}^{-1} + \mathbf{DA}^{-1}\mathbf{B})^{-1} + \mathbf{CDA}^{-1}\mathbf{B}(\mathbf{C}^{-1} + \mathbf{DA}^{-1}\mathbf{B})^{-1}\right]\mathbf{DA}^{-1}$$

$$= \mathbf{I} + \mathbf{BCDA}^{-1} - \mathbf{B}(\mathbf{I} + \mathbf{CDA}^{-1}\mathbf{B})(\mathbf{C}^{-1} + \mathbf{DA}^{-1}\mathbf{B})^{-1}\mathbf{DA}^{-1}$$

$$= \mathbf{I} + \mathbf{BCDA}^{-1} - \mathbf{BC}(\mathbf{I} + \mathbf{DA}^{-1}\mathbf{B})(\mathbf{C}^{-1} + \mathbf{DA}^{-1}\mathbf{B})^{-1}\mathbf{DA}^{-1}$$

$$= \mathbf{I} + \mathbf{BCDA}^{-1} - \mathbf{BCDA}^{-1}$$

$$= \mathbf{I}$$

This completes the proof.

Remark 3: If $\mathbf{D} = \mathbf{B}^{\mathrm{T}}$, then we have:

$$(\mathbf{A} + \mathbf{BCB}^{\mathrm{T}})^{-1} = \mathbf{A}^{-1} - \mathbf{A}^{-1}\mathbf{B}(\mathbf{C}^{-1} + \mathbf{B}^{\mathrm{T}}\mathbf{A}^{-1}\mathbf{B})^{-1}\mathbf{B}^{\mathrm{T}}\mathbf{A}^{-1} \tag{A.71}$$

Remark 4: If $\mathbf{C} = \mathbf{I}$, then we have:

$$(\mathbf{A} + \mathbf{BD})^{-1} = \mathbf{A}^{-1} - \mathbf{A}^{-1}\mathbf{B}(\mathbf{I} + \mathbf{DA}^{-1}\mathbf{B})^{-1}\mathbf{DA}^{-1} \tag{A.72}$$

A.2.4   Least Squares Estimator for Slow, Time-varying Parameters

The recursive least squares estimation method described in the previous section is not directly applicable when the parameters vary with time as new data are swamped by past data. There are two basic ways to modify the described recursive method to handle time-varying parameters.

1. Exponential window approach (exponentially weighted least squares)

It is obvious that the cost function (Eq. A.49) equally makes use of all observed data. However, if the system parameters are slowly time varying, the influence of old data on the parameters is gradually eliminated. The idea of the exponential window approach is to artificially emphasize the effect of current data by exponentially weighting past data values, and this is done by using a cost function with exponential weighting:

$$J(\theta) = \sum_{i=1}^{N} \lambda^{N-i} \varepsilon_i^2 = \sum_{i=1}^{N} \lambda^{N-i} (y_i - \hat{y}_i)^2 = \sum_{i=1}^{N} \lambda^{N-i} (y_i - \Phi(i)\theta)^2 = \min \tag{A.73}$$

Here, λ is called the forgetting factor, which is $0 < \lambda < 1$, and it is a measure of how fast the old data are forgotten. The recursive least squares estimation algorithm using the cost function (Eq. A.73) is given as follows:

$$\hat{\theta}(k+1) = \hat{\theta}(k) + \mathbf{K}(k+1)\left(y(k+1) - \phi(k+1)\hat{\theta}(k)\right) \tag{A.74}$$

$$\mathbf{K}(k+1) = \mathbf{P}(k)\phi^{\mathrm{T}}(k+1)\left[\lambda + \phi(k+1)\mathbf{P}(k)\phi^{\mathrm{T}}(k+1)\right]^{-1} \tag{A.75}$$

$$\mathbf{P}(k+1) = (\mathbf{I} - \mathbf{K}(k+1)\phi(k+1))\mathbf{P}(k)/\lambda \tag{A.76}$$

Note that $\lambda = 1$ gives the standard least squares estimation algorithm.

2. Rectangular window approach

The idea of the rectangular window approach is that the estimate at time k is only based on a finite number of past data, and all old data are completely discarded. To implement this idea, a rectangular window with fixed length N is first set, and whenever a new set of data is added at each time, a set of old data is discarded simultaneously, so the active number of data points is always kept to N. This approach requires that the last N estimate, $\hat{\theta}_{i+N,i+1}$, and covariance, $P_{i+N,i+1}$, be stored. For a more detailed description of the algorithm, the interested reader can refer to the book by Goodwin and Payne (1977) listed at the end of this appendix.

A.2.5   Generalized Least Squares Estimator

In previous sections, we discussed the statistical properties of the least squares estimation method and pointed out that the estimate is unbiased if the noise $\{\xi(k)\}$ in the model (Eq. A.54) is white noise or a sequence of uncorrelated zero-mean random variables with common variance σ^2. The white noise assumption is not a practical reality but is suitable for low-frequency control system analysis of practical systems. If the conditions of uncorrelatedness and zero mean of the noise sequence $\{\xi(k)\}$ cannot be satisfied in a system, the statistical properties of the least squares estimation will not be guaranteed in general. In this case, the generalized least squares method can be used and the estimate will be unbiased; this has been shown to work well in practice (Clarke, 1967; Söderström, 1974).

The idea of the generalized least squares estimation method is that the correlated sequence $\{\xi(k)\}$ is considered to be the output of a linear filter, which is driven by a white-noise sequence; that is:

$$\xi(k) + \sum_{i=1}^{p} c_i \xi(k-i) = w(k) \;\; \Rightarrow \;\; C(q^{-1})\xi(k) = w(k) \tag{A.77}$$

where $\{w(k)\}$ is a white-noise sequence, $C(q^{-1}) = 1 + c_1 q^{-1} + \cdots + c_p q^{-p}$.

For Eq. A.77, we know its z-transfer function is:

$$\frac{\xi(z)}{w(z)} = \frac{1}{C(z^{-1})} \tag{A.78}$$

Then, the system model may be written as:

$$A(q^{-1})y(k) = B(q^{-1})u(k) + C(q^{-1})w(k) \tag{A.79}$$

which can be further converted to:

$$A(q^{-1})y^*(k) = B(q^{-1})u^*(k) + w(k) \tag{A.80}$$

where

$$y^*(k) = \frac{y(k)}{C(q^{-1})} \;\; \text{and} \;\; u^*(k) = \frac{u(k)}{C(q^{-1})} \tag{A.81}$$

If $\{y^*(k)\}$ and $\{u^*(k)\}$ can be calculated, the parameters in $A(q^{-1})$ and $B(q^{-1})$ may be estimated by least squares, and the estimate is unbiased. However, the problem is that $C(q^{-1})$ is unknown. Thus, the parameters in $C(q^{-1})$ must be estimated along with $A(q^{-1})$ and $B(q^{-1})$, which results in the following generalized least squares estimation method:

1. Set $C(q^{-1}) = 1$, estimate parameter $\hat{\theta}$ in $A(q^{-1})$ and $B(q^{-1})$
2. Generate $\{\hat{\xi}(k)\}$ from $\xi(k) = \hat{A}(q^{-1})y^*(k) - \hat{B}(q^{-1})u^*(k)$
3. Estimate parameter $\hat{c}_i$ in $C(q^{-1})$ of Eq. A.77
4. Generate $\{y^*(k)\}$ and $\{u^*(k)\}$ based on the estimated $\hat{c}_i$ and Eq. A.78
5. Estimate parameter $\hat{\theta}$ in $A(q^{-1})$ and $B(q^{-1})$ based on the data points $\{y^*(k)\}$ and $\{u^*(k)\}$
6. If converged, stop; otherwise go to step 2.

A.3   State Estimation for Dynamic Systems

Some subsystems or components of a hybrid vehicle system are described by the state-space model given in Eq. A.44. In order to control an HEV subsystem properly, the system states sometimes need to be estimated based on observational data. In 1960, Kalman published his famous paper on the linear filtering problem, and his results were called the Kalman filter, which is a set of mathematical equations that provides a recursive computational method to estimate the states of a system in a way that minimizes the estimation error (Kalman, 1960a, 1960b). The Kalman filter techniques can be summarized as follows:

$$\text{State-space equations:} \quad \mathbf{x}(k+1) = A x(k) + \mathbf{B}u(k) + w(k)$$
$$\text{Output equations:} \quad y(k) = \mathbf{C}\mathbf{x}(k) + v(k) \tag{A.82}$$

$$\text{Noise:} \quad \mathrm{E}\{w(k)\} = 0, \quad \mathrm{E}\{v(k)\} = 0, \quad \mathrm{E}\{\mathbf{x}(0)\} = \mu, \quad \mathrm{E}\{w(j)w(k)\}_{j \neq k} = 0,$$
$$\mathrm{E}\{w(j)w(k)\}_{j=k} = Q, \quad \mathrm{E}\{v(j)v(k)\}_{j \neq k} = 0, \quad \mathrm{E}\{v(j)v(k)\}_{j=k} = R,$$
$$\mathrm{E}\{w(j)v(k)\} = 0, \quad \mathrm{E}\{\mathbf{x}(0)w(k)\} = 0, \quad \mathrm{E}\{\mathbf{x}(0)v(k)\} = 0,$$
$$\mathrm{E}\left\{[\mathbf{x}(0) - \mu][\mathbf{x}(0) - \mu]^{\mathrm{T}}\right\} = \mathbf{P}_0, \tag{A.83}$$

$$\text{Filter:} \quad \hat{\mathbf{x}}(k|k) = \hat{\mathbf{x}}(k|k-1) + \mathbf{K}(k)[y(k) - \mathbf{C}\hat{\mathbf{x}}(k|k-1)] \tag{A.84}$$

$$\text{Prediction:} \quad \hat{\mathbf{x}}(k|k-1) = \mathbf{A}\hat{\mathbf{x}}(k-1|k-1) \tag{A.85}$$

$$\text{Gain:} \quad \mathbf{K}(k) = \mathbf{P}(k,k-1)\mathbf{C}^{\mathrm{T}}\left[\mathbf{C}\mathbf{P}(k,k-1)\mathbf{C}^{\mathrm{T}} + R\right]^{-1} \tag{A.86}$$

$$\text{Filter error covariance:} \quad \mathbf{P}(k,k) = [\mathbf{I} - \mathbf{K}(k)\mathbf{C}]\mathbf{P}(k,k-1)[\mathbf{I} - \mathbf{K}(k)\mathbf{C}]^{\mathrm{T}} + \mathbf{K}(k)R\mathbf{K}^{\mathrm{T}}(k) \tag{A.87}$$

$$\text{Prediction error covariance:} \quad \mathbf{P}(k,k-1) = \mathbf{A}\mathbf{P}(k-1,k-1)\mathbf{A}^{\mathrm{T}} + Q \tag{A.88}$$

$$\text{Initial conditions:} \quad \hat{\mathbf{x}}(0,0) = \hat{\mathbf{x}}(0) = \mu, \quad \mathbf{P}(0,0) = \mathbf{P}_0 \tag{A.89}$$

For more details on the probabilistic origins and convergence properties of the Kalman filter, the interested reader should refer to Kalman's papers (Kalman, 1960a, 1960b; Kalman and Bucy, 1961) and Jazwinski's book (Jazwinski, 1970).

Example A.4
Consider the following system:

$$x(k+1) = \varphi x(k) + w(k)$$
$$y(k) = x(k) + v(k) \tag{A.90}$$

where $\{w(k)\}$ and $\{v(k)\}$ $k = 1, 2, \cdots$, are Gaussian noise sequences with zero mean and covariances Q and R, respectively. Estimate the state using the Kalman filter technique and list the several computational values if we assume that $\varphi = 1$, $P_0 = 100$, $Q = 25$, $R = 15$.

Table A.1   Computational results of Example A.4

k	$P(k, k-1)$	$K(k)$	$P(k, k)$
0	...	...	100
1	125	0.893	13.40
2	38.4	0.720	10.80
3	35.8	0.704	10.57
4	35.6	0.703	10.55

Solution:

The filtering equation is:

$$\hat{x}(k|k) = \hat{x}(k|k-1) + \mathbf{K}(k)[y(k) - C\hat{x}(k|k-1)] = \varphi\hat{x}(k-|k-1) + \mathbf{K}(k)[y(k) - \varphi\hat{x}(k-1|k-1)]$$

From Eq. A.88, we have the prediction error covariance, gain, and filter error covariance:

$$\mathbf{P}(k, k-1) = \varphi^2 \mathbf{P}(k-1, k-1) + Q$$

$$\mathbf{K}(k) = \left[\varphi^2 \mathbf{P}(k-1, k-1) + Q\right]\left[\varphi^2 \mathbf{P}(k-1, k-1) + Q + R\right]^{-1} = \frac{\varphi^2 \mathbf{P}(k-1, k-1) + Q}{\varphi^2 \mathbf{P}(k-1, k-1) + Q + R}$$

$$\mathbf{P}(k, k) = \frac{R[\varphi^2 \mathbf{P}(k-1, k-1) + Q]}{\varphi^2 \mathbf{P}(k-1, k-1) + Q + R}$$

For the given $\varphi = 1$, $P_0 = 100$, $Q = 25$, $R = 15$, the above equations will become:

$$\mathbf{P}(k, k-1) = \mathbf{P}(k-1, k-1) + 25$$

$$\mathbf{K}(k) = \frac{\mathbf{P}(k-1, k-1) + 25}{\mathbf{P}(k-1, k-1) + 40}$$

$$\mathbf{P}(k, k) = \frac{15[\mathbf{P}(k-1, k-1) + 25]}{[\mathbf{P}(k-1, k-1) + 40]} = 15\mathbf{K}(k)$$

The results of the first several steps are listed in Table A.1.

A.4   Joint State and Parameter Estimation for Dynamic Systems

In hybrid vehicle applications, it may also be necessary to estimate the states and parameters of a subsystem simultaneously. We devote this section to describing two approaches for joint state and parameter estimation of a dynamic system.

A.4.1   Extended Kalman Filter

While the Kalman filter provides a solution for estimating the states of a linear dynamic system, the extended Kalman filter gives a method for estimating the states of a nonlinear dynamic system. To obtain the extended Kalman filter, we consider the following nonlinear dynamic system:

$$\text{State-space equations}: \quad \mathbf{x}(k+1) = \mathbf{f}(\mathbf{x}(k), u(k)) + w(k)$$

$$\text{Output equations}: \quad y(k) = g(\mathbf{x}(k)) + v(k) \tag{A.91}$$

where $\mathbf{x}(k)$ is the state variable vector, $u(k)$ is the input variable, and $y(k)$ is the output variable, $\{w(k)\}$ and $\{v(k)\}$ again represent the system and measurement noise, which are assumed to have a Gaussian distribution and independent zero mean with covariances Q and R, respectively.

The extended Kalman filter algorithm is stated as:

$$\text{Filtering}: \quad \hat{\mathbf{x}}(k|k) = \hat{\mathbf{x}}(k|k-1) + \mathbf{K}(k)[y(k) - g(\hat{\mathbf{x}}(k|k-1))]$$

$$\mathbf{P}(k|k) = \left[\mathbf{I} - \mathbf{K}(k)\frac{\partial \mathbf{g}(x(k))}{\partial \mathbf{x}}\bigg|_{\hat{\mathbf{x}}(k|k-1)} \right] \mathbf{P}(k|k-1) \tag{A.92}$$

$$\text{Prediction}: \quad \hat{\mathbf{x}}(k|k-1) = \mathbf{f}(\hat{\mathbf{x}}(k-1|k-1), u(k-1))$$

$$\mathbf{P}(k|k-1) = \frac{\partial \mathbf{f}}{\partial \mathbf{x}}\bigg|_{\hat{\mathbf{x}}(k-1)} \mathbf{P}(k-1|k-1)\frac{\partial \mathbf{f}^{\mathrm{T}}}{\partial \mathbf{x}}\bigg|_{\hat{\mathbf{x}}(k-1)} + Q \tag{A.93}$$

$$\text{Gain}: \quad \mathbf{K}(k) = \mathbf{P}(k, k-1)\left(\frac{\partial \mathbf{g}(x(k))}{\partial \mathbf{x}}\bigg|_{\hat{\mathbf{x}}(k|k-1)} \right)^{\mathrm{T}}$$

$$\cdot \left[\left(\frac{\partial \mathbf{g}(x(k))}{\partial \mathbf{x}}\bigg|_{\hat{\mathbf{x}}(k|k-1)} \right) \mathbf{P}(k, k-1)\mathbf{C}\left(\frac{\partial \mathbf{g}(x(k))}{\partial \mathbf{x}}\bigg|_{\hat{\mathbf{x}}(k|k-1)} \right)^{\mathrm{T}} + R \right]^{-1} \tag{A.94}$$

$$\text{Initial conditions}: \quad \hat{\mathbf{x}}(0,0) = \hat{\mathbf{x}}(0) = \mu, \quad \mathbf{P}(0,0) = \mathbf{P}_0 \tag{A.95}$$

The iteration process of the extended Kalman filter algorithm is:

1. Obtain the last state estimate $\hat{\mathbf{x}}(k-1|k-1)$ and filter error covariance $\mathbf{P}(k-1|k-1)$
2. Compute $\hat{\mathbf{x}}(k|k-1)$ from $\hat{\mathbf{x}}(k|k-1) = \mathbf{f}(\hat{\mathbf{x}}(k-1|k-1), u(k-1))$
3. Compute $\mathbf{P}(k|k-1)$ from $\mathbf{P}(k|k-1) = \frac{\partial \mathbf{f}}{\partial \mathbf{x}}\big|_{\hat{\mathbf{x}}(k-1)} \mathbf{P}(k-1|k-1)\frac{\partial \mathbf{f}^{\mathrm{T}}}{\partial \mathbf{x}}\big|_{\hat{\mathbf{x}}(k-1)} + Q$
4. Compute $\mathbf{K}(k)$
5. Compute $\hat{\mathbf{x}}(k|k)$ from $\hat{\mathbf{x}}(k|k) = \hat{\mathbf{x}}(k|k-1) + \mathbf{K}(k)[y(k) - g(\hat{\mathbf{x}}(k|k-1))]$

6. Compute $\mathbf{P}(k|k)$ from $\mathbf{P}(k|k) = \left[\mathbf{I} - \mathbf{K}(k)\dfrac{\partial \mathbf{g}(x(k))}{\partial \mathbf{x}}\bigg|_{\hat{\mathbf{x}}(k|k-1)}\right]\mathbf{P}(k|k-1)$

7. Go to step 1.

The above extended Kalman filter algorithm can be applied to estimate states and parameters of a dynamic system simultaneously. Let us consider the nonlinear dynamic system described by the state-space equation given in Eq. A.91 and assume that there is an unknown parameter, $\mathbf{a}$, in $\mathbf{f}(\mathbf{x}(k), u(k))$ and an unknown parameter, $\mathbf{b}$, in $\mathbf{g}(\mathbf{x}(k))$, then, Eq. A.91 can be written as:

$$\text{State equations}: \quad \mathbf{x}(k+1) = \mathbf{f}(\mathbf{x}(k), u(k), \mathbf{a}) + w(k)$$
$$\text{Output equations}: \quad y(k) = g(\mathbf{x}(k), \mathbf{b}) + v(k) \tag{A.96}$$

In order to estimate parameters $\mathbf{a}$ and $\mathbf{b}$, we define the $\mathbf{a}$ and $\mathbf{b}$ as new states, described by the following equations:

$$\mathbf{a}(k+1) = \mathbf{a}(k)$$
$$\mathbf{b}(k+1) = \mathbf{b}(k) \tag{A.97}$$

Combining Eqs A.96 and A.97 and letting state variable $\mathbf{x}^*(k) = [\mathbf{x}(k), \mathbf{a}(k), \mathbf{b}(k)]^{\mathrm{T}}$, an augmented state-space equation is obtained as:

$$\text{State-space equation}: \quad \mathbf{x}^*(k+1) = \begin{pmatrix} \mathbf{f}(\mathbf{x}(k), u(k), \mathbf{a}(k)) \\ \mathbf{a}(k) \\ \mathbf{b}(k) \end{pmatrix} + \begin{pmatrix} w(k) \\ 0 \\ 0 \end{pmatrix} \tag{A.98}$$
$$\text{Output equation}: \quad y(k) = g(\mathbf{x}(k), \mathbf{b}(k)) + v(k)$$

By applying the extended Kalman filter algorithm to Eq. A.98, the augmented state $\mathbf{x}^*$ can be estimated; that is, the estimate of state $\mathbf{x}$ as well as parameters $\mathbf{a}$ and $\mathbf{b}$ is given. There are many good articles presenting practical examples of the extended Kalman filter; the interested reader should refer to the book by Grewal and Andrews (2008) listed at the end of this appendix.

A.4.2 Singular Pencil Model

The singular pencil (SP) model, possibly a new class of model for most control engineers, was first proposed by G. Salut and others and then developed by many researchers (Salut *et al.*, 1979, 1980; Aplevich 1981, 1985, 1991; Chen *et al.*, 1986). The SP model contains the input–output and state-space model as subsets. Models similar to this form have been called *generalized state-space models*, *descriptor systems*, *tableau equations*, *time-domain input–output models*, and *implicit linear systems*. An advantage when a system is described in the SP model format is that the simultaneous state and parameter estimation problem can be solved by an ordinary Kalman filter algorithm.

A dynamic system can be described in SP model form as follows:

$$[\mathbf{E} - \mathbf{DF} \ \ \mathbf{G}] \begin{bmatrix} \mathbf{x} \\ \mathbf{w} \end{bmatrix} = \mathbf{P}(\mathrm{D}) \begin{bmatrix} \mathbf{x} \\ \mathbf{w} \end{bmatrix} = 0 \tag{A.99}$$

where $\mathbf{x} \in R^n$ is the internal (auxiliary) variable vector and $\mathbf{w} \in R^{p+m}$ with known p and m is the external variable vector, usually consisting of input–output variables of the system. $\mathbf{E}$ and $\mathbf{F}$ are $(n + p)$ by n matrices, $\mathbf{G}$ is an $(n + p)$ by $(p + m)$ matrix, D is a linear operator, and the matrix $\mathbf{P}(\mathrm{D})$ is also called the system matrix.

SP models can describe general dynamic systems, which include special cases of two types of the most commonly used models – state-space models and input–output models.

If we define

$$\mathbf{E} - \mathbf{DF} = \begin{bmatrix} \mathbf{A} \\ \mathbf{C} \end{bmatrix} - \begin{bmatrix} \mathbf{I} \\ \mathbf{0} \end{bmatrix}, \mathbf{G} = \begin{bmatrix} 0 & \mathbf{B} \\ -\mathbf{I} & \mathbf{D} \end{bmatrix}, \mathbf{x} = \mathbf{x}(k),$$

And

$$\mathbf{w} = \begin{bmatrix} y_k \\ u_k \end{bmatrix},$$

Then the singular pencil model representation in Eq. A.99 can be written as:

$$\left\{ \begin{bmatrix} \mathbf{A} \\ \mathbf{C} \end{bmatrix} - \begin{bmatrix} \mathbf{I} \\ \mathbf{0} \end{bmatrix} z \right\} \mathbf{x}(k) + \begin{bmatrix} 0 & \mathbf{B} \\ -\mathbf{I} & \mathbf{D} \end{bmatrix} \begin{bmatrix} y(k) \\ u(k) \end{bmatrix} = 0 \tag{A.100}$$

Equation A.100 can be readily transformed to the standard state-space form as:

$$\begin{aligned} \mathbf{x}(k+1) &= \mathrm{A}\mathbf{x}(k) + \mathbf{B}u(k) \\ \mathbf{y}(k) &= \mathrm{C}\mathbf{x}(k) + \mathbf{D}u(k) \end{aligned} \tag{A.101}$$

where $\mathbf{u}(k)$ is the input vector, $\mathbf{y}(k)$ is the output vector, and $\mathbf{A}, \mathbf{B}, \mathbf{C}, \mathbf{D}$ are the state-space matrices in the usual notation.

The other special case is ARMAX models with the form:

$$\mathbf{y}(k) + \sum_{i=1}^{n} a_i \left(q^{-1} \right)^i \mathbf{y}(k) = \sum_{i=0}^{n} b_i \left(q^{-1} \right)^i \mathbf{u}(k) \tag{A.102}$$

a_i and b_i are parameters of the system and q^{-1} is the shift operator. Equation A.102 can also be written in first-order SP model form as:

$$\begin{bmatrix} 1 & & & \\ & \ddots & 0 & \\ & & 1 & \\ & & & 1 \\ & & & & 0 \end{bmatrix} \mathbf{x}(k+1) = \begin{bmatrix} 0 & & & \\ 1 & & & \\ & \ddots & & \\ & & & 1 \end{bmatrix} \mathbf{x}(k) + \begin{bmatrix} -a_n & b_n \\ \vdots & \vdots \\ -a_1 & b_1 \\ -1 & b_0 \end{bmatrix} \begin{bmatrix} \mathbf{y}(k) \\ \mathbf{u}(k) \end{bmatrix} \qquad (A.103)$$

When a system is described by the SP model, a Kalman filter algorithm can be used to estimate the state and parameters simultaneously. Let us take Eq. A.103 as an example to show how it works. First consider that state $\mathbf{x}(k)$ and parameters a_i and b_j are unknown and need to be estimated, and let $\boldsymbol{\gamma} = [a_n \ \cdots \ a_1 \ b_n \ \cdots \ b_0]^{\mathrm{T}}$ and $\mathbf{x}(k) = [x_1(k) \ \cdots \ x_n(k)]^{\mathrm{T}}$, then Eq. A.103 can be written as:

$$\begin{aligned} \mathbf{x}(k+1) &= \mathbf{E}_*\mathbf{x}(k) \\ 0 &= \mathbf{E}_0\mathbf{x}(k) \end{aligned} + \begin{bmatrix} -a_n & b_n \\ \vdots & \vdots \\ -a_1 & b_1 \\ -1 & b_0 \end{bmatrix} \begin{bmatrix} \mathbf{y}(k) \\ \mathbf{u}(k) \end{bmatrix} \begin{aligned} &= \mathbf{E}_*\mathbf{x}(k) + \mathbf{G}_*(k)\boldsymbol{\gamma} \\ &= \mathbf{E}_0\mathbf{x}(k) + \mathbf{G}_0(k)\boldsymbol{\gamma} - \mathbf{y}(k) \end{aligned} \qquad (A.104)$$

where

$$\mathbf{E}_* = \begin{bmatrix} 0 & 0 & \cdots & 0 & 0 \\ 1 & 0 & \cdots & 0 & 0 \\ \cdots & \cdots & \cdots & \cdots & \cdots \\ 0 & 0 & \cdots & 1 & 0 \end{bmatrix}_{n \times n}, \mathbf{E}_0 = \begin{bmatrix} 0 & 0 & \cdots & 1 \end{bmatrix}_{1 \times n},$$

$$G_*(k) = \begin{bmatrix} -y(k) & 0 & \cdots & 0 & u(k) & 0 & \cdots & 0 & 0 \\ 0 & -y(k) & \cdots & 0 & 0 & u(k) & \cdots & 0 & 0 \\ \cdots & \cdots & \cdots & \cdots & \cdots & \cdots & \cdots & \cdots & \cdots \\ 0 & 0 & 0 & -y(k) & 0 & 0 & \cdots & 0 & u(k) \end{bmatrix}_{n \times (2n+1)}$$

$$G_0(k) = \begin{bmatrix} 0 & \cdots & 0 & u(k) \end{bmatrix}_{1 \times (2n+1)}$$

In the presence of random disturbances, Eq. A.104 is modified to:

$$\begin{aligned}
\mathbf{x}(k+1) &= \mathbf{E}_*\mathbf{x}(k) + \mathbf{G}_*(k)\boldsymbol{\gamma} + \mathbf{C}_*e(k) \\
0 &= \mathbf{E}_0\mathbf{x}(k) + \mathbf{G}_0(k)\boldsymbol{\gamma} - \mathbf{y}(k) + \mathbf{e}(k)
\end{aligned}$$

(A.105)

where $\mathbf{e}(k)$ is a zero-mean, uncorrelated vector random variable and $\mathbf{C}_*$ is a matrix of noise parameters.

If we define the augmented states as:

$$\mathbf{s}(k) = \begin{bmatrix} \mathbf{x}(k) \\ \boldsymbol{\gamma}(k) \end{bmatrix}, \quad \mathbf{F} = \begin{bmatrix} \mathbf{E}_* & \mathbf{G}_* \\ \mathbf{0} & \mathbf{I} \end{bmatrix}, \quad \mathbf{H} = [\mathbf{E}_0 \ \ \mathbf{G}_0], \quad \mathbf{v}(k) = \begin{bmatrix} \mathbf{C}_*\mathbf{e}(k) \\ 0 \end{bmatrix}$$

(A.106)

then we have the following state-space model:

$$\begin{aligned}
\mathbf{s}(k+1) &= \mathbf{F}\mathbf{s}(k) + \mathbf{v}(k) \\
\mathbf{y}(k) &= \mathbf{H}\mathbf{s}(k) + \mathbf{e}(k)
\end{aligned}$$

(A.107)

By applying the ordinary Kalman filter algorithm described in Section A.3 to Eq. A.107, the states and parameters of the system in Eq. A.102 can be estimated simultaneously.

A.5 Enhancement of Numerical Stability in Parameter and State Estimation

In hybrid vehicle applications, parameter and state estimation algorithms are implemented in a microcontroller which normally has limited computational capability. Therefore, improving computational efficiency, enhancing numerical stability, and avoiding unnecessary storage are key factors influencing the implementation of real-time parameter and state estimation for vehicle applications. The following example shows what the numerical stability in real-time parameter/state estimation is, and two techniques are given to overcome the issue in this section.

First, let us consider the following single-parameter estimation by assuming that the model is:

$$\mathbf{y}(k) = \phi(k)\theta + \mathbf{e}(k)$$

(A.108)

where $\mathbf{y}$ and ϕ are the output and input, θ is the parameter, and $\{\mathbf{e}(k)\}$ is an uncorrelated white noise sequence with covariance R.

In order to present the issue seamlessly, let's rewrite the recursive least squares estimation equations A.67, A.68, and A.69:

$$\hat{\boldsymbol{\theta}}(k+1) = \hat{\boldsymbol{\theta}}(k) + \mathbf{K}(k+1)\left(y(k+1) - \phi(k+1)\hat{\boldsymbol{\theta}}(k)\right)$$

(A.109)

$$\mathbf{K}(k+1) = \mathbf{P}(k)\phi(k+1)\left[1+\mathbf{P}(k)\phi^2(k+1)\right]^{-1} \qquad (A.110)$$

$$\mathbf{P}(k+1) = (1-\mathbf{K}(k+1)\phi(k+1))\mathbf{P}(k) \qquad (A.111)$$

Since there is only one parameter in Eq. A.108, $\mathbf{P}$ and $\mathbf{K}$ are scalars in Eq. A.110 and A.111. The estimation error can be described as:

$$\widetilde{\theta}(k+1) = \theta - \hat{\mathbf{\theta}}(k+1)$$

$$= \theta - \hat{\mathbf{\theta}}(k) - \mathbf{K}(k+1)\left(y(k+1) - \phi(k+1)\hat{\mathbf{\theta}}(k)\right)$$

$$= \widetilde{\theta}(k) - \mathbf{K}(k+1) \cdot y(k+1) + \mathbf{K}(k+1) \cdot \phi(k+1)\hat{\mathbf{\theta}}(k) \qquad (A.112)$$

$$= \widetilde{\theta}(k) - \mathbf{K}(k+1) \cdot (\phi(k+1)\theta + \mathbf{e}(k+1)) + \mathbf{K}(k+1) \cdot \phi(k+1)\hat{\mathbf{\theta}}(k)$$

$$= (1 - \mathbf{K}(k+1)\phi(k+1))\widetilde{\theta} - \mathbf{K}(k+1)\mathbf{e}(k+1)$$

Equation A.112 is a stochastic difference equation, and it is unstable when $|1 - \mathbf{K}(k+1)\phi(k+1)| > 1$, or when $\mathbf{K}(k+1)\phi(k+1) < 0$, which will result in the estimated parameter blow-out.

From the gain equation (Eq. A.110), we know that:

$$\mathbf{K}(k+1)\phi(k+1) = \frac{\mathbf{P}(k)\phi^2(k+1)}{1+\mathbf{P}(k)\phi^2(k+1)} \qquad (A.113)$$

and from the above equation, we obviously find that the unstable condition $\mathbf{K}(k+1)\phi(k+1) < 0$ is equal to:

$$\begin{cases} \mathbf{P}(k) < 0 \\ \mathbf{P}(k)\phi^2(k+1) > -1 \end{cases} \qquad (A.114)$$

In other words, if the covariance matrix $\mathbf{P}$ turns negative, the estimate will be unstable until $\mathbf{P}(k)\phi^2(k+1)$ becomes less than -1. Also, from covariance equation A.111, we know that $\mathbf{P}(k+1)$ turns negative when $\mathbf{K}(k+1)\phi(k+1) > 1$. In addition, Eq. A.113 shows that $\mathbf{K}(k+1)\phi(k+1) \to 1$ when the value of $\mathbf{P}(k)$ is big enough; this results in the instability of the recursive calculation of the covariance matrix $\mathbf{P}(k)$.

In order to solve the unstable issue of the recursive least squares estimation, it is necessary to have the covariance matrix strictly positive definite. In this section, we introduce two common algorithms to achieve this goal.

A.5.1   Square-root Algorithm

The square-root algorithm is an effective technique for improving numerical stability in the parameter estimation (Peterka, 1975). Since the covariance matrix, $\mathbf{P}$, needs to be strictly positive definite, $\mathbf{P}$ can be factorized into the following form:

$$\mathbf{P}(k) = \mathbf{S}(k)\mathbf{S}^{\mathrm{T}}(k) \tag{A.115}$$

where $\mathbf{S}$ is a nonsingular matrix called the square root of $\mathbf{P}$.

In a recursive least squares estimation algorithm, if $\mathbf{S}$ is updated in real time rather than $\mathbf{P}$, the strictly positive definite form of $\mathbf{P}$ is ensured. The corresponding recursive algorithm for Eq. A.108 can be derived as follows.

If we have:

$$\mathbf{P}(k) = \mathbf{S}^2(k)$$

$$f(k+1) = \mathbf{S}(k)\phi(k+1)$$

$$\beta(k+1) = \lambda + f^2(k+1) \tag{A.116}$$

$$\mathbf{H}^2(k+1) = 1 - \frac{f^2(k+1)}{\beta(k+1)} = \frac{\lambda}{\lambda + f^2(k+1)}$$

the covariance matrix $\mathbf{P}$ can be expressed in the following recursive form:

$$\mathbf{S}^2(k+1) = \mathbf{P}(k+1) = \frac{1}{\lambda}(1 - \mathbf{K}(k+1)\phi(k+1))\mathbf{P}(k)$$

$$= \frac{1}{\lambda}\left(1 - \frac{\mathbf{S}^2(k)\phi^2(k+1)}{\lambda + \mathbf{S}^2(k)\phi^2(k+1)}\right)\mathbf{S}^2(k) \tag{A.117}$$

$$= \frac{1}{\lambda}\left(1 - \frac{f^2(k+1)}{\beta(k+1)}\right)\mathbf{S}^2(k) = \frac{1}{\lambda}\mathbf{H}^2(k+1)\mathbf{S}^2(k)$$

Or

$$\mathbf{S}(k+1) = \frac{1}{\sqrt{\lambda}}\mathbf{H}(k+1)\mathbf{S}(k)$$

Hence, the parameters can be estimated based on the following square-root recursive least squares estimation algorithm, and the numerical stability is ensured:

$$f(k+1) = \mathbf{S}^{\mathrm{T}}(k)\phi(k+1)$$

$$\beta(k+1) = \lambda + f^{\mathrm{T}}(k+1)f(k+1)$$

$$\alpha(k+1) = \frac{1}{\beta(k+1) + \sqrt{\lambda\beta(k+1)}}$$

$$\mathbf{K}(k+1) = \frac{\mathbf{S}(k)f(k+1)}{\beta(k+1)}$$ (A.118)

$$\mathbf{S}(k+1) = \frac{1}{\sqrt{\lambda}}\left[\mathbf{I} - \alpha(k+1)\beta(k+1)\mathbf{K}(k+1)\phi^{\mathrm{T}}(k+1)\right]\mathbf{S}(k)$$

$$\hat{\boldsymbol{\theta}}(k+1) = \hat{\boldsymbol{\theta}}(k) + \mathbf{K}(k+1)\left[y(k+1) - \phi^{\mathrm{T}}(k+1)\hat{\boldsymbol{\theta}}(k)\right]$$

A.5.2 UDU^T Covariance Factorization Algorithm

Since the square-root-based recursive least squares estimation algorithm requires extraction of the square root in real time, the computational efficiency is relatively low. Thornton and Bierman proposed an alternative approach to achieve numerical stability for the estimation, but it does not require square-root extraction. In their algorithm, both accuracy and computational efficiency are improved by using covariance factorization $\mathbf{P} = \mathbf{UDU}^{\mathrm{T}}$ where $\mathbf{U}$ is an upper triangular matrix with unit diagonals and $\mathbf{D}$ is diagonal. Compared with the general recursive least squares estimation algorithm, the $\mathbf{U}$–$\mathbf{D}$-factorization-based recursive least squares estimation algorithm updates covariance factors $\mathbf{U}$ and $\mathbf{D}$ together with gain $\mathbf{K}$ in real time (Thornton and Bierman, 1978).

If we assume that the computed output, $\mathbf{y}$, is given by the model:

$$\mathbf{y}(k) = \theta_1 x_1(k) + \theta_2 x_2(k) + \cdots + \theta_n x_n(k) + \mathbf{e}(k) = \phi(k)\boldsymbol{\theta} + \mathbf{e}(k)$$ (A.119)

where x_1, x_2, $\cdots$, x_n are inputs, $\mathbf{y}$ is the output, θ_1, θ_2, $\cdots$, θ_n are unknown parameters, $\{\mathbf{e}(k)\}$ again represents measurement noise following the Gaussian distribution with independent zero mean and covariance R. The pairs of observations $\{(x_i\ y_i), i = 1, 2, \cdots, N\}$ are obtained from measurements if we let $\boldsymbol{\theta} = [\theta_1\ \theta_2\ \cdots\ \theta_n]^{\mathrm{T}}$ and $\phi(k) = \left[x_1^{\mathrm{T}}(k)\ x_2^{\mathrm{T}}(k)\ \cdots\ x_n^{\mathrm{T}}(k)\right]^{\mathrm{T}}$

The parameter vector $\boldsymbol{\theta}$ can be estimated by the following $\mathbf{U}$–$\mathbf{D}$-factorization-based recursive least squares algorithm once the measured input and output and initial values of $\hat{\boldsymbol{\theta}}(0) = \theta_0$, $\mathbf{U}(0) = U_0$, and $\mathbf{D}(0) = D_0$ have been given:

$$\mathbf{K}(k+1) = \frac{\mathbf{P}(k)\phi(k+1)}{\lambda + \phi^{\mathrm{T}}(k+1)\mathbf{P}(k)\phi(k+1)} = \frac{\mathbf{U}(k)\mathbf{D}(k)\mathbf{U}^{\mathrm{T}}(k)\phi(k+1)}{\lambda + \phi^{\mathrm{T}}(k+1)\mathbf{U}(k)\mathbf{D}(k)\mathbf{U}^{\mathrm{T}}(k)\phi(k+1)}$$

$$= \frac{\mathbf{U}(k)\mathbf{D}(k)\mathbf{U}^{\mathrm{T}}(k)\phi(k+1)}{\beta(k+1)} = \frac{\mathbf{U}(k)\mathbf{G}(k+1)}{\beta(k+1)} \tag{A.120}$$

$$\mathbf{P}(k+1) = \frac{1}{\lambda}\left(\mathbf{I} - \frac{\mathbf{P}(k)\phi(k+1)\phi^{\mathrm{T}}(k+1)}{\lambda + \phi^{\mathrm{T}}(k+1)\mathbf{P}(k)\phi(k+1)}\right)\mathbf{P}(k) = \mathbf{U}(k+1)\mathbf{D}(k+1)\mathbf{U}^{\mathrm{T}}(k+1)$$

$$\hat{\boldsymbol{\theta}}(k+1) = \hat{\boldsymbol{\theta}}(k) + \mathbf{K}(k+1)\left[y(k+1) - \phi^{\mathrm{T}}(k+1)\hat{\boldsymbol{\theta}}(k)\right]$$

$$\mathbf{D}(k+1) = [d_{ij}(k+1)]_{n\times n}, \quad \Rightarrow \quad d_{ij}(k+1) = \begin{cases} d_i(k+1) = \dfrac{\beta_{i-1}(k+1)}{\lambda\beta_i(k+1)}d_i(k) & i=j \\ 0 & i\neq j \end{cases} \tag{A.121}$$

$$\mathbf{U}(k+1) = [u_{ij}(k+1)]_{n\times n},$$

$$\Rightarrow \quad u_{ij}(k+1) = \begin{cases} 1, & 1\leq i=j\leq n \\ u_{ij}(k) - \dfrac{f_j(k+1)}{\beta_{j-1}(k+1)}\displaystyle\sum_{k=i}^{j-1} u_{ik}(k)g_k(k+1), & 1\leq i<j\leq n \\ 0, & 1\leq j<i\leq n \end{cases} \tag{A.122}$$

$$\mathbf{F}(k+1) = \mathbf{U}^{\mathrm{T}}(k)\phi(k+1) = [f_i(k+1)]_{n\times 1}$$
$$\mathbf{G}(k+1) = \mathbf{D}(k)\mathbf{F}(k+1) = [g_i(k+1)]_{n\times 1} \tag{A.123}$$

$$\beta(k+1) = \lambda + \mathbf{F}^{\mathrm{T}}(k+1)\mathbf{G}(k+1) = \lambda + \sum_{i=1}^{n} f_i(k+1)g_i(k+1)$$

$$\Rightarrow \begin{cases} \beta_0(k+1) = \lambda \\ \beta_i(k+1) = \beta_{i-1}(k+1) + f_i(k+1)g_i(k+1) \end{cases} \tag{A.124}$$

where $\mathbf{U}(k) \in R^{n\times n}$ is an upper triangular matrix with unit diagonal, $\mathbf{D}(k) \in R^{n\times n}$ is a diagonal matrix, $\mathbf{F}(k)$ and $\mathbf{G}(k) \in R^n$ are vectors, β is a scalar and λ is the forgetting factor.

Example A.5
Given

$$\mathbf{P}(k) = \begin{bmatrix} 0.8 & 0.4 \\ 0.4 & 0.8 \end{bmatrix}, \quad \mathbf{U}(k) = \begin{bmatrix} 1 & 0.5 \\ 0 & 1 \end{bmatrix}, \quad \mathbf{D}(k) = \begin{bmatrix} 0.6 & 0 \\ 0 & 0.8 \end{bmatrix}, \quad \varphi(k+1) = \begin{bmatrix} 0.2 \\ 1 \end{bmatrix}, \quad \lambda = 1$$

Calculate the corresponding variables of the UDU^{T} covariance factorization algorithm.

$$F(k+1) = U^T(k)\varphi(k+1) = \begin{bmatrix} 1 & 0 \\ 0.5 & 1 \end{bmatrix} \begin{bmatrix} 0.2 \\ 1 \end{bmatrix} = \begin{bmatrix} 0.2 \\ 1.1 \end{bmatrix}$$

$$G(k+1) = D(k)F(k+1) = \begin{bmatrix} 0.6 & 0 \\ 0 & 0.8 \end{bmatrix} \begin{bmatrix} 0.2 \\ 1.1 \end{bmatrix} = \begin{bmatrix} 0.12 \\ 0.88 \end{bmatrix}$$

$\beta_0(k+1) = \lambda = 1;$

$\beta_1(k+1) = \beta_0(k+1) + f_1(k+1) \cdot g_1(k+1) = 1 + 0.2 \cdot 0.12 = 1.024$

$\beta_2(k+1) = \beta_1(k+1) + f_2(k+1) \cdot g_2(k+1) = 1.024 + 1.1 \cdot 0.88 = 1.992$

$$K(k+1) = \frac{U(k)G(k+1)}{\beta(k+1)} = \frac{1}{1.992} \begin{bmatrix} 1 & 0.5 \\ 0 & 1 \end{bmatrix} \begin{bmatrix} 0.12 \\ 0.88 \end{bmatrix} = \frac{1}{1.992} \begin{bmatrix} 0.56 \\ 0.88 \end{bmatrix}$$

$$D(k+1) = \begin{bmatrix} \dfrac{\beta_0(k+1)}{\lambda \cdot \beta_1(k+1)} d_1(k) & 0 \\ 0 & \dfrac{\beta_1(k+1)}{\lambda \cdot \beta_2(k+1)} d_1(k) \end{bmatrix}$$

$$= \begin{bmatrix} \dfrac{1}{1.024} 0.6 & 0 \\ 0 & \dfrac{1.024}{1.992} 0.8 \end{bmatrix} \cong \begin{bmatrix} 0.586 & 0 \\ 0 & 0.411 \end{bmatrix}$$

$$U(k+1) = \begin{bmatrix} 1 & u_{12}(k) - \dfrac{f_2(k+1)}{\beta_1(k+1)} u_{11}(k) g_1(k+1) \\ 0 & 1 \end{bmatrix} = \begin{bmatrix} 1 & 0.5 - \dfrac{1.1}{1.024} 0.12 \\ 0 & 1 \end{bmatrix} \cong \begin{bmatrix} 1 & 0.371 \\ 0 & 1 \end{bmatrix}$$

A.6 Procedure of Modeling and Parameter Identification

The modeling task is to build a mathematical relationship between inputs and outputs of a system; this is normally expressed by a set of dynamic equations, which are either ordinary differential equations or partial differential equations. In the steady-state case, these differential equations are reduced to algebraic equations or ordinary differential equations, respectively.

As mentioned earlier, there are two approaches to accomplishing the modeling task: by theoretically deriving and analyzing the system based on physical/chemical principles or by computing from observed (measured) data from special experiments or tests. The model achieved from the first approach is called a theory-based model, which takes in the basic physical and chemical rules as well as the continuity equations, reaction mechanisms (if known), diffusion theory etc. Major advantages of a full theoretical model are reliability and flexibility, which allows for major changes in system architecture and the prediction of behavior over a wide operating range. However, this ideal situation is seriously undermined by a lack of precise knowledge and the need to make assumptions which sometimes need to be verified. Even then, the extremely complex nature of many system models can make their solution impossible or economically unattractive.

The model established by the other approach is called a data-based model. Acceptance of data-based models is guided by their usefulness rather than truthfulness, which means that such models have good input–output relationships under certain operational ranges but one can never establish any exact, meaningful connection between them. In the absence of other knowledge or in difficult-to-define systems, this method is most important. The shortcoming of this approach is that it may lead to misleading conclusions resulting from missing essential data or errors in the data. In addition, although this identification method may yield fruitful relationships in tests, validation over wide operational ranges is necessary, and proper test techniques and facilities are also required. In general, building a system model needs to undergo the following six procedures.

1. **Determine the objectives of the modeling**

   First of all, the purpose of the model must be clearly defined; this plays a key role in determining the form of the model. It must be decided initially and it should be recognized that a model developed specifically for one purpose may not be the best, or even a suitable form, for another purpose. If the developed subsystem model is to support overall performance analysis of the whole hybrid system, the model should be small and relatively simple. An increase in overall system complexity may require a reduction in the detail of modeling of specific units or subsystems. In addition, the boundaries of the model variables also need to be taken into account.

2. **Gather prior knowledge of a system**

   It is very important to gather enough prior knowledge of a system in order to save modeling cost and time. Prior knowledge for modeling includes theoretical understandings, practical engineering intuition, and insight into the system, such as the main input variables (factors), their varying ranges and limits, and environmental noise mean and variance.

3. **Choose a candidate model set**

   This is a key step in modeling practice. Before a model has been developed, based on prior knowledge, a candidate model set and all necessary variables must be identified in order for the system behavior to be adequately described. The variables fall essentially into two categories: independent variables, also known as input variables, and dependent or output variables. Usually, a simple form of model should be used for a subsystem if the overall performances of a large system are of concern. Because a system interacts with others, boundaries of the model also need to be decided upon. To some extent, the boundaries determine the size and the complexity of the model.

4. **Design experiment and record data**

   In order to identify a dynamic system, the system must be excited enough to have sufficiently informative recorded data. Therefore, a design needs to be carried out for an identification experiment, which includes choice of input and measurement ports, test signals, sampling rate, total available time, and the availability of transducers and filters, etc. In addition, the experimental design must take account of the experimental conditions, such as constraints and limits, and the operating points of the system. The minimal requirement for the input signals is that the input signal must be able to excite all modes of the system; that is, the input signal has to meet the requirement of persistence of excitation. For more detailed experimental design problems, the interested reader should refer to the books by Goodwin and Payne (1977) and Ljung (1987).

5. Estimate model parameters

After selecting a set of candidate models and collecting enough data, estimation of the parameters of the model can be started. It needs to be emphasized that there are many methods of estimating the parameters and also different views of judging whether the estimated results are good. Common estimation methods, excluding the least squares and generalized least squares methods introduced in this appendix, are maximum likelihood, instrumental-variable, and stochastic approximation. Readers should select one of them to estimate the model parameters based on the purpose of the model and prior knowledge on noise.

6. Validate model

Validation tests whether the established model can meet the modeling requirements. One natural and important kind of validation is to evaluate the input–output behavior of the model to see if the model's outputs match actual, measured outputs resulting from the same input values. This can be carried out through a simulation, which simulates the system with actual inputs and compares measured outputs with the model outputs. Preferably, a different data set from the one that was used for the modeling should be used for the comparison. If the validation test is passed, the modeling task is complete; otherwise, the modeler needs to go back to procedure 2 above to refine the prior information on the system and redo procedures 3 to 6.

References

Aplevich, D. J. 'Time-domain input-output representations of linear systems,' *Automatica*, **17**, 509–521, 1981.

Aplevich, D. J. 'Minimal representations of implicit linear systems,' *Automatica*, **21**, 259–269, 1985.

Aplevich, D. J. *Implicit Linear Systems*, Lecture Notes in Control and Information Sciences, Vol. **152**, Springer-Verlag, Heidelberg, 1991.

Chen, Y., Aplevich, D. J., and Wilson, W. 'Simultaneous Estimation of State and Parameters for Multivariable Linear Systems with Singular Pencil Models,' *IEE Proceedings Pt. D*, **133**, 65–72, 1986.

Clarke, D. W. 'Generalized least-squares estimation of the parameters of a dynamic model,' *Proceedings of the IFAC Symposium on Identification in Automatic Control Systems*, Prague, 1967.

Goodwin, G. C. and Payne, R. L. *Dynamic System Identification: Experiment Design and Data Analysis*, Mathematics in Science and Engineering, Vol. **136**, Academic Press, New York, 1977.

Grewal, S. M. and Andrews, P. A. *Kalman Filtering: Theory and Practice Using MATLAB*, 3rd edition, John Wiley & Sons, Inc, 2008.

Jazwinski, A. H. *Stochastic Processes and Filtering Theory*, Academic Press, New York, 1970.

Kalman, R. E. 'On the General Theory of Control Systems,' *Proceedings of the First IFAC Congress*, Moscow, **1**, 481–492, Butterworths, London, 1960a.

Kalman, R. E. 'A New Approach to Linear Filtering and Prediction Problems,' *Transactions of ASME, Journal of Basic Engineering*, **82**, 34–45, 1960b.

Kalman, R. E. and Bucy, R. S. 'New Results in Linear Filtering and Prediction Theory,' *Transactions of ASME, Journal of Basic Engineering* (Ser. D), **83**, 95–108, 1961.

Ljung, L. *System Identification Theory for the User*, Prentice-Hall, Inc., Englewood Cliffs, New Jersey, 1987.

Peterka, V. 'A square root filter for real-time multivariable regression,' *Kybernetika*, **11**, 5A–67, 1975.

Salut, G. J., Aquilar-Martin, J., and Lefebvre, S. 'Canonical Input–Output Representation of Linear Multivariable Stochastic Systems and Joint Optimal Parameter and State Estimation,' *Stochastica*, **3**, 17–38, 1979.

Salut, G. J., Aquilar-Martin, J., and Lefebvre, S. 'New Results on Optimal Joint Parameter and State Estimation of Linear Stochastic Systems,' *Transactions of ASME, Journal of Dynamic System Measurement and Control*, **102**, 28–34, 1980.

Söderström, T. 'Convergence properties of the generalized least-squares identification method,' *Automatica*, **10**, 617–626, 1974.

Thornton, C. L. and Bierman, G. J. 'Filtering and Error Analysis via the $\mathbf{UDU}^T$ Covariance Factorization,' *IEEE Transactions on Automatic Control*, **AC-23**(5), 901–907, 1978.

Appendix B

Advanced Dynamic System Control Techniques

As stated in the introduction, the hybrid vehicle system is a complex electromechanical system, and many control problems are fundamentally multivariable, with many actuators, performance variables, and sensors; furthermore, they are generally nonlinear, exhibit fast parameter variation, and operate under uncertain and changing conditions. In addition, many of the control design objectives are very difficult to formalize, and many of the variables that are of great concern are not measurable.

Control theory and methodologies have advanced substantially since the 1950s. Pontryagin's maximum principle, Bellman's dynamic programming, and Kalman filtering set up the foundations of modern control theory, and the theories of controllability, observability, and feedback stabilization of linear state-space models have been the breakthroughs in terms of modern multivariable feedback control methodologies. Nowadays, a multifaceted but coherent body of control theory and methodology has been formed and is playing a crucial role in industrialized societies.

Typically, even a small improvement in a control algorithm can yield a significant enhancement in system performance. Advanced control theory and methodologies are providing new opportunities for hybrid vehicle systems from improvements in dynamic performance to enhancements in passenger safety levels and to the achievement of superior fuel economy. This appendix will briefly introduce some advanced control methodologies, including optimal control, adaptive control, model predictive control, robust control, and fault-tolerant control, for hybrid vehicle applications.

Hybrid Electric Vehicle System Modeling and Control, Second Edition. Wei Liu.
© 2017 John Wiley & Sons Ltd. Published 2017 by John Wiley & Sons Ltd.

B.1   Pole Placement in Control Systems

Pole location regulation is one of the first applications of the state-space approach. The problem of pole placement involves moving the poles of a given linear system to specific locations in the s-plane by means of state feedback. The pole-placement technique provides a method by which engineers can design a linear feedback controller that adjusts the system parameters so that the closed-loop system has the desired response characteristics. The design problem is formulated as follows.

Given a system described in state-space form as:

$$\dot{\mathbf{x}} = \mathbf{A}\mathbf{x} + \mathbf{B}u$$
$$y = \mathbf{C}\mathbf{x} \tag{B.1}$$

where u is the control variable, y is the measured output variable, and $\mathbf{x}$ is the state variable, and the corresponding transfer function is:

$$G(s) = \mathbf{C}(s\mathbf{I} - \mathbf{A})^{-1}\mathbf{B} \tag{B.2}$$

The design objective is to find the following admissible feedback control law such that the closed-loop system has a desired set of poles

$$u = r - \mathbf{K}\mathbf{x} \tag{B.3}$$

where r is the reference input and $\mathbf{K}$ is the feedback gain vector. The corresponding closed-loop system state-space equation and transfer function are:

$$\dot{\mathbf{x}} = (\mathbf{A} - \mathbf{B}\mathbf{K})\mathbf{x} + \mathbf{B}r \tag{B.4}$$

$$G_c(s) = \mathbf{C}(s\mathbf{I} - \mathbf{A} + \mathbf{B}\mathbf{K})^{-1}\mathbf{B} \tag{B.5}$$

In order to satisfy the condition that the poles in the system in Eq. B.5 can be placed in any desired location, the system must be controllable. As stated in Appendix A, if a system is controllable, the composite matrix $\begin{bmatrix} \mathbf{B} & \mathbf{A}\mathbf{B} & \cdots & \mathbf{A}^{n-1}\mathbf{B} \end{bmatrix}$ must be of rank n; furthermore, if the system in Eq. B.1 is controllable, there exists a nonsingular transformation matrix $\mathbf{T}$ that transforms Eq. B.1 into the phase-variable canonical form. Kuo gave the following method for constructing the transformation matrix $\mathbf{T}$ (Kuo, 1982).

$$\mathbf{T} = \begin{bmatrix} \mathbf{T}_1 \\ \mathbf{T}_1\mathbf{A} \\ \vdots \\ \mathbf{T}_1\mathbf{A}^{n-1} \end{bmatrix} \tag{B.6}$$

where

$$T_1 = [0 \ 0 \ \cdots \ 1] [B \ AB \ \cdots \ A^{n-1}B]^{-1} \tag{B.7}$$

The transformed phase-variable canonical form is:

$$\dot{x} = Ax + bu \xrightarrow{x = T^{-1}z} \dot{z} = \begin{bmatrix} 0 & 1 & 0 & \cdots & 0 \\ 0 & 0 & 1 & \cdots & 0 \\ \vdots & \vdots & \vdots & \ddots & \vdots \\ 0 & 0 & 0 & \cdots & 1 \\ -a_0 & -a_1 & -a_2 & \cdots & -a_{n-1} \end{bmatrix} z + \begin{bmatrix} 0 \\ 0 \\ \vdots \\ 0 \\ 1 \end{bmatrix} u \tag{B.8}$$

where a_i, $i = 0, \cdots n-1$ are coefficients of the following characteristic equation:

$$\det(sI - A) = s^n + a_{n-1}s^{n-1} + a_{n-2}s^{n-2} + \cdots + a_1 s + a_0 \tag{B.9}$$

If we assume that the state feedback is given by Eq. B.3 as:

$$u(z) = r - Kz = r - k_1 z_1 - k_2 z_2 - \cdots - k_n z_n \tag{B.10}$$

then the states of the closed-loop control system can be expressed as:

$$\dot{z} = \begin{bmatrix} 0 & 1 & 0 & \cdots & 0 \\ 0 & 0 & 1 & \cdots & 0 \\ \vdots & \vdots & \vdots & \ddots & \vdots \\ 0 & 0 & 0 & \cdots & 1 \\ -(a_0 + k_1) & -(a_1 + k_2) & -(a_2 + k_3) & \cdots & -(a_{n-1} + k_n) \end{bmatrix} z + \begin{bmatrix} 0 \\ 0 \\ \vdots \\ 0 \\ 1 \end{bmatrix} r \tag{B.11}$$

If the poles of the closed-loop system are placed in the desired locations, $s_1^*, s_2^*, \cdots, s_n^*$, then the desired characteristic equation is:

$$\det(sI - (A - BK)) = (s - s_1^*)(s - s_2^*) \cdots (s - s_n^*) = s^n + d_{n-1}s^{n-1} + d_{n-2}s^{n-2} + \cdots + d_1 s + d_0 \tag{B.12}$$

Thus, the required the feedback gain, $k_1, k_2, \cdots, k_n$, can be determined from the following equations:

$$\begin{aligned} a_0 + k_1 &= d_0 &\Rightarrow \quad k_1 &= d_0 - a_0 \\ a_1 + k_2 &= d_1 &\Rightarrow \quad k_2 &= d_1 - a_1 \\ \cdots \quad &\quad \cdots & \cdots \\ a_{n-1} + k_n &= d_{n-1} &\Rightarrow \quad k_n &= d_{n-1} - a_{n-1} \end{aligned} \tag{B.13}$$

In general, the following steps need to be taken to apply the pole-placement technique to the system described by state-space equation B.1.

1. Examine whether the rank of the controllability matrix $\begin{bmatrix} \mathbf{B} & \mathbf{AB} & \cdots & \mathbf{A}^{n-1}\mathbf{B} \end{bmatrix}$ is equal to n, otherwise the pole-placement technique cannot be applied.
2. Find the characteristic equation of the system

$$\det(s\mathbf{I}-\mathbf{A}) = s^n + a_{n-1}s^{n-1} + a_{n-2}s^{n-2} + \cdots + a_1 s + a_0 \tag{B.14}$$

3. Apply the inverse transformation matrix $\mathbf{T}^{-1}$ to the state-space equation B.1 to obtain the phase-variable canonical form as well as the corresponding open-loop control system diagram shown in Fig. B.1.

$$\mathbf{x} = \mathbf{T}^{-1}\mathbf{z} \quad \Rightarrow \quad \dot{\mathbf{z}} = \mathbf{A}_c \mathbf{z} + \mathbf{B}_c r \quad \text{and} \quad y = \mathbf{C}\mathbf{x} = \mathbf{C}_c \mathbf{z} \tag{B.15}$$

where

$$\mathbf{A}_c = \mathbf{T}\mathbf{A}\mathbf{T}^{-1}, \quad \mathbf{B}_c = \mathbf{T}\mathbf{B}, \quad \mathbf{C}_c = \mathbf{C}\mathbf{T}^{-1}, \mathbf{A}_c = \begin{bmatrix} 0 & 1 & 0 & \cdots & 0 \\ 0 & 0 & 1 & \cdots & 0 \\ \vdots & \vdots & \vdots & \ddots & \vdots \\ 0 & 0 & 0 & \cdots & 1 \\ -a_0 & -a_1 & -a_2 & \cdots & -a_{n-1} \end{bmatrix}, \mathbf{B}_c = \begin{bmatrix} 0 \\ 0 \\ \vdots \\ 0 \\ 1 \end{bmatrix}$$

4. Set up a feedback control law and write down the state-space equation of the closed-loop control system. The corresponding closed-loop system is shown in Fig. B.2.

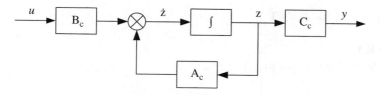

Figure B.1 Open-loop control system diagram

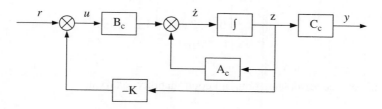

Figure B.2 Closed-loop control system diagram

$$u(\mathbf{z}) = r - k_1 z_1 - k_2 z_2 - \cdots - k_{n-1} z_{n-1} - k_n z_n = r - \mathbf{Kz} \tag{B.16}$$

$$\dot{\mathbf{z}} = (\mathbf{A_c} - \mathbf{B_c K})\mathbf{z} + \mathbf{B_c}r = \begin{bmatrix} 0 & 1 & 0 & \cdots & 0 \\ 0 & 0 & 1 & \cdots & 0 \\ \vdots & \vdots & \vdots & \ddots & \vdots \\ 0 & 0 & 0 & \cdots & 1 \\ -a_0 - k_1 & -a_1 - k_2 & -a_2 - k_3 & \cdots & -a_{n-1} - k_n \end{bmatrix} + \begin{bmatrix} 0 \\ 0 \\ \vdots \\ 0 \\ 1 \end{bmatrix} r \tag{B.17}$$

$$y = \mathbf{C_c z}$$

5. Place all closed-loop poles in the desired locations

$$\det(s\mathbf{I} - (\mathbf{A_c} - \mathbf{B_c K})) = (s - s_1^*)(s - s_2^*) \cdots (s - s_n^*)$$
$$= s^n + d_{n-1}s^{n-1} + d_{n-2}s^{n-2} + \cdots + d_1 s + d_0 \tag{B.18}$$

6. Equate to Eq. B.11 to solve for the feedback gain, $k_1, k_2, \cdots, k_n$.
7. Apply the transformation matrix $\mathbf{T}$ to Eq. B.17 to obtain the feedback gain in the original coordinates:

$$\dot{\mathbf{x}} = \mathbf{T}^{-1}(\mathbf{A_c} - \mathbf{B_c K})\mathbf{Tx} + \mathbf{T}^{-1}\mathbf{B_c}r = \mathbf{Ax} - \mathbf{BFx} + \mathbf{B}r$$
$$y = \mathbf{C_c Tx} = \mathbf{Cx} \tag{B.19}$$

That is, $\mathbf{z} = \mathbf{Tx}$ and $\mathbf{F} = \mathbf{KT}$

Example B.1
Consider the following system:

$$\dot{\mathbf{x}} = \begin{bmatrix} 1 & 0 & 0 \\ 0 & 2 & 1 \\ 0 & 0 & 2 \end{bmatrix} \mathbf{x} + \begin{bmatrix} 1 \\ 0 \\ 1 \end{bmatrix} u \tag{B.20}$$
$$y = \begin{bmatrix} 1 & 1 & 0 \end{bmatrix} \mathbf{x}$$

Where the characteristic equation and controllability matrix are:

$$\det(s\mathbf{I} - \mathbf{A}) = s^3 - 5s^2 + 8s - 4 \tag{B.21}$$

$$\begin{bmatrix} \mathbf{B} & \mathbf{AB} & \mathbf{A^2B} \end{bmatrix} = \begin{bmatrix} \begin{bmatrix} 1 \\ 0 \\ 1 \end{bmatrix} & \begin{bmatrix} 1 & 0 & 0 \\ 0 & 2 & 1 \\ 0 & 0 & 2 \end{bmatrix} \begin{bmatrix} 1 \\ 0 \\ 1 \end{bmatrix} & \begin{bmatrix} 1 & 0 & 0 \\ 0 & 2 & 1 \\ 0 & 0 & 2 \end{bmatrix}^2 \begin{bmatrix} 1 \\ 0 \\ 1 \end{bmatrix} \end{bmatrix} = \begin{bmatrix} 1 & 1 & 1 \\ 0 & 1 & 4 \\ 1 & 2 & 4 \end{bmatrix}$$

(B.22)

$$\text{rank}\left(\begin{bmatrix} \mathbf{B} & \mathbf{AB} & \mathbf{A^2B} \end{bmatrix} \right) = \text{rank} \left(\begin{bmatrix} 1 & 1 & 1 \\ 0 & 1 & 4 \\ 1 & 2 & 4 \end{bmatrix} \right) = 3$$

The transformation matrix $\mathbf{T}$ and transformed matrices $\mathbf{A}_c, \mathbf{B}_c, \mathbf{C}_c$ are:

$$\mathbf{T}_1 = \begin{bmatrix} 0 & 0 & 1 \end{bmatrix} \begin{bmatrix} 1 & 1 & 1 \\ 0 & 1 & 4 \\ 1 & 2 & 4 \end{bmatrix}^{-1} = \begin{bmatrix} 0 & 0 & 1 \end{bmatrix} \begin{bmatrix} 4 & 2 & -3 \\ -4 & -3 & 4 \\ 1 & 1 & -1 \end{bmatrix}$$

(B.23)

$$= \begin{bmatrix} 1 & 1 & -1 \end{bmatrix}, \quad \mathbf{T} = \begin{bmatrix} 1 & 1 & -1 \\ 1 & 2 & -1 \\ 1 & 4 & 0 \end{bmatrix}, \quad \mathbf{T}^{-1} = \begin{bmatrix} 4 & -4 & 1 \\ -1 & 1 & 0 \\ 2 & -3 & 1 \end{bmatrix}$$

$$\mathbf{A}_c = \mathbf{TAT}^{-1} = \begin{bmatrix} 1 & 1 & -1 \\ 1 & 2 & -1 \\ 1 & 4 & 0 \end{bmatrix} \begin{bmatrix} 1 & 0 & 0 \\ 0 & 2 & 1 \\ 0 & 0 & 2 \end{bmatrix} \begin{bmatrix} 4 & -4 & 1 \\ -1 & 1 & 0 \\ 2 & -3 & 1 \end{bmatrix} = \begin{bmatrix} 0 & 1 & 0 \\ 0 & 0 & 1 \\ 4 & -8 & 5 \end{bmatrix},$$

$$\mathbf{B}_c = \mathbf{TB} = \begin{bmatrix} 1 & 1 & -1 \\ 1 & 2 & -1 \\ 1 & 4 & 0 \end{bmatrix} \begin{bmatrix} 1 \\ 0 \\ 1 \end{bmatrix} = \begin{bmatrix} 0 \\ 0 \\ 1 \end{bmatrix}$$

(B.24)

$$\mathbf{C}_c = \mathbf{CT}^{-1} = \begin{bmatrix} 1 & 1 & 0 \end{bmatrix} \begin{bmatrix} 4 & -4 & 1 \\ -1 & 1 & 0 \\ 2 & -3 & 1 \end{bmatrix} = \begin{bmatrix} 3 & -3 & 1 \end{bmatrix}$$

If we intend to find the feedback gain for the system such that the poles of the closed-loop system are located at -1, -2, and -3, the characteristic equation of the closed-loop system can be obtained as:

$$\det(s\mathbf{I} - (\mathbf{A}_c - \mathbf{B}_c\mathbf{K})) = (s+1)(s+2)(s+3) = s^3 + 6s^2 + 11s + 6 \qquad (B.25)$$

From Eqs B.13 and B.17, we have:

$$
\begin{aligned}
a_0 + k_1 &= d_0 & \Rightarrow \quad k_1 &= d_0 - a_0 = 6 + 4 = 10 \\
a_1 + k_2 &= d_1 & \Rightarrow \quad k_2 &= d_1 - a_1 = 11 - 8 = 3 \\
a_2 + k_3 &= d_2 & \Rightarrow \quad k_3 &= d_2 - a_2 = 6 + 5 = 11
\end{aligned}
\qquad (B.26)
$$

The closed-loop control system state and output equations in the transformed coordinates are:

$$
\dot{\mathbf{z}} = \begin{bmatrix} 0 & 1 & 0 \\ 0 & 0 & 1 \\ -6 & -11 & -6 \end{bmatrix} \mathbf{z} + \begin{bmatrix} 0 \\ 0 \\ 1 \end{bmatrix} r
$$
$$
y = \begin{bmatrix} 3 & -3 & 1 \end{bmatrix} \mathbf{z}
\qquad (B.27)
$$

In the original coordinates, the feedback gains can be obtained from:

$$
\mathbf{F} = \mathbf{KT} = \begin{bmatrix} k_1 & k_2 & k_3 \end{bmatrix} \begin{bmatrix} 1 & 1 & -1 \\ 1 & 2 & -1 \\ 1 & 4 & 0 \end{bmatrix} = \begin{bmatrix} 10 & 3 & 11 \end{bmatrix} \begin{bmatrix} 1 & 1 & -1 \\ 1 & 2 & -1 \\ 1 & 4 & 0 \end{bmatrix} = \begin{bmatrix} 24 & 60 & -13 \end{bmatrix} \qquad (B.28)
$$

Finally, the closed-loop control system is achieved in the original coordinates:

$$
\dot{\mathbf{x}} = \begin{bmatrix} -23 & -60 & 13 \\ 0 & 2 & 1 \\ -24 & -60 & 15 \end{bmatrix} \mathbf{x} + \begin{bmatrix} 1 \\ 0 \\ 1 \end{bmatrix} u
$$
$$
y = \begin{bmatrix} 1 & 1 & 0 \end{bmatrix} \mathbf{x}
\qquad (B.29)
$$

B.2   Optimal Control

The fundamental control objective of a hybrid vehicle system is to maximize fuel economy and minimize emissions while abiding by the drivability and safety requirements under various driving scenarios. Unlike optimization, the optimal control is to find a control law for a given dynamic system such that a certain optimal objective is achieved with physical constraints satisfied at the same time. This section intends to introduce the basic principle of optimal control design, outline brief design procedures, and illustrate results with simple examples.

B.2.1   The Optimal Control Problem Formulation

The optimal control problem is formulated as follows:

1. **The given system**

   It is assumed that the system is described by the following state-space model:

$$\dot{\mathbf{x}} = \mathbf{f}(\mathbf{x}, \mathbf{u}, t) \tag{B.30}$$

   where $\mathbf{x} = [x_1 \ x_2 \ \cdots \ x_n]^T$ is the state vector, and $\mathbf{u} = [u_1 \ u_2 \ \cdots \ u_m]^T$ is the control vector.

2. **Admissible control region**

   In practical problems, the control variables, $u_1, u_2, \cdots, u_m$, are normally constrained between upper and lower bounds; that is:

$$u_i^{\text{lower}} \le u_i \le u_i^{\text{upper}} \tag{B.31}$$

   The region where all control variables satisfy the constraints is called the *admissible control region*.

3. **Initial and end conditions**

   Initial conditions are the starting points $\mathbf{x}(t_0)$ of the state variable, $\mathbf{x}$, and the end conditions are the ending point, $\mathbf{x}(t_f)$, of the state variable, $\mathbf{x}$. If $\mathbf{x}(t_0)$ is arbitrary for any value, the control problem is called a free initial control problem; if $\mathbf{x}(t_0)$ is given, the control problem is called a fixed initial control problem. Similarly, if $\mathbf{x}(t_f)$ is arbitrary for any value, the control problem is called a free endpoint control problem; if $\mathbf{x}(t_f)$ is given, the problem is called a fixed endpoint control problem. The initial and end conditions can be expressed as:

$$\begin{aligned} \mathbf{x}(t_0) &= \mathbf{x}_0 \\ \mathbf{x}(t_f) &= \mathbf{x}_f \end{aligned} \tag{B.32}$$

4. **Objective function**

   The objective function, also known as the cost function, is a function of the state and control variables. The principal form of the objective function is the quadratic function, which is normally a sum of weightings of the squares of the individual state variables and control variables, as follows:

$$J = \int_0^t \left(\mathbf{x}^T \mathbf{Q} \mathbf{x} + \mathbf{u}^T \mathbf{R} \mathbf{u} \right) dt \tag{B.33}$$

This form will lead to a suitable mathematical treatment of qualitative consideration of penalizing both large deviations from the desired state and increased control actions more severely.

B.2.2 Pontryagin's Maximum Method

Pontryagin presented his maximum principle to solve the optimal control problem in 1962 (Pontryagin *et al.*, 1962). The basic design procedure with an example is given in this section.

For a given system:

$$\dot{\mathbf{x}} = \mathbf{f}(\mathbf{x}, u, t), \quad \mathbf{x}(t_0) = \mathbf{x}_0, \quad \mathbf{x}(t_f) \text{ is free} \tag{B.34}$$

if the objective function and Hamilton function are defined as:

$$J = \Phi(\mathbf{x}(t_f)) + \int_{t_0}^{t_f} L(\mathbf{x}(t), u(t), t) dt \tag{B.35}$$

$$H(\mathbf{x}(t), \lambda(t), u(t), t) = L(\mathbf{x}(t), \lambda(t), u(t), t) + \lambda^T \mathbf{f}(\mathbf{x}(t), u(t), t)$$

the optimal control $u^*(t)$, which drives the system from the initial states, $\mathbf{x}(t_0)$, to the end states, $\mathbf{x}(t_f)$, with the maximal J, can be found based on the following steps extracted from Pontryagin's maximum principle:

1. Introduce a co-state vector, $\lambda(t)$; $\lambda(t)$ and $\mathbf{x}(t)$ shall satisfy the following differential equation system if there exists the optimal control, $u^*(t)$

$$\dot{\mathbf{x}} = \frac{\partial H}{\partial \lambda} = \mathbf{f}(\mathbf{x}, u, t)$$

$$\dot{\lambda} = -\frac{\partial H}{\partial \mathbf{x}} = -\frac{\partial L}{\partial \mathbf{x}} - \left[\frac{\partial f}{\partial \mathbf{x}} \right]^T \lambda(t) \tag{B.36}$$

2. Determine the initial and end conditions of the state variable $\mathbf{x}(t)$ and co-state variable $\lambda(t)$ based on the detailed application scenarios. For the case given by Eqs B.34 and B.35, the optimal control problem has fixed initial and free end states, and the terminal conditions of Eq. B.36 are:

$$\mathbf{x}(t_0) = \mathbf{x}_0, \quad \lambda(t_f) = \left[\frac{\partial \Phi}{\partial \mathbf{x}} \right]_{t_f} \tag{B.37}$$

3. In practical applications, if the optimal states $\mathbf{x}^*$ and co-states λ^* are determined or known, then the Hamilton function $H(\mathbf{x}^*(t), \lambda^*(t), u(t), t)$ is the only function of control

variable, $u(t)$, which means that the optimal control, $u^*(t)$, will result in the value of the Hamilton function being maximized. Therefore, the optimal control, $u^*(t)$, can be found in the admissible control region by the following equation:

$$\max_{u \in U} H(\mathbf{x}^*(t), \boldsymbol{\lambda}^*(t), u(t), t) = H(\mathbf{x}^*(t), \boldsymbol{\lambda}^*(t), u^*(t), t), \quad t \in [t_0, t_f] \tag{B.38}$$

where U is the admissible region of control variable $u(t)$.

Example B.2

Consider the following system:

$$\dot{x} = ax + bu \tag{B.39}$$

The objective function is defined as:

$$J = \int_0^\infty x^2 dt \tag{B.40}$$

and the admissible control region is:

$$-M \le u \le M \tag{B.41}$$

The optimal control action, $u^*(t)$, can be found by the following steps:

1. Set up the Hamilton function:

$$H = x^2 + \lambda(ax + bu) \tag{B.42}$$

2. Set up the co-state equation:

$$\dot{\lambda} = -\frac{\partial H}{\partial x} = -2x + a\lambda \tag{B.43}$$

3. Find the optimal control $u^*(t)$ which has the value of Hamilton function at its minimum:

$$u^*(t) = \begin{cases} -M, & \text{if } \lambda b > 0 \\ M, & \text{if } \lambda b \le 0 \end{cases} \quad \text{i.e., } u^*(t) = -M \cdot \text{sign}(\lambda b) \tag{B.44}$$

4. The optimal state trace is:

$$x^*(t) = ax^* - M \cdot \text{sign}(\lambda b) \tag{B.45}$$

B.2.3 Dynamic Programming

The fundamental principle of the dynamic programming method is based on Bellman's principle of optimality, which states that an optimal policy has the property that whatever the initial state and the initial decision, the remaining decision must form an optimal policy with regard to the state resulting from the first decision (Bellman, 1957). Bellman's principle of optimality immediately leads to the following computational algorithm to find the optimal control law.

Consider a discrete system described by the following difference equation:

$$\mathbf{x}(k+1) = \mathbf{f}(\mathbf{x}(k), \mathbf{u}(k)), \quad \mathbf{x}(0) = \mathbf{x}_0, \quad k = 0, 1, \cdots, N-1 \tag{B.46}$$

The objective function is:

$$J(\mathbf{x}(0), N) = \sum_{k=0}^{N-1} L(\mathbf{x}(k), \mathbf{u}(k)) + \Phi(\mathbf{x}(N)) \tag{B.47}$$

The dynamic programming method finds the sequence of decision, $\mathbf{u}(0)$, $\mathbf{u}(1), \cdots, \mathbf{u}(N-1)$, which maximizes the objective function (Eq. B.47). Dynamic programming can be used either backwards or forwards. The backward version is the normal method and it starts from the final point to the initial point where $\mathbf{x}(0)$ is specified as $\mathbf{x}_0$, and the sequence of optimal controls $\mathbf{u}(N-1), \mathbf{u}(N-2), \cdots, \mathbf{u}(0)$ is thus generated. A set of recursive relationships is provided, which enables the establishment of the sequence $J(\mathbf{x}(N-1), 1)$, $J(\mathbf{x}(N-2), 2)$, $\cdots$, $J(\mathbf{x}(0), N)$ and the optimal control sequence $\mathbf{u}^*(N-1), \mathbf{u}^*(N-2), \cdots, \mathbf{u}^*(0)$.

At present, if the final state $\mathbf{x}(N)$ is specified as $\mathbf{x}(N) = \mathbf{x}_f$, the final state need not be included in the summation of the objective function (Eq. B.47), and this is a fixed end condition problem. In practice, an optimal decision is made using the following procedures:

1. The first step is to find the optimal control, $\mathbf{u}^*(N-1)$, which denotes the optimal objective function, $J(\mathbf{x}(N-1), 1) = L(\mathbf{x}(N-1), \mathbf{u}(N-1)) + \Phi(\mathbf{x}(N))$, i.e.

$$J^*(\mathbf{x}(N-1), 1) = \min_{\mathbf{u}(N-1)} \{L(\mathbf{x}(N-1), \mathbf{u}(N-1)) + \Phi(\mathbf{x}(N))\} \tag{B.48}$$

Based on the system state equation (Eq. B.46), the final state $\mathbf{x}(N) = \mathbf{x}_f$ is reached from the previous state $\mathbf{x}(N-1)$ driven by the optimal control, $\mathbf{u}^*(N-1)$, i.e.:

$$\mathbf{x}(N) = \mathbf{f}(\mathbf{x}(N-1), \mathbf{u}^*(N-1)) = \mathbf{x}_f \tag{B.49}$$

2. The second step is to find the optimal control, $\mathbf{u}^*(N-2)$, which minimizes the objective function $J(\mathbf{x}(N-2), 2) = \min_{\mathbf{u}(N-2)} \{L(\mathbf{x}(N-2), \mathbf{u}(N-2)) + J^*(\mathbf{x}(N-1), 1)\}$ and drives the state to $\mathbf{x}(N-1)$ from $\mathbf{x}(N-2)$.

3. Repeating the above procedures, assume that the stage now goes backwards to stage $N-2$. At this stage, the initial state is $\mathbf{x}(1)$ and the objective is to find the optimal control sequence $\mathbf{u}^*(N-1),\mathbf{u}^*(N-2),\cdots,\mathbf{u}^*(1)$ to make the objective function, $J(\mathbf{x}(1),N-1)$, minimal.

$$J^*(\mathbf{x}(1),N-1) = \min_{\mathbf{u}(1)} \{L(\mathbf{x}(1),\mathbf{u}(1)) + J^*(\mathbf{x}(2),N-2)\} \qquad (B.50)$$

where the notation $J^*(\mathbf{x}(2),N-2)$ signifies the objective function evaluated over $N-1$ times and the end state, $\mathbf{x}_\mathrm{f}$, is reached by the optimal control sequence $\mathbf{u}^*(N-1),\mathbf{u}^*(N-2),\cdots,\mathbf{u}^*(2)$.

4. Finally, the whole optimal control sequence, $\mathbf{u}^*(N-1),\mathbf{u}^*(N-2),\cdots,\mathbf{u}^*(1)$ is determined; this drives the system from the initial state $\mathbf{x}(0)=\mathbf{x}_0$ to the specified end state, $\mathbf{x}(N)=\mathbf{x}_\mathrm{f}$, and the optimal objective function value is achieved over the whole control process.

$$J^*(\mathbf{x}(0),N) = \min_{\mathbf{u}(0)} \{L(\mathbf{x}(0),\mathbf{u}(0)) + J^*(\mathbf{x}(1),N-1)\} \qquad (B.51)$$

The dynamic programming method is an important optimal control technique to improve hybrid vehicle fuel economy and other performance measures. Many HEV control problems, such as energy management, can be treated as an optimal control problem and solved by dynamic programming. The following two examples are given to help readers understand the principle of dynamic programming, and two detailed engineering examples are presented in Chapters 6 and 8.

Example B.3

Given a discrete event process with the incurred cost for each path labeled in Fig. B.3, use the dynamic programming method to determine the lowest cost path from A to F.

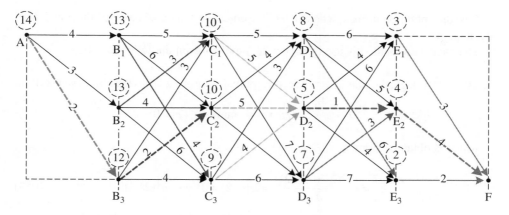

Figure B.3   Working diagram of the dynamic programming method

Solution:

As all costs incurred on alternative stage paths have been given, it is easy to use the backward method to solve the problem. At stage E, there are three paths to reach end point F, and the cost of each path is circled with a dashed line on the diagram based on the principles of optimality. Then, if we go back to stage D, for each of the event points D_1, D_2, and D_3, the optimal paths are determined by minimizing the cost to the final point F. The calculated minimal cost is circled above the point, and the corresponding path is highlighted and colored on the diagram. That is, the optimal path from D_1 to F is $D_1 \rightarrow E_3 \rightarrow F$ and the cost of this path is **8**; the optimal path from D_2 to F is $D_2 \rightarrow E_2 \rightarrow F$ and the cost with this path is **5**; the optimal path from D_3 to F is $D_3 \rightarrow E_2 \rightarrow F$ and the cost is **7** from this path. Now, we are back to stage C. Similar to the previous step, the lowest paths are determined for each event point, C_1, C_2, and C_3, the costs of which are circled above the point and highlighted. If we start from C_1, the optimal path is $C_1 \rightarrow D_2 \rightarrow E_2 \rightarrow F$ and the cost is **10** for this path; if we start from C_2, the optimal path is $C_2 \rightarrow D_2 \rightarrow E_2 \rightarrow F$ and the cost with this path is **10**; and if we start from C_3, the optimal path is $C_3 \rightarrow D_2 \rightarrow E_2 \rightarrow F$ and the cost is **9** for this path. Returning to stage B, the optimal path starting from B_1 is $B_1 \rightarrow C_3 \rightarrow D_2 \rightarrow E_2 \rightarrow F$ and the cost is **13**; the optimal path from B_2 is $B_2 \rightarrow C_1 \rightarrow D_2 \rightarrow E_2 \rightarrow F$ and the cost is **13**; and the optimal path from B_3 is $B_3 \rightarrow C_2 \rightarrow D_2 \rightarrow E_2 \rightarrow F$ with a cost of **12**. Finally, we are back to the first stage, A; the optimal path is to go to the point B_3. Therefore, the determined overall optimal path is $A \rightarrow B_3 \rightarrow C_2 \rightarrow D_2 \rightarrow E_2 \rightarrow F$, highlighted by the dashed line in Fig. B.3, and the overall cost is **14,** which is the minimum.

Example B.4

Consider a discrete linear system with the following objective function:

$$x(k+1) = ax(k) + bu(k)$$

$$J = \sum_{k=0}^{2} \left(x^2(k) + ru^2(k) \right) \tag{B.52}$$

Find the optimal control, u, which makes the objective J minimal based on the principle of dynamic programming.

The backward method is used to solve the problem from the following steps.

1. At the last stage, the state $x(2)$ is free; the optimal control $u^*(2)$ has the objective function:

$$J^* = x^2(2) + ru^2(2) = \quad \min \text{ is } u^*(2) = 0$$

which is obtained from:

$$\frac{\partial J}{\partial u} = \frac{(x^2(2) + ru^2(2))}{\partial u} = 2ru^*(2) = 0 \quad \Rightarrow \quad u^*(2) = 0 \tag{B.53}$$

and the achieved optimal value of the objective function is $J^* = x^{*2}(2)$

2. At the stage $x = x(1)$, assume that the state $x(1)$ is on the optimal trajectory, i.e., $x(1) = x^*(1)$, then the optimal control $u^*(1)$ is solved from:

$$J^*(x(1)) = \min_{u(1)} \left\{ \left(x^{*2}(1) + ru^2(1) \right) + J^*(x(2)) \right\}$$

$$= \min_{u(1)} \left\{ \left(x^{*2}(1) + ru^2(1) \right) + x^{*2}(2) \right\} \tag{B.54}$$

$$= \min_{u(1)} \left\{ x^{*2}(1) + ru^2(1) + \left(ax^*(1) + bu(1) \right)^2 \right\}$$

and

$$\frac{\partial (J^*(x(1)))}{\partial u(1)} = \frac{\partial \left(x^{*2}(1) + ru^2(1) + \left(ax^*(1) + bu(1) \right)^2 \right)}{\partial u(1)} \tag{B.55}$$

$$= 2ru(1) + 2b(ax^*(1) + bu(1)) = 0$$

The optimal control, $u^*(1)$, and the corresponding value of the objective function are:

$$u^*(1) = -\left(\frac{abx^*(1)}{r + b^2} \right), \quad J^*(x(1)) = \left(1 + \frac{ra^2}{r + b^2} \right) x^{*2}(1) \tag{B.56}$$

3. Now go back to the stage $x = x(0)$; the optimal control, $u^*(0)$, is solved by:

$$J^*(x(0)) = \min_{u(1)} \left\{ \left(x^{*2}(0) + ru^2(0) \right) + J^*(x(1)) \right\}$$

$$= \min_{u(1)} \left\{ \left(x^{*2}(0) + ru^2(0) \right) + \left(1 + \frac{ra^2}{r + b^2} \right) x^{*2}(1) \right\} \tag{B.57}$$

$$= \min_{u(1)} \left\{ \left(x^{*2}(0) + ru^2(0) \right) + \left(1 + \frac{ra^2}{r + b^2} \right) \left(ax^*(0) + bu(0) \right)^2 \right\}$$

Solve to $u(0)$ from the following equation:

$$\frac{\partial (J^*(x(0)))}{\partial u} = 2ru(0) + 2b \left(1 + \frac{ra^2}{r + b^2} \right) (ax^*(0) + bu(0)) = 0$$

$$\Rightarrow \quad u(0) = -\frac{ab(r + b^2 + ra^2)}{(r + b^2)^2 + ra^2 b^2} x(0) \tag{B.58}$$

Since the obtained optimal control sequence $u^*(2), u^*(1), u^*(0)$ is a function of states $x(1)$ and $x(0)$, the optimal sequence can be implemented in the closed-loop feedback control form.

B.2.4   Linear Quadratic Control

Linear quadratic control (LQC) is another important type of optimal problem, which assumes that the system is linear and the objective function has quadratic form, as follows:

$$\dot{\mathbf{x}} = \mathbf{A}\mathbf{x} + \mathbf{B}\mathbf{u}$$

$$\mathbf{y} = \mathbf{C}\mathbf{x}$$

$$J = \frac{1}{2}\int_{t_0}^{t_f} \left(\mathbf{x}^{\mathrm{T}}\mathbf{Q}\mathbf{x} + \mathbf{u}^{\mathrm{T}}\mathbf{R}\mathbf{u}\right)dt + \frac{1}{2}\mathbf{x}_f^{\mathrm{T}}\mathbf{S}\mathbf{x}_f \tag{B.59}$$

where $\mathbf{Q}$ is symmetric positive semi-definite, and $\mathbf{R}$ and $\mathbf{S}$ are symmetric positive definite, and they are normally chosen as diagonal matrices in practice.

Based on Pontryagin's maximum principle, the optimal LQC problem can be solved using the following steps:

1. Define the Hamilton function

$$H(\mathbf{x},\mathbf{u},\boldsymbol{\lambda}) = \frac{1}{2}\left(\mathbf{x}^{\mathrm{T}}\mathbf{Q}\mathbf{x} + \mathbf{u}^{\mathrm{T}}\mathbf{R}\mathbf{u}\right) + \boldsymbol{\lambda}^{\mathrm{T}}(\mathbf{A}\mathbf{x} + \mathbf{B}\mathbf{u}) \tag{B.60}$$

then the system state and co-state can be described by:

$$\dot{\mathbf{x}} = \frac{\partial H}{\partial \boldsymbol{\lambda}} = \mathbf{A}\mathbf{x} + \mathbf{B}\mathbf{u}$$

$$\dot{\boldsymbol{\lambda}} = -\frac{\partial H}{\partial \mathbf{x}} = -\frac{\partial\left(\frac{1}{2}\left(\mathbf{x}^{\mathrm{T}}\mathbf{Q}\mathbf{x} + \mathbf{u}^{\mathrm{T}}\mathbf{R}\mathbf{u}\right)\right)}{\partial \mathbf{x}} - \frac{\partial(\mathbf{A}\mathbf{x} + \mathbf{B}\mathbf{u})}{\partial \mathbf{x}}\boldsymbol{\lambda} = -\mathbf{Q}\mathbf{x} - \mathbf{A}^{\mathrm{T}}\boldsymbol{\lambda} \tag{B.61}$$

with the specified initial conditions of $\mathbf{x}(t_0) = \mathbf{x}_0$ and boundary conditions $\boldsymbol{\lambda}(t_f) = \mathbf{S}\mathbf{x}(t_f)$.

2. The optimal control, $\mathbf{u}^*$, is the admissible control that maximizes the Hamilton function, H. If we assume that $\mathbf{u}$ is unbounded, differentiating H with respect to the input $\mathbf{u}$ yields the following equation for a maximum in H:

$$\frac{\partial H}{\partial u} = \mathbf{R}\mathbf{u} + \mathbf{B}^{\mathrm{T}}\boldsymbol{\lambda} = 0 \;\Rightarrow\; \mathbf{u}^* = -\mathbf{R}^{-1}\mathbf{B}^{\mathrm{T}}\boldsymbol{\lambda} \tag{B.62}$$

3. From Eq. B.62, we know that $\mathbf{u}^*$ is a linear function of co-state, $\boldsymbol{\lambda}$. In order to form feedback control, it is necessary to obtain the transformation matrix $\mathbf{P}$ which links $\mathbf{x}$ and $\boldsymbol{\lambda}$:

$$\boldsymbol{\lambda}(t) = \mathbf{P}(t)\mathbf{x}(t) \tag{B.63}$$

4. Substituting $\boldsymbol{\lambda}(t) = \mathbf{P}(t)\mathbf{x}(t)$ into $\mathbf{u}^* = -\mathbf{R}^{-1}\mathbf{B}^{\mathrm{T}}\boldsymbol{\lambda}$ results in the feedback control law:

$$\mathbf{u}^* = -\mathbf{R}^{-1}\mathbf{B}^{\mathrm{T}}\mathbf{P}(t)\mathbf{x}(t) \tag{B.64}$$

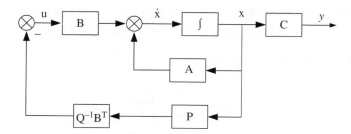

Figure B.4   Diagram of the optimal LQC feedback control system

5. Taking $\mathbf{u}^* = -\mathbf{R}^{-1}\mathbf{B}^T\mathbf{P}(t)\mathbf{x}(t)$ into Eq. B.61, we have:

$$\dot{\mathbf{x}} = \mathbf{A}\mathbf{x} - \mathbf{B}\mathbf{R}^{-1}\mathbf{B}^T\mathbf{P}(t)\mathbf{x}(t) = \left(\mathbf{A} - \mathbf{B}\mathbf{R}^{-1}\mathbf{B}^T\mathbf{P}(t)\right)\mathbf{x}(t)$$
$$\dot{\boldsymbol{\lambda}} = -\mathbf{Q}\mathbf{x} - \mathbf{A}^T\boldsymbol{\lambda} = -\left(\mathbf{Q} + \mathbf{A}^T\mathbf{P}(t)\right)\mathbf{x}(t) \tag{B.65}$$

6. Applying derivation to Eq. B.63, we get:

$$\dot{\boldsymbol{\lambda}}(t) = \dot{\mathbf{P}}(t)\mathbf{x}(t) + \dot{\mathbf{P}}(t)\dot{\mathbf{x}}(t) \tag{B.66}$$

7. Eliminating $\dot{\boldsymbol{\lambda}}(t)$, $\dot{\mathbf{x}}(t)$, and $\mathbf{x}(t)$ in Eqs B.65 and B.66, we have:

$$\dot{\mathbf{P}}(t) = -\mathbf{P}(t)\mathbf{A} - \mathbf{A}^T\mathbf{P}(t) + \mathbf{P}(t)\mathbf{B}\mathbf{R}^{-1}\mathbf{B}^T\mathbf{P}(t) - \mathbf{Q}, \quad \mathbf{P}(t_f) = \mathbf{S} \tag{B.67}$$

This equation is called the matrix Riccati equation, and it is a nonlinear differential equation. Once $\mathbf{P}$ has been solved for, the feedback control law is formed. The optimal feedback system may be represented by the diagram shown in Fig. B.4.

8. If $t_f \to \infty$, $\mathbf{P}(t)$ becomes a constant matrix and the above Riccati equation is reduced to the matrix algebra equation:

$$\mathbf{P}(t)\mathbf{A} + \mathbf{A}^T\mathbf{P}(t) - \mathbf{P}(t)\mathbf{B}\mathbf{R}^{-1}\mathbf{B}^T\mathbf{P}(t) + \mathbf{Q} = 0 \tag{B.68}$$

where $\mathbf{P}$ is a symmetric positive matrix.

Example B.5
A linear dynamic system is described by the equation:

$$\begin{bmatrix} \dot{x}_1 \\ \dot{x}_2 \end{bmatrix} = \begin{bmatrix} 0 & 1 \\ 0 & 0 \end{bmatrix} \begin{bmatrix} x_1 \\ x_2 \end{bmatrix} + \begin{bmatrix} 0 \\ 1 \end{bmatrix} u$$
$$J = \frac{1}{2}\int_0^\infty \left(x_1^2 + 2bx_1x_2 + ax_2^2 + u^2\right)dt \quad \text{where } a > b^2 > 0 \tag{B.69}$$

Find the optimal feedback control law that has the objective function J minimized.

Solution:
From the given J, we know:

$$\mathbf{Q} = \begin{bmatrix} 1 & b \\ b & a \end{bmatrix}, \quad \mathbf{R} = 1$$

From the following controllability matrix, we know that the given system is controllable:

$$\text{Rank}([\mathbf{B} \quad \mathbf{AB}]) = \text{Rank}\left(\begin{bmatrix} 0 \\ 1 \end{bmatrix} \begin{bmatrix} 0 & 1 \\ 0 & 0 \end{bmatrix} \begin{bmatrix} 0 \\ 1 \end{bmatrix} \right) = \text{Rank}\left(\begin{bmatrix} 0 \\ 1 \end{bmatrix} \begin{bmatrix} 1 \\ 0 \end{bmatrix} \right) = 2 \qquad \text{(B.70)}$$

From Eq. B.64, the optimal feedback control law is:

$$u^* = -\mathbf{R}^{-1}\mathbf{B}^{\mathsf{T}}\mathbf{P}(t)\mathbf{x}(t) = -[0 \quad 1] \begin{bmatrix} p_{11} & p_{12} \\ p_{21} & p_{22} \end{bmatrix} \begin{bmatrix} x_1 \\ x_2 \end{bmatrix} = -p_{12}x_1 - p_{22}x_2 \qquad \text{(B.71)}$$

where p_{12} and p_{22} are the solution of the following Riccati equation:

$$\begin{bmatrix} p_{11} & p_{12} \\ p_{21} & p_{22} \end{bmatrix} \begin{bmatrix} 0 & 1 \\ 0 & 0 \end{bmatrix} + \begin{bmatrix} 0 & 0 \\ 1 & 0 \end{bmatrix} \begin{bmatrix} p_{11} & p_{12} \\ p_{21} & p_{22} \end{bmatrix} - \begin{bmatrix} p_{11} & p_{12} \\ p_{21} & p_{22} \end{bmatrix} \begin{bmatrix} 0 \\ 1 \end{bmatrix} [0 \quad 1] \begin{bmatrix} p_{11} & p_{12} \\ p_{21} & p_{22} \end{bmatrix} + \begin{bmatrix} 1 & b \\ b & a \end{bmatrix} = \begin{bmatrix} 0 & 0 \\ 0 & 0 \end{bmatrix}$$
$$\text{(B.72)}$$

Equating the above matrices, we obtain the following three equations:

$$p_{12}^2 = 1, \quad p_{11} - p_{12}p_{22} + b = 0, \quad 2p_{12} - p_{22}^2 + a = 0 \qquad \text{(B.73)}$$

Solving the equations, we obtain:

$$p_{12} = \pm 1, \quad p_{22} = \pm\sqrt{a + 2p_{12}}, \quad p_{11} = p_{12}p_{22} - b \qquad \text{(B.74)}$$

Since the matrix $\mathbf{P}$ has to be symmetric positive definite, the solutions are determined as:

$$p_{12} = 1, \quad p_{22} = \sqrt{a + 2}, \quad p_{11} = \sqrt{a + 2} - b$$

Thus, the optimal control law is:

$$u^* = -x_1 - \sqrt{a + 2}\, x_2 \qquad \text{(B.75)}$$

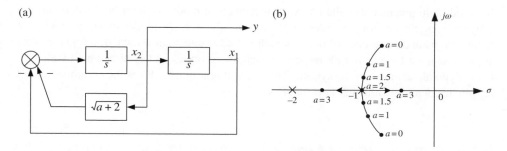

Figure B.5 Obtained optimal feedback control system

If we assume that the system output is x_2, the closed-loop control system is:

$$\begin{bmatrix} \dot{x}_1 \\ \dot{x}_2 \end{bmatrix} = \begin{bmatrix} 0 & 1 \\ -1 & -\sqrt{a+2} \end{bmatrix} \begin{bmatrix} x_1 \\ x_2 \end{bmatrix}$$

$$y = \begin{bmatrix} 0 & 1 \end{bmatrix} \begin{bmatrix} x_1 \\ x_2 \end{bmatrix}$$

(B.76)

The diagram of the closed-loop system is shown in Fig. B.5(a). The transfer function and the pole locations are described by the following equation:

$$G_c(s) = \frac{1}{s^2 + \sqrt{a+2}\,s + 1} \quad \Rightarrow \quad s_{1,2} = -\frac{\sqrt{a+2}}{2} \pm j\frac{\sqrt{2-a}}{2}$$

(B.77)

If we take parameter a as the changeable variable, we can obtain the root locus of the optimal feedback system, and the root locus diagram is shown in Fig. B.5(b).

From the root locus diagram, we can summarize the observations as:

1. When $a = 0$, which means there is no demand on the state x_2, the system poles are located at $s_{1,2} = -\frac{\sqrt{2}}{2} \pm j\frac{\sqrt{2}}{2}$.

2. With an increase in system parameter, a, the poles move toward the real axis, resulting in less oscillation and a longer response time.

3. The original system is not stable, but the optimal closed-loop system is asymptotically stable.

B.3   Stochastic and Adaptive Control

So far, the presented control methods have been for systems in which the dynamics are known and equations have been established. However, a hybrid vehicle system normally operates in a dynamic environment with different road, weather, and load conditions.

It would be impractical to build an exact mathematical model to describe the dynamics of a hybrid vehicle system. In order to ensure that the vehicle model meets performance requirements in the presence of model and disturbance uncertainties, the developed control algorithms must be robust with respect to such uncertainties and even be able to adapt to slow, dynamic changes in the system. This section briefly introduces the common control techniques dealing with such uncertainties.

B.3.1   Minimal Variance Control

B.3.1.1   Minimal Variance Prediction

Consider a single-input, single-output linear system described by the following ARMAX model:

$$A\left(q^{-1}\right)y(k) = q^{-d}B\left(q^{-1}\right)u(k) + C\left(q^{-1}\right)\xi(k) \tag{B.78}$$

where:

$$A(q^{-1}) = 1 + a_1 q^{-1} + \cdots + a_n q^{-n}$$

$$B(q^{-1}) = b_0 + b_1 q^{-1} + \cdots + b_m q^{-m} \quad (b_0 \neq 0, m \leq n)$$

$$C(q^{-1}) = 1 + c_1 q^{-1} + \cdots + c_l q^{-l}$$

$$E\{\xi(k)\} = 0$$

$$E\{\xi(i)\xi(j)\} = \begin{cases} \sigma^2 & i=j \\ 0 & i \neq j \end{cases}$$

and $\{\xi(k)\}$ is an uncorrelated random variable with zero mean and standard deviation σ. It is also assumed that all the zeros of the polynomial $C(q^{-1})$ are inside the unit disc.

Now the problem is to predict $y(k+d)$ based on the observed input and output variables, $u(k), u(k-1), \cdots, y(k), y(k-1), \cdots$, at time k in the presence of measurement noise. It can be proved that the optimal (minimal variance) prediction of $y(k+d)$ can be achieved by the following equation, see Åstrom and Witternmark (1984).

Let us rewrite Eq. B.78 as:

$$A\left(q^{-1}\right)y(k+d) = B\left(q^{-1}\right)u(k) + C\left(q^{-1}\right)\xi(k+d) \tag{B.79}$$

That is:

$$y(k+d) = \frac{B(q^{-1})}{A(q^{-1})}u(k) + \frac{C(q^{-1})}{A(q^{-1})}\xi(k+d) = \frac{B(q^{-1})}{A(q^{-1})}u(k) + \frac{q^d C(q^{-1})}{A(q^{-1})}\xi(k) \tag{B.80}$$

The last term of the equation can be expressed as:

$$\frac{q^d C(q^{-1})}{A(q^{-1})}\xi(k+d) = \left(1 + e_1 q^{-1} + \cdots + e_{d-1} q^{-d+1}\right)\xi(k+d) + \frac{G(q^{-1})}{A(q^{-1})}\xi(k)$$

$$= E(q^{-1})\xi(k+d) + \frac{G(q^{-1})}{A(q^{-1})}\xi(k) \tag{B.81}$$

where the polynomials $E(q^{-1})$ and $G(q^{-1})$ are the quotient and the remainder of the division $\dfrac{q^d C(q^{-1})}{A(q^{-1})}$.

Thus, Eq. B.80 has the following form:

$$y(k+d) = \frac{B(q^{-1})}{A(q^{-1})}u(k) + E(q^{-1})\xi(k+d) + \frac{G(q^{-1})}{A(q^{-1})}\xi(k) \tag{B.82}$$

Equation B.78 can also be rewritten as:

$$\xi(k) = \frac{A(q^{-1})y(k)}{C(q^{-1})} - \frac{q^{-d}B(q^{-1})u(k)}{C(q^{-1})} \tag{B.83}$$

Substituting Eq. B.83 into B.82, we have:

$$y(k+d) = \frac{B(q^{-1})}{A(q^{-1})}u(k) + E(q^{-1})\xi(k+d) + \frac{G(q^{-1})}{A(q^{-1})}\left(\frac{A(q^{-1})y(k)}{C(q^{-1})} - \frac{q^{-d}B(q^{-1})u(k)}{C(q^{-1})}\right)$$

$$= E(q^{-1})\xi(k+d) + \frac{G(q^{-1})y(k)}{C(q^{-1})} + \frac{B(q^{-1})\left(C(q^{-1}) - q^{-d}G(q^{-1})\right)}{A(q^{-1})C(q^{-1})}u(k)$$

$$= E(q^{-1})\xi(k+d) + \frac{G(q^{-1})y(k)}{C(q^{-1})} + \frac{B(q^{-1})E(q^{-1})}{C(q^{-1})}u(k) \tag{B.84}$$

If we let $\hat{y}(k+d)$ be the prediction of $y(k+d)$, then the minimal variance prediction of $y(k+d)$ will be:

$$J = E\left\{[y(k+d) - \hat{y}(k+d)]^2\right\}$$

$$= E\left\{\left[E(q^{-1})\xi(k+d) + \frac{G(q^{-1})y(k)}{C(q^{-1})} + \frac{B(q^{-1})E(q^{-1})}{C(q^{-1})}u(k) - \hat{y}(k+d)\right]^2\right\} \tag{B.85}$$

Since the observed input and output data, $u(k), u(k-1), \cdots, y(k), y(k-1), \cdots$ are independent with the further noise sequence, $\xi(k+1), \xi(k+2), \cdots, \xi(k+d)$, Eq. B.85 is equal to:

$$J = E\left\{ \left[E(q^{-1})\xi(k+d) \right]^2 \right\} + E\left\{ \left[\frac{G(q^{-1})y(k)}{C(q^{-1})} + \frac{B(q^{-1})E(q^{-1})}{C(q^{-1})}u(k) - \hat{y}(k+d) \right]^2 \right\} \quad (B.86)$$

Since the first term in Eq. B.86 is not predictable, the second term must be zero to have J at a minimum.

Therefore, the minimal variance prediction of $y(k+d)$ is:

$$\hat{y}(k+d) = \frac{G(q^{-1})y(k)}{C(q^{-1})} + \frac{B(q^{-1})E(q^{-1})}{C(q^{-1})}u(k) = \frac{G(q^{-1})y(k)}{C(q^{-1})} + \frac{F(q^{-1})}{C(q^{-1})}u(k) \quad (B.87)$$

and the variance of the optimal prediction error is:

$$J^* = E\left\{ \left[E(q^{-1})\xi(k+d) \right]^2 \right\} = \left(1 + e_1^2 + \cdots + e_{d-1}^2 \right)\sigma^2 \quad (B.88)$$

where:

$$\begin{aligned}
E(q^{-1}) &= 1 + e_1 q^{-1} + \cdots + e_{d-1}q^{-d+1} \\
C(q^{-1}) &= A(q^{-1})E(q^{-1}) + q^{-d}G(q^{-1}) \\
F(q^{-1}) &= B(q^{-1})E(q^{-1}) = f_0 + f_1 q^{-1} + \cdots + f_{m+d-1}q^{-m-d+1} \\
G(q^{-1}) &= g_0 + g_1 q^{-1} + \cdots + g_{n-1}q^{-n+1}
\end{aligned} \qquad (B.89)$$

Example B.6

Consider a system described by:

$$y(k) + a_1 y(k-1) = b_0 u(k-2) + \xi(k) + c_1 \xi(k-1) \quad (B.90)$$

where parameters $a = 1-0.9$, $b_0 = 0.5$, $c_1 = 0.7$, and $\{\xi(k)\}$ is a sequence of independent random variables. Determine the minimal variance prediction and the variance of the prediction error over two steps; that is, $\hat{y}^*(k+2|k)$.

Solution:

From the system input–output equation, we have the following polynomials:

$$A(q^{-1}) = 1 + a_1 q^{-1}, \quad B(q^{-1}) = b_0, \quad C(q^{-1}) = 1 + c_1 q^{-1}, \quad d = 2 \quad (B.91)$$

Based on the above polynomials, we can assume the following equations:

$$E(q^{-1}) = 1 + e_1 q^{-1}, \quad G(q^{-1}) = g_0, \quad F(q^{-1}) = f_0 + f_1 q^{-1} \tag{B.92}$$

Solving the following Diophantine equation, we can obtain the quotient and remainder as:

$$\frac{C(q^{-1})}{A(q^{-1})} = \frac{1 + c_1 q^{-1}}{1 + a_1 q^{-1}} = \left(1 + (c_1 - a_1)q^{-1}\right) + q^{-2}\frac{a_1(a_1 - c_1)}{1 + a_1 q^{-1}} \Rightarrow$$

$$E(q^{-1}) = 1 + (c_1 - a_1)q^{-1}, \quad G(q^{-1}) = a_1(a_1 - c_1) \tag{B.93}$$

Equating the coefficients of Eqs B.92 and B.93, we have:

$$e_1 = c_1 - a_1, \quad g_0 = a_1(a_1 - c_1) \quad \text{and} \quad f_0 = b_0, \quad f_1 = b_0(c_1 - a_1) \tag{B.94}$$

Thus, the minimal variance prediction and the variance of the prediction error are:

$$\hat{y}^*(k+2|k) = \frac{g_0 y(k) + (f_0 + f_1 q^{-1})u(k)}{1 + c_1 q^{-1}} \tag{B.95}$$

$$E\{[\tilde{y}^*(k+2|k)]\} = \left(1 + e_1^2\right)\sigma^2$$

Taking parameters $a_1 = -0.9$, $b_0 = 0.5$, $c_1 = 0.7$ into Eq. B.94, we get $e_1 = -1.6$, $g_0 = 1.44$, $f_0 = 0.5$, and $f_1 = 0.8$. The variance of the solved prediction error, $E\{[\tilde{y}^*(k+2|k)]\} = \left(1 + e_1^2\right)\sigma^2 = \left(1 + 1.6^2\right)\sigma^2 = 3.56\sigma^2$.

B.3.1.2   Minimal Variance Control

To derive the minimal variance control law, we assume that the system in Eq. B.78 is a minimal phase system in which all zeros are located inside the unit disc. The objective of minimal variance control is to determine a control law which has the following objective function at its minimum:

$$J = E\left\{ [y(k+d) - y_{\text{desired}}(k+d)]^2 \right\} = \min \tag{B.96}$$

Based on the minimal variance prediction, the system in Eq. B.78 can be expressed by:

$$y(k+d) = E(q^{-1})\xi(k+d) + \frac{G(q^{-1})y(k)}{C(q^{-1})} + \frac{B(q^{-1})E(q^{-1})}{C(q^{-1})}u(k) \tag{B.97}$$

Substituting Eq. B.97 into the objective function (Eq. B.96), we have:

$$
\begin{aligned}
J &= E\left\{\left[y(k+d) - y_{\text{desired}}(k+d)\right]^2\right\} \\
&= E\left\{\left[E(q^{-1})\xi(k+d) + \frac{G(q^{-1})y(k)}{C(q^{-1})} + \frac{B(q^{-1})E(q^{-1})}{C(q^{-1})}u(k) - y_{\text{desired}}(k+d)\right]^2\right\}
\end{aligned}
\tag{B.98}
$$

Since the term $E(q^{-1})\xi(k+d) = \xi(k) + e_1\xi(k+1) + \cdots + e_{d-1}\xi(k+d)$ in the equation is a random sequence and independent with the observed input and output data at time k, $k-1$, $k-2$, $\cdots$, Eq. B.98 is equal to:

$$
J = E\left\{\left[E(q^{-1})\xi(k+d)\right]^2\right\} + E\left\{\left[\frac{G(q^{-1})y(k)}{C(q^{-1})} + \frac{B(q^{-1})E(q^{-1})}{C(q^{-1})}u(k) - y_{\text{desired}}(k+d)\right]^2\right\}
\tag{B.99}
$$

Without losing generality, we can set $y_{\text{desired}}(k+d) = 0$, so the minimal variance control, $u^*(k)$, can be obtained from the following equation:

$$
E\left\{\left[\frac{G(q^{-1})y(k)}{C(q^{-1})} + \frac{B(q^{-1})E(q^{-1})}{C(q^{-1})}u(k)\right]^2\right\} = 0
\tag{B.100}
$$

That is:

$$
B(q^{-1})E(q^{-1})u(k) = -G(q^{-1})y(k)
\tag{B.101}
$$

Since the considered system is a minimal phase system, the polynomial $B(q^{-1})$ has a stable inverse, the optimal control to minimize the variance in the control error is:

$$
u^*(k) = -\frac{G(q^{-1})y(k)}{B(q^{-1})E(q^{-1})}
\tag{B.102}
$$

The minimal variance control law has the feedback form. A control system diagram with minimal variance controller is shown in Fig. B.6.

Example B.7

Consider a system given by:

$$
y(k) = \frac{1}{1+0.5q^{-1}}u(k-1) + \frac{1+0.7q^{-1}}{1+0.2q^{-1}}\xi(k)
\tag{B.103}
$$

where $\{\xi(k)\}$ is a sequence of independent random variables with zero mean. Determine the minimal variance control law, $u^*(k)$.

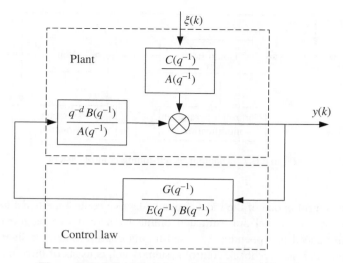

Figure B.6 Minimal variance control system diagram

Solution:

From the system's input–output equation, we have the following polynomials:

$$A(q^{-1}) = (1 + 0.5q^{-1})(1 + 0.2q^{-1}) = 1 + 0.7q^{-1} + 0.1q^{-2}, \quad B(q^{-1}) = 1 + 0.2q^{-1}$$
$$C(q^{-1}) = (1 + 0.5q^{-1})(1 + 0.7q^{-1}) = 1 + 1.2q^{-1} + 0.35q^{-2}, \quad d = 1 \tag{B.104}$$

Solving the following Diophantine equation, we can obtain the quotient and remainder:

$$\frac{C(q^{-1})}{A(q^{-1})} = \frac{1 + 1.2q^{-1} + 0.35q^{-2}}{1 + 0.7q^{-1} + 0.1q^{-2}} = 1 + q^{-1}\frac{0.5 + 0.25q^{-1}}{1 + 0.7q^{-1} + 0.1q^{-2}} \Rightarrow$$
$$E(q^{-1}) = 1, \quad G(q^{-1}) = 0.5 + 0.25q^{-1} \tag{B.105}$$

Thus, the minimal variance control law is:

$$u^*(k) = -\frac{0.5(1 + 0.5q^{-1})y(k)}{1 + 0.2q^{-1}} \tag{B.106}$$

B.3.2 Self-tuning Control

Minimal variance control provides an effective control method for a system in process and in the presence of measurement noise, while adaptive control techniques are the relevant control methods for dealing with variations in model parameters and operating environments. The fundamental principle of adaptive control is to assess such variations online and then change the control strategy correspondingly to maintain satisfactory control performance.

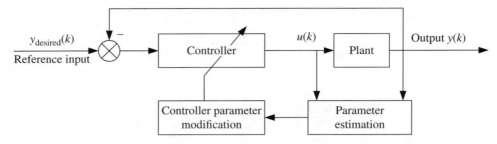

Figure B.7   Diagram of self-tuning control system

Self-tuning control is one of the most applied adaptive control methods; here, the parameters of the system model are estimated online by a recursive parameter estimation method, as introduced in Appendix A. A self-tuning control system, a diagram of which is shown in Fig. B.7, is a feedback control system which consists of three main functions: parameter estimation, control law modification, and control decision. Since such an adaptive control strategy requires considerable online computations and may also include various aspects of stochastic analysis, the controller must adapt faster than the parameter change rates of the plant and/or environment. In addition, since the shown adaptive control system is an actual feedback control system, the entire system has to be subject to the stability consideration, which usually results in complex analysis on the system stability, convergence, and performance. This section briefly outlines the widely used minimal variance self-tuning control.

Assume that we have the system described by Eq. B.79; the minimal variance self-tuning adaptive controller can be designed based on the following procedures:

1. Set up a prediction model of the system as:

$$\hat{y}(k+d) = G\left(q^{-1}\right)y(k) + F\left(q^{-1}\right)u(k) + E\left(q^{-1}\right)\xi(k+d) \qquad \text{(B.107)}$$

2. The corresponding minimal variance prediction of the system is:

$$
\begin{aligned}
\hat{y}^*(k+d) &= G\left(q^{-1}\right)y(k) + F\left(q^{-1}\right)u(k) \\
&= g_0 y(k) + g_1 y(k-1) + \cdots + g_{n-1} y(k-n+1) \\
&\quad + f_0 u(k) + f_1 u(k-1) + \cdots + f_{m+d-1} u(k-m-d+1)
\end{aligned}
\qquad \text{(B.108)}
$$

3. Predetermine $f_0 = b_0$ based on tests or on knowledge of the system, and establish the following relationship between inputs and outputs:

$$y(k) - b_0 u(k-d) = \varphi(k-d)\theta \qquad \text{(B.109)}$$

where

$$\theta = [g_0 \ \ g_1 \ \ \cdots \ \ g_{n-1} \ \ f_1 \ \ f_2 \ \ \cdots \ \ f_{m+d-1}]^{\mathrm{T}}$$
$$\varphi(k) = [y(k) \ \ y(k-1) \ \ \cdots \ \ y(k-n+1) \ \ u(k-1) \ \ u(k-2) \ \ \cdots \ \ u(k-m-d+1)]$$

4. Carry out the parameter estimation by a recursive estimation method based on the observed input and output data; recursive least squares is the most commonly used method to perform this task.
5. Establish the minimal variance controller based on the estimated parameter $\hat{\theta}$ as:

$$u(k) = -\frac{1}{f_0}\varphi(k)\theta(k) \tag{B.110}$$

B.3.3 Model Reference Adaptive Control

Model reference adaptive control (MRAC) is another method of adapting the controller to maintain satisfactory system performance according to changes in the plant or environment. The model reference adaptive control system was originally developed at MIT for aerospace applications, and the parameter adjustment method is called the MIT rule, where the parameter adjustment mechanism minimizes the error between the model output and the actual plant output. The control action, $u(k)$, needs to be within the admissible boundary, and normally it is a linear combination of the model output, $y_m(k)$, the reference input, $r(k)$, and the system output, $y_p(k)$. The MIT-rule-based MRAC design technique is introduced in this section with the control system diagram shown in Fig. B.8.

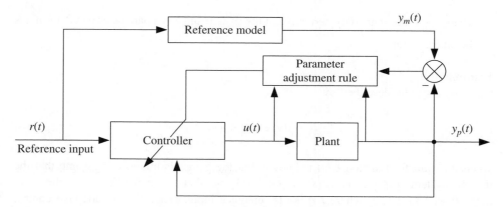

Figure B.8 Diagram of model reference adaptive control system

A model reference adaptive controller can be designed based on the following procedures:

1. Consider a single-input, single-output linear system described by:

$$y_p(s) = G(s)u(s) \tag{B.111}$$

where the reference model has the following form:

$$y_m(s) = G_m(s)r(s) \tag{B.112}$$

2. Set up an error function between the system output and the reference model output:

$$e(t) = y_p(t) - y_m(t) \tag{B.113}$$

3. Define the objective function, for example, the objective function can be defined as:

$$J(\theta) = \frac{1}{2}e^2(t) \tag{B.114}$$

4. Determine the adaptive control law. If the objective function is defined as in Eq. B.114, the parameter of the controller can be updated in the direction of the negative gradient of J; that is:

$$\frac{d\theta}{dt} = -\lambda\frac{\partial J}{\partial\theta} = -\lambda\frac{\partial(e(t))}{\partial\theta}e(t) \tag{B.115}$$

where $\dfrac{\partial e}{\partial\theta}$ is the sensitivity derivative of the system and λ is a parameter that determines the adaption rate.

Example B.8

Consider a system described by:

$$y_p(s) = kG(s)u(s) \tag{B.116}$$

where the transfer function $G(s)$ is known, but the gain k is unknown. Assume that the desired system response is $y_m(s) = k_m G(s)r(s)$ and that the controller is of the form $u(t) = \theta_1 \cdot r(t) + \theta_2 \int r(t)$, where $r(t)$ is the reference input. Determine an adaptive control law based on the MIT rule.

Solution:

The error between the reference model and the actual output can be expressed as:

$$e(s) = y_p(s) - y_m(s) = kG(s)u(s) - k_mG(s)r(s) = kG(s)\left(\theta_1 \cdot r(s) + \frac{\theta_2 r(s)}{s}\right) - k_mG(s)r(s)$$

(B.117)

The sensitivity to the controller parameter is:

$$\frac{\partial(e)}{\partial\theta_1} = kG(s)r(s) = \frac{k}{k_m}y_m$$

$$\frac{\partial(e)}{\partial\theta_2} = \frac{kG(s)r(s)}{s} = \frac{k}{k_m}\frac{y_m}{s} \quad\Rightarrow\quad \frac{\partial(e)}{\partial\theta_2} = \frac{k}{k_m}\int y_m dt$$

(B.118)

Based on the MIT rule, in order to make the system response follow the desired response, the controller's parameter, θ, is adjusted as follows:

$$\frac{d\theta_1}{dt} = -\lambda_1\frac{\partial J}{\partial\theta_1} = -\lambda_1\frac{\partial(e)}{\partial\theta_1}e = -\lambda_1\frac{k}{k_m}y_m e = -\beta_1 y_m e$$

$$\frac{d\theta_2}{dt} = -\lambda_2\frac{\partial J}{\partial\theta_2} = -\lambda_2\frac{\partial(e)}{\partial\theta_2}e = -\lambda_2\frac{k}{k_m}\left(\int y_m dt\right)e = -\beta_2 e\int y_m dt$$

(B.119)

Thus,

$$\theta_1 = -\beta_1\int y_m e dt$$

$$\theta_2 = -\beta_2\int\left(\int y_m dt\right)e dt$$

(B.120)

The corresponding control system diagram is shown in Fig. B.9.

MRAC is a broad subject area covering many different methods and applications. Since the 1970s, MRAC designs have mainly been based on Lyapunov's stability theory and Popov's hyper-stability theory, as the stability of the closed-loop system is guaranteed from these approaches. The main drawback of Lyapunov's design approach is that there is no systematic way of finding a suitable Lyapunov function that can be used to specify the adaptive control law. Popov's hyper-stability theory is concerned with finding conditions that must be satisfied to make the feedback system globally asymptotically stable. These conditions collectively produce the Popov integral inequality. Based on Popov's hyper-stability theory, the feedback control system is stable if all controls satisfy the Popov integral inequality. Thus, the approach with Popov hyper-stability is much more flexible than the Lyapunov approach for designing adaptive control law. For more detailed materials

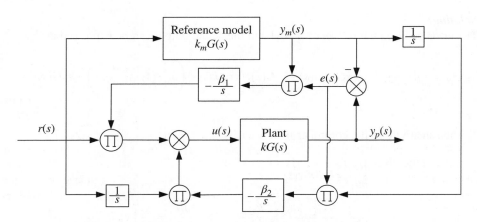

Figure B.9 Implemented adaptive control system diagram based on the MIT rule

regarding these two approaches, the interested reader should refer to the texts by Åstrom and Witternmark (1995), Landau (1979), and Butler (1992).

B.3.4 Model Predictive Control

Model predictive control (MPC), referred to as moving horizon control or receding horizon control, is an advanced control technology developed in the 1980s, which has been widely implemented in practical applications (Garcia, *et al.*, 1989). MPC can be formularized as follows.

Assume that the system is described by the single-input, single-output model shown in Eq. B.78 or the following state-space model:

$$x(k+1) = Ax(k) + B_m u(k) + B_d w(k)$$
$$y(k) = Cx(k) + v(k)$$

(B.121)

where $y(k)$ is the measurable system output, $u(k)$ is the manipulated system input, $w(k)$ is the measurable disturbance, and $v(k)$ is assumed to be sequences of white noise.

The aim of model predictive control is to select the manipulated input variables, $u(k+i|k)$, $i = 1, \cdots, p$, at time k such that they minimize the following objective function:

$$J = \sum_{i=1}^{p} \left(y(k+i|k) - y_{\text{sp}}(k) \right)^2 + \sum_{i=1}^{m} r_i \Delta u(k+i|k)^2 = \min$$

(B.122)

subject to:

$$u_{\max} \geq u(k+i-1|k) \geq u_{\min}, \quad i = 1, \cdots, m$$

(B.123)

$$\Delta u_{\max} \geq \Delta u(k+i-1|k) \geq -\Delta u_{\max}, \quad i=1,\cdots,m \qquad \text{(B.124)}$$

$$y_{\max} \geq y(k+i|k) \geq y_{\min}, \quad i=1,\cdots,p \qquad \text{(B.125)}$$

where p and $m<p$ are the lengths of the system output prediction and the manipulated input horizons, $u(k+i|k)$, $i=1,\cdots,p$ is the set of future manipulated input values which makes the objective function minimal, $y_{sp}(k)$ is the setpoint, and Δ is the difference operator; that is, $\Delta u(k+i|k)=u(k+i|k)-u(k+i-1|k)$.

It needs to be emphasized that although the manipulated variables are determined by optimizing the objective function (Eq. B.122) over the horizons m, the control action only takes the first step. Therefore, the optimization of MPC is a rolling optimization, and the amount of computation is one of the concerns when the MPC strategy is implemented in practical applications. The MPC control scheme is shown in Fig. B.10.

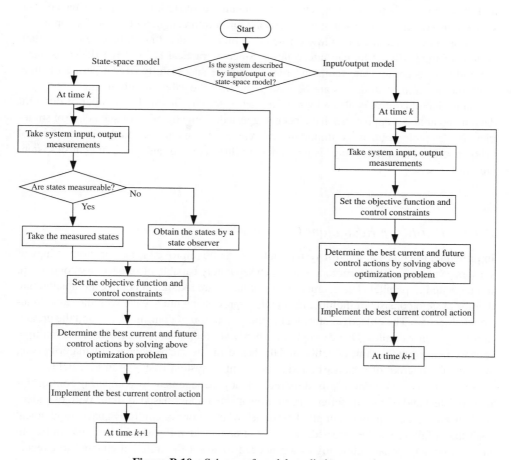

Figure B.10   Scheme of model predictive control

B.4   Fault-tolerant Control

Fault-tolerant control is a set of advanced control methodologies which admit that one or more key components of a physical feedback system will fail; this can have a significant impact on system stability and other performance factors. At the simplest level, sensor or actuator failure may be considered, while at a more complex level, a subsystem failure needs to be tolerated. In the same vein, engineers can also worry about computer hardware and software failure in the system. The idea of a fault-tolerant control design is to retain stability and safety in the system while losing some level of performance in a graceful manner. To implement such an idea, it may be necessary to reconfigure the control system in real time following the detection of such failures.

Advances in hybrid vehicle systems are leading to increasingly complex systems with ever-more-demanding performance goals. A modern hybrid vehicle system will require fault detection, isolation, and control reconfiguration to be completed in a short time, and the controller will allow the vehicle to maintain adequate levels of performance even with failures in one or more actuators or sensors; furthermore, higher degrees of autonomous operation may be required to allow for health monitoring and fault tolerance over certain periods of time without human intervention. Since a practical HEV/PHEV/EV must have the ability to accommodate faults in order to operate successfully over a period of time, fault-tolerant control strategies are necessary to fulfill the safety requirements.

Fault-tolerant control is also a topic of immense activity in which a number of different design and analysis approaches have been suggested. Some mention of such control strategies is desirable, but it is important to realize the nature of fault-tolerant control problems. The briefest coverage is given here to illustrate two fault-tolerant control system architectures.

B.4.1   Hardware Redundant Control

In principle, tolerance to control system failures can be improved if two or more strings of sensors, actuators, and microprocessors, each separately capable of satisfactory control, are implemented in parallel. The characteristics of hardware redundant control architecture are that the system consists of multiple information-processing units, each with the same functional objective. The objective may include the generation of control signals and the prediction of system variables. One example of a hardware redundant control system is multiple sensors measuring the same quantity, and the best estimate can be obtained by majority vote. A principle diagram of a hardware redundant control system is shown in Fig. B.11.

A voting scheme is normally used for redundancy management by comparing control signals to detect and overcome failures in a hardware-redundant-based system. With two identical channels, a comparator simply determines whether or not control signals are identical such that a failure can be detected; however, it cannot identify which string has failed. In most cases, additional online logics are needed to select the unfailed channel to execute the control task. The design tasks of a hardware-redundant-based fault-tolerant control

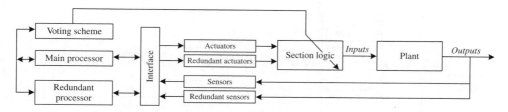

Figure B.11   Diagram of a hardware redundant control system

system usually solve the following problems: selection logic, nuisance trips, generic failures, reliability of voting/selection units, control action contention, cross-strapping, increased cost of operation and maintenance, and the number of operating channels required for dispatch, etc. Since the fault-detection algorithm is the core functional algorithm, it has to meet higher standard requirements; it must be sensitive to failures yet insensitive to small operational errors, including data lost or interrupted due to non-collocation of sensors or actuators. In addition, false indications of failure must be minimized to ensure that useful resources are kept online and missions are not aborted prematurely.

B.4.2   Software Redundant Control

A hardware redundant control strategy can protect against control system component failures, but it increases the cost, makes maintenance more difficult, and does not address failures in plant components such as the battery subsystem, transmission, and motors. Furthermore, in the general case, this form of parallelism implies that fault tolerance can only be improved by physically isolating the processors from each other. On the other hand, a software redundant control strategy provides an ability to improve fault tolerance in the control system on the above three aspects with fewer additional components.

Although hardware redundancy is at the core of most solutions for reliability, uninterrupted operation can also be achieved in control systems through redistribution of the control authority between different functioning actuators rather than through duplication or triplication of those actuators. The idea of a software redundant strategy is to utilize different subsets of measurements along with different system models and control algorithms. In other words, if a variety of system fault patterns and corresponding fault situations are able to be characterized and some possible algorithms can be selected and simulated based on the particular fault pattern, a decision unit will determine the appropriate control algorithm or appropriate combination of algorithms. When the correct algorithm is determined, the control system can be reconstructed in time against the failures. If failures occur within the system, the control algorithm can be changed with time. Software-based redundancy is also called functional redundancy, and it consists mainly of fault detection or a diagnosis unit identifying the failed components, controller change, and selection logic units selecting

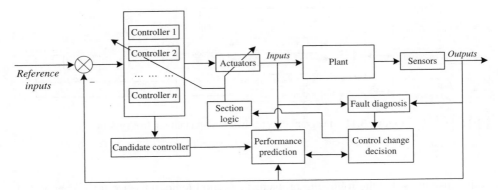

Figure B.12   Diagram of a software redundant control system

the controller and control channels based on the performance prediction of the selected controller so that the system will adapt to the failure situation. A diagram of a software redundant control system is shown in Fig. B.12.

References

Åstrom, K. J. and Witternmark, B. *Computer Controlled Systems – Theory and Design*, Prentice-Hall Inc., Englewood Cliffs, N.J., 1984.

Åstrom, K. J. and Witternmark, B. *Adaptive Control*, 2nd edition, Addison-Wesley Publishing Company Inc., 1995.

Bellman, R. E. *Dynamic Programming*, Princeton University Press, Princeton, NJ, 1957.

Butler, H. *Model-Reference Adaptive Control – From Theory to Practice*, Prentice-Hall, 1992.

Garcia, E. C., Prett, M. D., and Morari, M. 'Model Predictive Control: Theory and Practice – A Survey,' *Automatica*, **25**(3), 335–348, 1989.

Kuo, B. C. *Automatic Control Systems*, 4th edition, Prentice-Hall Inc., Englewood Cliffs, N.J., 1982.

Landau, Y. D. *Adaptive Control: The Model Reference Approach*, Marcel Dekker, New York, 1979.

Pontryagin, L. S., Boltyanskii, V., Gamkrelidze, R., and Mishchenko, E. *The Mathematical Theory of Optimal Processes*, Interscience, NY, 1962.

Index

Hybrid Electric Vehicle System Modeling and Control, Second Edition. Wei Liu.
© 2017 John Wiley & Sons Ltd. Published 2017 by John Wiley & Sons Ltd.